THE URBAN SOCIOLOGY READER

Second edition

The urban world is an exciting terrain for investigating the central institutions, structures, and problems of the social world and how they have transformed through the last two hundred years. This reader comprises sections on urban social theory, racial and social difference in the city, culture in everyday life, culture and the urban economy, globalization and transnational social relations, and the regulation of urban space.

Drawing together seminal selections covering the nineteenth to the twenty-first centuries, this reader includes forty-three significant writings from eminent names such as Simmel, Wirth, Park, Burgess, Du Bois, Zukin, Sassen, and Harvey. The second edition illuminates more recent urban issues, such as sprawl, sustainability, immigration, and urban protest. Selections are predominantly sociological, but some readings cross disciplinary boundaries.

Providing an essential resource for students of urban studies, this book brings together important but, until now, widely dispersed writings. Editorial commentaries precede each entry, introducing the text, demonstrating its significance, and outlining the issues surrounding its topic, whilst the associated bibliography enables deeper investigations.

Jan Lin is Professor of Sociology at Occidental College, Los Angeles.

Christopher Mele is Associate Professor of Sociology at State University of New York at Buffalo.

THE ROUTLEDGE URBAN READER SERIES

Series editors

Richard T. LeGates

Professor Emeritus of Urban Studies and Planning, San Francisco State University

Frederic Stout

Lecturer in Urban Studies, Stanford University

The Routledge Urban Reader Series responds to the need for comprehensive coverage of the classic and essential texts that form the basis of intellectual work in the various academic disciplines and professional fields concerned with cities and city planning.

The readers focus on the key topics encountered by undergraduates, graduate students, and scholars in urban studies, geography, sociology, political science, anthropology, economics, culture studies, and professional fields such as city and regional planning, urban design, architecture, environmental studies, international relations, and landscape architecture. They discuss the contributions of major theoreticians, practitioners, and other individuals, groups, and organizations that study the city or practice in a field that directly affects the city.

As well as drawing together the best of classic and contemporary writings on the city, each reader features extensive introductions to the book, sections, and individual selections prepared by the volume editors to place the selections in context, illustrate relations among topics, provide information on the author, and point readers towards additional related bibliographic material.

Each reader contains:

- Between thirty-five and sixty *selections* divided into six sections to eight sections. Almost all of the selections are previously published works that have appeared as journal articles or portions of books.
- A *general introduction* describing the nature and purpose of the reader.
- *Section introductions* for each section of the reader to place the readings in context.
- *Selection introductions* for each selection describing the author, the intellectual background and context of the selection, competing views of the subject matter of the selection and bibliographic references to other readings by the same author and other readings related to the topic.
- One or more plate sections and illustrations at the beginning of each section.
- An index.

The series consists of the following titles:

THE CITY READER

The City Reader: Fifth edition – an interdisciplinary urban reader aimed at urban studies, urban planning, urban geography and urban sociology courses – is the *anchor urban reader*. Routledge published a first edition of *The City Reader* in 1996, a second edition in 2000, a third edition in 2003, and a fourth edition in 2007. *The City Reader* has become one of the most widely used anthologies in urban studies, urban geography, urban sociology and urban planning courses in the world.

URBAN DISCIPLINARY READERS

The series contains *urban disciplinary readers* organized around social science disciplines and professorial fields: urban sociology, urban geography, urban politics, urban and regional planning, and urban design. The urban disciplinary readers include both classic writings and recent, cutting-edge contributions to the respective disciplines. They are lively, high-quality, competitively priced readers which faculty can adopt as course texts and which also appeal to a wider audience.

TOPICAL URBAN ANTHOLOGIES

The urban series includes *topical urban readers* intended both as primary and supplemental course texts and for the trade and professional market. The topical titles include readers related to sustainable urban development, global cities, cybercities, and city cultures.

INTERDISCIPLINARY ANCHOR TITLE

The City Reader: fifth edition
Richard T. LeGates and Frederic Stout (eds)

URBAN DISCIPLINARY READERS

The Urban Geography Reader
Nick Fyfe and Judith Kenny (eds)

The Urban Politics Reader
Elizabeth Strom and John Mollenkopf (eds)

The Urban and Regional Planning Reader
Eugenie Birch (ed.)

The Urban Sociology Reader, second edition
Jan Lin and Christopher Mele (eds)

The Urban Design Reader, second edition
Michael Larice and Elizabeth Macdonald (eds)

TOPICAL URBAN READERS

The City Cultures Reader, second edition
Malcolm Miles, Tim Hall with Iain Borden (eds)

The Cybercities Reader
Stephen Graham (ed.)

The Sustainable Urban Development Reader, second edition
Stephen M. Wheeler and Timothy Beatley (eds)

The Global Cities Reader
Neil Brenner and Roger Keil (eds)

Forthcoming:

Cities of the Global South Reader
Faranak Miraftab and Neema Kudva (eds)

For further information on The Routledge Urban Reader Series please visit our website:

http://www.routledge.com/articles/featured_series_routledge_urban_reader_series/

or contact

Andrew Mould
Routledge
2 Park Square, Milton Park,
Abingdon, Oxon, OX14 4RN
England
andrew.mould@routledge.co.uk

Richard T. LeGates
Department of Urban Studies and Planning
San Francisco State University
1600 Holloway Avenue
San Francisco, CA 94132
(510) 642-3256
dlegates@sfsu.edu

Frederic Stout
Urban Studies Program
Stanford University
Stanford, California 94305-2048
fstout@stanford.edu

The Urban Sociology Reader

Second edition

Edited by

Jan Lin

and

Christopher Mele

LONDON AND NEW YORK

First published 2005
by Routledge
This edition published 2013
by Routledge
2 Park Square, Milton Park, Abingdon, Oxon OX14 4RN

Simultaneously published in the USA and Canada
by Routledge
711 Third Avenue, New York, NY 10017

Routledge is an imprint of the Taylor & Francis Group, an informa business

British Library Cataloguing in Publication Data
A catalogue record for this book is available from the British Library

Library of Congress Cataloging in Publication Data
The urban sociology reader / [edited by] Jan Lin and Christopher Mele.
p. cm.
Rev. ed. of: The urban sociology reader. 2005.
Includes bibliographical references and index.
1. Sociology, Urban. I. Lin, Jan. II. Mele, Christopher.
HT108.U733 2012
307.76—dc23

2012005536

ISBN: 978-0-415-66530-8 (hbk)
ISBN: 978-0-415-66531-5 (pbk)
ISBN: 978-0-203-10333-3 (ebk)

Typeset in 9.5/12pt Amasis MT
by Graphicraft Limited, Hong Kong

Printed and bound in Great Britain by
TJ International Ltd, Padstow, Cornwall

Contents

Plates

Acknowledgments

The second edition of this anthology is rededicated to Janet Abu-Lughod, Emeritus Professor of Sociology at the New School University in New York City. We were among the first cohort of sociologists that Janet trained at the Graduate Faculty of the New School upon her move from Northwestern University in 1987. We became ardent urbanists under her guidance, stimulated by her deep comparative knowledge of cities, communities, and civilizations, and her passionate commitment to public intellectualism as an urban planner and a proponent of research with an action focus in the neighboring communities of Lower Manhattan. She supervised our dissertation research, Jan Lin working in Chinatown and Christopher Mele in the East Village. We have maintained a lively partnership through the years, in part through our association with the Community and Urban Sociology Section (CUSS) of the American Sociological Association (ASA).

We gratefully acknowledge Richard T. LeGates, the pioneer and general editor (along with Frederic Stout) of The Routledge Urban Readers Series, for inviting us to produce a second edition of this volume. His active encouragement and valuable insights helped to improve our book proposal and manuscript immensely along every stage of its development. Andrew Mould, our Publisher at Routledge, and Faye Leerink, Senior Editorial Assistant, were also of instrumental importance in shepherding our project through its various phases. We would also like to thank the anonymous reviewers for their very helpful critiques and comments during the proposal phase of our project.

Rob Kappel, Gregory Hall, Wendy Hilleren, and Ryan Wymyczak at the University at Buffalo offered invaluable assistance in the production of the volume.

GENERAL INTRODUCTION

Cities are focal arenas for the contemplation of the human condition and man's struggle for self-expression. Cities are landscapes of cultural diversity and subcultural differentiation, what Robert Park called a "mosaic of social worlds." The bohemias, bright light districts, and red light districts of the city are crucibles for the exploration of artistic, cultural, and sexual expression. The city contains our workplaces, our residences, and our commercial marketplaces. The metropolis is a terrain of social inequality, from the decline and deterioration of marginal places like the Southside of Chicago and New York's Bowery to the affluence of prime spaces like Midtown Manhattan and Rodeo Drive in Los Angeles. Cities are landscapes of gender inequality and social differences in sexuality. Cities are key sites in the transformation of the global economy. There is a new cultural economy of cities that gives us an analytic window on the character of our postindustrial society. Rising inequality has led to a climate of fear in cities, which have become high-security fortresses. Urban social movements have arisen to articulate the demands of the socially and economically disenfranchised in our cities. Urban protests erupted in 2011 with renewed fervor in the Arab Spring and the Occupy Wall Street movement.

In this second edition of *The Urban Sociology Reader*, we have retooled our array of selections that span the nineteenth to the twenty-first century. We dropped some outmoded early twentieth-century readings and added more contemporary selections that reflect the economic, political, and cultural changes affecting cities in the new millennium. We offer readings to spark humanistic discourse in the classic liberal arts tradition as well as addressing advanced theoretical debates in urban sociology. *The Urban Sociology Reader* is aimed at graduate urban sociologists as well as undergraduate students. We target readers drawn to urban sociology while contemplating or involved in careers such as education, urban policy and planning, public administration and government, community organizing, arts, and the cultural economy. We promote the exploration of urban sociological practice as well as theory.

The urban world is a provocative terrain to contemplate central experiences, structures, and problems of the social world, and how they have transformed over the last two hundred years. Our reader traverses that terrain through central themes of urban social theory (classic and contemporary), inequality and social difference (by class, race/ethnicity, gender, and sexuality), globalization and the world-system, culture and the urban economy, and urban social movements. Our selections are predominantly sociological, but some readings cut across disciplinary boundaries, reflecting underlying movements in the social sciences and social changes in the real world since the 1960s. These movements include those under the broad rubrics of "multiculturalism," "globalization," "postmodernism," and "neoliberalism." Our selections are primarily American, but we have enhanced our attention to European writing and cities in the developing world. We have attempted to offer a balanced mix of political-economic as well as cultural perspectives.

Ethnographic studies represent a strong tradition in urban sociology, beginning with the many community studies published in the early years of the Chicago School. Undergraduates find ethnographies enjoyable to read; they are drawn to their clear, jargon-free presentation and story-telling plotlines. The readings we have chosen represent the tradition of "thick description," and are written in a lively style.

These include readings by James Duncan, Donald Donham, Sirpa Tani, Paul Stoller and Jasmin McConatha, Peggy Levitt, Karin Aguilar-San Juan, Teresa Caldeira, Christopher Mele, and Setha Low.

URBANIZATION AND COMMUNITY

Our anthology begins with Part One, Urbanization and Community, drawing from classic European as well as American texts in urban sociology spanning the nineteenth and twentieth centuries. The selections by Ferdinand Tönnies and Georg Simmel consider the major social and psychological changes that accompanied urbanization and the development of capitalism in Europe. Tönnies was comparable to Emile Durkheim in linking the decline of primary ties and community life with increasing specialization in the division of labor and social life. Simmel saw the capitalist city as a sensorium that assaulted the urbanite with a cacophony of sights and sounds, including advertising, commodities, pedestrians, and vehicular traffic. The decline of traditional mores and small-town prejudices had fostered greater freedom and cosmopolitanism for the individual. Simmel believed the experience of modernity was somewhat paradoxical; the urban commercial sensorium fed the self while starving the spirit. The liberated individual was also a restless one. There was a rootlessness that came with participation in urban society and the modern marketplace.

Louis Wirth carried the perspectives of the nineteenth-century European theorists into the American city of the early twentieth century. He drew greater attention than Simmel to the negative consequences of modernity, especially the status of Durkheimian *anomie* and the onset of urban social problems and personality disorders such as crime, delinquency, and mental illness. He felt that differentiation and the "mosaic of social worlds" fostered social pathology and social distance between people. Claude Fischer accepted the concept of general urban effects, but interpreted the "mosaic of social worlds" as a more positive phenomenon of subcultural differentiation. Fischer saw social differentiation as a creative process rather than a symptom of moral drift or social decline. Traditional communities gave way to subcultural communities with the growth of the metropolis. The growth of subcultural communities counterbalanced the general decline of community with the ongoing course of urbanization.

Other writers, like Jane Jacobs, Barry Wellman, and Barry Leighton, articulated similar views recognizing the renaissance and transformation of community. Jane Jacobs decried the destruction of community by misguided urban renewal policies promoted by the bureaucratic-rational state in the period after World War II. These authors gave intellectual voice to the growing neighborhood preservation and community action movements of the 1960s and 1970s, with the architectural critic and neighborhood mobilization leader Jacobs gaining recognition as a public spokesperson for their views. Wellman and Leighton pushed us to move beyond the idea of place-based community being saved or resurrected to the notion of community being "liberated" from spatial attachments through new innovations in communication and transportation technology. They bring attention to the rise of communities without propinquity through the Internet and bring attention to how these ramified and branching social networks give participants access to greater outside resources.

Robert Putnam is more skeptical about the ability of ramified social networks to create new forms of community. He recognizes that non-profit organizations representing interests such as the environment, women, or retired people, create a sense of dues-paying membership and interest-group politics. But he contends that they don't generate the same kind of civic engagement as traditional organizations like the Boy Scouts, Parent–Teacher Associations, and the Red Cross. He perceives there has been a long-term decline in trust, reciprocity, and social capital in America. The loss of civic engagement and social capital in America has drawn increasing interest as variables contributing to urban economic and social decline. There is growing attention to ramified social networks on the Internet and their ability to foster sociability, community, and civic engagement. The role of Internet social networking in fostering political engagement has become increasingly prominent in the new millennium, as evident in urban-based protests such as the Arab Spring and Occupy Wall Street movement in 2011.

UNDERSTANDING URBAN GROWTH IN THE CAPITALIST CITY

The essays in Part Two help address the dynamics of growth in the capitalist city, with attention to the transition from the human ecology theory to Marxist and political/economy approaches. The founder of the Chicago School, Robert Park, applied the ideas of Charles Darwin to justify the presence of urban social inequality, though he ultimately diverged from social Darwinists like Herbert Spencer, who were advocates of imperialism and racial eugenics. Ernest Burgess promulgated the Chicago School's famous concentric zone theory of urban development, which was based on the city of Chicago during the railroad phase of its development. Burgess believed that increasingly geographic and social mobility led to urban social disorganization. Human ecology held a negative view of the consequences of human mobility and cultural change.

The human ecologists increasingly gave way to Marxist and urban political economy perspective that emerged in the 1960s against the backdrop of urban social movements. The British geographer David Harvey offers a path-breaking Marxist account of urban economic development that recognizes features of contradiction, overaccumulation, and mobility of different circuits of capital in production, finance, and the built environment. His essay has growing significance in the new millennium against the backdrop of economic recession and crisis in the financial and housing sectors.

John Logan and Harvey Molotch represent an American perspective on urban political economy that is highly critical of social inequality and urban elites. They promote a political economy of place that repudiates the human ecology perspective. Their concept of the urban power elite as a "growth machine" has influenced a new generation of urban scholars. Gregory Squires gives a different view on the urban growth machine by addressing the growth of urban policies of "public/private partnership," which he recognizes as a trope masking an underlying ideology of privatism. He questions whether the U.S. government really acted in the public interest when providing subsidies to private capital through successive historical episodes of canal and railway building, freeway construction and urban renewal, and downtown redevelopment. Michael Dear extends the critique of human ecology with his discussion of the L.A. School of urban studies. Dear suggests that the modernist hegemony of urban elites has given way to a polycentric, polyglot, and polycultural pastiche of urban development. Dear suggests that Los Angeles has superseded Chicago as the paradigm of urban growth in the twenty-first century.

Neil Brenner and Nik Theodore address urban political and economic transitions in global context through their analysis of neoliberal doctrines in urban planning and politics. They describe how international banking organizations have pushed for open and deregulated markets and implemented structural adjustment policies that lead to the scaling back of Fordist-Keynesian social programs in national and urban context. They describe neoliberalism as a volatile and contested transition. They draw special attention to cities as political flashpoints where neoconservative political and social movements that vilify the poor as threats to public order and morality have been met with social movements of the disenfranchised and powerless. They give a good political and economic framework to comprehend and debate the growth of recent political and urban social movements such as the Tea Party and Occupy Wall Street movements.

Peter Dreier and his co-authors turn to problem of urban sprawl, which they blame for hurting the inner city, aggravating city–suburb divides, and degrading the environment. They see the promise for cities to be "engines of prosperity," and argue for "smart growth" initiatives and regional planning policies to bridge the growing economic and political gaps between inner cities and suburbs. They give concrete suggestions for building effective political coalitions to achieve a "metropolitics" to implement more regional planning policies that foster more livable and social equitable environments while promoting economic growth.

The problems of urban sprawl and unsustainable economic development are put in more global context by William Rees and Mathis Wackernagel. They are pioneers in developing estimation models for "ecological footprint" analysis that measure the human load on the urban and regional environment. They argue that cities are entropic and dissipative structures, "black holes" of goods and energy consumption and waste generation. They give policy proposals for making cities more sustainable.

RACIAL AND SOCIAL INEQUALITY

Part Three turns to matters of racial and social inequality, with attention first to the problems of segregation that have confronted the inner city poor and the African American population and persisted in changing forms over the last one hundred years. We turn to examine the more positive impacts of segregation in immigrant enclaves that generate jobs, revenues, and chances for social mobility for ethnic participants. We close with selections on the social experiences of homeless populations and questions of race and class effects on the experiences of people displaced by Hurricane Katrina.

W. E. B. Du Bois starts with a selection that examines the housing problems confronting blacks who were segregated in the Seventh Ward of Philadelphia at the turn of the twentieth century. He draws attention to severe overcrowding, lack of fresh water and adequate sanitation, and building of residential tenements in rear lots. Black residents tolerated relatively high rents near the central city to be near jobs in the white community. Sub-letting and predatory rents were commonly found. He identified a four-tiered social stratification center in the African American community. His study mixes the intentions of social science with social reformism.

Loïc Wacquant and William Wilson begin with a chapter that considers the "hyperghettoization" of the ghetto underclass in the US despite the passage of civil rights legislation in the 1960s. Suburbanization and economic change has left the underclass socially isolated in the inner city, bereft of access to good jobs, schools, and housing. Douglas Massey and Nancy Denton place greater emphasis on racial issues with their outrage at the "hypersegregation" that continues to plague the black ghetto with the failure of civil rights legislation in America. They decry the condition of "American Apartheid" that is the result of discriminatory lending, racial steering by realtors, and redlining practices by banks and mortgage companies. White flight leads to the disenfranchisement of the white suburbs from the needs and concerns of the black inner cities. The spatial isolation of the underclass intensifies the presence of an "oppositional culture" in the ghetto that further marginalizes the residents from mainstream society. Massey and Denton point to a very clear dialectic between spatial segregation and social exclusion.

Alejandro Portes and Robert Manning examine the phenomenon of immigrant enclaves, which are alternative sub-economies separate from the US primary sector economy and the low-wage secondary sector where poor minorities proliferate. Immigrant enclaves comprise a protected sector offering job opportunities in often exploitative working conditions. While some immigrant enterprises primarily serve co-ethnic clientele, other immigrant businesses serve as "middlemen" between white elites and poor minorities. These middleman minorities act as a social buffer between the rich and the poor and may sometimes bear the brunt of underclass anger against the broader system that marginalizes them.

The Chinese enclave, or "ethnoburb," of the suburban San Gabriel Valley cities of Los Angeles are addressed by Jan Lin and Paul Robinson. They observe that ethnoburbs challenge traditional assumptions of human ecology that immigrant colonies would wither with the spatial and social mobility of immigrants into American society. They found evidence in the ethnoburb of a lower-class core and two middle-class fringe districts, with some evidence of cultural assimilation, but broader trends of persisting Chinese language use. Limited English-speaking ability, they found, was not a barrier to socioeconomic mobility. White flight and racial/ethnic succession to an immigrant Chinese population in some San Gabriel Valley cities was accompanied by nativist politics and opposition to the rapid economic growth, traffic congestion, and Chinese signage that was brought by immigrant capital.

James Duncan discusses the spatial and social exclusion of the homeless population of cities, who are seen as a threat to social and moral order. He describes how they are segregated into marginal spaces and jurisdictional voids of the city. They manage to occupy some prime public spaces while observing certain behavioral etiquettes that satisfy the law enforcement and public gaze. The people displaced by Hurricane Katrina are the subject of the selection by James Elliot and Jeremy Pais, who consider how race and class variables affected the ability of the population to evacuate in advance of and after the storm hit. They found that white residents were more likely to leave before the storm hit than black residents.

GENDER AND SEXUALITY

The selections in Part Four consider the social positions of women and sexual minorities in the city. Markusen considers the idea that women are "segregated" in the suburbs. The spatial entrapment of women in suburban housing is culturally reinforced, and this segregation impedes their job opportunities and social mobility as long as employment is based in the central city. Ann Markusen asserts that the suburbs were a creation of patriarchy, rather than traditional explanations that the suburbs were the benign outcome of consumer choice, highway lobbyists, suburban developers, or the postwar Federal Housing Administration. She believes that the widespread separation of residential suburbs from central city workplaces leads to gross inefficiencies due to commuting times and energy consumption, as well as the alienation of many individuals and families. The privatizing of family life in the suburbs fosters the decline of extended family networks and community social capital. Markusen calls for urban social policies that re-collectivize child-care facilities, housing, job opportunities, and recreational activities for families in America. We invite speculation on the impact of the Internet on the economic and cultural life of women in the suburbs.

Melissa Gilbert considers a different kind of spatial entrapment, that of poor women of color in the inner city. She finds that poor women can adapt to this spatial segregation, effectively generating opportunity out of constraint, by exploiting the social capital inherent in dense networks of kin and friends to procure child-care assistance, housing, education, and other resources. Sy Adler and Johanna Brenner similarly explore the interpersonal social capital networks that operate among lesbian women in the city. Lesbian colonies and social networks are relatively invisible compared to the high visibility of gay men in the city. Donald Donham explores the new visibility and more public sexual culture available to gay men in South Africa in the post-apartheid period.

Sirpa Tani turns to an examination of the informal sex economy in Helsinki, Finland. Tani describes how spaces of prostitution are socially constructed as well as contested. She recounts how public officials and the media represented women as either non-sexual victims of prostitution or as prostitutes. In response to this stigmatization, women residents fought the stigmatization of their neighborhood and gender identity by waging campaigns against prostitution.

GLOBALIZATION AND TRANSNATIONALISM

Part Five moves to the macro-level to explore the phenomenon of globalization and transnational identity. John Friedman is the seminal writer on this subject, and his chapter considers the rise since the 1960s of world cities as command centers in the new international division of labor. Saskia Sassen considers the strategic advantages of global city positioning in advancing the interests of political and economic elites. At the same time, global cities require a low-wage labor force to satisfy key operations. New forms of politics are arising in global cities, where the disadvantaged labor force composed primarily of immigrants and people of color has found the social space to make new claims. Brenda Yeoh and T. C. Chang examine how state managers in Singapore have balanced policies attractive to maintaining the presence of transnational elites while being sensitive to the presence of local heritage and communities. They address four categories of transnational population in Singapore, including business elites, the "third world" population, expressive specialists, and tourists.

Paul Stoller and Jasmin Tahmaseb McConatha take a closer look at processes of globalization from the ground up through their study of West African street vendors in New York City as a transnational community network. Their experience of transnational community is enhanced through the intermediation of Muslim fellowship, informal and formal mutual aid associations for West African immigrants in New York. Peggy Levitt turns to the presence of transnational community among Dominican immigrants to Boston. She draws attention to the importance of social remittances (of ideas, behaviors, and social capital) in maintaining immigrant ties to their home communities. She addresses the transfer of innovative practices in education, business, and health care between sending and receiving societies. The experience of transnational villagers, she notes, transforms earlier assumptions on the nature of immigrant assimilation to U.S. society.

CULTURE AND THE CITY

Part Six examines the growth of the cultural economy, a phenomenon that has counterbalanced some of the decline in manufacturing employment in many cities. The cultural economy is partly related to the sectors known as the "information economy" or "high technology" but includes a wider array of activities related to the consumption of culture and cultural products. Sharon Zukin uses the term "symbolic economy" to describe the ensemble of activities that includes the arts, entertainment, sports, fashion, restaurants, and tourism. Public officials and other growth machine interests have begun to perceive the advantages of linking these kinds of activities to the redevelopment and gentrification process in central cities, provoking the displacement of marginal players such as the homeless, the poor, and mom-and-pop businesses. As the inner city is restructured and "revitalized," the "other" is simultaneously evicted in a physical sense, and appropriated in a symbolic sense for middle-class consumers. A series of conflicts and controversies are provoked over the ownership and control of culture, in its aesthetic, historical, and ethnic dimensions. Richard Florida offers a more boosterish perspective on the cultural economy of cities with his concept of "creative capital," which is a variant of the "social capital" concept of Robert Putnam and others. He has provoked great interest from scholars, planners, and public officials through his attention to creative capital as a strategy for promoting local economic innovation and regional growth.

Edward LiPuma and Thomas Koelble examine how culture has been used to frame place identity through the construction of an "urban imaginary." As a quintessential site of racial and ethnic diversity, Miami reveals some expected interethnic tensions but also many surprising opportunities for intergroup connections and hybridized identities. Despite its complex and conflicted cultural composition, Miami manages to sustain a stable place-identity for the global marketplace that idealizes many aspects of its cultural diversity while downplaying others. Karin Aguilar San-Juan examines the use of culture in the framing of place identity in the Vietnamese enclaves of Little Saigon in Southern California's Orange County and Fields Corner in Boston. She finds differences in the role of ethnic entrepreneurs and the extent of collaboration with other ethnic communities across the two districts.

REGULATION AND RIGHTS IN URBAN SPACE

Part Seven examines the growing use of legal controls and zoning regulations by state managers, business and social elites, and planners to exclude socially undesirable populations from urban spaces and limit their full participation in society. Sally Engle Merry describes the rise of tactics of spatial governmentality that include use of "zero tolerance" policies to regulate certain prime urban spaces as "off limits" to unwanted social groups. These mechanisms of crime control foster social exclusion and protect investments in the upscale redevelopment of city centers. Setha Low addresses the growth of social regulation as evident in the long-run decline in urban public spaces, including streets, sidewalks, parks, plazas, and open areas for public use. She addresses the growing redefinition of urban public space to serve the interests of risk-averse homeowners, elites, business, governments, and planners. She examines in particular the growing privatization of ownership and management of public spaces in Lower Manhattan, and the growing exclusion of youth subcultures, the poor, and other undesirable groups. Her essay takes on sharper relevance with the emergence of the Occupy Wall Street movement at Zuccotti Park in 2011 and its subsequent repression by the City of New York.

Teresa Caldeira paints a portrait of social inequality and spatial segregation in São Paolo, where middle- and upper-class residents barricade themselves in gated communities for fear of the dangerous "other" beyond high-security gates. She finds comparable trends occurring in Los Angeles, where gated communities promote security consciousness along with status enhancement. Christopher Mele finds social regulation and exclusion at work in Chester, Pennsylvania, where neoliberal reinvestment strategies have been at work to attract investment capital in an eclectic mix of power and waste management industries as well as sports and entertainment complexes. As municipal power has been ceded to private interests, black residents and other disenfranchised groups have lost control over decisions that shape their city.

The last two readings elicit the spirit of the great French critical urban theorist Henri Lefebvre, who was inspired by the social protests of the 1960s to advocate for a form of urban politics that promoted the inherent rights of the inhabitants. James Holston applies this perspective to cities in the developing world, such as Brasilia, where modernist planning failed to produce enlightened public spaces and engaged civic life. He reveals the possibilities for more democratic and engaged public life in a diverse assortment of spaces of insurgent citizenship, including squatter settlements and spatial occupations by dispossessed, disentitled, and stateless population groups. David Harvey contends that our contemporary concept of rights has been normalized to concerns for private property and profit making. He contends that we have to revise our comprehension of natural rights for any kind of emancipatory urban politics to take hold. Harvey believes we cannot separate ourselves from the kind of society we want to (re)invent. It is a fundamental human right to re-imagine and re-make the city, as the city delivers us the opportunity to change ourselves.

PART ONE

Urbanization and Community

Plate 1 A transatlantic arriving in the harbor, New York City, USA 1959 by Henri Cartier-Bresson. Reproduced by permission of Magnum Photos.

Plate 2 Skateboarder and biker. Reproduced by permission of Corbis.

INTRODUCTION TO PART ONE

A central intellectual question since the original days of urban sociology is: "What are the effects of urbanization on community?" The European founders of the field of sociology such as Emile Durkheim and Ferdinand Tönnies were witness to rapid urbanization coupled with the locomotive onrush of industrialization in the capitalist societies of the late nineteenth century. The decline of primary, sentimental relationships and small group solidarities with the absorption of the rural into the urban world seemed quite inevitable at the time. Sociology originated as a positivistic science that compared the progress from rural societies to urban societies, to the biological maturation of natural organisms from infancy to maturity. The belief in the Enlightenment and the idea of progress underlay the theory of evolution and species development from simple to complex organisms. Classic urban sociology linked the phenomenon of specialization in the division of labor to the phenomenon of social differentiation in urban society.

Understanding the character of urbanism as a way of life is another seminal subject of urban sociology. Ferdinand Tönnies described the rural–urban shift through the conceptual categories of *Gemeinschaft* (community) and *Gesellschaft* (urban society). These concepts are good illustrations of *ideal types* in sociological analysis. The ideal type functions as an analytical paradigm or model that can be analyzed and tested for its validity through comparison. Tönnies did not consider these societal types as mutually exclusive polar opposites, but as two categories in a continuum of societies undergoing social change. The shift from *Gemeinschaft* to *Gesellschaft* may be compared with Emile Durkheim's conception of society undergoing a transition from mechanical to organic solidarity. What Tönnies described as *kurville*, or collective will, is similar to what Durkheim described as collective consciousness, a collective soul or conscience that guides group behavior. The state was seen to act fairly and judiciously as the will of the people.

Both Tönnies and Durkheim recognized the fading of primary bonds of kinship, sentiment, and community life, with the ascendance of secondary bonds of occupational, legal, and political association. Tönnies somewhat romanticized the loss of *Gemeinschaft* but in fact he saw *Gesellschaft* as a rational and necessary vehicle for guiding a more specialized and diverse society. The governmental state guaranteed that urbanism as way of life would guarantee rights, civility, and security to urban residents. Tönnies' outlook on a rational and specialized urban society led by a legitimate state is a contrast from the Marxian view on class struggle and inequality in the division of labor. Tönnies was concerned that *Gesellschaft* be kept honest and not be sabotaged by corruption or kidnapped by totalitarian political interests. Durkheim, in contrast, was more concerned with the moral consequences of the rise of *anomie* caused by *Gesellschaft* society.

Georg Simmel had a less sentimental view of the decline of *Gemeinschaft.* He recognized factors of intensification that assaulted the psychological life of urbanites, fostering anonymity and impersonality in urban life. The importance of money in a capitalist society, he furthermore believed, contributed to a calculating and discriminating nature to the urban personality. Simmel viewed metropolitan man as blasé, jaded, and materialistic. Yet urbanism also promoted cosmopolitanism, which fostered greater social tolerance for unconventional behaviors and freedom from provinciality and prejudice. The oversaturation of our social life with materialism, superficiality, and objective values, however, has suppressed our subjectivity, spirituality, and social life. The urban personality is both bombarded and liberated by the sensory commercial marketplaces of modern capitalism. For Simmel, the experience of modern urban life is suffused with the experience of a money economy where quality has been reduced to quantity and consumers are materially rewarded but spiritually deprived. There is a loneliness that is brought about by an affluent society that has freed people to explore their individualism but left their souls in a state of restlessness and flux.

Louis Wirth updated the Durkheimian view on the decline of group solidarity to analysis of the modern American city. Wirth perceived that factors of size, density, and heterogeneity fostered role segmentation through the emancipation of the individual from traditional rules and mores. He clearly articulated the resulting normlessness, or *anomie*, the social void, which contributed to a spectrum of urban social problems, such as crime, delinquency, mental breakdown, and other forms of psychological and social disorganization. He updated Robert Park's famous quote regarding the city (see "The City: Suggestions for the Investigation of Human Behavior in the City Environment," *American Journal of Sociology* 20, 5 [March 1915]: 577–612) as "a mosaic of social worlds which touch but do not interpenetrate" to the concept of the city as a "mosaic of social worlds in which the transition from one to the other is abrupt." Geographic mobility, the growing decline of traditional norms and mores, and social heterogeneity were breeding social and personality disorders in the city. Wirth felt that sociologists had a mission to analyze and ameliorate urban social problems.

Claude Fischer reformulated Wirthian urbanism, applying the urban factors of size, density, and heterogeneity to the idea of creating rather than destroying communities. He argued that size and density of population in cities created "critical mass" sufficient to formulate new subcultural communities. The increasingly heterogeneous "mosaic of social worlds" further intensified subcultures through his concept that they touch, but then "recoil, with sparks flying." His concept of subculture includes an eclectic assortment of special hobbyists, interest groups, artists, innovative thinkers, ethnic groups, religious subcultures, homosexuals, and others commonly classified as "deviant." That they congregate socially and spatially as communities reverses the traditional thinking that urbanism leads to the decline of community and the growth of social disorganization. Fischer sees cities as diverse mosaics of heterogeneous neighborhoods that are crucibles for the exploration of subcultural diversity and social difference.

Fischer contributes to a growing view voiced by other writers such as Jacobs, and Wellman and Leighton, that there has been a popular renaissance and transformation of what we understand as "community" in the contemporary city. There is a kaleidoscopic array of new community forms in the city of the new millennium. Some revive the traditional enclaves of the old *Gemeinschaft*, like the "urban villages" that are nodes for the incorporation of international immigrants to the global city. Subcultural communities are more emergent phenomena that are formed out of new social networks of friendship and association, sometimes with an outsider status against the cultural mainstream. New technology, including the Internet, further widens the opportunities for social networking. The growth of new communities is also strongly connected with the rise of neighborhood-based mobilizations and other "urban social movements" that since the 1960s have risen to contest urban power brokers and the political establishment. The community resurgence has achieved growing public support, and promoted neighborhood planning as an antidote to the callousness of large, centralized planning bureaucracies.

The spirit of popular insurgency was codified through the writing and activism of Jane Jacobs, a fierce architectural critic of modernist planning that also stood up to New York power broker Robert Moses and helped organize a neighborhood movement and save Greenwich Village against the plan for an intercity freeway through Lower Manhattan. Jacobs gave voice to a scathing critique of the rational-bureaucratic state that promoted misguided urban renewal policies and destroyed vibrant neighborhoods throughout the nation for freeway building or new construction of architecturally dull housing towers surrounded by indefensible spaces. In her selection she expands on the importance of neighborhoods as "organs of self-government" that possess a natural ability to guide the quality of urban life, but have been marginalized and disempowered by powerful and insensitive centralized institutions. She asserts the most successful neighborhood-based political districts possess dense, territorially bounded social leadership networks with powers of communicative and political mobilization. The most powerful and stable neighborhood districts, she asserts, are socioeconomically and culturally diverse.

Wellman and Leighton similarly note the rise of "saved" and "liberated" communities that have challenged the older determinism of "community lost," but contest the importance of neighborhoods in the ongoing mobilization and experience of community in our society. They draw attention to the growth of new kinds of communities liberated from the attachment to space that are constructed through more sparsely knit and loosely bounded social networks than traditional place-based communities. Innovations in communications and transport technology have enabled the rise of more "communities without propinquity" and the growth of the Internet has enabled new forms of social networking in the public sphere of the media. Though loose networks are more weakly sentimental or intimate, they contend that "liberated communities" are constructed of ramified, branching networks giving access to greater outside resources.

Robert Putnam, by contrast, downplays the sociological force of some of the new voluntary and membership organizations in our society based on ramified networks, such as the Sierra Club, the National Organization for Women, and the American Association of Retired Persons. Though these non-profit, "third sector" organizations might have significant political clout, he contrasts the growth of dues-paying interest group politics from the kinds of civic engagement that was mobilized by traditional organizations like Parent–Teacher Associations, the Boy Scouts, and the Red Cross. The decline of these traditional organizations has also meant the loss of generalized trust, reciprocity, and social capital in our society, he asserts, hobbling civic engagement and fostering greater individualism and political apathy in America. The decline of recreational bowling leagues in a nation now intent on "bowling alone" is his metaphor for the loss of intimate dense networks of sociability and civic engagement.

While some of Putnam's assertions may be debated, there is growing interest in the relevance of the social capital concept as a way of understanding how economic decline and the out-movement of people from the inner city may be correlated to the loss of vital social networks and civic institutions. We may ponder to what degree ramified social networks and liberated communities are more concentrated in the affluent classes, which possess greater access to new technologies, resources, and other economic and political interests. While white middle-class communities may have been some of the pioneering participants in urban social movements and ramified social networking, racial/ethnic minorities and other subcultural communities have growing clout in neighborhood empowerment movements and participation in Internet social networking. The resurgence of revolutionary political activity in the Arab world in the spring of 2011 has certainly increased public and intellectual interest in the power of social networking as a tool for promoting protest politics among oppressed minorities. To what degree social networking can sustain a secure civic life, durable political institutions, and judicious democracies is a question for our continuing consideration.

"Community and Society"

from C. P. Loomis (ed.), *Community and Society* (1963) [1887]

Ferdinand Tönnies

Editors' Introduction

Ferdinand Tönnies (1855–1936) was born into a wealthy farming family in Schleswig-Holstein, Germany, in an era in which the peasant culture of the rural province was being transformed by mechanization and the money economy. His oldest brother was engaged in a thriving trade with English merchants, exposing Tönnies first-hand to the world of English capitalism. In 1881 he became a lecturer at the University of Kiel, where he remained until ousted by the Nazis in 1933 because of his social democratic political associations. Though less influential than his contemporaries Max Weber and Emile Durkheim, Tönnies may be recognized as a founding father of sociology.

His enduring contribution to urban sociology is the distinction between two basic types of social formations, *Gemeinschaft* (community) and *Gesellschaft* (society), with a general historical trend from the former to the latter. Societies of the earlier form are organized around family, village, and town, with a mainly agricultural economy and local political culture. The latter form of society, by contrast, is exemplified by larger-level social units of metropolis and nation-state, and based on complex trade and industry. Primary sentimental relationships predominate in *Gemeinschaft*, while secondary associational relationships proliferate in *Gesellschaft*. While some of his interpreters proliferated the impression that Tönnies sentimentalized *Gemeinschaft* while criticizing *Gesellschaft*, he disclaimed such intention. For him, the shift was a normal developmental process of the body social, comparable to the transition from youth to adulthood.

Tönnies was strongly influenced by English thinkers, including the political philosopher Thomas Hobbes, Sir Henry Maine, and the Social Darwinist Herbert Spencer. The concept of will was central to his theory. Tönnies argued that there are two basic forms of human volition, or will. *Gemeinschaft* is formed around *Wesenwille*, or essential will, which is the underlying, organic, self-fulfilling or instinctive driving force, while *Gesellschaft* is characterized by *Kurwille*, or arbitrary will, which is deliberative, purposive, instrumental, and future (goal) oriented. *Wesenwille* is that which springs intrinsically from a person's temper and character. *Kurwille* is the capacity to distinguish means from ends and to act practically out of rational self-interest.

Tönnies decried totalitarianism (including the Nazism that emerged in Germany), but he was intrigued by the force of "public opinion" that enforces the communal will of society and may involve the use of sanctions against dissidents. He dealt with these ideas in other publications, including *Die Sitte* (1909) and *Critique of Public Opinion* (Lanham: Rowman & Littlefield, 2002, edited and translated by Hanno Hardt and Slavko Splichal from *Kritik der Offentlichen Meinung*, 1922). His concept of *Kurwille* can thus be related to the Hobbesian social contract, whereby citizens control the state through deliberation and reasoned discussion to counter tyrannical authority and avaricious despotism.

Tönnies developed his concepts *Gemeinschaft/Gesellschaft* as "ideal types," which are paradigms or models that may not fully conform to social reality, but are useful for purposes of analytical comparison. Rather than being polar extremes, the two ideal types can be seen as being on opposite ends of a continuum. Tönnies conceived of any society as always to some degree possessing characteristics of both ideal types. The original concept of ideal types may be credited to the German sociologist Max Weber. *Gemeinschaft* may be compared with the traditional society conceived by the French sociologist Emile Durkheim (*The Division of Labor in Society* [1893], translated by George Simpson. New York: Free Press, 1933) through his notion of mechanical solidarity, characterized by a simple division of labor and a morally homogeneous population bound by similar values and beliefs. *Gesellschaft* corresponds with Durkheim's notion of organic solidarity, found in the modern society that has a complex division of labor and a heterogeneous population held together by interdependency, laws, and contracts. The American sociologist Robert Redfield, on the basis of fieldwork in rural Mexico, later characterized the traditional society as the "folk society" ("The Folk Society," *American Journal of Sociology* 52 [1947], 293–308).

Gemeinschaft und Gesellschaft is also available in an earlier edition, which also contained some of Tönnies' later essays, as *Fundamental Concepts of Sociology* (Oxford: American Book Co., 1940). Tönnies' ten other books, of which the major work dealing with sociology is his 1931 *Einführung in die Soziologie* (*An Introduction to Sociology*; Stuttgart: Ferdinand Enke), plus most of his essays, still await English translations. A full bibliography of Tönnies' work can be found in *American Journal of Sociology*, 42 (1937), 100–101. A brief critique of Tönnies' works can be found in Louis Wirth, "The Sociology of Ferdinand Tönnies," *American Journal of Sociology*, 32 (1927), 412–422.

ORDER – LAW – MORES

There is a contrast between a social order which – being based upon consensus of wills – rests on harmony and is developed and ennobled by folkways, mores, and religion, and an order which – being based upon a union of rational wills – rests on convention and agreement, is safeguarded by political legislation, and finds its ideological justification in public opinion.

There is, further, in the first instance a common and binding system of positive law, of enforcible norms regulating the interrelation of wills. It has its roots in family life and is based on land ownership. Its forms are in the main determined by the code of the folkways and mores. Religion consecrates and glorifies these forms of the divine will, i.e., as interpreted by the will of wise and ruling men. This system of norms is in direct contrast to a similar positive law which upholds the separate identity of the individual rational wills in all their interrelations and entanglements. The latter derives from the conventional order of trade and similar relations but attains validity and binding force only through the sovereign will and power of the state. Thus, it becomes one of the most important instruments of policy; it sustains, impedes, or furthers social trends; it is defended or contested publicly by doctrines and opinions and thus is changed, becoming more strict or more lenient.

There is, further, the dual concept of morality as a purely ideal or mental system of norms for community life. In the first case, it is mainly an expression and organ of religious beliefs and forces, by necessity intertwined with the conditions and realities of family spirit and the folkways and mores. In the second case, it is entirely a product and instrument of public opinion, which encompasses all relations arising out of contractual sociableness, contacts, and political intentions.

Order is natural law, law as such = positive law, mores = ideal law. Law as the meaning of what may or ought to be, of what is ordained or permitted, constitutes an object of social will. Even the natural law, in order to attain validity and reality, has to be recognized as positive and binding. But it is positive in a more general or less definite way. It is general in comparison with special laws. It is simple compared to complex and developed law.

DISSOLUTION

The substance of the body social and the social will consists of concord, folkways, mores, and religion, the manifold forms of which develop under favorable conditions during its lifetime. Thus, each individual receives his share from this common center, which is manifest in his own sphere, i.e., in his sentiment, in his mind and heart, and in his conscience as well as in his environment, his possessions, and his activities. This is also true of each group. It is in this center that the individual's strength is rooted, and his rights derive, in the last instance, from the one original law which, in its divine and natural character, encompasses and sustains him, just as it made him and will carry him away. But under certain conditions and in some relationships, man appears as a free agent (person) in his self-determined activities and has to be conceived of as an independent person. The substance of the common spirit has become so weak or the link connecting him with the others worn so thin that it has to be excluded from consideration. In contrast to the family and co-operative relationship, this is true of all relations among separate individuals where there is no common understanding, and no time-honored custom or belief creates a common bond. This means war and the unrestricted freedom of all to destroy and subjugate one another, or, being aware of possible greater advantage, to conclude agreements and foster new ties. To the extent that such a relationship exists between closed groups or communities or between their individuals or between members and non members of a community, it does not come within the scope of this study. In this connection we see a community organization and social conditions in which the individuals remain in isolation and veiled hostility toward each other so that only fear of clever retaliation restrains them from attacking one another, and, therefore, even peaceful and neighborly relations are in reality based upon a warlike situation. This is, according to our concepts, the condition of Gesellschaft-like civilization, in which peace and commerce are maintained through conventions and the underlying mutual fear. The state protects this civilization through legislation and politics. To a certain extent science and public opinion, attempting to conceive it as necessary and eternal, glorify it as progress toward perfection.

But it is in the organization and order of the Gemeinschaft that folk life and folk culture persist. The state, which represents and embodies Gesellschaft, is opposed to these in veiled hatred and contempt, the more so the further the state has moved away from and become estranged from these forms of community life. Thus, also in the social and historical life of mankind there is partly close interrelation, partly juxtaposition and opposition of natural and rational will.

THE PEOPLE (VOLKSTUM) AND THE STATE (STAATSTUM)

In the same way as the individual natural will evolves into pure thinking and rational will, which tends to dissolve and subjugate its predecessors, the original collective forms of Gemeinschaft have developed into Gesellschaft and the rational will of the Gesellschaft. In the course of history, folk culture has given rise to the civilization of the state.

The main features of this process can be described in the following way. The anonymous mass of the people is the original and dominating power which creates the houses, the villages, and the towns of the country. From it, too, spring the powerful and self-determined individuals of many different kinds: princes, feudal lords, knights, as well as priests, artists, scholars. As long as their economic condition is determined by the people as a whole, all their social control is conditioned by the will and power of the people. Their union on a national scale, which alone could make them dominant as a group, is dependent on economic conditions. And their real and essential control is economic control, which before them and with them and partly against them the merchants attain by harnessing the labor force of the nation. Such economic control is achieved in many forms, the highest of which is planned capitalist production or large-scale industry. It is through the merchants that the technical conditions for the national union of independent individuals and for capitalistic production are created. This merchant class is by nature, and mostly also by origin, international as well as national and urban, i.e., it belongs to Gesellschaft, not Gemeinschaft. Later all social groups and dignitaries and, at least in tendency, the whole people acquire the characteristics of the Gesellschaft.

Men change their temperaments with the place and conditions of their daily life, which becomes hasty and changeable through restless striving. Simultaneously, along with this revolution in the social order, there takes place a gradual change of the law, in meaning as well as in form. The contract as such becomes the basis of the entire system, and rational will of Gesellschaft, formed by its interests, combines with authoritative will of the state to create, maintain and change the legal system. According to this conception, the law can and may completely change the Gesellschaft in line with its own discrimination and purpose; changes which, however, will be in the interest of the Gesellschaft, making for usefulness and efficiency. The state frees itself more and more from the traditions and customs of the past and the belief in their importance. Thus, the forms of law change from a product of the folkways and mores and the law of custom into a purely legalistic law, a product of policy. The state and its departments and the individuals are the only remaining agents, instead of numerous and manifold fellowships, communities, and commonwealths which have grown up organically. The characters of the people, which were influenced and determined by these previously existing institutions, undergo new changes in adaptation to new and arbitrary legal constructions. These earlier institutions lose the firm hold which folkways, mores, and the conviction of their infallibility gave to them.

Finally, as a consequence of these changes and in turn reacting upon them, a complete reversal of intellectual life takes place. While originally rooted entirely in the imagination, it now becomes dependent upon thinking. Previously, all was centered around the belief in invisible beings, spirits and gods; now it is focalized on the insight into visible nature. Religion, which is rooted in folklife or at least closely related to it, must cede supremacy to science, which derives from and corresponds to consciousness. Such consciousness is a product of learning and culture and, therefore, remote from the people. Religion has an immediate contact and is moral in its nature because it is most deeply related to the physical–spiritual link which connects the generations of men. Science receives its moral meaning only from an observation of the laws of social life, which leads it to derive rules for an arbitrary and reasonable order of social organization. The intellectual attitude of the individual becomes gradually less and less influenced by religion and more and more influenced by science. Utilizing the research findings accumulated by the preceding industrious generation, we shall investigate the tremendous contrasts which the opposite poles of this dichotomy and these fluctuations entail. For this presentation, however, the following few remarks may suffice to outline the underlying principles.

TYPES OF REAL COMMUNITY LIFE

The exterior forms of community life as represented by natural will and Gemeinschaft were distinguished as house, village, and town. These are the lasting types of real and historical life. In a developed Gesellschaft, as in the earlier and middle stages, people live together in these different ways. The town is the highest, viz., the most complex, form of social life. Its local character, in common with that of the village, contrasts with the family character of the house. Both village and town retain many characteristics of the family; the village retains more, the town less. Only when the town develops into the city are these characteristics almost entirely lost. Individuals or families are separate identities, and their common locale is only an accidental or deliberately chosen place in which to live. But as the town lives on within the city, elements of life in the Gemeinschaft, as the only real form of life, persist within the Gesellschaft, although lingering and decaying. On the other hand, the more general the condition of Gesellschaft becomes in the nation or a group of nations, the more this entire "country" or the entire "world" begins to resemble one large city. However, in the city and therefore where general conditions characteristic of the Gesellschaft prevail, only the upper strata, the rich and the cultured, are really active and alive. They set up the standards to which the lower strata have to conform. These lower classes conform partly to supersede the others, partly in imitation of them in order to attain for themselves social power and independence. The city consists, for both groups (just as in the case of the "nation" and the "world"), of free persons who stand in contact with each other, exchange with each other and cooperate without any Gemeinschaft or will thereto developing among them except as such might develop sporadically or as a leftover from former conditions. On the contrary, these numerous external contacts, contracts, and contractual relations only cover up as many inner hostilities and antagonistic interests. This is especially true of the antagonism between the rich or the so-called cultured class and the poor or the servant class, which try to obstruct and destroy each other. It is this contrast which, according to Plato, gives the "city" its dual character and makes it divide in itself. This itself, according to our concept, constitutes the city, but the same contrast is also manifest in every large-scale relationship between capital and labor. The common town life remains within the Gemeinschaft of family and rural life; it is devoted to some agricultural pursuits but concerns itself especially with art and handicraft which evolve from these natural needs and habits. City life, however, is sharply distinguished from that; these basic activities are used only as means and tools for the special purposes of the city.

The city is typical of Gesellschaft in general. It is essentially a commercial town and, in so far as commerce dominates its productive labor, a factory town. Its wealth is capital wealth which, in the form of trade, usury, or industrial capital, is used and multiplies. Capital is the means for the appropriation of products of labor or for the exploitation of workers. The city is also the center of science and culture, which always go hand in hand with commerce and industry. Here the arts must make a living; they are exploited in a capitalistic way. Thoughts spread and change with astonishing rapidity. Speeches and books through mass distribution become stimuli of far-reaching importance.

The city is to be distinguished from the national capital, which, as residence of the court or center of government, manifests the features of the city in many respects although its population and other conditions have not yet reached that level. In the synthesis of city and capital, the highest form of this kind is achieved: the metropolis. It is the essence not only of a national Gesellschaft, but contains representatives from a whole group of nations, i.e., of the world. In the metropolis, money and capital are unlimited and almighty. It is able to produce and supply goods and science for the entire earth as well as laws and public opinion for all nations. It represents the world market and world traffic; in it world industries are concentrated. Its newspapers are world papers, its people come from all corners of the earth, being curious and hungry for money and pleasure.

COUNTERPART OF GEMEINSCHAFT

Family life is the general basis of life in the Gemeinschaft. It subsists in village and town life. The village community and the town themselves can be considered as large families, the various clans and houses representing the elementary organisms of its body; guilds, corporations, and offices, the tissues and organs of the town. Here original kinship and inherited status remain an essential, or at least the most important, condition of participating fully in common property and other rights. Strangers may be accepted and protected as serving-members or guests either temporarily or permanently. Thus, they can belong to the Gemeinschaft as objects, but not easily as agents and representatives of the Gemeinschaft. Children are, during minority, dependent members of the family, but according to Roman custom they are called free because it is anticipated that under possible and normal conditions they will certainly be masters, their own heirs. This is true neither of guests nor of servants, either in the house or in the community. But honored guests can approach the position of children. If they are adopted or civic rights are granted to them, they fully acquire this position with the right to inherit. Servants can be esteemed or treated as guests or even, because of the value of their functions, take part as members in the activities of the group. It also happens sometimes that they become natural or appointed heirs. In reality there are many gradations, lower or higher, which are not exactly met by legal formulas. All these relationships can, under special circumstances, be transformed into merely interested and dissolvable interchange between independent contracting parties. In the city such change, at least with regard to all relations of servitude, is only natural and becomes more and more widespread with its development. The difference between natives and strangers becomes irrelevant. Everyone is what he is, through his personal freedom, through his wealth and his contracts. He is a servant only in so far as he has granted certain services to someone else, master in so far as he receives such services. Wealth is, indeed, the only effective and original differentiating characteristic; whereas in Gemeinschaften property it is considered as participation in the common ownership and as a specific legal concept is entirely the consequence and result of freedom or ingenuity, either original or acquired. Therefore, wealth, to the extent that this is possible, corresponds to the degree of freedom possessed.

In the city as well as in the capital, and especially in the metropolis, family life is decaying. The more and the longer their influence prevails the more the residuals of family life acquire a purely accidental character. For there are only few who will confine their energies within such a narrow circle; all are attracted outside by business, interests, and pleasures, and thus separated from one another. The great and mighty, feeling free and independent, have always felt a strong inclination to break through the barriers of the folkways and mores. They know that they can do as they please. They have the power to bring about changes in their favor, and this is positive proof of individual arbitrary power. The mechanism of money, under usual conditions and if working under high pressure, is means to overcome all resistance, to obtain everything wanted and desired, to eliminate all dangers and to cure all evil. This does not hold always. Even if all controls of the Gemeinschaft are eliminated, there are nevertheless controls in the Gesellschaft to which the free and independent individuals are subject. For Gesellschaft (in the narrower sense), convention takes to a large degree the place of the folkways, mores, and religion. It forbids much as detrimental to the common interest which the folkways, mores, and religion had condemned as evil in and of itself.

The will of the state plays the same role through law courts and police, although within narrower limits. The laws of the state apply equally to everyone; only children and lunatics are not held responsible to them. Convention maintains at least the appearance of morality; it is still related to the folkways, mores, and religious and aesthetic feeling, although this feeling tends to become arbitrary and formal. The state is hardly directly concerned with morality. It has only to suppress and punish hostile actions which are detrimental to the common weal or seemingly dangerous for itself and society. For as the state has to administer the common weal, it must be able to define this as it pleases. In the end it will probably realize that no increase in knowledge and culture alone will make people kinder, less egotistic, and more content and that dead folkways, mores, and religions cannot be revived by coercion and teaching. The state will then arrive at the conclusion that in order to create moral forces and moral beings it must prepare the ground and fulfill the necessary conditions, or at least it must eliminate counteracting forces. The state, as the reason of Gesellschaft, should decide to destroy Gesellschaft or at least to reform or renew it. The success of such attempts is highly improbable.

THE REAL STATE

Public opinion, which brings the morality of Gesellschaft into rules and formulas and can rise above the state, has nevertheless decided tendencies to urge the state to use its irresistible power to force everyone to do what is useful and to leave undone what is damaging. Extension of the penal code and the police power seems the right means to curb the evil impulses of the masses. Public opinion passes easily from the demand for freedom (for the upper classes) to that of despotism (against the lower classes). The makeshift convention has but little influence over the masses. In their striving for pleasure and entertainment they are limited only by the scarcity of the means which the capitalists furnish them as price for their labor, which condition is as general as it is natural in a world where the interests of the capitalists and merchants anticipated all possible needs and in mutual competition incite to the most varied expenditures of money. Only through fear of discovery and punishments, that is, through fear of the state, is a special and large group, which encompasses far more people than the professional criminals, restrained in its desire to obtain the key to all necessary and unnecessary pleasures. The state is their enemy. The state, to them, is an alien and unfriendly power; although seemingly authorized by them and embodying their own will, it is nevertheless opposed to all their needs and desires, protecting property which they do not possess, forcing them into military service for a country which offers them hearth and altar only in the form of a heated room on the upper floor or gives them, for native soil, city streets where they may stare at the glitter and luxury in lighted windows forever beyond their reach! Their own life is nothing but a constant alternative between work and leisure, which are both distorted into factory routine and the low pleasure of the saloons. City life and Gesellschaft down the common people to decay and death; in vain they struggle to attain power through their own multitude, and it seems to them that they can use their power only for a revolution if they want to free themselves from their fate. The masses become conscious of this social position through the education in schools and through newspapers. They proceed from class consciousness to class struggle. This class struggle may destroy society and the state which it is its purpose to reform. The entire culture has been transformed into a civilization of state and Gesellschaft, and this transformation means the doom of culture itself if none of its scattered seeds remain alive and again bring forth the essence and idea of Gemeinschaft, thus secretly fostering a new culture amidst the decaying one.

"The Metropolis and Mental Life"

from Kurt H. Wolff (ed.), *The Sociology of Georg Simmel* (1950) [1903]

Georg Simmel

Editors' Introduction

Georg Simmel (1858–1918) was born to a prosperous Jewish family, at the intersection of Friedrichstrasse and Leipzigerstrasse, the very heart of Berlin's commercial and theatrical bright light district, the equivalent of New York's Times Square or London's Piccadilly Circus. Simmel obtained his doctorate in philosophy in 1881 at the University of Berlin. Marginalized by the German academic system because of Jewish ancestry and intellectual radicalism, Simmel did not obtain a regular academic appointment until the last four years of his life. For most of his career, he maintained a recurring lecturing position at the University of Berlin, where his lectures influenced an extraordinary legacy of students, including Georg Lukacs, Ernst Bloch, Karl Mannheim, and Robert Park. Despite being somewhat an academic outsider, he was nevertheless an engaged public intellectual who frequented fashionable salons and enjoyed the friendship of eminent sociologists like Max Weber and the German poet Rainer Maria Rilke. As a nonobservant Jew in Weimar Berlin, he was a rootless cosmopolitan while being a public figure.

Simmel's seminal essay ("Die Grossstätde und das Geistesleben") was originally delivered as a lecture within a series during the winter of 1902–03 connected to an exhibition held in Dresden by the Gehe Foundation on the emergence of the modern metropolis. This First German City Exhibit (*Erste Deutsche Städte-Ausstellung*) was following upon the 1896 Berlin Trade Exhibition, part of a historical vogue in world city expositions, such as those held in Paris in 1886 and Chicago in 1893. The lectures and exhibits examined the intellectual, economic, and political dimensions of German urbanism, and addressed planning problems and social issues related to public transportation, housing, employment, health, welfare, and cultural institutions. Simmel's essay focused more upon the philosophical and psychological implications of these transitions. Simmel was interested in the social construction of the modern urban self.

The commercial emporium of the world city expositions framed Simmel's view of the metropolis as the nexus point for the circulation of capital, commodities, and people. That commerce was central to the great transformations of modernity was not lost upon Marx in his writings on the political economy of capital, but what Simmel explored in this essay as well as his magnum opus, *The Philosophy of Money* (1909), were the philosophical and psychological dimensions of money in modern culture. Simmel discussed the triumph of the money economy over traditional society, the rise of objectification and quantification. He saw the capitalist city as a sensorium of psychic overstimulations and commodity temptations. The decline of fixed, ancient, and venerable traditions with the rise of flux, transitoriness, and arbitrary value is what Marx described with his famed adage: "all that is solid melts into air" (see *The Communist Manifesto*).

Simmel's detached and capricious urban cosmopolitan is much similar to the "flaneur" of philosopher Walter Benjamin (*Paris, Capital of the Nineteenth Century*) and poet Charles Baudelaire, the Parisian pedestrian who sumptuously wandered the shopping arcades and boulevards, intoxicated by the spectacle of commerce and the anonymity of the urban crowd. The notion of the loneliness of life in the crowd reflects the particular reserve and impersonality that is displayed by the pedestrian dandy, the bourgeois shopper, or the urban commuter. The barrage of lures, stimulations, and choices in the modern city of commerce has induced a kind of monkish self-reflection that can be seen as transcendence as much as retreat. Freed from the prejudices and obligations of family and community, the bourgeois urbanist experiences the restlessness of liberation, a new condition of self-consciousness and inner emotional development. For all his liberation from the communal society, the urban modernist is now embedded in the iron cage of a world of work and bureaucracy, and the consumer's dilemma of a search for identity in a soulless mass society.

Simmel's seminal essay on "The Stranger" further elucidates his interpretation of the soul of the metropolitan man who is more marginal to the axes of power. The short but powerful essay expresses some of the outsider status Simmel experienced in the academy, while communicating some general properties of Simmel's thought regarding the dialectic between the individual and the society. Simmel's wandering Jewish trader bears the stigmata of the quintessential outsider who is not regarded as an individual, but as a social type or category. This status may be extended to other social types such as the black underclass, immigrant foreigner, the social or sexual deviant as Simmel suggests: "the stranger, like the poor and like sundry 'inner enemies,' is an element of the group itself." The presence of the stranger establishes spatial rules and social etiquettes of social distance. Differentiation of the "other" as well as the "deviant" establishes rules of conduct in a secular society, sustaining the solidarity of the in-group.

Simmel's interest in micro-sociological realms, the minutiae of everyday life, has attracted sociologists associated with the "cultural turn" in sociology since the 1960s. The British sociologist David Frisby has recently done much in the translating and popularizing of Simmel in this way. His writings include *Georg Simmel* (London and Chichester: Tavistock/Ellis Horwood, 1984), *Simmel and Since: Essays on Georg Simmel's Social Theory* (London: Routledge, 1992), *Sociological Impressionism: A Reassessment of Georg Simmel's Social Theory* (London: Routledge, 1992), and a reader of original translated writings titled *Simmel on Culture: Selected Writings*, edited by David Frisby and Mike Featherstone (Thousand Oaks, CA: Sage Publications, 1997). An excellent biography of Simmel can be found in Lewis A. Coser, *Masters of Sociological Thought: Ideas in Historical and Social Context*, Second Edition (New York: Harcourt Brace Jovanovich, 1977).

The deepest problems of modern life derive from the claim of the individual to preserve the autonomy and individuality of his existence in the face of overwhelming social forces, of historical heritage, of external culture, and of the technique of life. The fight with nature which primitive man has to wage for his bodily existence attains in this modern form its latest transformation. The eighteenth century called upon man to free him of all the historical bonds in the state and in religion, in morals and in economics. Man's nature, originally good and common to all, should develop unhampered. In addition to more liberty, the nineteenth century demanded the functional specialization of man and his work; this specialization makes one individual incomparable to another, and each of them indispensable to the highest possible extent. However, this specialization makes each man the more directly dependent upon the supplementary activities of all others. Nietzsche sees the full development of the individual conditioned by the most ruthless struggle of individuals; socialism believes in the suppression of all competition for the same reason. Be that as it may, in all these positions the same basic motive is at work: the person resists to being leveled down and worn out by a social–technological mechanism. An inquiry into the inner meaning of specifically modern life and its products, into the soul of the cultural body, so to speak, must seek to solve the equation that structures like the metropolis set up between the individual and the super-individual contents of life. Such an inquiry must answer the question of how the personality accommodates itself in the adjustments to external forces. This will be my task today.

The psychological basis of the metropolitan type of individuality consists in the *intensification of nervous stimulation* that results from the swift and uninterrupted change of outer and inner stimuli. Man is a differentiating creature. His mind is stimulated by the difference between a momentary impression and the one that preceded it. Lasting impressions, impressions which differ only slightly from one another, impressions which take a regular and habitual course and show regular and habitual contrasts – all these use up, so to speak, less consciousness than does the rapid crowding of changing images, the sharp discontinuity in the grasp of a single glance, and the unexpectedness of onrushing impressions. These are the psychological conditions that the metropolis creates. With each crossing of the street, with the tempo and multiplicity of economic, occupational and social life, the city sets up a deep contrast with small town and rural life with reference to the sensory foundations of psychic life. The metropolis exacts from man as a discriminating creature a different amount of consciousness than does rural life. Here the rhythm of life and sensory mental imagery flows more slowly, more habitually, and more evenly. Precisely in this connection the sophisticated character of metropolitan psychic life becomes understandable – as over against small town life, which rests more upon deeply felt and emotional relationships. These latter are rooted in the more unconscious layers of the psyche and grow most readily in the steady rhythm of uninterrupted habituations. The intellect, however, has its locus in the transparent, conscious, higher layers of the psyche; it is the most adaptable of our inner forces. In order to accommodate to change and to the contrast of phenomena, the intellect does not require any shocks and inner upheavals; it is only through such upheavals that the more conservative mind could accommodate to the metropolitan rhythm of events. Thus the metropolitan type of man – which, of course, exists in a thousand individual variants – develops an organ protecting him against the threatening currents and discrepancies of his external environment which would uproot him. He reacts with his head instead of his heart. In this an increased awareness assumes the psychic prerogative. Metropolitan life, thus, underlies a heightened awareness and a predominance of intelligence in metropolitan man. The reaction to metropolitan phenomena is shifted to that organ which is least sensitive and quite remote from the depth of the personality. Intellectuality is thus seen to preserve subjective life against the overwhelming power of metropolitan life, and intellectuality branches out in many directions and is integrated with numerous discrete phenomena.

The metropolis has always been the seat of the money economy. Here the multiplicity and concentration of economic exchange gives an importance to the means of exchange that the scantiness of rural commerce would not have allowed. Money economy and the dominance of the intellect are intrinsically connected. They share a matter-of-fact attitude in dealing with men and with things; and, in this attitude, a formal justice is often coupled with an inconsiderate hardness. The intellectually sophisticated person is indifferent to all genuine individuality, because relationships and reactions result from it that cannot be exhausted with logical operations. In the same manner, the individuality of phenomena is not commensurate with the pecuniary principle. Money is concerned only with what is common to all: it asks for the exchange value, it reduces all quality and individuality to the question: How much? All intimate emotional relations between persons are founded in their individuality, whereas in rational relations man is reckoned with like a number, like an element that is in itself indifferent. Only the objective measurable achievement is of interest. Thus metropolitan man reckons with his merchants and customers, his domestic servants and often even with persons with whom he is obliged to have social intercourse. These features of intellectuality contrast with the nature of the small circle in which the inevitable knowledge of individuality as inevitably produces a warmer tone of behavior, a behavior which is beyond a mere objective balancing of service and return. In the sphere of the economic psychology of the small group it is of importance that under primitive conditions production serves the customer who orders the good, so that the producer and the consumer are acquainted. The modern metropolis, however, is supplied almost entirely by production for the market, that is, for entirely unknown purchasers who never personally enter the producer's actual field of vision. Through this anonymity the interests of each party acquire an unmerciful matter-of-factness; and the intellectually calculating economic egoisms of both parties need not fear any deflection because of the imponderables of personal relationships. The money economy dominates the metropolis; it has displaced the last survivals of domestic production and the direct barter of goods; it minimizes, from day to day, the amount of work ordered by customers. The matter-of-fact attitude is obviously so intimately interrelated with the money economy, which is dominant in the metropolis, that nobody can say whether the intellectualistic mentality first promoted the money economy or whether the latter determined the former. The metropolitan way of life is certainly the most fertile soil for this reciprocity, a point which I shall document merely by citing the dictum of the most eminent English constitutional historian: throughout the whole course of English history, London has never acted as England's heart but often as England's intellect and always as her moneybag!

In certain seemingly insignificant traits, which lie upon the surface of life, the same psychic currents characteristically unite. Modern mind has become more and more calculating. The calculative exactness of practical life that the money economy has brought about corresponds to the ideal of natural science: to transform the world into an arithmetic problem, to fix every part of the world by mathematical formulas. Only money economy has filled the days of so many people with weighing, calculating, with numerical determinations, with a reduction of qualitative values to quantitative ones. Through the calculative nature of money a new precision, a certainty in the definition of identities and differences, unambiguousness in agreements and arrangements has been brought about in the relations of life-elements – just as externally this precision has been affected by the universal diffusion of pocket watches. However, the conditions of metropolitan life are at once cause and effect of this trait. The relationships and affairs of the typical metropolitan usually are so varied and complex that without the strictest punctuality in promises and services the whole structure would break down into an inextricable chaos. Above all, this necessity is brought about by the aggregation of so many people with such differentiated interests, who must integrate their relations and activities into a highly complex organism. If all clocks and watches in Berlin would suddenly go wrong in different ways, even if only by one hour, all economic life and communication of the city would be disrupted for a long time. In addition an apparently mere external factor, long distances, would make all waiting and broken appointments result in an ill-afforded waste of time. Thus, the technique of metropolitan life is unimaginable without the most punctual integration of all activities and mutual relations into a stable and impersonal time schedule. Here again the general conclusions of this entire task of reflection become obvious, namely, that from each point on the surface of existence – however closely attached to the surface alone – one may drop a sounding into the depth of the psyche so that all the most banal externalities of life finally are connected with the ultimate decisions concerning the meaning and style of life. Punctuality, calculability, exactness are forced upon life by the complexity and extension of metropolitan existence and are not only most intimately connected with its money economy and intellectualist character. These traits must also color the contents of life and favor the exclusion of those irrational, instinctive, sovereign traits and impulses which aim at determining the mode of life from within, instead of receiving the general and precisely schematized form of life from without. Even though sovereign types of personality, characterized by irrational impulses, are by no means impossible in the city, they are nevertheless opposed to typical city life. The passionate hatred of men like Ruskin and Nietzsche for the metropolis is understandable in these terms. Their natures discovered the value of life alone in the unschematized existence that cannot be defined with precision for all alike. From the same source of this hatred of the metropolis surged their hatred of money economy and of the intellectualism of modern existence.

The same factors that have thus coalesced into the exactness and minute precision of the form of life have coalesced into a structure of the highest impersonality; on the other hand, they have promoted a highly personal subjectivity. There is perhaps no psychic phenomenon that has been so unconditionally reserved to the metropolis as has the blasé attitude. The blasé attitude results first from the rapidly changing and closely compressed contrasting stimulations of the nerves. From this, the enhancement of metropolitan intellectuality, also, seems originally to stem. Therefore, stupid people who are not intellectually alive in the first place usually are not exactly blasé. A life in boundless pursuit of pleasure makes one blasé because it agitates the nerves to their strongest reactivity for such a long time that they finally cease to react at all. In the same way, through the rapidity and contradictoriness of their changes, more harmless impressions force such violent responses, tearing the nerves so brutally hither and thither that their last reserves of strength are spent; and if one remains in the same milieu they have no time to gather new strength. Incapacity thus emerges to react to new sensations with the appropriate energy. This constitutes that blasé attitude which, in fact, every metropolitan child shows when compared with children of quieter and less changeable milieus.

This physiological source of the metropolitan blasé attitude is joined by another source that flows from the money economy. The essence of the blasé attitude consists in the blunting of discrimination. This does not mean that the objects are not perceived, as is the case with the half-wit, but rather that the meaning and differing values of things, and thereby the things themselves, are experienced as insubstantial. They appear to the blasé person in an evenly flat and gray tone; no one object deserves preference over any other. This mood is the faithful subjective reflection of the completely internalized money economy. By being the equivalent to all the manifold things in one and the same way, money becomes the most frightful leveler. For money expresses all qualitative differences of things in terms of "how much?" Money, with all its colorlessness and indifference, becomes the common denominator of all values; irreparably it hollows out the core of things, their individuality, their specific value, and their incomparability. All things float with equal specific gravity in the constantly moving stream of money. All things lie on the same level and differ from one another only in the size of the area that they cover. In the individual case this coloration, or rather discoloration, of things through their money equivalence may be unnoticeably minute. However, through the relations of the rich to the objects to be had for money, perhaps even through the total character that the mentality of the contemporary public everywhere imparts to these objects, the exclusively pecuniary evaluation of objects has become quite considerable. The large cities, the main seats of the money exchange, bring the purchasability of things to the fore much more impressively than do smaller localities. That is why cities are also the genuine locale of the blasé attitude. In the blasé attitude the concentration of men and things stimulates the nervous system of the individual to its highest achievement so that it attains its peak. Through the mere quantitative intensification of the same conditioning factors this achievement is transformed into its opposite and appears in the peculiar adjustment of the blasé attitude. In this phenomenon the nerves find in the refusal to react to their stimulation the last possibility of accommodating to the contents and forms of metropolitan life. The self-preservation of certain personalities is bought at the price of devaluing the whole objective world, a devaluation that in the end unavoidably drags one's own personality down into a feeling of the same worthlessness.

Whereas the subject of this form of existence has to come to terms with it entirely for himself, his self-preservation in the face of the large city demands from him a no less negative behavior of a social nature. This mental attitude of metropolitans toward one another we may designate, from a formal point of view, as reserve. If so many inner reactions were responses to the continuous external contacts with innumerable people as are those in the small town, where one knows almost everybody one meets and where one has a positive relation to almost everyone, one would be completely atomized internally and come to an unimaginable psychic state. Partly this psychological fact, partly the right to distrust that men have in the face of the touch-and-go elements of metropolitan life, necessitates our reserve. As a result of this reserve we frequently do not even know by sight those who have been our neighbors for years. And it is this reserve that in the eyes of the small-town people makes us appear to be cold and heartless. Indeed, if I do not deceive myself, the inner aspect of this outer reserve is not only indifference but, more often than we are aware, it is a slight aversion, a mutual strangeness and repulsion, which will break into hatred and fight at the moment of a closer contact, however caused. The whole inner organization of such an extensive communicative life rests upon an extremely varied hierarchy of sympathies, indifferences, and aversions of the briefest as well as of the most permanent nature. The sphere of indifference in this hierarchy is not as large as might appear on the surface. Our psychic activity still responds to almost every impression of somebody else with a somewhat distinct feeling. The unconscious, fluid and changing character of this impression seems to result in a state of indifference. Actually this indifference would be just as unnatural as the diffusion of indiscriminate mutual suggestion would be unbearable. From both these typical dangers of the metropolis, indifference and indiscriminate suggestibility, antipathy protects us. A latent antipathy and the preparatory stage of practical antagonism affect the distances and aversions without which this mode of life could not at all be led. The extent and the mixture of this style of life, the rhythm of its emergence and disappearance, the forms in which it is satisfied – all these, with the unifying motives in the narrower sense – form the inseparable whole of the metropolitan style of life. What appears in the metropolitan style of life directly as dissociation is in reality only one of its elemental forms of socialization.

This reserve with its overtone of hidden aversion appears in turn as the form or the cloak of a more general mental phenomenon of the metropolis: it grants to the individual a kind and an amount of personal freedom which has no analogy whatsoever under other conditions. The metropolis goes back to one of the large developmental tendencies of social life as such, to one of the few tendencies for which an approximately universal formula can be discovered. The earliest phase of social formations found in historical as well as in contemporary social structures is this: a relatively small circle firmly closed against neighboring, strange, or in some way antagonistic circles. However, this circle is closely coherent and allows its individual members only a narrow field for the development of unique qualities and free, self-responsible movements. Political and kinship groups, parties and religious associations begin in this way. The self-preservation of very young associations requires the establishment of strict boundaries and a centripetal unity. Therefore they cannot allow the individual freedom and unique inner and outer development. From this stage social development proceeds at once in two different, yet corresponding, directions. To the extent to which the group grows – numerically, spatially, in significance and in content of life – to the same degree the group's direct, inner unity loosens, and the rigidity of the original demarcation against others is softened through mutual relations and connections. At the same time, the individual gains freedom of movement, far beyond the first jealous delimitation. The individual also gains a specific individuality to which the division of labor in the enlarged group gives both occasion and necessity. The state and Christianity, guilds and political parties, and innumerable other groups have developed according to this formula, however much, of course, the special conditions and forces of the respective groups have modified the general scheme. This scheme seems to me distinctly recognizable also in the evolution of individuality within urban life. The small-town life in Antiquity and in the Middle Ages set barriers against movement and relations of the individual toward the outside, and it set up barriers against individual independence and differentiation within the individual self. These barriers were such that under them modern man could not have breathed. Even today a metropolitan man who is placed in a small town feels a restriction similar, at least, in kind. The smaller the circle which forms our milieu is, and the more restricted those relations to others are which dissolve the boundaries of the individual, the more anxiously the circle guards the achievements, the conduct of life, and the outlook of the individual, and the more readily a quantitative and qualitative specialization would break up the framework of the whole little circle.

The ancient *polis* in this respect seems to have had the very character of a small town. The constant threat to its existence at the hands of enemies from near and afar effected strict coherence in political and military respects, a supervision of the citizen by the citizen, a jealousy of the whole against the individual whose particular life was suppressed to such a degree that he could compensate only by acting as a despot in his own household. The tremendous agitation and excitement, the unique colorfulness of Athenian life, can perhaps be understood in terms of the fact that a people of incomparably individualized personalities struggled against the constant inner and outer pressure of a deindividualizing small town. This produced a tense atmosphere in which the weaker individuals were suppressed and those of stronger natures were incited to prove themselves in the most passionate manner. This is precisely why it was that there blossomed in Athens what must be called, without defining it exactly, "the general human character" in the intellectual development of our species. For we maintain factual as well as historical validity for the following connection: the most extensive and the most general contents and forms of life are most intimately connected with the most individual ones. They have a preparatory stage in common, that is, they find their enemy in narrow formations and groupings the maintenance of which places both of them into a state of defense against expanse and generality lying without and the freely moving individuality within. Just as in the feudal age, the "free" man was the one who stood under the law of the land, that is, under the law of the largest social orbit, and the unfree man was the one who derived his right merely from the narrow circle of a feudal association and was excluded from the larger social orbit – so today metropolitan man is "free" in a spiritualized and refined sense, in contrast to the pettiness and prejudices which hem in the small-town man. For the reciprocal reserve and indifference and the intellectual life conditions of large circles are never felt more strongly by the individual in their impact upon his independence than in the thickest crowd of the big city. This is because the bodily proximity and narrowness of space makes the mental distance only the more visible. It is obviously only the obverse of this freedom if, under certain circumstances, one nowhere feels as lonely and lost as in the metropolitan crowd. For here as elsewhere it is by no means necessary that the freedom of man be reflected in his emotional life as comfort.

It is not only the immediate size of the area and the number of persons that, because of the universal historical correlation between the enlargement of the circle and the personal inner and outer freedom, has made the metropolis the locale of freedom. It is rather in transcending this visible expanse that any given city becomes the seat of cosmopolitanism. The horizon of the city expands in a manner comparable to the way in which wealth develops; a certain amount of property increases in a quasi-automatical way in ever more rapid progression. As soon as a certain limit has been passed, the economic, personal, and intellectual relations of the citizenry, the sphere of intellectual predominance of the city over its hinterland, grow as in geometrical progression. Every gain in dynamic extension becomes a step, not for an equal, but for a new and larger extension. From every thread spinning out of the city, ever-new threads grow as if by themselves, just as within the city the unearned increment of ground rent, through the mere increase in communication, brings the owner automatically increasing profits. At this point, the quantitative aspect of life is transformed directly into qualitative traits of character. The sphere of life of the small town is, in the main, self-contained and autarchic. For it is the decisive nature of the metropolis that its inner life overflows by waves into a far-flung national or international area. Weimar is not an example to the contrary, since its significance was hinged upon individual personalities and died with them; whereas the metropolis is indeed characterized by its essential independence even from the most eminent individual personalities. This is the counterpart to the independence, and it is the price the individual pays for the independence, which he enjoys in the metropolis. The most significant characteristic of the metropolis is this functional extension beyond its physical boundaries. And this efficacy reacts in turn and gives weight, importance, and responsibility to metropolitan life. Man does not end with the limits of his body or the area comprising his immediate activity. Rather is the range of the person constituted by the sum of effects emanating from him temporally and spatially. In the same way, a city consists of its total effects that extend beyond its immediate confines. Only this range is the city's actual extent in which its existence is expressed. This fact makes it obvious that individual freedom, the logical and historical complement of such extension, is not to be understood only in the negative sense of mere freedom of mobility and elimination of prejudices and petty philistinism. The essential point is that the particularity and incomparability, which ultimately every human being possesses, be somehow expressed in the working-out of a way of life. That we follow the laws of our own nature – and this after all is freedom – becomes obvious and convincing to ourselves and to others only if the expressions of this nature differ from the expressions of others. Only our unmistakability proves that our way of life has not been superimposed by others.

Cities are, first of all, seats of the highest economic division of labor. They produce thereby such extreme phenomena as in Paris the remunerative occupation of the *quatorzième*. They are persons who identify themselves by signs on their residences and who are ready at the dinner hour in correct attire, so that they can be quickly called upon if a dinner party should consist of thirteen persons. In the measure of its expansion, the city offers more and more the decisive conditions of the division of labor. It offers a circle that through its size can absorb a highly diverse variety of services. At the same time, the concentration of individuals and their struggle for customers compel the individual to specialize in a function from which he cannot be readily displaced by another. It is decisive that city life has transformed the struggle with nature for livelihood into an inter-human struggle for gain, which here is not granted by nature but by other men. For specialization does not flow only from the competition for gain but also from the underlying fact that the seller must always seek to call forth new and differentiated needs of the lured customer. In order to find a source of income that is not yet exhausted, and to find a function that cannot readily be displaced, it is necessary to specialize in one's services. This process promotes differentiation, refinement, and the enrichment of the public's needs, which obviously must lead to growing personal differences within this public.

All this forms the transition to the individualization of mental and psychic traits that the city occasions in proportion to its size. There is a whole series of obvious causes underlying this process. First, one must meet the difficulty of asserting his own personality within the dimensions of metropolitan life. Where the quantitative increase in importance and the expense of energy reach their limits, one seizes upon qualitative differentiation in order somehow to attract the attention of the social circle by playing upon its sensitivity for differences. Finally, man is tempted to adopt the most tendentious peculiarities, that is, the specifically metropolitan extravagances of mannerism, caprice, and preciousness. Now, the meaning of these extravagances does not at all lie in the contents of such behavior, but rather in its form of "being different," of standing out in a striking manner and thereby attracting attention. For many character types, ultimately the only means of saving for themselves some modicum of self-esteem and the sense of filling a position is indirect, through the awareness of others. In the same sense a seemingly insignificant factor is operating, the cumulative effects of which are, however, still noticeable. I refer to the brevity and scarcity of the inter-human contacts granted to the metropolitan man, as compared with social intercourse in the small town. The temptation to appear "to the point," to appear concentrated and strikingly characteristic, lies much closer to the individual in brief metropolitan contacts than in an atmosphere in which frequent and prolonged association assures the personality of an unambiguous image of himself in the eyes of the other.

The most profound reason, however, why the metropolis conduces to the urge for the most individual personal existence – no matter whether justified and successful – appears to me to be the following: the development of modern culture is characterized by the preponderance of what one may call the "objective spirit" over the "subjective spirit." This is to say, in language as well as in law, in the technique of production as well as in art, in science as well as in the objects of the domestic environment, there is embodied a sum of spirit. The individual in his intellectual development follows the growth of this spirit very imperfectly and at an ever-increasing distance. If, for instance, we view the immense culture that for the last hundred years has been embodied in things and in knowledge, in institutions and in comforts, and if we compare all this with the cultural progress of the individual during the same period – at least in high status groups – a frightful disproportion in growth between the two becomes evident. Indeed, at some points we notice retrogression in the culture of the individual with reference to spirituality, delicacy, and idealism. This discrepancy results essentially from the growing division of labor. For the division of labor demands from the individual an ever more one-sided accomplishment, and the greatest advance in a one-sided pursuit only too frequently means death to the personality of the individual. In any case, he can cope less and less with the overgrowth of objective culture. The individual is reduced to a negligible quantity, perhaps less in his consciousness than in his practice and in the totality of his obscure emotional states that are derived from this practice. The individual has become a mere cog in an enormous organization of things and powers which tear from his hands all progress, spirituality, and value in order to transform them from their subjective form into the form of a purely objective life. It needs merely to be pointed out that the metropolis is the genuine arena of this culture that outgrows all personal life. Here in buildings and educational institutions, in the wonders and comforts of space-conquering technology, in the formations of community life, and in the visible institutions of the state, is offered such an overwhelming fullness of crystallized and impersonalized spirit that the personality, so to speak, cannot maintain itself under its impact. On the one hand, life is made infinitely easy for the personality in that stimulations, interests, uses of time and consciousness are offered to it from all sides. They carry the person as if in a stream, and one needs hardly to swim for oneself. On the other hand, however, life is composed more and more of these impersonal contents and offerings that tend to displace the genuine personal colorations and incomparabilities. This results in the individual's summoning the utmost in uniqueness and particularization, in order to preserve his most personal core. He has to exaggerate this personal element in order to remain audible even to himself. The atrophy of individual culture through the hypertrophy of objective culture is one reason for the bitter hatred that the preachers of the most extreme individualism, above all Nietzsche, harbor against the metropolis. But it is, indeed, also a reason why these preachers are so passionately loved in the metropolis and why they appear to the metropolitan man as the prophets and saviors of his most unsatisfied yearnings.

If one asks for the historical position of the two forms of individualism that are nourished by the quantitative relation of the metropolis, namely, individual independence and the elaboration of individuality itself, then the metropolis assumes an entirely new rank order in the world history of the spirit. The eighteenth century found the individual in oppressive bonds that had become meaningless – bonds of a political, agrarian, guild, and religious character. They were restraints that, so to speak, forced upon man an unnatural form and outmoded, unjust inequalities. In this situation the cry for liberty and equality arose, the belief in the individual's full freedom of movement in all social and intellectual relationships. Freedom would at once permit the noble substance common to all to come to the fore, a substance which nature had deposited in every man and which society and history had only deformed. Besides this eighteenth-century ideal of liberalism, in the nineteenth century, through Goethe and Romanticism, on the one hand, and through the economic division of labor, on the other hand, another ideal arose: individuals liberated from historical bonds now wished to distinguish themselves from one another. The carrier of man's values is no longer the "general human being" in every individual, but rather man's qualitative uniqueness and irreplaceability. The external and internal history of our time takes its course within the struggle and in the changing entanglements of these two ways of defining the individual's role in the whole of society. It is the function of the metropolis to provide the arena for this struggle and its reconciliation. For the metropolis presents the peculiar conditions which are revealed to us as the opportunities and the stimuli for the development of both these ways of allocating roles to men. Therewith these conditions gain a unique place, pregnant with inestimable meanings for the development of psychic existence. The metropolis reveals itself as one of those great historical formations in which opposing streams that enclose life unfold, as well as join one another with equal right. However, in this process the currents of life, whether their individual phenomena touch us sympathetically or antipathetically, entirely transcend the sphere for which the judge's attitude is appropriate. Since such forces of life have grown into the roots and into the crown of the whole of the historical life in which we, in our fleeting existence, as a cell, belong only as a part, it is not our task either to accuse or to pardon, but only to understand.

"Urbanism as a Way of Life"

from *American Journal of Sociology* (1938)

Louis Wirth

Editors' Introduction

Published in 1938, Wirth's essay on urbanism, and the factors of size, density, and heterogeneity, is one of the foundational statements of the Chicago School of urban sociology. It is clearly influenced by Ferdinand Tönnies, Georg Simmel, and Robert E. Park. Like Tönnies, he views the theory of urbanism as an ideal type. Wirth's concept of the "schizoid" urban personality, beset by "segmental roles," is akin to Simmel's blasé and reserved metropolitan man. Simmel felt, however, that the cosmopolitanism of city life liberated urbanites from the prejudices and provincialities of rural life. Wirth was less impressed by the positive benefits of this emancipation from primary group controls. He drew our attention to the growth of Durkheimian *anomie*, which consequently engendered a host of modern social problems, including crime, deviance, and various kinds of mental illness that were seen to proliferate in the city. Wirth also informed our understanding of Robert Park's concept of the city as a "mosaic of social worlds" that increases social distance between people. He viewed this as an outcome of urban density and specialization. He was more sensitive to the practical implications of a theory of urbanism than Tönnies or Simmel, as he suggested that knowledge of the causes of urban social problems were important to apply to a range of social policy and urban planning practices.

Louis Wirth was born August 28, 1897, in Gemünden, a small village in the Rhineland district of Germany to a Jewish rural cattle farming family. He followed his maternal uncle to Omaha, Nebraska, in the United States, to take advantage of educational opportunities. He was a successful high school debater and eventually won a scholarship to the University of Chicago. He flirted for a while with leftist anti-war causes during World War I, and then worked with delinquent boys with the Jewish Charities of Chicago after college. He obtained a Ph.D. in Sociology at the University of Chicago in 1925. His doctoral thesis on the Jewish quarter of Chicago was published as *The Ghetto* (Chicago: University of Chicago Press, 1928). After various teaching posts and fellowships, he joined the Chicago faculty under the chairmanship of Robert E. Park in 1931. In *The Ghetto*, Wirth examined the consequences of centuries of discrimination on Jewish community life, ranging from Renaissance Italy to Chicago's Maxwell Street. The book served as a model for the university's researchers in ethnicity, many of whom later studied under Wirth when he joined the university's faculty.

As a professor at the University of Chicago, Wirth blended empirical research and theory in his work and contributed to the emergence of sociology as a profession. The advent of the Roosevelt administration gave many opportunities for sociologists to work with government in congressional testimony, consulting, and funded research. Wirth also played a significant role in organizing an introductory course in the social sciences and was popularly known as a persuasive lecturer. During the late 1930s he grew involved in community affairs in Chicago and was often invited to make public addresses on urban planning and race relations. He became a well-known radio speaker, acting as a moderator and discussant on a series of 62 University of Chicago "round tables" broadcast between 1937 and 1952.

As an academic committed to social action, Louis Wirth became involved in numerous groups, committees, and associations concerned with the effects of racial prejudice on community life. He was a founder and president of the Chicago-based American Council on Race Relations, which sponsored research into problems of fair employment, education, housing, and integration. In 1947, with funds from the Carnegie and Rockefeller Foundations, Wirth also established the Committee on Education, Training, and Research in Race Relations at the University of Chicago. Led by Wirth, demographer Philip Hauser, and anthropologist Sol Tax, the committee played a key role in addressing the social and political factors underlying racial discrimination in the city of Chicago.

Wirth was President of the American Sociological Society and his Presidential Address, "Consensus and Mass Communication," was delivered at the organization's annual meeting in New York City in December 1947. He was also the first President (1949–52) of the International Sociological Association. Wirth died suddenly and unexpectedly one spring day in 1952 in Buffalo, New York at the young age of 55. He had been in Buffalo to speak at a conference on community relations; he collapsed and died following his presentation.

Wirth also published a book on the selected writings of Karl Mannheim, entitled *Ideology and Utopia*, which he co-edited with Edward Shils. A useful book on Louis Wirth's legacy is by Albert J. Reiss, Jr., *Louis Wirth: On Cities and Social Life* (Chicago: University of Chicago Press, 1964). This book includes an excellent biographical memorandum by Elizabeth Wirth Marvick.

THE CITY AND CONTEMPORARY CIVILIZATION

Just as the beginning of Western civilization is marked by the permanent settlement of formerly nomadic peoples in the Mediterranean basin, so the beginning of what is distinctively modern in our civilization is best signalized by the growth of great cities. Nowhere has mankind been farther removed from organic nature than under the conditions of life characteristic of these cities. The contemporary world no longer presents a picture of small isolated groups of human beings scattered over a vast territory as Sumner described primitive society. The distinctive feature of man's mode of living in the modern age is his concentration into gigantic aggregations around which cluster lesser centers and from which radiate the ideas and practices that we call civilization.

[. . .]

Since the city is the product of growth rather than of instantaneous creation, it is to be expected that the influences which it exerts upon the modes of life should not be able to wipe out completely the previously dominant modes of human association. To a greater or lesser degree, therefore, our social life bears the imprint of an earlier folk society, the characteristic modes of settlement of which were the farm, the manor, and the village. This historic influence is reinforced by the circumstances that the population of the city itself is in large measure recruited from the countryside, where a mode of life reminiscent of this earlier form of existence persists. Hence we should not expect to find abrupt and discontinuous variation between urban and rural types of personality. The city and the country may be regarded as two poles in reference to one or the other of which all human settlements tend to arrange themselves. In viewing urban-industrial and rural-folk society as ideal types of communities, we may obtain a perspective for the analysis of the basic models of human association as they appear in contemporary civilization.

SOCIOLOGICAL DEFINITION OF THE CITY

Despite the preponderant significance of the city in our civilization, our knowledge of the nature of urbanism and the process of urbanization is meager, notwithstanding many attempts to isolate the distinguishing characteristics of urban life. Geographers, historians, economists, and political scientists have incorporated the points of view of their respective disciplines into diverse definitions of the city. While in no sense intended to supersede these, the formulation of a sociological approach to the city may incidentally serve to call attention to the interrelations between them by emphasizing the peculiar characteristics of the city as a particular form of human association. A sociologically significant definition of the city seeks to select those elements of urbanism which mark it as a distinctive mode of human group life.

[. . .]

While urbanism, or that complex of traits which makes up the characteristic mode of life in cities, and urbanization, which denotes the development and extensions of these factors, are thus not exclusively found in settlements which are cities in the physical and demographic sense, they do, nevertheless, find their most pronounced expression in such areas, especially in metropolitan cities. In formulating a definition of the city it is necessary to exercise caution in order to avoid identifying urbanism as a way of life with any specific locally or historically conditioned cultural influences which, though they may significantly affect the specific character of the community, are not the essential determinants of its character as a city.

[. . .]

For sociological purposes a city may be defined as a relatively large, dense, and permanent settlement of socially heterogeneous individuals. On the basis of the postulates which this minimal definition suggests, a theory of urbanism may be formulated in the light of existing knowledge concerning social groups.

A THEORY OF URBANISM

In the rich literature on the city we look in vain for a theory systematizing the available knowledge concerning the city as a social entity. We do indeed have excellent formulations of theories on such special problems as the growth of the city viewed as a historical trend and as a recurrent process, and we have a wealth of literature presenting insights of social relevance and empirical studies offering detailed information on a variety of particular aspects of urban life. But despite the multiplication of research and textbooks on the city, we do not as yet have a comprehensive body of compendent hypotheses which may be derived from a set of postulates implicitly contained in a sociological definition of the city. Neither have we abstracted such hypotheses from our general sociological knowledge which may be substantiated through empirical research. The closest approximations to a systematic theory of urbanism are to be found in a penetrating essay, "Die Stadt," by Max Weber and in a memorable paper by Robert E. Park on "The City: Suggestions for the Investigation of Human Behavior in the Urban Environment." But even these excellent contributions are far from constituting an ordered and coherent framework of theory upon which research might profitably proceed.

[. . .]

To say that large numbers are necessary to constitute a city means, of course, large numbers in relation to a restricted area or high density of settlement. There are, nevertheless, good reasons for treating large numbers and density as separate factors, because each may be connected with significantly different social consequences. Similarly the need for adding heterogeneity to numbers of population as a necessary and distinct criterion of urbanism might be questioned, since we should expect the range of differences to increase with numbers. In defense, it may be said that the city shows a kind and degree of heterogeneity of population which cannot be wholly accounted for by the law of large numbers or adequately represented by means of a normal distribution curve. Because the population of the city does not reproduce itself, it must recruit its migrants from other cities, the countryside, and – in the United States until recently – from other countries. The city has thus historically been the melting-pot of races, peoples, and cultures, and a most favorable breeding-ground of new biological and cultural hybrids. It has not only tolerated but rewarded individual differences. It has brought together people from the ends of the earth *because* they are different and thus useful to one another, rather than because they are homogeneous and like-minded.

There are a number of sociological propositions concerning the relationship between (a) numbers of population, (b) density of settlement, (c) heterogeneity of inhabitants and group life can be formulated on the basis of observation and research.

Size of the population aggregate

Ever since Aristotle's *Politics*, it has been recognized that increasing the number of inhabitants in a settlement beyond a certain limit will affect the relationships between them and the character of the city. Large numbers involve, as has been pointed out, a greater range of individual variation. Furthermore, the greater the number of individuals participating in a process of interaction, the greater is the *potential* differentiation between them. The personal traits, the occupations, the cultural life, and the ideas of the members of an urban community may, therefore, be expected to range between more widely separated poles than those of rural inhabitants.

That such variations should give rise to the spatial segregation of individuals according to color, ethnic heritage, economic and social status, tastes and preferences, may readily be inferred. The bonds of kinship, of neighborliness, and the sentiments arising out of living together for generations under a common folk tradition are likely to be absent or, at best, relatively weak in an aggregate the members of which have such diverse origins and backgrounds. Under such circumstances competition and formal control mechanisms furnish the substitutes for the bonds of solidarity that are relied upon to hold a folk society together.

[. . .]

The multiplication of persons in a state of interaction under conditions which make their contact as full personalities impossible produces that segmentalization of human relationships which has sometimes been seized upon by students of the mental life of the cities as an explanation for the "schizoid" character of urban personality. This is not to say that the urban inhabitants have fewer acquaintances than rural inhabitants, for the reverse may actually be true; it means rather that in relation to the number of people whom they see and with whom they rub elbows in the course of daily life, they know a smaller proportion, and of these they have less intensive knowledge.

Characteristically, urbanites meet one another in highly segmental roles. They are, to be sure, dependent upon more people for the satisfactions of their life-needs than are rural people and thus are associated with a great number of organized groups, but they are less dependent upon particular persons, and their dependence upon others is confined to a highly fractionalized aspect of the other's round of activity. This is essentially what is meant by saying that the city is characterized by secondary rather than primary contacts. The contacts of the city may indeed be face to face, but they are nevertheless impersonal, superficial, transitory, and segmental. The reserve, the indifference, and the blasé outlook which urbanites manifest in their relationships may thus be regarded as devices for immunizing themselves against the personal claims and expectations of others.

The superficiality, the anonymity, and the transitory character of urban social relations make intelligible, also, the sophistication and the rationality generally ascribed to city-dwellers. Our acquaintances tend to stand in a relationship of utility to us in the sense that the role which each one plays in our life is overwhelmingly regarded as a means for the achievement of our own ends. Whereas the individual gains, on the one hand, a certain degree of emancipation or freedom from the personal and emotional controls of intimate groups, he loses, on the other hand, the spontaneous self-expression, the morale, and the sense of participation that comes with living in an integrated society. This constitutes essentially the state of *anomie*, or the social void, to which Durkheim alludes in attempting to account for the various forms of social disorganization in technological society.

The segmental character and utilitarian accent of interpersonal relations in the city find their institutional expression in the proliferation of specialized tasks which we see in their most developed form in the professions. The operations of the pecuniary nexus lead to predatory relationships which tend to obstruct the efficient functioning of the social order unless checked by professional codes and occupational etiquette. The premium put upon utility and efficiency suggests the adaptability of the corporate device for the organization of enterprises in which individuals can engage only in groups. The advantage that the corporation has over the individual entrepreneur and the partnership in the urban-industrial world derives not only from the possibility it affords of centralizing the resources of thousands of individuals or from the legal privilege of limited liability and perpetual succession, but from the fact that the corporation has no soul.

The specialization of individuals, particularly in their occupations, can proceed only, as Adam Smith pointed out, upon the basis of an enlarged market, which in turn accentuates the division of labor. This enlarged market is only in part supplied by the city's hinterland; in large measure it is found among the larger numbers that the city itself contains. The dominance of the city over the surrounding hinterland becomes explicable in terms of the division of labor which urban life occasions and promotes. The extreme degree of interdependence and the unstable equilibrium of urban life are closely associated with the division of labor and the specialization of occupations. This interdependence and this instability are increased by the tendency of each city to specialize in those functions in which it has the greatest advantage.

[. . .]

Density

As in the case of numbers, so in the case of concentration in limited space, certain consequences of relevance in sociological analysis of the city emerge. Of these only a few can be indicated.

As Darwin pointed out for flora and fauna and as Durkheim noted in the case of human societies, an increase in numbers when area is held constant (i.e., an increase in density) tends to produce differentiation and specialization, since only in this way can the area support increased numbers. Density thus reinforces the effect of numbers in diversifying men and their activities and in increasing the complexity of the social structure.

On the subjective side, as Simmel has suggested, the close physical contact of numerous individuals necessarily produces a shift in the media through which we orient ourselves to the urban milieu, especially to our fellowmen. Typically, our physical contacts are close but our social contacts are distant. The urban world puts a premium on visual recognition. We see the uniform which denotes the role of the functionaries, and are oblivious to the personal eccentricities hidden behind the uniform. We tend to acquire and develop a sensitivity to a world of artifacts, and become progressively farther removed from the world of nature.

We are exposed to glaring contrasts between splendor and squalor, between riches and poverty, intelligence and ignorance, order and chaos. The competition for space is great, so that each area generally tends to be put to the use which yields the greatest economic return. Place of work tends to become dissociated from place of residence, for the proximity of industrial and commercial establishments makes an area both economically and socially undesirable for residential purposes.

Density, land values, rentals, accessibility, healthfulness, prestige, aesthetic consideration, absence of nuisances such as noise, smoke, and dirt determine the desirability of various areas of the city as places of settlement for different sections of the population. Place and nature of work, income, racial and ethnic characteristics, social status, custom, habit, taste, preference, and prejudice are among the significant factors in accordance with which the urban population is selected and distributed into more or less distinct settlements. Diverse population elements inhabiting a compact settlement thus become segregated from one another in the degree in which their requirements and modes of life are incompatible and in the measure in which they are antagonistic. Similarly, persons of homogeneous status and needs unwittingly drift into, consciously select, or are forced by circumstances into the same area. The different parts of the city thus acquire specialized functions. The city consequently tends to resemble a mosaic of social worlds in which the transition from one to the other is abrupt. The juxtaposition of divergent personalities and modes of life tends to produce a relativistic perspective and a sense of toleration of differences which may be regarded as prerequisites for rationality and which lead toward the secularization of life.

The close living together and working together of individuals who have no sentimental and emotional ties foster a spirit of competition, aggrandizement, and mutual exploitation. Formal controls are instituted to counteract irresponsibility and potential disorder. Without rigid adherence to predictable routines a large compact society would scarcely be able to maintain itself. The clock and the traffic signal are symbolic of the basis of our social order in the urban world. Frequent close physical contact, coupled with great social distance, accentuates the reserve of unattached individuals toward one another and, unless compensated by other opportunities for response, gives rise to loneliness. The necessary frequent movement of great numbers of individuals in a congested habitat causes friction and irritation. Nervous tensions which derive from such personal frustrations are increased by the rapid tempo and the complicated technology under which life in dense areas must be lived.

Heterogeneity

The social interaction among such a variety of personality types in the urban milieu tends to break down the rigidity of caste lines and to complicate the class structure, and thus induces a more ramified and differentiated framework of social stratification than is found in more integrated societies. The heightened mobility of the individual, which brings him within the range of stimulation by a great number of diverse individuals and subjects him to fluctuating status in the differentiated social groups that compose the social structure of the city, brings him toward the acceptance of instability and insecurity in the world at large as a norm. This fact helps to account too for the sophistication and cosmopolitanism of the urbanite. No single group has the undivided allegiance of the individual. The groups with which he is affiliated do not lend themselves readily to a simple hierarchical arrangement. By virtue of his different interests arising out of different aspects of social life, the individual acquires membership in widely divergent groups, each of which functions only with reference to a certain segment of his personality. Nor do these groups easily permit of a concentric arrangement so that the narrower ones fall within the circumference of the more inclusive ones, as is more likely to be the case in the rural community or in primitive societies. Rather, the groups with which the person typically is affiliated are tangential to each other or intersect in highly variable fashion.

Partly as a result of the physical footlooseness of the population and partly as a result of their social mobility, the turnover in group membership generally is rapid. Place of residence, place and character of employment, income, and interests fluctuate, and the task of holding organizations together and maintaining and promoting intimate and lasting acquaintanceship between the members is difficult. This applies strikingly to the local areas within the city into which persons become segregated more by virtue of differences in race, language, income, and social status than through choice or positive attraction to people like themselves. Overwhelmingly the city-dweller is not a home-owner, and since a transitory habitat does not generate binding traditions and sentiments, only rarely is he a true neighbor. There is little opportunity for the individual to obtain a conception of the city as a whole or to survey his place in the total scheme. Consequently he finds it difficult to determine what is to his own "best interests" and to decide between the issues and leaders presented to him by the agencies of mass suggestion. Individuals who are thus detached from the organized bodies which integrate society comprise the fluid masses that make collective behavior in the urban community so unpredictable and hence so problematical.

Although the city, through the recruitment of variant types to perform its diverse tasks and the accentuation of their uniqueness through competition and the premium upon eccentricity, novelty, efficient performance, and inventiveness, produces a highly differentiated population, it also exercises a leveling influence. Wherever large numbers of differently constituted individuals congregate, the process of depersonalization also enters. This leveling tendency inheres in part in the economic basis of the city. The development of large cities, at least in the modern age, was largely dependent upon the concentrative force of steam. The rise of the factory made possible mass production for an impersonal market. The fullest exploitation of the possibilities of the division of labor and mass production, however, is possible only with standardization of processes and products. A money economy goes hand in hand with such a system of production. Progressively as cities have developed upon a background of this system of production, the pecuniary nexus which implies the purchasability of services and things has displaced personal relations as the basis of association. Individuality under these circumstances must be replaced by categories. When large numbers have to make common use of facilities and institutions, those facilities and institutions must serve the needs of the average person rather than those of particular individuals. The services of the public utilities, of the recreational, educational, and cultural institutions, must be adjusted to mass requirements. Similarly, the cultural institutions, such as the schools, the movies, the radio, and the newspapers, by virtue of their mass clientele, must necessarily operate as leveling influences. The political process as it appears in urban life could not be understood unless one examined the mass appeals made through modern propaganda techniques. If the individual would participate at all in the social, political, and economic life of the city, he must subordinate some of his individuality to the demands of the larger community and in that measure immerse himself in mass movements.

THE RELATION BETWEEN A THEORY OF URBANISM AND SOCIOLOGICAL RESEARCH

By means of a body of theory such as that illustratively sketched above, the complicated and many-sided phenomena of urbanism may be analyzed in terms of a limited number of basic categories. The sociological approach to the city thus acquires an essential unity and coherence enabling the empirical investigator not merely to focus more distinctly upon the problems and processes that properly fall in his province but also to treat his subject matter in a more integrated and systematic fashion. A few typical findings of empirical research in the field of urbanism, with special reference to the United States, may be indicated to substantiate the theoretical propositions set forth in the preceding pages, and some of the crucial problems for further study may be outlined.

On the basis of the three variables, number, density of settlement, and degree of heterogeneity, of the urban population, it appears possible to explain the characteristics of urban life and to account for the differences between cities of various sizes and types.

Urbanism as a characteristic mode of life may be approached empirically from three interrelated perspectives: (1) as a physical structure comprising a population base, a technology, and an ecological order; (2) as a system of social organization involving a characteristic social structure, a series of social institutions, and a typical pattern of social relationships; and (3) as a set of attitudes and ideas, and a constellation of personalities engaging in typical forms of collective behavior and subject to characteristic mechanisms of social control.

Urbanism in ecological perspective

Since in the case of physical structure and ecological process we are able to operate with fairly objective indices, it becomes possible to arrive at quite precise and generally quantitative results. The dominance of the city over its hinterland becomes explicable through the functional characteristics of the city which derive in large measure from the effect of numbers and density. Many of the technical facilities and the skills and organizations to which urban life gives rise can grow and prosper only in cities where the demand is sufficiently great. The nature and scope of the services rendered by these organizations and institutions and the advantage which they enjoy over the less developed facilities of smaller towns enhance the dominance of the city, making ever wider regions dependent upon the central metropolis.

The composition of an urban population shows the operation of selective and differentiating factors. Cities contain a larger proportion of persons in the prime of life than rural areas, which contain more old and very young people. In this, as in so many other respects, the larger the city the more this specific characteristic of urbanism is apparent. With the exception of the largest cities, which have attracted the bulk of the foreign-born males, and a few other special types of cities, women predominate numerically over men. The heterogeneity of the urban population is further indicated along racial and ethnic lines. The foreign-born and their children constitute nearly two-thirds of all the inhabitants of cities of one million and over. Their proportion in the urban population declines as the size of the city decreases, until in the rural areas they comprise only about one-sixth of the total population. The larger cities similarly have attracted more Negroes and other racial groups than have the smaller communities. Considering that age, sex, race, and ethnic origin are associated with other factors such as occupation and interest, one sees that a major characteristic of the urban-dweller is his dissimilarity from his fellows. Never before have such large masses of people of diverse traits as we find in our cities been thrown together into such close physical contact as in the great cities of America. Cities generally, and American cities in particular, comprise a motley of peoples and cultures of highly differentiated modes of life between which there often is only the faintest communication, the greatest indifference, the broadest tolerance, occasionally bitter strife, but always the sharpest contrast.

The failure of the urban population to reproduce itself appears to be a biological consequence of a combination of factors in the complex of urban life, and the decline in the birth rate generally may be regarded as one of the most significant signs of the urbanization of the Western world. While the proportion of deaths in cities is slightly greater than in the country, the outstanding difference between the failure of present-day cities to maintain their population and that of cities of the past is that in former times it was due to the exceedingly high death rates in cities, whereas today, since cities have become more livable from a health standpoint, it is due to low birth rates. These biological characteristics of the urban population are significant sociologically, not merely because they reflect the urban mode of existence but also because they condition the growth and future dominance of cities and their basic social organization. Since cities are the consumers rather than the producers of men, the value of human life and the social estimation of the personality will not be unaffected by the balance between births and deaths. The pattern of land use, of land values, rentals, and ownership, the nature and functioning of the physical structures, of housing, of transportation and communication facilities, of public utilities – these and many other phases of the physical mechanism of the city are not isolated phenomena unrelated to the city as a social entity but are affected by and affect the urban mode of life.

Urbanism as a form of social organization

The distinctive features of the urban mode of life have often been described sociologically as consisting of the substitution of secondary for primary contacts, the weakening of bonds of kinship, and the declining social significance of the family, the disappearance of the neighborhood, and the undermining of the traditional basis of social solidarity. All these phenomena can be substantially verified through objective indices. Thus, for instance, the low and declining urban-reproduction rates suggest that the city is not conducive to the traditional type of family life, including the rearing of children and the maintenance of the home as the locus of a whole round of vital activities. The transfer of industrial, educational, and recreational activities to specialized institutions outside the home has deprived the family of some of its most characteristic historical functions. In cities mothers are more likely to be employed, lodgers are more frequently part of the household, marriage tends to be postponed, and the proportion of single and unattached people is greater. Families are smaller and more frequently without children than in the country. The family as a unit of social life is emancipated from the larger kinship group characteristic of the country, and the individual members pursue their own diverging interests in their vocational, educational, religious, recreational, and political life.

Such functions as the maintenance of health, the methods of alleviating the hardships associated with personal and social insecurity, the provisions for education, recreation, and cultural advancement have given rise to highly specialized institutions on a community-wide, statewide, or even national basis. The same factors which have brought about greater personal insecurity also underlie the wider contrasts between individuals to be found in the urban world. While the city has broken down the rigid caste lines of preindustrial society, it has sharpened and differentiated income and status groups. Generally, a larger proportion of the adult urban population is gainfully employed than is the case with the adult-rural population. The white-collar class, comprising those employed in trade, in clerical, and in professional work, are proportionately more numerous in large cities and in metropolitan centers and in smaller towns than in the country.

On the whole, the city discourages an economic life in which the individual in time of crisis has a basis of subsistence to fall back upon, and it discourages self-employment. While incomes of city people are on the average higher than those of country people, the cost of living seems to be higher in the larger cities. Home-ownership involves greater burdens and is rarer. Rents are higher and absorb a larger proportion of the income. Although the urban-dweller has the benefit of many communal services, he spends a large proportion of his income for such items as recreation and advancement and a smaller proportion for food. What the communal services do not furnish, the urbanite must purchase, and there is virtually no human need which has remained unexploited by commercialism. Catering to thrills and furnishing means of escape from drudgery, monotony, and routine thus become one of the major functions of urban recreation, which at its best furnishes means for creative self-expression and spontaneous group association, but which more typically in the urban world results in passive spectatorism, on the one hand, or sensational record-smashing feats, on the other.

Reduced to a state of virtual impotence as an individual, the urbanite is bound to exert himself by joining with others of similar interest into groups organized to obtain his ends. This results in the enormous multiplication of voluntary organizations directed toward as great a variety of objectives as there are human needs and interests. While, on the one hand, the traditional ties of human association are weakened, urban existence involves a much greater degree of interdependence between man and man and a more complicated, fragile, and volatile form of mutual interrelations over many phases of which the individual as such can exert scarcely any control. Frequently there is only the most tenuous relationship between the economic position or other basic factors that determine the individual's existence in the urban world and the voluntary groups with which he is affiliated. In a primitive and in a rural society it is generally possible to predict on the basis of a few known factors who will belong to what and who will associate with whom in almost every relationship of life, but in the city we can only project the general pattern of group formation and affiliation, and this pattern will display many incongruities and contradictions.

Urban personality and collective behavior

It is largely through the activities of the voluntary groups, be their objectives economic, political, educational, religious, recreational, or cultural, that the urbanite expresses and develops his personality, acquires status, and is able to carry on the round of activities that constitutes his life. It may easily be inferred, however, that the organizational framework which these highly differentiated functions call into being does not of itself insure the consistency and integrity of the personalities whose interests it enlists. Personal disorganization, mental breakdown, suicide, delinquency, crime, corruption, and disorder might be expected under these circumstances to be more prevalent in the urban than in the rural community. This has been confirmed in so far as comparable indexes are available, but the mechanisms underlying these phenomena require further analysis.

Since for most group purposes it is impossible in the city to appeal individually to the large number of discrete and differentiated citizens, and since it is only through the organizations to which men belong that their interests and resources can be enlisted for a collective cause, it may be inferred that social control in the city should typically proceed through formally organized groups. It follows, too, that the masses of men in the city are subject to manipulation by symbols and stereotypes managed by individuals working from afar or operating invisibly behind the scenes through their control of the instruments of communication. Self-government either in the economic, or political, or the cultural realm is under these circumstances reduced to a mere figure of speech, or, at best, is subject to the unstable equilibrium of pressure groups. In view of the ineffectiveness of actual kinship ties, we create fictional kinship groups. In the face of the disappearance of the territorial unit as a basis of social solidarity, we create interest units. Meanwhile the city as a community resolves itself into a series of tenuous segmental relationships superimposed upon a territorial base with a definite center but without a definite periphery, and upon a division of labor which far transcends the immediate locality and is world-wide in scope. The larger the number of persons in a state of interaction with another, the lower is the level of communication and the greater is the tendency for communication to proceed on an elementary level, i.e., on the basis of those things which are assumed to be common or to be of interest to all.

It is obviously, therefore, to the emerging trends in the communication system and to the production and distribution technology that has come into existence with modern civilization that we must look for the symptoms which will indicate the probable development of urbanism as a mode of social life. The direction of the ongoing changes in urbanism will for good or ill transform not only the city but the world.

It is only in so far as the sociologist, with a workable theory of urbanism, has a clear conception of the city as a social entity that he can hope to develop a unified body of reliable knowledge – which what passes as "urban sociology" is certainly not at the present time. By taking his point of departure from a theory of urbanism such as that sketched in the foregoing pages, a theory to be elaborated, tested, and revised, in the light of further analysis and empirical research, the sociologist can hope to determine the criteria of relevance and validity of factual data. The miscellaneous assortment of disconnected information which has hitherto found its way into sociological treatises on the city may thus be sifted and incorporated into a coherent body of knowledge. Incidentally, only by means of some such theory will the sociologist escape the futile practice of voicing in the name of sociological science a variety of often unsupportable judgments about poverty, housing, city-planning, sanitation, municipal administration, policing, marketing, transportation, and other technical issues. Though the sociologist cannot solve any of these practical problems – at least not by himself – he may, if he discovers his proper function, have an important contribution to make to their comprehension and solution. The prospects for doing this are brightest through a general, theoretical, rather than through an *ad hoc* approach.

"Theories of Urbanism"

from *The Urban Experience*, second edition (1984) [1976]

Claude S. Fischer

Editors' Introduction

Claude Fischer offers a seminal codification of academic thought on the theory of urbanism, and in the process makes a valuable reformulation of the theory through his attention to the emergence of subcultures. Fischer accepts some of the precepts of the prevailing determinism in urban sociology that acknowledges the primary and independent impact of urban effects, which Louis Wirth had identified as size, density, and heterogeneity. He differs from the compositional approach of Oscar Lewis and Herbert Gans, who rebuke urban effects and look to the cultural, demographic, and class characteristics of urbanites. Fischer believes that size and density of the population, or what he calls *critical mass* in cities, have independent effects in fostering subcultures. The emergence of subcultures fosters the further creation of more subcultures through the touch and recoil of more intensive interactions between more diverse populations and heterogeneous communities. That is to say that increasing size and density fosters greater heterogeneity. This recalls Robert Park's concept that the modern city becomes a *mosaic of social worlds*. The larger the city, the greater there is the potential to produce subcultural communities.

Fischer reconstructs the theory of urbanism by downplaying the negative effects of Durkheimian *anomie* with reference to the crime, mental health, and social problems that are found in the metropolis. He gives another perspective on *differentiation* as a cultural process linked with specialization in the division of labor. Fischer sees the city and its subcultures as a vital force for the amplifying of cultural experience and human creativity. Subcultures mark the emancipation of the individual from traditional controls and conventions, while providing a new set of subgroup identities and communities. In this way, they counterbalance some of the alienation and normlessness, the spiritual anxieties and social disorders found in our cities and marketplaces, which result from the breakdown of traditional customs and primary relationships.

Fischer makes an intriguing contribution to subcultural theory, which has also been explored from the standpoint of media and cultural studies, including Dick Hebdige (*Subculture: The Meaning of Style*. London: Methuen, 1979) and an edited reader by Ken Gelder and Sarah Thornton (*The Subcultures Reader*. London: Routledge, 1997). Subcultures, in these cultural studies, are seen as a creative force of communication or *bricolage*, which provide youth, sexual and racial/ethnic minorities with a means of defying and criticizing the established cultural hegemony. Fischer's understanding of subculture is less associated with identity politics and more eclectic and wide-ranging in its definition, comprising groups as diverse as delinquents, criminals, artists, bohemians, new religious sects, hobbyists, dance aficionados, hippies, and construction workers.

Understanding the subcultural life of the city helps understand the impact of the social movements of the 1960s and 1970s, which mobilized youthful, racial/ethnic, and gender/sexual minorities in movements of political resistance and empowerment. Fischer enlightens our understanding of the emergence of artistic, bohemian, and gay/lesbian neighborhoods in American cities, such as New York's Greenwich Village and East Village, the Castro district of San Francisco, and the Hollywood and West Hollywood districts of Los Angeles. These subcultural neighborhoods are often a lure for the children of suburbia who are drawn to the central city in search of the authenticity, excitement of what is unfamiliar. This is a distinct contrast from the earlier generation of the postwar period, which escaped the city in search of privacy and open space. Understanding urban subcultures also connects with the growing interest in the creative and cultural life of cities (see Part Six of this volume) with the onset of widespread gentrification and the emergence of an urban cultural economy.

This selection is extracted from Claude Fischer's book, *The Urban Experience* (1984) [1976]. The subcultural theory of urbanism is further articulated in the articles "Toward a Subcultural Theory of Urbanism," *American Journal of Sociology* 80, 6 (1975): 1319–1341 and "The Subcultural Theory of Urbanism: A Twentieth-Year Assessment," *American Journal of Sociology* 101, 3 (November 1995): 543–577.

Claude Fischer teaches in the Sociology Department at the University of California, Berkeley. He is currently the Executive Editor of *Contexts*, an official publication of the American Sociological Association. Fischer served as Chair of the Community and Urban Sociology Section in 1991–92. In 1996, he won the Robert and Helen Lynd Award for lifetime contribution to community and urban sociology from the American Sociological Association. Along with Michael Hout, Claude Fischer directs a project funded by the Russell Sage Foundation at the University of California, Berkeley, called "USA: A Century of Difference." Drawing upon a century of data up to the 2000 Census, this project will report on how Americans live, work, consume, and pray at the beginning of the twenty-first century. Among his books are *To Dwell Among Friends: Personal Networks in Town and City* (Chicago, IL: University of Chicago Press, 1982) and *America Calling: A Social History of the Telephone to 1940* (Berkeley, CA: University of California Press, 1992).

In *To Dwell Among Friends*, Fischer considers the thesis of declining community by comparing the differences in peoples' personal relations in urban versus non-urban areas. He takes the view that people exercise considerable individual agency in building their personal ties and networks. Initial relations are given to us, such as parents and other kin, but as we grow into adults, we select which ties are maintained and which are dropped. He concludes that people living in large cities versus small towns have roughly the same number of social ties; neither group is any more likely to be isolated. Small-town residents tend to be more involved with kin, city respondents with non-kin. Urbanites also tend to have less dense networks, but more intense interpersonal relations that involve multiple exchanges with given individuals. Urban dwellers tend to display tendencies similar to those of the young and the educated. Cities also tend to contain younger, more educated, and more diverse populations than small towns.

We begin our inquiry into the nature of the urban experience by considering the theorizing of social scientists about the consequences of urbanism. The purpose of developing such theories before looking at the "real world" is to provide the investigator with a set of concepts needed to organize his or her perceptions of what would otherwise be a bewildering complexity. Properly developed, these concepts focus attention on the most critical features of the "real world." To begin a major study without a good theory or theories is like being dropped into a dark jungle with neither map nor compass.

But before reviewing those theories, we must consider, once again and more exactly, the problem of defining "urban." For it turns out that some of the disagreement and confusion about the nature of the urban experience stems from differences in interpretation of the terms "urban" and "city." Surely, little progress can be made in understanding city life if we do not understand what the word "city" means. The four broad types of definitions are: demographic, institutional, cultural, and behavioral.

Demographic definitions involve essentially the size and density of population. In the present book, a community is more or less "urban" depending on the size of its population; a "city," therefore, is a place with a relatively large population. *Institutional* definitions reserve the term "city" for communities with certain specific institutions. For example, to be a city, a community must have its own autonomous political elite; or, it must have specific economic institutions, such as a commercial market. *Cultural* definitions require that a community possess particular cultural features, such as a group of literate people. And *behavioral* definitions require certain distinctive and typical behavioral styles among the people of a community – for example, an impersonal style of social interaction – before the community is labeled a "city."

The demographic definition has at least three advantages: One, the numerical criterion is common to virtually all definitions of "urban" or "city"; even those focusing on other variables employ size. Two, the purely demographic definition does not beg the question as to whether any other factor is necessarily associated with size; that remains an open issue. And three, the demographic definition implies that "urban" and "city" refer to matters of degree; they are not all-or-nothing variables.

What theories are there about the social–psychological consequences of urbanism in the sense of the demographic definition–population concentration? Here and throughout the rest of the book, we shall center on three major theories of urbanism, two of which confront each other directly, and a third which attempts their synthesis:

1 *Determinist theory* (also called *Wirthian* theory or the *theory of urban anomie*) argues that urbanism increases social and personality disorders over those found in rural places.
2 *Compositional* (or *nonecological*) *theory* denies such effects of urbanism; it attributes differences between urban and rural behavior to the composition of the different populations.
3 *Subcultural theory* adopts the basic orientation of the compositional school but holds that urbanism does have certain effects on the people of the city, with consequences much like the ones determinists see as evidence of social disorganisation.

Before discussing each theory in detail, we should consider the history of social thought from which they all emerged.

The most influential and historically significant theory of urbanism received its fullest exposition in a 1938 paper by Louis Wirth (thus the term, "Wirthian") entitled "Urbanism as a Way of Life." This essay, one of the most often quoted, reprinted, and cited in the whole sociological literature, needs to be examined carefully. It is heir to a long tradition of sociological theory.

The events that formed the focal concern of social philosophers during the nineteenth and early twentieth centuries have been termed the "Great Transformation" (Karl Polanyi (1944) *The Great Transformation*. New York: Farrar and Rinehart). Western society was undergoing vast and dramatic changes as a result of the Industrial Revolution and its accompanying processes of urbanization, nationalization, and bureaucratization. These early social scientists (Marx, Durkheim, Weber, Simmel, Tönnies, and others) sought to understand the forms of social life and the psychological character of the emerging civilization – our civilization.

The analysis they developed greatly emphasized the matter of *scale*. Innovations in transportation and communication, together with rapid increases in population, meant that many more individuals than ever before were able to interact and trade with each other. Instead of a person's daily life being touched at most by only the few hundred people of one village, in modern society an individual is in virtually direct contact with thousands, and in indirect contact with millions.

This "dynamic density," to use Durkheim's term, in turn produces *social differentiation*, or diversification, the most significant aspect of which is an increased division of labor. In the preindustrial society, most workers engaged in similar activities; in modern society, they have very different and specialized occupations. In a small, undifferentiated population, where people know each other, perform the same sort of work, and have the same interests – where they look, act, and think alike – it is relatively easy to maintain a consensus on proper values and appropriate behavior. But in a large, differentiated society, where people differ in their work and do not know each other personally, they have divergent interests, views, and styles. A pipefitter and a ballet dancer have little in common. And so, there can be little consensus or cohesion in such a society, and the social order is precarious. Further ramifications of social differentiation, it was thought, included the development of formal institutions, such as contracts and bureaucracy; the rise of rational, scientific modes of understanding the world; an increase in individual freedom, at the cost of interpersonal estrangement; and a rise in the rate of deviant behavior and social disorganization.

The essence of this classic sociological analysis is the connection of the structural characteristics of a society, particularly its scale, to the quality of its "moral order." That turns out, not coincidentally, to parallel the focal interest of urban sociology: the interest between structural features of communities – particularly scale – and their moral orders. In fact, the city has long played a significant role in classic sociological theories. It was seen as modern society in microcosm, so that the ways of life in urban places were viewed as harbingers of life in the emerging civilization. At the same time, the classic theories have had a significant role in influencing the study of cities, having been borrowed from liberally in the formation of the determinist approach.

The development of urban theory moved from Europe to the University of Chicago during the first third of this century. There, the Department of Sociology, under the leadership of Robert Ezra Park, a former journalist and student of the classical German sociologist Georg Simmel, produced a vast and seminal array of theoretical and empirical studies of urban life, based on research conducted chiefly in the city of Chicago. In an influential essay (Robert Park (1915) "The City: Suggestions for the Investigation of Human Behavior in the City," *American Journal of Sociology* 20, 5: 577–612), Park followed the lead of the classic theorists by arguing that urbanism produced new ways of life and new types of people, and that sociologists should venture out to explore these new forms in their own cities, much in the style of anthropologists studying primitive tribes. Another strong motivation for such research was the social turmoil then accompanying the rapid growth and industrialization of Western cities, a realm of civic activity in which Chicago about 1910 was no doubt a leader. The serious social problems accompanying these developments demanded study and explanation.

The varied studies of Chicago resulted in a remarkable series of descriptions of urban ways of life. The "natural histories" depicted many different groups and areas: taxi-hall dancers, hobos, Polish-Americans, juvenile gangs, the Jewish ghetto, pickpockets, police, and so on. A theme running through the findings of these various studies was that the groups, whether "normal" or "deviant," formed their own "social worlds." That is, they tended to be specialized social units in which the members associated mainly with each other, held their own rather distinctive set of beliefs and values, spoke in a distinctive argot, and displayed characteristic styles of behavior. Together these studies described a city that was, to quote Park's famous phrase, "a mosaic of social worlds which touch but do not interpenetrate." As we shall see, the explanation for the urban phenomena observed by Chicago's sociologists was drawn largely from the classic theories of the Great Transformation.

DETERMINIST THEORY

Some leads to a determinist theory of urbanism can be found in Park's 1915 paper, but the full exposition of this theory was achieved in Wirth's essay 23 years later. Wirth begins with a definition of the city as "a relatively large, dense, and permanent settlement of socially heterogeneous individuals" – an essentially demographic definition. He then seeks to demonstrate how these inherent, essential features of urbanism produce social disorganization and personality disorders – the dramatic aspects of the city scene that had captured the attention of the Chicago School. Wirth's analysis operates on essentially two levels, one a psychological argument, the other an argument of social structure.

The psychological analysis draws heavily upon a 1905 paper by Georg Simmel, a teacher of both Park and Wirth. In his essay "The Metropolis and Mental Life" Simmel centered on the ways that living in the city altered individuals' minds and personalities. The key, he thought, lay in the sensations which life in the city produces: "The psychological basis of the metropolitan type of individuality consists in the *intensification of nervous stimulation* which results from the swift and uninterrupted change of inner and outer stimuli." The city's most profound effects, Simmel maintained, are its profusion of sensory stimuli – sights, sounds, smells, actions of others, their demands and interferences. The onslaught is stressful; individuals must protect themselves, they must adapt. Their basic mode of adaptation is to react with their heads instead of their hearts. This means that urban dwellers tend to become intellectual, rationally calculating, and emotionally distant from one another. At the same time, these changes promote freedom for self-development and creativity. . . .

Wirth's treatment of this process follows Simmel's and begins with the assumption that the large, dense, and heterogeneous environment of the city assaults the hapless city dweller with profuse and varied stimuli. Horns blare, signs flash, solicitors tug at coattails, poll-takers telephone, newspaper headlines try to catch the eye, strange-looking and strange-behaving persons distract attention – all these features of the urban milieu claim a different response from the individual. Adaptations to maintain mental equilibrium are necessary and they appear. These adaptations liberate urbanites from the claims being pressed upon them. They also insulate them from the other people. City dwellers become aloof, brusque, impersonal in their dealings with others, emotionally buffered in their human relationships. Even these protective devices are not enough, so that "psychic overload" exacts at least a partial toll in irritation, anxiety, and nervous strain.

The interpersonal estrangement that follows from urbanites' adaptations produces further consequences. The bonds that connect people to one another are loosened – even sundered – and without them people are left both unsupported and unrestrained. At the worst, they must suffer through material and emotional crises without assistance, must deal with them alone; being alone, they are more likely to fail, to suffer physical deterioration or mental illness, or both. The typical picture is one of an elderly pensioner living in a seedy hotel without friends or kin, suffering loneliness, illness, and pain. But this same estrangement permits people in the city to spin the wildest fantasies – and to act upon those fantasies, whether they result in feats of genius or deeds of crime and depravity. The typical picture here is one of a small-town boy suddenly unshackled by conventional constraints, and possessing unlimited options including a life of creative art or a life of crime. Ultimately, interpersonal estrangement produces a decline of community cohesion and a corresponding loss of "sense of community." These are the psychological changes and further consequences, Wirth argued, that follow from increases in urbanism.

In his analysis of social structure, Wirth reaches essentially the same conclusion as he does in his psychological analysis, but he posits different processes. Through economic processes of competition, comparative advantage, and specialization, the size, density, and heterogeneity of a population produce the multi-faceted community differentiation mentioned earlier. This is manifested most significantly in the division of labor, but it exists in other forms as well: in the diversity of locales – business districts, residential neighborhoods, "bright-lights" areas, and so on; in people's places of activity, with work conducted in one place, family life in another, recreation in yet a third; in people's social circles, with one set of persons co-workers, another set neighbors, another friends, and still another kin; in institutions, with the alphabetized diversity of government agencies, specialized school systems, and media catering to every taste. An important aspect of this community differentiation is that it is reflected in people's activities. Their time and attention come to be divided among different and disconnected places, and people. For example, a business executive might move from breakfast with her family, to discussions with office co-workers, to lunch with business contacts, to a conference with clients, to golf with friends from the club, and finally to dinner with neighbors.

The differentiation of the social structure and of the lives of individuals living within that structure weakens social bonds in two ways. At the community level, people differ so much from each other in such things as their jobs, their neighborhoods, and their life-styles that moral consensus becomes difficult. With divergent interests, styles, and views of life, groups in the city cannot agree on values or beliefs, on ends or on means. As community-wide cohesion is weakened, so is the cohesion of the small, intimate, "primary" groups of society, such as family, friends, and neighbors – the ones on which social order and individual balance depend. These groups are weakened because, as a result of the differentiation of urban life, each encompasses less of an individual's time or needs. For instance, people work outside the family and increasingly play outside the family, so that the family becomes less significant in their lives. Similarly, they can leave the neighborhood for shopping or recreation, so that the neighbors become less important. Claiming less of people's attention, controlling less of their lives, the primary groups become debilitated. Thus, by dividing the community and by weakening its primary groups, differentiation produces a general loosening of social ties.

This situation in turn results in *anomie*, a social condition in which the norms – the rules and conventions of proper and permissible behavior – are feeble. People do not agree about the norms, do not endorse them, and tend to challenge or ignore them. Yet some degree of social order must be, indeed is, maintained even in the largest cities. Since personal means of providing order have been weakened, other means must be used. These other means – rational and impersonal procedures that arise to prevent or to moderate anomie – are called *formal integration*. For example, instead of controlling the behavior of unruly teenagers by talking to them or their parents personally, neighbors call in the police. Instead of settling a community problem through friendly and informal discussions, people organize lobbying groups and campaign in formal elections.

This sort of formal integration avoids chaos and can even maintain a well-functioning social order. However, according to the classic theories that Wirth applied in his analysis of cities, such an order can never fully replace a communal order based on consensus and the moral strength of small, primary groups. Consequently, more anomie must develop in urban than in nonurban places.

The behavioral consequences of anomie and of the shedding of social ties are similar to those eventually resulting from overstimulation. People are left unsupported to suffer their difficulties alone; and they are unrestrained by social bonds or rules from committing all sorts of acts, from the simply "odd" to the dangerously criminal.

These, then, are the arguments with which Wirth explained what seemed to the Chicago School to be peculiarly urban phenomena – stress, estrangement, individualism, and especially social disorganization. On the psychological level, urbanism produces threats to the nervous system that then lead people to separate themselves from each other. On the level of social structure, urbanism induces differentiation, which also has the consequences of isolating people. A society in which social relationships are weak provides freedom for individuals, but it also suffers from a debilitated moral order, a weakness that permits social disruption and promotes personality disorders.

COMPOSITIONAL THEORY

The determinist approach has been challenged on a number of fronts. The most significant challenge has been posed by compositional theory, perhaps best represented by the work of Herbert Gans and Oscar Lewis. . . . Compositionalists emerged from the same Chicago School tradition as the determinists, but they derived their inspiration largely from that part of the Chicago orientation that describes the city as a "mosaic of social worlds." These "worlds" are intimate social circles based on kinship, ethnicity, neighborhood, occupation, life-style, or similar personal attributes. They are exemplified by enclaves such as immigrant neighborhoods ("Little Italy") and upper-class colonies ("Nob Hill"). . . . The crux of the compositional argument is that these private milieus endure even in the most urban of environments.

In contrast to determinists, social scientists such as Gans and Lewis do not believe that urbanism weakens small, primary groups. They maintain that these groups persist undiminished in the city. Not that people are torn apart because they must live simultaneously in different social worlds, but instead that people are enveloped and protected by their social worlds. This point of view denies that ecological factors – particularly the size, density, and heterogeneity of the wider community – have any serious, direct consequences for personal social worlds. In this view, it matters little to the average kith-and-kin group whether there are 100 people in the town or 100,000; in either case the basic dynamics of that group's social relationships and its members' personalities are unaffected.

In compositionalist terms, the dynamics of social life depend largely on the nonecological factors of social class, ethnicity, and stage in the life-cycle. Individuals' behavior is determined by their economic position, cultural characteristics, and by their marital and family status. The same attributes also determine who their associates are and what social worlds they live in. It is these attributes – not the size of the community or its density – that shape social and psychological experience.

Compositionalists do not suggest that urbanism has *no* social-psychological consequences, but they do argue that both the *direct* psychological effects on the individual and the *direct* anomic effects on social worlds are insignificant. If community size does have any consequences, these theorists stipulate, they result from ways in which size affects positions of individuals in the economic structure, the ethnic mosaic, and the life-cycle. For example, large communities may provide better-paying jobs, and the people who obtain those jobs will be deeply affected. But they will be affected by their new economic circumstances, not directly by the urban experience itself. Or, a city may attract a disproportionate number of males, so that many of them cannot find wives. This will certainly affect their behavior, but not because the city has sundered their social ties. Thus, the compositional approach can acknowledge urban–rural social–psychological differences, and can account for them insofar as these differences reflect variations in class, ethnicity, or life-cycle. But the compositional approach does not expect such differences to result from the psychological experience of city life or from an alteration in the cohesion of social groups.

The contrast between the determinist and compositional approaches can be expressed this way: Both emphasize the importance of social worlds in forming the experiences and behaviors of individuals, but they disagree sharply on the relationship of urbanism to the viability of those personal milieus. Determinist theory maintains that urbanism has a direct impact on the coherence of such groups, with serious consequences for individuals. Compositional theory maintains that these social worlds are largely impervious to ecological factors, and that urbanism thus has no serious, *direct* effects on groups or individuals.

SUBCULTURAL THEORY

The third approach, *subcultural theory* (Claude Fischer (1975) "Toward a Subcultural Theory of Urbanism," *American Journal of Sociology* 80, 6: 1319–1341), contends that urbanism independently affects social life – not, however, by destroying social groups as determinism suggests, but instead by helping to create and strengthen them. The most significant social consequence of community size is the promotion of diverse *subcultures* (culturally distinctive groups, such as college students or Chinese-Americans). Like compositional theory, subcultural theory maintains that intimate social circles persist in the urban environment. But, like determinism, it maintains that ecological factors do produce significant effects in the social orders of communities, precisely by supporting the emergence and vitality of distinctive subcultures.

Like the Chicago School in certain of its works and like compositionalists, the subcultural position holds that people in cities live in meaningful social worlds. These worlds are inhabited by persons who share relatively distinctive traits (like ethnicity or occupation), and who tend to interact especially with one another, and who manifest a relatively distinct set of beliefs and behaviors. Social worlds and subcultures are roughly synonymous. Obvious examples of subcultures include ones like those described by the Chicago School: the country club set in Grosse Pointe, Michigan; the Chicano community in East Los Angeles; and hippies in urban communes. There are more complex subcultures as well. For example, on the south side of Chicago is an area heavily populated by workers in the nearby steel mills. These workers together form a community and occupational subculture, with particular habits, interests, and attitudes. But they are further divided into even more specific subcultures by ethnicity and neighborhood; thus there are, for example, the recently immigrated Serbo-Croatian steelworkers in one area and the earlier-generation ones elsewhere, each group somewhat different from the other. In both subcultural and compositional theory, these subcultures persist as meaningful environments for urban residents.

However, in contrast to the compositional analysis, which discounts any effects of urbanism, subcultural theory argues that these groups *are* affected directly by urbanism, particularly by the effects of "critical mass." Increasing scale on the rural-to-urban continuum creates new subcultures, modifies existing ones, and brings them into contact with each other. Thus urbanism has unique consequences, including the production of "deviance," but not because it destroys social worlds – as determinism argues – but more often because it creates them.

The subcultural theory holds, first, that there are two ways in which urbanism produces Park's "mosaic of little worlds which touch but do not interpenetrate": 1) Large communities attract migrants from wider areas than do small towns, migrants who bring with them a great variety of cultural backgrounds, and thus contribute to the formation of a diverse set of social worlds. And 2), large size produces the structural differentiation stressed by the determinists – occupational specialization, the rise of specialized institutions, and of special interest groups. To each of these structural units are usually attached subcultures. For example, police, doctors, and longshoremen tend to form their own milieus – as do students, or people with political interests or hobbies in common. In these ways, urbanism generates a variety of social worlds.

But urbanism does more: It intensifies subcultures. Again, there are two processes. One is based on *critical mass*, a population size large enough to permit what would otherwise be only a small group of individuals to become a vital, active subculture. Sufficient numbers allow them to support institutions – clubs, newspapers, and specialized stores, for example – that serve the group; allow them to have a visible and affirmed identity, to act together on their own behalf, and to interact extensively with each other. For example, let us suppose that one in every thousand persons is intensely interested in modern dance. In a small town of 5,000 that means there would be, on the average, five such persons, enough to do little else than engage in conversation about dance. But in a city of one million, there would be a thousand – enough to support studios, occasional ballet performances, local meeting places, and a special social milieu. Their activity would probably draw other people beyond the original thousand into the subculture (those quintets of dance-lovers migrating from the small towns). The same general process of critical mass operates for artists, academics, bohemians, corporate executives, criminals, computer programmers – as well as for ethnic and racial minorities.

The other process of intensification results from contacts between these subcultures. People in different social worlds often do "touch," in Park's language. But in doing so, they sometimes rub against one another only to recoil, with sparks flying upward. Whether the encounter is between blacks and Irish, hard-hats and hippies, or town and gown, people from one subculture often find people in another subculture threatening, offensive, or both. A common reaction is to embrace one's own social world all the more firmly, thus contributing to its further intensification. This is not to deny that there are often positive contacts between groups. There are; and there is a good deal of mutual influence – for example, the symbolism of young construction workers growing beards, or middle-class white students using black ghetto slang. It is, however, the contrast and recoil that intensify and help to define urban subcultures.

Among the subcultures spawned or intensified by urbanism are those which are considered to be either downright deviant by the larger society – such as delinquents, professional criminals, and homosexuals; or to be at least "odd" – such as artists, missionaries of new religious sects, and intellectuals; or to be breakers of tradition – such as life-style experimenters, radicals, and scientists. These flourishing subcultures, together with the conflict that arises among them and with mainstream subcultures, are both effects of urbanism, and they both produce what the Chicago School thought of as social "disorganization." According to subcultural theory, these phenomena occur not because social worlds break down, and people break down with them, but quite the reverse – because social worlds are formed and nurtured.

Subcultural theory is thus a synthesis of the determinist and compositional theories: like the compositional approach, it argues that urbanism does not produce mental collapse, anomie, or interpersonal estrangement; that urbanites at least as much as ruralites are integrated into viable social worlds. However, like the determinist approach, it also argues that cities *do* have effects on social groups and individuals – that the differences between rural and urban persons have other causes than the economic, ethnic, or life-style circumstances of those persons. Urbanism does have *direct* consequences.

"The Uses of City Neighborhoods"

from *The Death and Life of Great American Cities* (1961)

Jane Jacobs

Editors' Introduction

Contemporary urban sociology owes much to the legacy of Jane Jacobs, a staff writer at *Architectural Forum* and a community activist who helped change our views about architectural modernism and urban renewal policy in postwar America. She assaulted the misguided policies of slum clearance bureaucrats who devastated vital urban neighborhoods in favor of expressways and oppressively dull housing blocks surrounded by seas of indefensible open spaces. Dense networks of lively streets and sociologically diverse neighborhoods are important touchstones to urban quality of life for Jacobs. She championed the cause of neighborhood preservation and helped save Greenwich Village from the bulldozers of New York power broker Robert Moses and his plan for an urban expressway through Lower Manhattan.

This chapter is excerpted from her 1961 book, *The Death and Life of Great American Cities*. In this book she also describes the lively daily ballet of life on Hudson Street where she lived, the significance of public characters, and the informal surveillance networks of "eyes on the street" that helped regulate public security for residents living among urban strangers. She celebrated the messiness and spontaneity of urban street life that was so full of serendipitous encounters. She sought to shift urban planners away from slum clearance toward a more enlightened policy that promoted rather than destroyed neighborhoods.

In this selection she urges us to understand and better mobilize the powers of neighborhoods as mundane "organs of self-government" that comprise the larger city. She warns us from sentimentalizing neighborhood preservation into nostalgic evocations of town life. Neighborhoods are not introverted or self-contained simulations of the rural village life, but better understood as naturally extroverted constituent units of urbanity. City planners cannot revive neighborhoods simply through location of streets, parks, and housing. Jacobs says city planners also need to create a social life and local identity through additional use of public buildings and landmarks. Community assets and social fabric were destroyed in many urban neighborhoods by urban renewal.

But Jacobs says we must also learn the secrets of self-governing neighborhoods that are able to mobilize assets and resources for their constituent residents. Effective urban planning and governance structures for the good and just city can be built out of mobilizing communities of interest into effective political districts. She warns against island-like fiefdoms that cater to insular interests in favor of joint committees and organizational coalitions of leaders from inside and outside the district. She identifies a range of civic, neighborhood improvement, and protest organizations from which the leadership of neighborhood districts may be drawn. A network of leaders with "hop-and-skip" links to a variety of organizations forms the core of effective neighborhood districts, including key cross-link communicators that are efficient mobilizers of trust and reciprocity. These are the kinds of social networks and social capital that were lost in many inner city communities devastated by slum clearance, depopulation, and disinvestment.

She also speculates on the proper size of neighborhood districts, and finds that smaller districts like the North End are common in smaller cities like Boston, while larger districts prevail in larger cities like Back of the Yards in Chicago and the Lower East Side in New York. She also cites the common incidence of socially cohesive ethnic neighborhoods, although she thinks they are more powerful if made more sociologically diverse by newcomers and networks of cross-link leaders. She believes that sameness can handicap residents from access to more opportunities.

Jane Jacobs moved from New York City to Toronto, Canada in 1968, where she was just as influential in the cancelling of the Spadina Expressway. She published several more important books in urban studies, including *The Economy of Cities* (New York: Vintage, 1969) and *Cities and the Wealth of Nations* (New York: Vintage, 1984). She died in 2006. In addition to leaving an intellectual legacy, her city planning ideas had a significant impact on the architecture and urban planning movement known as the New Urbanism. This movement promotes many city planning principles that Jacobs celebrated, such as pedestrian walkability, neotraditional architecture, mixed-uses, civic identity, local heritage, and "smart growth" versus urban sprawl.

Neighborhood is a word that has come to sound like a Valentine. As a sentimental concept, "neighborhood" is harmful to city planning. It leads to attempts at warping city life into imitations of town or suburban life. Sentimentality plays with sweet intentions in place of good sense.

A successful city neighborhood is a place that keeps sufficiently abreast of its problems so it is not destroyed by them. An unsuccessful neighborhood is a place that is overwhelmed by its defects and problems and progressively more helpless before them. Our cities contain all degrees of success and failure. But on the whole we Americans are poor at handling city neighborhoods, as can be seen by the long accumulations of failures in our great gray belts on the one hand, and by the Turfs of rebuilt city on the other hand.

It is fashionable to suppose that certain touchstones of the good life will create good neighborhoods – schools, parks, clean housing and the like. How easy life would be if this were so! How charming to control a complicated and ornery society by bestowing upon it rather simple physical goodies. In real life, cause and effect are not so simple.

[...]

To hunt for city neighborhood touchstones of success in high standards of physical facilities, or in supposedly competent and nonproblem populations, or in nostalgic memories of town life, is a waste of time. It evades the meat of the question, which is the problem of what city neighborhoods do, if anything, that may be socially and economically useful in cities themselves, and how they do it.

We shall have something solid to chew on if we think of city neighborhoods as mundane organs of self-government. Our failures with city neighborhoods are, ultimately, failures in localized self-government. And our successes are successes at localized self-government. I am using self-government in its broadest sense, meaning both the informal and formal self-management of society.

Both the demands on self-government and the techniques for it differ in big cities from the demands and techniques in smaller places. For instance, there is the problem of all those strangers. To think of city neighborhoods as organs of city self-government or self-management, we must first jettison some orthodox but irrelevant notions about neighborhoods which may apply to communities in smaller settlements but not in cities. We must first of all drop any ideal of neighborhoods as self-contained or introverted units.

[...]

Whatever city neighborhoods may be, or may not be, and whatever usefulness they may have, or may be coaxed into having, their qualities cannot work at cross-purposes to thoroughgoing city mobility and fluidity of use, without economically weakening the city of which they are a part. The lack of either economic or social self-containment is natural and necessary to city neighborhoods – simply because they are part of cities...

But for all the more innate extroversion of city neighborhoods, it fails to follow that city people can therefore get along magically without neighborhoods. Even the most urbane citizen does care about the atmosphere of the street and district where he lives, no matter how much choice he has of pursuits outside it; and the common run of city people do depend greatly on their neighborhoods for the kind of everyday lives they lead.

Let us assume (as is often the case) that city neighbors have nothing more fundamental in common with each other than that they share a fragment of geography. Even so, if they fail at managing the fragment decently, the fragment will fail. There exists no inconceivably energetic and all-wise "They" to take over and substitute for localized self-management. Neighborhoods in cities need not supply for their people an artificial town or village life, and to aim at this is both silly and destructive. But neighborhoods in cities do need to supply some means for civilized self-government. This is the problem.

Looking at city neighborhoods as organs of self-government, I can see evidence that only three kinds of neighborhoods are useful: a) the city as a whole, b) street neighborhoods, c) districts of large, subcity size, composed of 100,000 people or more in the case of the largest cities.

Each of these kinds of neighborhoods has different functions, but the three supplement each other in complex fashion. It is impossible to say that one is more important than the others. For success with staying power at any spot, all three are necessary. But I think that other neighborhoods than these three kinds just get in the way, and make successful self-government difficult or impossible.

The most obvious of the three, although it is seldom called a neighborhood, is the city as a whole. We must never forget or minimize this parent community while thinking of a city's smaller parts. This is the source from which most public money flows, even when it comes ultimately from the federal or state coffers. This is where most administrative and policy decisions are made, for good or ill. This is where general welfare often comes into direst conflict, open or hidden, with illegal or other destructive interests.

Moreover, up on this plane we find vital special-interest communities and pressure groups. The neighborhood of the entire city is where people especially interested in the theater or in music or in other arts find one another and get together, no matter where they may live. This is where people immersed in specific professions or businesses or concerned about particular problems exchange ideas and sometimes start action.

[...]

A city's very wholeness in bringing together people with communities of interest is one of its greatest assets, possibly the greatest. And, in turn, one of the assets a city district needs is people with access to the political, the administrative, and the special-interest communities of the city as a whole.

[...]

When Greenwich Village fought to prevent its park, Washington Square, from being bisected by a highway, for example, majority opinion was overwhelmingly against the highway. But not unanimous opinion; among those for the highway were numerous people of prominence, with leadership positions in smaller sections of the district. Naturally they tried to keep the battle on a level of sectional organization, and so did the city government. Majority opinion would have frittered itself away in these tactics, instead of winning. Indeed, it was frittering itself away in these tactics, instead of winning. Indeed, it was frittering itself away until this truth was pointed out by Raymond Rubinow, a man who happened to work in the district, but did not live there. Since Rubinow helped form a *Joint* Emergency Committee, a true district organization cutting through other organizational lines. Effective districts operate as Things in their own right, and most particularly must their citizens who are in agreement with each other on controversial questions act together at district scale, or they get nowhere. Districts are not groups of petty principalities, working in federation. If they work, they work as integral units of power and opinion, large enough to count.

Our cities possess many islandlike neighborhoods too small to work as districts, and these include not only the neighborhoods inflicted by planning, but also many unplanned neighborhoods. These unplanned, too small units have grown up historically, and often are enclaves of distinctive ethnic groups. They frequently perform well and strongly the neighborhood functions of streets and thus keep marvelously in hand the kinds of neighborhood social problems and rot that develop from without. They are shortchanged on public improvements and services because they lack power to get them. They are helpless to reverse the slow-death warrants of area credit-blacklisting by mortgage lenders, a problem terribly difficult to fight even with impressive district power. If they develop conflicts with people in adjoining neighborhoods, both they and the adjoining people are apt to be helpless at improving relationships. Indeed, insularity makes these relationships deteriorate further.

[. . .]

If the only kinds of city neighborhoods that demonstrate useful functions in real-life self-government are the city as a whole, streets, and districts, then effective neighborhood physical planning for cities should aim at these purposes:

First, to foster lively and interesting streets.

Second, to make the fabric of these streets as continuously a network as possible *throughout* a district of potential subcity size and power.

Third, to use parks and squares and public buildings as part of this street fabric; use them to intensify and knit together the fabric's complexity and multiple use. They should not be used to island off different uses from each other, or to island off subdistrict neighborhoods.

Fourth, to emphasize the functional identity of areas large enough to work as districts.

If the first three aims are well pursued, the fourth will follow. Here is why: Few people, unless they live in a world of paper maps, can identify with an abstraction called a district, or care much about it. Most of us identify with a place in the city because we use it, and get to know it reasonably intimately. We take our two feet and move around in it and come to count on it. The only reason anyone does this much is that useful and interesting or convenient differences fairly nearby exert an attraction.

[. . .]

Differences, *not duplications*, make for cross-use and hence for a person's identification with an area greater than his immediate street network. As for Turf, planned or unplanned, nobody outside the Turf can possibly feel a natural identity of interest with it or with what it contains.

Centers of use grow up in lively, diverse districts, just as centers of use occur on a smaller scale in parks, and such centers count especially in district identification if they contain also a landmark that comes to stand for the place symbolically and, in a way, for the district. But centers cannot carry the load of district identification by themselves; differing commercial and cultural facilities, and different-looking scenes, must crop up all through. Within this fabric, physical barriers, such as huge traffic arteries, too large parks, big institutional groupings, are functionally destructive because they block cross-use.

How big, in absolute terms, must an effective district be? I have given a functional definition of size: big enough to fight city hall, but not so big that street neighborhoods are unable to draw district attention and to count.

In absolute terms, this means different sizes in different cities, depending partly on the size of the city as a whole. In Boston, when the North End had a population upward of 30,000 people, it was strong in district power. Now its population is about half that, partly from the salutary process of uncrowding its dwellings as its people have unslummed, and partly from the unsalutary process of being ruthlessly amputated by a new highway. Cohesive though the North End is, it has lost an important sum of district power. In a city like Boston, Pittsburgh or possibly even Philadelphia, as few as 30,000 people may be sufficient to form a district. In New York or Chicago, however, a district as small as 30,000 amounts to nothing. Chicago's most effective district, the Back-of-the-Yards, embraces about 100,000 people, according to the director of the district Council, and is building up its population further. In New York, Greenwich Village is on the small side for an effective district, but is viable because it manages to make up for this with other advantages. It contains approximately 80,000 residents, along with a working population (perhaps a sixth of them the same people) of approximately 125,000. East Harlem and the Lower East Side of New York, both struggling to create effective districts, each contain about 200,000 residents, and need them.

Of course other qualities than sheer population size count in effectiveness – especially good communication and good morale. But population size is vital because it represents, if most of the time only by implication, votes. There are only two ultimate public powers shaping and running American cities: votes and control of the money. To sound nicer, we may call these "public opinion" and "disbursement of funds," but they are still votes and money. An effective district – and through its mediation, the street neighborhoods – possesses one of these powers: the power of votes. Through this, and this alone, can it effectively influence the power brought to bear on it, for good or for ill, by public money.

Robert Moses, whose genius at getting things done largely consists in understanding this, has made an art of using control of public money to get his way with those whom the voters elect and depend on to represent their frequently opposing interests. This is, of course, in other guises, an old, sad story of democratic government. The art of negating the power of votes with the power of money can be practiced just as effectively by honest public administrators as by dishonest representatives of purely private interests. Either way, seduction or subversion of the elected is easiest when the electorate is fragmented into ineffective units of power.

On the maximum side, I know of no district larger than 200,000 which operates like a district. Geographical size imposes empirical population limits in any case. In real life, the maximum size of naturally evolved, effective districts seems to be roughly about a mile and a half square. Probably this is because anything larger gets too inconvenient for sufficient local cross-use and for the functional identity that underlies district political identity. In a very big city, populations must therefore be dense to achieve successful districts; otherwise, sufficient political power is never reconciled with viable geographic identity.

This point on geographic size does not mean a city can be mapped out in segments of about a square mile, the segments defined with boundaries, and districts thereby brought to life. It is not boundaries that make a district, but the cross-use and life. The point in considering the physical size and limits of a district is this: the kinds of objects, natural or man-made, that form physical barriers to easy cross-use must be somewhere. It is better that they be at the edges of areas large enough to work as districts than that they cut into the continuity of otherwise feasible districts. The fact of a district lies in what it is internally, and in the internal continuity and overlapping with which it is used, not in the way it ends or in how it looks in an air view. Indeed, in many cases very popular city districts spontaneously extend their edges, unless prevented from doing so by physical barriers. A district too thoroughly buffered off also runs the danger of losing economically stimulating visitors from other parts of the city.

Neighborhood planning units that are significantly defined only by their fabric and the life and intricate cross-use they generate, rather than by formalistic boundaries, are of course at odds with orthodox planning conceptions. The difference is the difference between dealing with living, complex organisms, capable of shaping their own destinies, and dealing with fixed and inert settlements, capable merely of custodial care (if that) of what has been bestowed upon them.

In dwelling on the necessity for districts, I do not want to give the impression that an effective city district is self-contained either economically, politically or socially. Of course it is not and cannot be, any more than a street can be. Nor can districts be duplicates of one another; they differ immensely, and should. A city is not a collection of repetitious towns. An interesting district has a character of its own and specialties of its own. It draws users from outside (it has little truly economic variety unless it does), and its own people go forth.

Nor is there necessity for district self-containment. In Chicago's Back-of-the-Yards, most of the breadwinners used to work, until the 1940s, at the slaughterhouses within the district. This did have a bearing on district formation in this case, because district organization here was a sequel to labor union organization. But as these residents and their children have graduated from the slaughterhouse jobs, they have moved, into the working life and public of the greater city. Most, other than teenagers with after-school jobs, now work outside the district. This movement has not weakened the district; coincident with it, the district has grown stronger.

The constructive factor that has been operating here meanwhile is time. Time, in cities, is the substitute for self-containment. Time, in cities, is indispensable.

The cross-links that enable a district to function as a Thing are neither vague nor mysterious. They consist of working relationships among specific people, many of them without much else in common than that they share a fragment of geography.

The first relations to form in city areas, given any neighborhood stability, are those in street neighborhoods and those among people who do have something else in common and belong to organizations with one another – churches, PTAs, businessmen's associations, political clubs, local civic leagues, fundraising committees for health campaigns or other public causes, sons of such-and-such a village (common clubs among Puerto Ricans today, as they have been with Italians), property owners' associations, block improvement associations, protestors against injustices, and so on, ad infinitum.

To look into almost any relatively established area of a big city turns up so many organizations, mostly little, as to make one's head swim . . . Small organizations and special-interest organizations grow in our cities like leaves on the trees, and in their own way are just as awesome a manifestation of the persistence and doggedness of life.

The crucial stage in the formation of an effective district goes much beyond this, however. An interweaving, but different, set of relationships must grow up; these are working relationships among people, usually leaders, who enlarge their local public life beyond the neighborhoods of streets and specific organizations or institutions and form relationships with people whose roots and backgrounds are in entirely different constituencies, so to speak. These hop-and-skip relationships are more fortuitous in cities than are the analogous, almost enforced, hop-and-skip links among people from different small groupings within self-contained settlements. Perhaps because we are typically more advanced at forming whole-city neighborhoods of interest than at forming districts, hop–skip district relationships sometimes originate fortuitously among people from a district who meet in a special-interest neighborhood of the whole city, and then carry over this relationship into their district. Many district networks in New York, for instance, start in this fashion.

It takes surprisingly few hop–skip people, relative to a whole population, to weld a district into a real Thing. A hundred or so people do it in a population a thousand times their size. But these people must have time to find each other, time to try expedient cooperation – as well as time to have rooted themselves, too, in various smaller neighborhoods of place or special interest.

When my sister and I first came to New York from a small city, we used to amuse ourselves with a game we called Messages. I suppose we were trying, in a dim way, to get a grip on the great, bewildering world into which we had come from our cocoon. The idea was to pick two wildly dissimilar individuals – say a headhunter in the Solomon Islands and a cobbler in Rock Island, Illinois – and assume that one had to get a message to the other by word of mouth; then we would each silently figure out a plausible, or at least possible, chain of persons to whom the message could go. The one who could make the shortest plausible chain of messages won. The headhunter would speak to the headman of his village, who would speak to the trader who came to buy copra, who would speak to the Australian patrol officer when he came through, who would tell the man who was next slated to go to Melbourne on leave, etc. Down at the other end, the cobbler would hear from his priest, who got it from the mayor, who got it from a state senator, who got it from the governor, etc. We soon had these close-to-home messages down to a routine for almost everybody we could conjure up, but we would get tangled in long chains at the middle until we began employing Mrs. Roosevelt. Mrs. Roosevelt made it suddenly possible to skip whole chains of intermediate connections. She knew the most unlikely people. The world shrank remarkably. It shrank us right out of our game, which became too cut and dried.

A district requires a small quota of its own Mrs. Roosevelts – people who know unlikely people, and therefore eliminate the necessity for long chains of communication (which in real life would not occur at all).

Settlement-house directors are often the ones who begin such systems of district hop–skip links, but they can only begin them and work at opportune ways to extend them; they cannot carry the load. These links require the growth of trust, the growth of cooperation that is, at least at first, apt to be happenstance and tentative; and they require people who have considerable self-confidence, or sufficient concern about local public problems to stand them in the stead of self-confidence.

[. . .]

Once a good strong network of these hop–skip links does get going in a city district, the net can enlarge relatively swiftly and weave all kinds of resilient new patterns. One sign that it is doing so, sometimes, is the growth of a new kind of organization, more or less district-wide, but impermanent, formed specifically for *ad hoc* purposes. But to get going, a district needs these three requisites: a start of some kind; a physical area with which sufficient people can identify as users; and Time.

The people who form hop–skip links, like the people who form the smaller links in streets and special-interest organizations, are not at all the statistics that are presumed to represent people in planning and housing schemes. Statistical people are a fiction for many reasons, one of which is that they are treated as infinitely interchangeable. Real people are unique, they invest years of their lives in significant relationships with other unique people, and are not interchangeable in the least. Severed from their relationships, they are destroyed as effective social beings – sometimes for a little while, sometimes forever.

In city neighborhoods, whether streets or districts, if too many slowly grown public relationships are disrupted at once, all kinds of havoc can occur – so much havoc, instability and helplessness that it sometimes seems time will never again get in its licks.

Harrison Salisbury, in a series of *New York Times* articles, "The Shook-Up Generation," put well this vital point about city relationships and their disruption: "Even a ghetto [he quoted a pastor as saying], after it has remained a ghetto for a period of time builds up its social structure and this makes for more stability, more leadership, more agencies for helping the solution of public problems."

But when slum clearance enters an area [Salisbury went on], it does not merely rip out slatternly houses. It uproots the people. It tears out the churches. It destroys the local business man. It sends the neighborhood lawyer to new offices downtown and it mangles the tight skein of community friendships and group relationships beyond repair.

It drives the old-timers from their broken-down flats or modest homes and forces them to find new and alien quarters. And it pours into a neighborhood hundreds and thousands of new faces.

[. . .]

Renewal planning, which is largely aimed at saving buildings, and incidentally some of the population, but at strewing the rest of a locality's population, has much the same result. So does too heavily concentrated private building, capitalizing in a rush on the high values created by a stable city neighborhood. From Yorkville, in New York, an estimated 15,000 families have been driven out between 1951 and 1960 by this means; virtually all of them left unwillingly. In Greenwich Village, the same thing is happening. Indeed, it is a miracle that our cities have any functioning districts, not that they have so few. In the first place, there is relatively little city territory at present which is, by luck, well suited physically to forming districts with good cross-use and identity. And within this, incipient or slightly too weak districts are forever being amputated, bisected and generally shaken up by misguided planning policies. The districts that are effective enough to defend themselves from planned disruption are eventually trampled in an unplanned gold rush by those who aim to get a cut of these rare social treasures.

To be sure, a good city neighborhood can absorb newcomers into itself, both newcomers by choice and immigrants settling by expediency, and it can protect a reasonable amount of transient population too. But these increments or displacements have to be gradual. If self-government in the place is to work, underlying any float of population must be a continuity of people who have forged neighborhood networks. These networks are a city's irreplaceable social capital. Whenever the capital is lost, from whatever cause, the income from it disappears, never to return until and unless new capital is slowly and chancily accumulated.

Some observers of city life, noting that strong city neighborhoods are so frequently ethnic communities – especially communities of Italians, Poles, Jews or Irish – have speculated that a cohesive ethnic base is required for a city neighborhood that works as a social unit . . . these ethnically cohesive communities are not always as naturally cohesive as they may look to outsiders . . . Ethnic cohesiveness may have played a part in the formation of these sections, but it has been no help in welding district cross-links . . . Today many streets in these old ethnic communities have assimilated into their neighborhoods a fantastic ethnic variety from almost the whole world. They have also assimilated a great sprinkling of middle-class professionals and their families, who prove to do very well at city street and district life, in spite of the planning myth that such people need protective islands of pseudosuburban "togetherness." Some of the streets that function best in the Lower

East Side (before they were wiped out) were loosely called "Jewish," but contained, as people actually involved in the street neighborhoods, individuals of more than forty differing ethnic origins.

[...]

Here is a seeming paradox: To maintain in a neighborhood sufficient people who stay put, a city must have . . . fluidity and mobility of use . . . Over intervals of time, many people change their jobs and the locations of their jobs, shift or enlarge their outside friendships and interests, change their family sizes, change their incomes up or down, even change many of their tastes. In short they live, rather than just exist. If they live in diversified, rather monotonous, districts – in districts, particularly, where many details of physical changes can constantly be accommodated – and if they like the place, they can stay put despite changes in the locales or natures of their other pursuits or interests. Unlike the people who must move from a lower-middle to a middle-middle to an upper-middle suburb as their incomes and leisure activities change (or be very outré indeed), or the people of a little town who must move to another town or to a city to find different opportunities, city people need not pull up stakes for such reasons.

A city's collection of opportunities of all kinds, and the fluidity with which these opportunities and choices can be used, is an asset – not a detriment – for encouraging city–neighborhood stability.

However, this asset has to be capitalized upon. It is thrown away where districts are handicapped by sameness and are suitable, therefore, to only a narrow range of incomes, tastes and family circumstances. Neighborhood accommodations for fixed, bodiless, statistical people are accommodations for instability. The people in them, as statistics, may stay the same. But the people in them, as people, do not. Such places are forever way stations.

I have been emphasizing assets and strengths peculiar to big cities, and weaknesses peculiar to them also. Cities, like anything else, succeed only by making the most of their assets. I have tried to point out the kinds of places in cities that do this, and the way they work. My idea, however, is not that we should therefore try to reproduce, routinely and in a surface way, the streets and districts that do display strength and success as fragments of city life. This would be impossible, and sometimes would be an exercise in architectural antiquarianism. Moreover, even the best streets and districts can stand improvement, especially amenity.

But if we understand the principles behind the behavior of cities, we can build on potential assets and strengths, instead of acting at cross-purposes to them. First we have to know the general results we want – and know because of knowing how life in cities works. We have to know, for instance, that we want lively, well-used streets and other public spaces, and why we want them. But knowing what to want, although it is a first step, is far from enough. The next step is to examine some of the workings of cities at another level; the economic workings that produce those lively streets and districts for city users.

"Networks, Neighborhoods, and Communities: Approaches to the Study of the Community Question"

from *Urban Affairs Quarterly* (1979)

Barry Wellman and Barry Leighton

Editors' Introduction

Barry Wellman and Barry Leighton were collaborators at the University of Toronto Center for Urban and Community Research doing field interview research in 1978 on community, interpersonal, and social network ties of residents of East York, a Toronto borough. It was a follow-up to research Barry Wellman had initiated on the "community question" in 1968 at the same site. In this selection, Wellman and Leighton present their research on social networks and their relation to traditional research on communities in urban sociology. These social networks are analogous to the cross-links and "hop-and-skip" activist links that Jane Jacobs describes in her strong neighborhood political districts. But Wellman and Leighton comprehend social networks in the wider context of secondary relationships that include friendship, interpersonal, and other ties of interest.

The authors begin with a discussion of the central place of the neighborhood in community and urban sociology. They move on to a discussion of the "community question" and identify the traditional prevalence of the story of "community lost." They conclude that, though there is revived interest in the status of neighborhood-based communities (community saved), it is more interesting to consider the ways that community is being fostered (community liberated) without the attachment to neighborhood. While neighborhood has traditionally been a "crucial nexus" for normative integration of the individual with the larger social system, they argue for freedom from place attachment to comprehend the growth of more sparsely knit social networks that offer the rewards of community without propinquity.

Wellman and Leighton explain the revival of community and the "community saved" perspective as responses and alternatives to the depersonalizing effects of large rational-bureaucratic institutions in our society. They note the usefulness of the community saved argument in neighborhood-based government anti-poverty programs that engendered maximum feasible participation of marginalized residents. But rather than retreating into the havens and safe spaces of traditional neighborhood communities, they favor growth in liberated communities that are enabled by new transport and communications technologies, ongoing separation of workplace and home, and increasing geographic mobility in America.

While saved communities offer dense closely knit networks that are strong in sentiment, intimacy, or group solidarity, liberated communities are constructed of looser, sparsely knit, branching, and ramifying networks that bring access to greater external resources. They contend there are policy implications for liberated social networks in a variety of areas, including: a) therapeutic healing networks in physical and mental health,

b) political coalition-building among marginalized communities, and c) integration of Internet social networking technology into residential communities in urban planning.

Barry Wellman began increasing studies of the Internet in the 1990s, and collaborated with Keith Hampton in field research in a digitally wired suburb in Toronto they called "Netville." Their research has continuing relevance with the explosion in Internet social networking in many economic, cultural, and political arenas of life in the new millennium. He also explored social research in the international context in the edited volume *Networks in the Global Village* (Boulder, CO: Westview 1999). Also he published *The Internet in Everyday Life* (with Caroline Haythornthwaite; Oxford: Blackwell 2002).

NEIGHBORHOOD OR COMMUNITY?

Urban sociology has tended to be *neighborhood sociology*. This has meant that analyses of large-scale urban phenomena (such as the fiscal crisis of the state) have been neglected in favor of small-scale studies of communities. It has also meant that the study of such communities has been firmly rooted in the study of neighborhoods, be they the "symbiotic" communities of Park (1936) or the "street corners" of Liebow (1967). It is to the sorting out of this second tendency, the merger of "neighborhood" and "community" that we address this paper.

There are a number of reasons why the concept of "neighborhood" has come to be substituted for that of "community."

First, urban researchers have to start somewhere. The neighborhood is an easily identifiable research site, while the street corner is an obvious and visible place for mapping small-scale interaction.

Second, many scholars have interpreted the neighborhood as the microcosm of the city and the city as an aggregate of neighborhoods. They have emphasized the local rather than the cosmopolitan in a building block approach to analysis which has given scant attention to large-scale urban structure.

Third, administrative officials have imposed their own definitions of neighborhood boundaries upon urban maps in attempts to create bureaucratic units. Spatial areas, labeled and treated as coherent neighborhoods, have come to be regarded as natural phenomena.

Fourth, urban sociology's particular concern with spatial distributions has tended to be translated into local area concerns. Territory has come to be seen as the inherently most important organizing factor in urban social relations rather than just one potentially important factor.

Fifth, and the most importantly, many analysts have been preoccupied with the conditions under which solidary sentiments can be maintained. Their preoccupation reflects a persistent overarching sociological concern with normative integration and consensus. The neighborhood has been studied as an apparently obvious container of normative solidarity.

For these reasons at least, the concentration on the neighborhood has had a strong impact on definitions of, research on, and theorizing about community. Neighborhood studies have produced hundreds of finely wrought depictions of urban life, and they have given us powerful ideas about how small-scale social systems operate in a variety of social contexts. But does the concept of "neighborhood" equal the concept of "community"? Are the two really one and the same?

Definitions of community tend to include three ingredients: *networks of interpersonal ties* (outside of the household) which *provide sociability and support* to members, residence in a *common locality*, and *solidarity sentiments and activities*. It is principally the emphasis on common locality, and to a lesser extent the emphasis on solidarity, which has encouraged the identification of "community" with "neighborhood."

Yet the paramount concern of sociologists is social structure, and concerns about the spatial location of social structures and their normative integration must necessarily occupy secondary positions. To sociologists, unlike geographers, spatial distributions are not inherently important variables, but assume importance only as they affect such social structural questions as the formation of interpersonal networks and the flow of resources through such networks.

[. . .]

The community question

With its manifest concerns for the activities of populations in territories, urban sociology has often seemed to stand apart from broader theoretical concerns. Yet its concentration on the study of the neighborhood-as-community is very much a part of a fundamental sociological issue. This fundamental issue, which has occupied much sociological thinking, is the *community question*: the study of how large-scale divisions of labor in social systems affect the organization and content of interpersonal ties.

Sociologists have been particularly concerned with that form of the community question which investigates the impact of the massive industrial bureaucratic transformations of North America and Europe during the past two hundred years have had on a variety of primary ties: in the home, the neighborhood, the workplace, with kin and friends, and among interest groups. Have such ties attenuated or flourished in contemporary societies? In what sort of networks are they organized? Have the contents of such ties remained as holistic as alleged to be in preindustrial societies or have they become narrowly specialized and instrumental?

The community question thus forms a crucial nexus between macroscopic and microscopic analysis. It directly addresses the structural integration of a social system and the interpersonal means by which its members can gain access to scarce resources. We urge, therefore, that the study of the community question be freed from its identification with the study of neighborhoods.

THE NETWORK PERSPECTIVE

We suggest that the *network analysis perspective* is a more appropriate response to the community question in urban studies than the traditional focus on the neighborhood. A network analysis of community takes as its starting point the search for social linkages and flows of resources. Only then does it enquire into the spatial distribution and solidary sentiments associated with the observed linkages. Such an approach largely frees the study of community from spatial and normative bases. It makes possible the discovery of network-based communities which are neither linked to a particular neighborhood nor to a set of solidary sentiments.

However, the network perspective is not inherently anti-neighborhood. By leaving the matter of spatial distributions initially open, this perspective makes it equally possible to discover an "urban village" (Gans 1962) as it is to discover a "community without propinquity" (Webber 1963). A network analysis might also tell us that strong ties remain abundant and important, but that they rarely are located in the neighborhood. With this approach we are then better able to assess the position of neighborhood ties within the context of overall structures of social relationships.

The community question has been extensively debated by urban scholars. In this paper, we evaluate three competing scholarly arguments about the community question from a network perspective. The first two arguments to be discussed both focus on the neighborhood: the *community lost*, asserting the absence of local solidarities, and the *community saved* argument, asserting their persistence. The *community liberated* argument, in contrast, denies any neighborhood basis to community.

COMMUNITY LOST

The community lost argument contends that the transformation of Western societies to centralized, industrial bureaucratic structures has gravely weakened primary ties and communities, making the individual more dependent on formal organizational resources for sustenance. The first attempts to deal with the community question were, at the turn of the [twentieth] century, closely associated with broader sociological concerns about the impact of the Industrial Revolution on communal ties and normative integration.

[. . .]

Lost networks

The community lost argument makes a number of specific assertions about the kinds of primary ties, social networks, and community structures that will tend to be present under its assumptions. By casting the lost argument in network analytic terms, we shall be better able to evaluate it in comparison with the community saved and community liberated arguments.

a) Rather than being a full member of a solidary community, urbanites are now *limited members* (in terms of amount, intensity, and commitment of interaction) of *several social networks.*
b) Primary ties are *narrowly defined*: there are *fewer strands* in the relationship.
c) The narrowly defined ties tend to be *weak in intensity.*
d) Ties tend to be *fragmented* into isolated *two-person* relationships rather than being parts of extensive networks.
e) Those networks that do exist tend to be *sparsely knit* (a low proportion of all potential links between members actually exists) rather than being densely knit (a high proportion of potential links exist).
f) The networks are *loosely bounded*; there are few discrete clusters or primary groups.
g) Sparse density, loose boundaries and narrowly defined ties provide *little* structural basis for *solidary activities or sentiments.*
h) The narrowly defined ties dispersed among a number of networks create *difficulties in mobilizing assistance* from network members.

[. . .]

Policy implications

The community lost argument has significantly affected urban policy in North America and Western Europe. There have been extensive "community development" programs designed to end alienation and to grow urban roots, such as the putative War on Poverty. The desired community ideal in such programs has been the regeneration of the densely knit, tightly bounded, solidary neighborhood community. When, despite the programs, a return to the pastoral ideal has not seemed achievable, then despair about social disorganization has led to elaborate social control policies, designed to keep in check the supposedly alienated, irrational, violence-prone masses. When even the achievement of social control has not seemed feasible, policies of neglect – benign or otherwise – have been developed. Administrators have removed services from inner-city neighborhoods, asserting their inability to cope with socially disorganized behavior and leaving the remaining inhabitants to fend for themselves. The residents of such inner-city American areas as Pruitt-Igoe and the South Bronx have become to be regarded as unredeemably "sinful" as they suffer the supposed war of all against all.

[. . .]

COMMUNITY SAVED

The community saved argument maintains that neighborhood communities have persisted in industrial bureaucratic social systems as important sources of support and sociability. It argues that the very formal, centralizing tendencies of bureaucratic institutions have paradoxically encouraged the maintenance of primary ties as more flexible sources of sociability and support. The saved argument contends that urbanites continue to organize safe communal havens, with neighborhood, kinship and work solidarities mediating and coping with bureaucratic institutions.

The saved argument shares with the lost argument the identification of "community" with "neighborhood." However, saved scholars have reacted against the tendency of some lost scholars to write secondary analyses *about* the neighborhood community rather than primary analyses *of* neighborhood communities.

[. . .]

Saved networks

The saved argument, cast into network analytic terms, is quite different from the lost argument:

a) Urbanites tend to be *heavily involved members* of a *single neighborhood community*, although they may combine this with membership in other social networks.
b) There are *multiple strands* of relationships between the members of these neighborhood communities.
c) While network ties vary in intensity, many of them are *strong.*
d) Neighborhood ties tend to be organized into *extensive networks.*
e) Networks tend to be *densely knit.*

f) Neighborhood networks are *tightly bounded*, with few external linkages. Ties tend to loop back into the same cluster of network members.
g) High density, tight boundaries, and multistranded ties provide a structural basis for a good deal of *solidary activities and sentiments*. The multistranded strong ties clustered in densely knit networks *facilitate the mobilization* of assistance for dealing with routine and emergency matters.

Saved scholars have tended to regard human beings as fundamentally good and inherently gregarious. They are viewed as apt to organize self-regulating communities under all circumstances, even extreme conditions of poverty, oppression, or catastrophe.

Hence the saved argument has shared the neighborhood community ideal with the lost argument, but it has seen this ideal as attainable and often already existing. Neighborhood communities are valued precisely because they can provide small-scale loci of interaction and can effectively mediate urbanites' dealings with large-scale institutions. Densely knit, tightly bounded communities are valued as structures particularly suited to the tenacious conservation of its internal resources, the maintenance of local autonomy, and the social control of members (and intruders) in the face of powerful impinging external forces.

Policy implications

Public acceptance of the saved argument has greatly increased during the past two decades. Active neighborhood communities are now valued as antidotes to industrial bureaucratic societies' alleged impersonality, specialized relationships, and loss of comprehensible scale.

[. . .]

The neighborhood unit has been the twentieth-century planning ideal for new housing. Saved ideologies have also argued the necessity for preserving existing neighborhoods against the predations of ignorant and rapacious institutions. The saved argument has been the ideological foundation of the neighborhood movement, which seeks to stop expressways, demolish developers, and renovate old areas. Some neighborhoods have been successfully rescued from "urban renewal."

In political analyses, rioters, far from being socially disorganized, are now seen to be rooted, well-connected community members. Their motivations tend to be in defense of existing communal interests or claims to new ones, rather than the irrational, individualistic, psychologistic responses claimed by the lost argument. Indeed, the means by which urbanites get involved in a riot are very much associated with the competitions, coalitions, and solidary ties of their social networks.

Many saved social pathologists have encouraged the nurturance of densely knit, bounded communities as a structural salve for the stresses of poverty, ethnic segregation, and physical and mental diseases. Getting help informally through neighborhood communities is alleged to be more sensitive to peculiar local needs and protective of the individual against bureaucratic claims. Furthermore, such programs have been welcomed by administrators as more cost-effective (or, as some critics allege, merely cheaper to operate) than the formal institutional intervention implied by the lost argument.

In the early 1960s the saved argument became the new orthodoxy in communities studies with the publication of such works as Gans' *The Urban Villagers* (1962), Greer's (1962) synthesis of postwar survey research, and Jacob's (1961) assertion of the vitality of dense, diverse central cities. Such case studies as Young and Willmott's (1957) study of a working-class London neighborhood, Gans' (1967) account of middle-class, new suburban networks, and Liebow's (1967) portrayal of inner-city blacks' heavy reliance on network ties helped clinch the case.

The rebuttal of the lost argument's assertion of urban social disorganization has therefore been accomplished, theoretically and empirically, by studies emphasizing the persistence of neighborhood communities. In the process, though, the lost argument's useful starting point may have come to be neglected: that the industrial bureaucratic division of labor has strongly affected the structure of primary ties. Saved scholars have tended to look on for – and at – the persistence of functioning neighborhood communities. Consequently we now know that neighborhood communities persist and often flourish, but we do not know the position of neighborhood-based ties within overall social networks.

Many recent saved analyses have recognized this difficulty by introducing the "community limited liability" concept, which treats the neighborhood as just one of a series of communities among which urbanites divide their membership (see Janowitz 1952, Greer 1962, Suttles 1972, et al.). Hunter and Suttles (1972: 61), for example, portray such communities as a set of concentric zones radiating out from the block to "entire sectors of the city." However, while such analyses recognize the possibilities for urbanites to be members of diverse networks with limited involvement in each network, the "limited liability community" formulation is still predicated on the neighborhood concept, seeing urban ties as radiating out from a local, spatially defined base.

COMMUNITY LIBERATED

The third response to the community question, the liberated argument, agrees with the lost argument's contention that the industrial bureaucratic nature of social systems has caused the weakening of neighborhood communities. But the liberated argument also agrees with the saved argument's contention that primary ties have remained viable, useful, and important. It shares the saved argument's contention that communities still flourish in the city, but it maintains that such communities are rarely organized within neighborhoods.

The liberated argument contends that a variety of structural and technological developments have liberated communities from the confines of neighborhoods and dispersed network ties from all-embracing solidary communities to more narrowly based ones: a) cheap, effective transportation and communication facilities; b) the separation of workplace and kinship ties into nonlocal, nonsolidary networks; c) high rates of social and residential mobility.

The liberated argument, like the other two arguments, begins with the concept of space. Yet where the other arguments see communities as resident in neighborhoods, the liberated argument confronts spatial restrictions only in order to transcend them. Although harkening back to some of the more optimistic writings of Simmel about the liberating effects of urban life . . . the argument has become prominent only in the past two decades following the proliferation of personal automotive and airplane travel and telecommunications in the Western world. It contends that there is now the possibility of "community without propinquity" (Webber 1964) in which distance and travel time are minimal constraints.

Liberated networks

With its emphasis on aspatial communities, the liberated argument has been methodologically associated with network analytic techniques. However, it must be emphasized that network analysis does not necessarily share the liberated argument's ideological bias and can be used to evaluate the existence of all *three* community patterns: lost, saved, and liberated.

In network terms, the liberated argument contends that:

a) Urbanites now tend to be *limited members of several social networks*, possibly including one located in their neighborhood.
b) There is *variation in the breadth of the strands* of relationships between network members; there are multistranded ties with some, single-stranded ties with many others, and relationships of intermediate breadth with the rest.
c) The ties range in intensity; *some* of them are *strong*, while others are weak but nonetheless useful.
d) An individual's ties tend to be organized into a *series of networks with few connections* between them.
e) Networks tends to be *sparsely knit* although certain portions of the networks, such as those based on kinship, may be more densely knit.
f) The networks are *loosely bounded, ramifying* structures, branching out extensively to form linkages to additional people and resources.
g) Sparse density, loose boundaries, and narrowly defined ties provide *little structural basis for solidary activities and sentiments* in the overall networks of urbanites, although some solidary clusters of ties are often present.
h) *Some network ties can be mobilized* for the general purpose of specific assistance in dealing with routine or emergency matters. The likelihood of mobilization depends more on the quality of the two-person ties than on the nature of the larger network structure.

The liberated argument is fundamentally optimistic about urban life. It is appreciative of urban diversity; imputations of social disorganization and pathology find little place within it. The argument's view of human behavior emphasizes its entrepreneurial and manipulative aspects. People are seen as having a propensity to form primary ties, not out of inherent good or evil, but in order to accomplish specific, utilitarian ends.

The liberated argument, as does the lost argument, minimizes the importance of neighborhood communities. But where the lost argument sees this as throwing the urbanite upon the resources of formal organizations, the liberated argument contends that sufficient ties are available in non-neighborhood networks to provide critical social support and sociability. Furthermore, it argues that the diverse links between these networks organize the city as a "network of networks" (Craven and Wellman 1973) to provide a flexible coordinating structure not possible through a lost formal bureaucratic hierarchy or a saved agglomeration of neighborhoods.

The liberated argument recoils from the lost and saved arguments' village-like community norm. The argument celebrates the structural autonomy of being able to move among various social networks. It perceives solidary communities as fostering stifling social control and of causing isolation from outside contact and resources. Multiple social networks are valued because the cross-cutting commitments and alternative escape routes limit the claims that any one community can make upon its members.

Policy implications

Liberated analysts have called for the reinforcement of other social networks in addition to the traditional ones of the neighborhood and the family. Whereas industrial power considerations have worked against the development of solidary networks in the workplace, much attention has been paid recently to fostering "helping networks" that would prevent or heal the stress of physical and mental diseases. No longer is the neighborhood community seen as the safe, supportive haven; no longer are formal institutions to be relied on for all healing attempts. Instead, networks are to be mobilized, and where they do not exist they can be constructed so that urbanites may find supportive places. However, the efficacy of such deliberately constructed "natural support systems" has not yet been adequately demonstrated.

The liberated argument has had an important impact on thinking about political phenomena, especially that related to collective disorders. Research by Charles Tilly (e.g. 1975, 1978) and associates, in particular, has shown such collective disorders to be integral parts of broader contentions for power by competing interest groups. In addition to the internal solidarity emphasized by the saved argument, a contending group's chances for success have been shown to be strongly associated with the capacity for making linkages to external coalitions that cross-cutting ties between networks can provide (e.g. Gans 1974a, 1974b; Granovetter, 1974b).

Recent British New Town planning (e.g. Milton Keynes) has been predicated on the high rates of personal automotive mobility foreseen by the liberated argument. However, the argument's contention that there are minimal costs to spatial separation has come up against the increase in the monetary costs of such separation associated with the significant rise in the price of oil within the last decade. One response has been to advocate increased reliance on telecommunications to maintain community ties over large distances. New developments in computer technology foreshadow major increases in telecommunications capabilities, such as "electronic mail" and "computer conferencing." Yet the strength of the liberated argument does not necessarily depend on technological innovations. Recent research in preindustrial social systems has indicated that long-distance ties can be maintained without benefit of telephone or private automobiles, as long as such ties are structurally embedded in kinship systems of common local origins.

COMMUNITIES: LOST, SAVED, OR LIBERATED?

Are communities lost, saved, or liberated? Too often, the three arguments have been presented as: a) competing alternative to depictions of the "true" nature of Western industrial bureaucratic social systems, or b) evolutionary successors, with preindustrial saved communities giving way to industrial lost, only to be superseded by post-industrial liberated.

In contrast, we believe that all three arguments have validity when stripped of their ideological paraphernalia down to basic network structures. Indeed their structural character might be highlighted by thinking of them as sparse, dense, and ramified network patterns. Different network patterns tend to have different consequences for the acquisition and control of resources. We might then expect to find the prevalence of lost, saved, and liberated communities to vary according to the kinds of societal circumstances in which they are located.

Saved communities/dense networks

In saved networks, densely knit ties and tight boundaries tend to occur together. This may be because network members have a finite lump of sociability, so that if they devote most of their energies to within-network ties, they do not have much scope for maintaining external linkages. Conversely, tight boundaries may also foster the creation of new ties within the community, as internal links become the individual's principal hopes of gaining access to resources.

Such dense, bounded saved networks, be they neighborhood, kinship, or otherwise based, are apt to be solidary in sentiments and activities. They are well-structured for maintaining informal social control over members and intruders. The dense ties and communal solidarity should facilitate the ready mobilization of the community's resources for the aid of members in good standing. But because solidarity does not necessarily mean egalitarianism, not all of the community's resources may be gathered or distributed equally.

Community studies have shown the saved pattern to be quite prevalent in situations in which community members do not have many individual personal resources and where there are unfavorable conditions for forming external ties. Certain ethnic minority and working-class neighborhoods clearly follow this pattern. In such situations, concerns about conserving, controlling, and efficiently pooling those resources the beleaguered community possesses also resonate with its members' inability to acquire additional resources elsewhere. A heavy load consequently is placed on ties within the saved community.

Liberated communities/ramified networks

If saved network patterns are particularly suited to conditions of resource scarcity and conservation, liberated network patterns are particularly suited in conditions of resource abundance and acquisition. Such sparsely knit, loosely bounded networks are not structurally well-equipped for internal social control. Implicit assurance in the security of one's home base is necessary before one can reach out into new areas.

Loose boundaries and sparse density foster networks that extensively branch out to link with new members. These ramifying liberated networks are well-structured for acquiring additional resources through a larger number of direct and indirect external connections. Their structure is apt to connect liberated network members with a more diverse array of resources than saved networks are apt to encounter, although the relative lack of solidarity in such liberated networks may well mean that a lower proportion of resources will be available to other network members.

It may well be that the liberated pattern is peculiarly suited to affluent sectors of contemporary Western societies. It places a premium on a base of individual security, entrepreneurial skills in moving between networks, and the ability to function without the security of membership in a solidary community. However, its appearance in other social contexts indicates that it reflects a more fundamental alternative to the saved community pattern.

But the saved or liberated community patterns can appear as desirable alternatives to those enmeshed in the other pattern. To those unsatisfied with the uncertain multiplicities of liberated networks, holistic, solidary saved communities can appear as a welcome retreat. To those who feel trapped in all-embracing saved networks, the availability of alternative liberated primary networks may offer a welcome escape route. Much migration from rural areas may follow this tendency.

Lost communities/sparse networks

What of circumstances where no alternative network sources of escape or retreat are possible? It is in such situations that the lost pattern of direct affiliation with formal institutions can become attractive: the army, the church, the firm, and the university. However, the lost pattern may always be unstable for individuals and communities as formal institutional ties devolve into complex primary network webs. Therefore, as primary ties develop between organizations, we may expect to find networks taking on the pattern of saved or liberated communities.

Personal communities

When studying neighborhoods and communities, we are likely to find diversity rather than a universal pattern to either local or personal networks. We have proposed that dense saved network patterns are better suited for internal control of resources while ramified liberated patterns are better suited for obtaining access to external resources. Although we have suggested that each of these patterns should be more prevalent in one sort of a society than another, it is quite likely that the total network of a community will comprise a mixture of these two patterns in varying proportions. That is, some of the ties within a network will be densely knit and tightly bounded, while others will be sparsely knit and ramified. The different patterns are useful for different things. As Merton (1957) early pointed out, most communities have some network members for exchanging resources with the outside world ("cosmopolitans") and some for allocating them internally ("locals").

Our own research in the Borough of East York, Toronto, has revealed that individuals, too may be simultaneous members of both saved and liberated pattern networks.[1] Some of an urbanite's ties tend to be clustered into densely knit, tightly bounded networks, their solidarity often reinforced by either kinship structures or residential or workplace propinquity. Such saved networks are better able to mobilize help in emergencies through efficient communication and structurally enforced norms. Their density and boundedness tend to give these clusters more of a tangible collective image, so that network members have a sense of solidary attachment.

Yet we have found (Wellman 1979) that such clusters are likely to comprise only a minority of one's important network ties. The other ties tend to be much less densely connected. Instead of looping back into one another within boundaries, they tend to be ramified, branching out to encounter new members to whom the original network members are not directly connected. These sparsely knit, loosely bounded liberated networks are structurally not as efficient in mobilizing collective assistance for their members, but their branching character allows additional resources to be reached. Furthermore, the liberated ties, while not as conducive to internal solidarity as the saved clusters, better facilitate coalition building between networks.

Neighborhood and community

Almost all of the people we studied have many strong ties and they are able to obtain assistance through a number of close relationships. Yet only a small proportion of these "intimate" ties are located in the same neighborhood. While neighboring ties are still prevalent and important in East York, they rarely achieve the intensity of intimacy.

Neighborhood relationships persist but only as specialized components of the overall primary networks. The variety of ties in which an urbanite can be involved – with distant parents, intimate friends, less intimate friends, coworkers, and so on – and the variety of networks in which these are organized can provide flexible structural bases for dealing with routine and emergency matters.

In sum, we must be concerned with neighborhood and community rather than neighborhood *or* community. We have suggested that the two are separate concepts which may or may not be closely associated. In some situations we can observe the saved pattern of community as solidary neighborhood. In many other situations, if we go out and look for neighborhood-based networks, we are apt to find them. They can be heavily used for the advantages of quick accessibility. But if we broaden our field of view to include other primary relations, then the apparent neighborhood solidarities may now be seen as clusters in the rather sparse, loosely bounded structures of urbanites' total networks.

NOTE

1 The data collected in a 1968 random-sample, closed-ended survey of 845 adult East Yorkers, directed by Donald B. Coates, with Barry Wellman as coordinator. East York (1971 population – 104,646) is an upper working-class, lower middle-class, predominantly British-Canadian inner-city suburb of Toronto. It has the reputation of being one of the most solidary areas of the city.

"Bowling Alone: America's Declining Social Capital"

from *Journal of Democracy* (1995)

Robert Putnam

Editors' Introduction

In this selection, Robert Putnam explores the long-term decline in civic engagement in America and the correlation with the loss in trust, generalized reciprocity, and social capital in our society. He cites Alexis de Tocqueville and his classic *Democracy in America*, in asserting the exceptional quality of democratic civil society in the United States in its founding days. He reinforces the importance of addressing the more recent declines of civic engagement if America is still to be a democratic model for the rest of world, emulated by developing and postcommunist societies.

He measures the trends in U.S. civic engagement of declining membership in traditional voluntary organizations, including Parent–Teacher Associations, the Boy Scouts, and the Red Cross. Civic engagement and social connectedness are important factors in addressing education, urban poverty, crime, unemployment, and health outcomes. Trust and generalized reciprocity can help reduce opportunism and corruption in public life. But people have become disaffiliated from public life and distrustful of government, through such experiences as prominent assassinations, social unrest over the Vietnam War, and the Watergate scandal. Declines can also be seen in many fraternal organizations, like the Lions, Shriners, and Elks. Putnam also offers his most famous statistic, that of declining participation in recreational bowling leagues. While Americans are still bowling in growing numbers, they are more frequently "bowling alone."

He does acknowledge a countertrend to the decline in social capital, created by a new "tertiary sector" of nonprofit organizations addressing environmental or social issues, like the Sierra Club, the National Organization for Women, and the American Association of Retired Persons (AARP). He characterizes these as mainly dues-paying membership groups and downplays their relative power to generate social connectedness, reciprocity, and social capital like the traditional fraternal and civic organizations. He has greater hope in the force of self-help support groups like Alcoholics Anonymous.

Putnam suggests a variety of reasons for declining social capital in America, including the entry of women into the workforce, greater geographic mobility from home communities, the erosion of the traditional family, and growth of privatizing and individualizing leisure and media technologies such as television and the Internet. He also addresses practical steps towards reversing the decline in social capital and civic engagements, such as through electronic forums, rooting out social intolerance and corruption in our civic life, and deploying government programs like the county agricultural-agent system, community colleges, and tax deductions for charitable organizations.

Unlike Jacobs and Wellman/Leighton, Putnam doesn't give much attention to the resurgence of social change and community activism in cities since the 1960s against urban renewal and the rational-bureaucratic centralized state. He offers a better explanation for the growing social and political disaffection of the residents, civic and fraternal organizations that historically populated small towns and mid-twentieth-century American cities that were traditionally upholders rather than critics of the establishment and the mass society. He's more skeptical than Wellman/Leighton about the power of the branching, ramified social networks of the nonprofit sector to effectively mobilize new civic engagement. Like Wellman/Leighton, however, he has some faith in the potential of electronic forums to foster greater social and civic connectedness. Internet social networks were powerful tools in communicating and mobilizing revolutionary protests in the Arab world in the spring of 2011. Putnam would ask to what degree social networking can help us build sustainable civic institutions and democratic systems for the long term.

Social capital has been explored by a number of sociologists. Some classic articles include James Coleman (1988), "Social Capital in the Creation of Human Capital," *American Journal of Sociology* Supplement 94: S95, and A. Portes (1998), "Social Capital: Its Origins and Applications in Modern Sociology," *Annual Review of Sociology*, 24: 1–24.

Putnam is a Professor of Political Science and Public Policy at Harvard University. He is also the author of *Making Democracy Work: Civic Traditions in Modern Italy* (Princeton, NJ: Princeton University Press, 1995) and *Better Together: Restoring the American Community* (with Lewis Feldstein and Don Cohen; New York: Simon and Schuster, 2003).

Many students of the new democracies that have emerged over the past decade and a half have emphasized the importance of a strong and active civil society to the consolidation of democracy. Especially with regard to the postcommunist countries, scholars and democratic activists alike have lamented the absence or obliteration of traditions of independent civic engagement and a widespread tendency toward passive reliance on the state. To those concerned with the weakness of civil societies in the developing or postcommunist world, the advanced Western democracies and above all the United States have typically been taken as models to be emulated. There is striking evidence, however, that the vibrancy of American civil society has notably declined over the past several decades.

Ever since the publication of Alexis de Tocqueville's *Democracy in America*, the United States has played a central role in systematic studies of the links between democracy and civil society. Although this is in part because trends in American life are often regarded as harbingers of social modernization, it is also because America has traditionally been considered unusually "civic" (a reputation that, as we shall later see, has not been entirely unjustified).

When Tocqueville visited the United States in the 1830s, it was the Americans' propensity for civic association that most impressed him as the key to their unprecedented ability to make democracy work. "Americans of all ages, all stations in life, and all types of disposition," he observed, "are forever forming associations. There are not only commercial and industrial associations in which all take part, but others of a thousand different types – religious, moral, serious, futile, very general and very limited, immensely large and very minute . . . Nothing, in my view, deserves more attention than the intellectual and moral associations in America."

Recently, American social scientists of a neo-Tocquevillean bent have unearthed a wide range of empirical evidence that the quality of public life and the performance of social institutions (and not only in America) are indeed powerfully influenced by norms and networks of civic engagement. Researchers in such fields as education, urban poverty, unemployment, the control of crime and drug abuse, and even health have discovered that successful outcomes are more likely in civically engaged communities. Similarly, research on the varying economic attainments of different ethnic groups in the United States has demonstrated the importance of social bonds within each group. These results are consistent with research in a wide range of settings that demonstrates the vital importance of social networks for job placement and many other economic outcomes.

Meanwhile, a seemingly unrelated body of research on the sociology of economic development has also focused attention on the role of social networks. Some of this work is situated in the developing countries, and some of it elucidates the peculiarly successful "network capitalism" of East Asia. Even in less exotic Western economies, however, researchers have discovered highly efficient, highly flexible "industrial districts" based on networks of collaboration among workers and small entrepreneurs. Far from being paleoindustrial anachronisms, these dense interpersonal and interorganizational networks undergird ultramodern industries, from the high tech of Silicon Valley to the high fashion of Benetton.

The norms and networks of civic engagement also powerfully affect the performance of representative government. That, at least, was the central conclusion of my own 20-year, quasi-experimental study of subnational governments in different regions of Italy. Although all these regional governments seemed identical on paper, their levels of effectiveness varied dramatically. Systematic inquiry showed that the quality of governance was determined by longstanding traditions of civic engagement (or its absence). Voter turnout, newspaper readership, membership in choral societies and football clubs – these were the hallmarks of a successful region. In fact, historical analysis suggested that these networks of organized reciprocity and civic solidarity, far from being an epiphenomenon of socioeconomic modernization, were a precondition for it.

No doubt the mechanisms through which civic engagement and social connectedness produce such results – better schools, faster economic development, lower crime, and more effective government – are multiple and complex. While these briefly recounted findings require further confirmation and perhaps qualification, the parallels across hundreds of empirical studies in a dozen disparate disciplines and subfields are striking. Social scientists in several fields have recently suggested a common framework for understanding these phenomena, a framework that rests on the concept of social capital. By analogy with notions of physical capital and human capital – tools and training that enhance individual productivity – "social capital" refers to features of social organization such as networks, norms, and social trust that facilitate coordination and cooperation for mutual benefit.

For a variety of reasons, life is easier in a community blessed with a substantial stock of social capital. In the first place, networks of civic engagement foster sturdy norms of generalized reciprocity and encourage the emergence of social trust. Such networks facilitate coordination and communication, amplify reputations, and thus allow dilemmas of collective action to be resolved. When economic and political negotiation is embedded in dense networks of social interaction, incentives for opportunism are reduced. At the same time, networks of civic engagement embody past success at collaboration, which can serve as a cultural template for future collaboration. Finally, dense networks of interaction probably broaden the participants' sense of self, developing the "I" into the "we," or (in the language of rational-choice theorists) enhancing the participants' "taste" for collective benefits.

I do not intend here to survey (much less contribute to) the development of the theory of social capital. Instead, I use the central premise of that rapidly growing body of work – that social connections and civic engagement pervasively influence our public life, as well as our private prospects – as the starting point for an empirical survey of trends in social capital in contemporary America. I concentrate here entirely on the American case, although the developments I portray may in some measure characterize many contemporary societies.

WHATEVER HAPPENED TO CIVIC ENGAGEMENT?

We begin with familiar evidence on changing patterns of political participation, not least because it is immediately relevant to issues of democracy in the narrow sense. Consider the well-known decline in turnout in national elections over the last three decades. From a relative high point in the early 1960s, voter turnout had by 1990 declined by nearly a quarter; tens of millions of Americans had forsaken their parents' habitual readiness to engage in the simplest act of citizenship. Broadly similar trends also characterize participation in state and local elections.

It is not just the voting booth that has been increasingly deserted by Americans. A series of identical questions posed by the Roper Organization to national samples ten times each year over the last two decades reveals that since 1973 the number of Americans who report that "in the past year" they have "attended a public meeting on town or school affairs" has fallen by more than a third (from 22 percent in 1973 to 13 percent in 1993). Similar (or even greater) relative declines are evident in responses to questions about attending a political rally or speech, serving on a committee of some local organization, and working for a political party. By almost every measure, Americans' direct engagement in politics and government has fallen steadily and sharply over the last generation, despite the fact that average levels of education – the best individual-level predictor of political participation – have risen sharply throughout this period. Every year over the last decade or two, millions more have withdrawn from the affairs of their communities.

Not coincidentally, Americans have also disengaged psychologically from politics and government over this era. The proportion of Americans who reply that they "trust the government in Washington" only "some of the time" or "almost never" has risen steadily from 30 percent in 1966 to 75 percent in 1992.

These trends are well known, of course, and taken by themselves would seem amenable to a strictly political explanation. Perhaps the long litany of political tragedies and scandals since the 1960s (assassinations, Vietnam, Watergate, Irangate, and so on) has triggered an understandable disgust for politics and government among Americans, and that in turn has motivated their withdrawal. I do not doubt that this common interpretation has some merit, but its limitations become plain when we examine trends in civic engagement of a wider sort.

Our survey of organizational membership among Americans can usefully begin with a glance at the aggregate results of the General Social Survey, a scientifically conducted, national-sample survey that has been repeated 14 times over the last two decades. Church-related groups constitute the most common type of organization joined by Americans; they are especially popular with women. Other types of organizations frequently joined by women include school-service groups (mostly parent–teacher associations), sports groups, professional societies, and literary societies. Among men, sports clubs, labor unions, professional societies, fraternal groups, veterans' groups, and service clubs are all relatively popular.

Religious affiliation is by far the most common associational membership among Americans. Indeed, by many measures America continues to be (even more than in Tocqueville's time) an astonishingly "churched" society. For example, the United States has more houses of worship per capita than any other nation on Earth. Yet religious sentiment in America seems to be becoming somewhat less tied to institutions and more self-defined.

How have these complex crosscurrents played out over the last three or four decades in terms of Americans' engagement with organized religion? The general pattern is clear: The 1960s witnessed a significant drop in reported weekly churchgoing – from roughly 48 percent in the late 1950s to roughly 41 percent in the early 1970s. Since then, it has stagnated or (according to some surveys) declined still further. Meanwhile, data from the General Social Survey show a modest decline in membership in all "church-related groups" over the last 20 years. It would seem, then, that net participation by Americans, both in religious services and in church-related groups, has declined modestly (by perhaps a sixth) since the 1960s.

For many years, labor unions provided one of the most common organizational affiliations among American workers. Yet union membership has been falling for nearly four decades, with the steepest decline occurring between 1975 and 1985. Since the mid-1950s, when union membership peaked, the unionized portion of the nonagricultural work force in America has dropped by more than half, falling from 32.5 percent in 1953 to 15.8 percent in 1992. By now, virtually all of the explosive growth in union membership that was associated with the New Deal has been erased. The solidarity of union halls is now mostly a fading memory of aging men.

The parent–teacher association (PTA) has been an especially important form of civic engagement in twentieth-century America because parental involvement in the educational process represents a particularly productive form of social capital. It is, therefore, dismaying to discover that participation in parent–teacher organizations has dropped drastically over the last generation, from more than 12 million in 1964 to barely 5 million in 1982 before recovering to approximately 7 million now.

Next, we turn to evidence on membership in (and volunteering for) civic and fraternal organizations. These data show some striking patterns. First, membership in traditional women's groups has declined more or less steadily since the mid-1960s. For example, membership in the national Federation of Women's Clubs is down by more than half (59 percent) since 1964, while membership in the League of Women Voters (LWV) is off 42 percent since 1969.

Similar reductions are apparent in the numbers of volunteers for mainline civic organizations, such as the Boy Scouts (off by 26 percent since 1970) and the Red Cross (off by 61 percent since 1970). But what about the possibility that volunteers have simply switched their loyalties to other organizations? Evidence on "regular" (as opposed to occasional or "drop-by") volunteering is available from the Labor Department's Current Population Surveys of 1974 and 1989. These estimates suggest that serious volunteering declined by roughly one-sixth over these 15 years, from 24 percent of adults in 1974 to 20 percent in 1989. The multitudes of Red Cross aides and Boy Scout troop leaders now missing in action have apparently not been offset by equal numbers of new recruits elsewhere.

Fraternal organizations have also witnessed a substantial drop in membership during the 1980s and 1990s. Membership is down significantly in such groups as the Lions (off 12 percent since 1983), the Elks (off 18 percent since 1979), the Shriners (off 27 percent since 1979), the Jaycees (off 44 percent since 1979), and the Masons (down 39 percent since 1959). In sum, after expanding steadily throughout most of this century, many major civic organizations have experienced a sudden, substantial, and nearly simultaneous decline in membership over the last decade or two.

The most whimsical yet discomfiting bit of evidence of social disengagement in contemporary America that I have discovered is this: more Americans are bowling today than ever before, but bowling in organized leagues has plummeted in the last decade or so. Between 1980 and 1993 the total number of bowlers in America increased by 10 percent, while league bowling decreased by 40 percent. (Lest this be thought a wholly trivial example, I should note that nearly 80 million Americans went bowling at least once during 1993, nearly a third more than voted in the 1994 congressional elections and roughly the same number as claim to attend church regularly. Even after the 1980s' plunge in league bowling, nearly 3 percent of American adults regularly bowl in leagues.) The rise of solo bowling threatens the livelihood of bowling-lane proprietors because those who bowl as members of leagues consume three times as much beer and pizza as solo bowlers, and the money in bowling is in the beer and pizza, not the balls and shoes. The broader social significance, however, lies in the social interaction and even occasionally civic conversations over beer and pizza that solo bowlers forgo. Whether or not bowling beats balloting in the eyes of most Americans, bowling teams illustrate yet another vanishing form of social capital.

COUNTERTRENDS

At this point, however, we must confront a serious counterargument. Perhaps the traditional forms of civic organization whose decay we have been tracing have been replaced by vibrant new organizations. For example, national environmental organizations (like the Sierra Club) and feminist groups (like the National Organization for Women) grew rapidly during the 1970s and 1980s and now count hundreds of thousands of dues-paying members. An even more dramatic example is the American Association of Retired Persons (AARP), which grew exponentially from 400,000 card-carrying members in 1960 to 33 million in 1993, becoming (after the Catholic Church) the largest private organization in the world. The national administrators of these organizations are among the most feared lobbyists in Washington, in large part because of their massive mailing lists of presumably loyal members.

These new mass-membership organizations are plainly of great political importance. From the point of view of social connectedness, however, they are sufficiently different from classic "secondary associations" that we need to invent a new label – perhaps "tertiary associations." For the vast majority of their members, the only act of membership consists in writing a check for dues or perhaps occasionally reading a newsletter. Few ever attend any meetings of such organizations, and most are unlikely ever (knowingly) to encounter any other member. The bond between any two members of the Sierra Club is less like the bond between any two members of a gardening club and more like the bond between any two Red Sox fans (or perhaps any two devoted Honda owners): they root for the same team and they share some of the same interests, but they are unaware of each other's existence. Their ties, in short, are to common symbols, common leaders, and perhaps common ideals, but not to one another. The theory of social capital argues that associational membership should, for example, increase social trust, but this prediction is much less straightforward with regard to membership in tertiary associations. From the point of view of social connectedness, the Environmental Defense Fund and a bowling league are just not in the same category.

If the growth of tertiary organizations represents one potential (but probably not real) counterexample to my thesis, a second countertrend is represented by the growing prominence of nonprofit organizations, especially nonprofit service agencies. This so-called third sector includes everything from Oxfam and the Metropolitan Museum of Art to the Ford Foundation and the Mayo Clinic. In other words, although most secondary associations are nonprofits, most nonprofit agencies are not secondary associations. To identify trends in the size of the nonprofit sector with trends in social connectedness would be another fundamental conceptual mistake.

A third potential countertrend is much more relevant to an assessment of social capital and civic engagement. Some able researchers have argued that the last few decades have witnessed a rapid expansion in "support groups" of various sorts . . . Many of these groups are religiously affiliated, but many others are not . . . All three of these potential countertrends – tertiary organizations, nonprofit organizations, and support groups – need somehow to be weighed against the erosion of conventional civic organizations. One way of doing so is to consult the General Social Survey.

Within all educational categories, total associational membership declined significantly between 1967 and 1993. Among the college-educated, the average number of group memberships per person fell from 2.8 to 2.0 (a 26-percent decline); among high-school graduates, the number fell from 1.8 to 1.2 (32 percent); and among those with fewer than 12 years of education, the number fell from 1.4 to 1.1 (25 percent). In other words, at all educational (and hence social) levels of American society, and counting all sorts of group memberships, the average number of associational memberships has fallen by about a fourth over the last quarter-century. Without controls for educational levels, the trend is not nearly so clear, but the central point is this: more Americans than ever before are in social circumstances that foster associational involvement (higher education, middle age, and so on), but nevertheless aggregate associational membership appears to be stagnant or declining.

Broken down by type of group, the downward trend is most marked for church-related groups, for labor unions, for fraternal and veterans' organizations, and for school-service groups. Conversely, membership in professional associations has risen over these years, although less than might have been predicted, given sharply rising educational and occupational levels. Essentially the same trends are evident for both men and women in the sample. In short, the available survey evidence confirms our earlier conclusion: American social capital in the form of civic associations has significantly eroded over the last generation.

GOOD NEIGHBORLINESS AND SOCIAL TRUST

I noted earlier that most readily available quantitative evidence on trends in social connectedness involves formal settings, such as the voting booth, the union hall, or the PTA. One glaring exception is so widely discussed as to require little comment here: the most fundamental form of social capital is the family, and the massive evidence of the loosening of bonds within the family (both extended and nuclear) is well known. This trend, of course, is quite consistent with – and may help to explain – our theme of social decapitalization.

A second aspect of informal social capital on which we happen to have reasonably reliable time-series data involves neighborliness. In each General Social Survey since 1974 respondents have been asked, "How often do you spend a social evening with a neighbor?" The proportion of Americans who socialize with their neighbors more than once a year has slowly but steadily declined over the last two decades, from 72 percent in 1974 to 61 percent in 1993. (On the other hand, socializing with "friends who do not live in your neighborhood" appears to be on the increase, a trend that may reflect the growth of workplace-based social connections.)

Americans are also less trusting. The proportion of Americans saying that most people can be trusted fell by more than a third between 1960, when 58 percent chose that alternative, and 1993, when only 37 percent did. The same trend is apparent in all educational groups; indeed, because social trust is also correlated with education and because educational levels have risen sharply, the overall decrease in social trust is even more apparent if we control for education.

Our discussion of trends in social connectedness and civic engagement has tacitly assumed that all the forms of social capital that we have discussed are themselves coherently correlated across individuals. This is in fact true. Members of associations are much more likely than nonmembers to participate in politics, to spend time with neighbors, to express social trust, and so on.

The close correlation between social trust and associational membership is true not only across time and across individuals, but also across countries. Evidence from the 1991 World Values Survey demonstrates the following:

1 Across the 35 countries in this survey, social trust and civic engagement are strongly correlated; the greater the density of associational membership in a society, the more trusting its citizens. Trust and engagement are two facets of the same underlying factor – social capital.
2 America still ranks relatively high by cross-national standards on both these dimensions of social capital. Even in the 1990s, after several decades' erosion, Americans are more trusting and more engaged than people in most other countries of the world.
3 The trends of the past quarter-century, however, have apparently moved the United States significantly lower in the international rankings of social capital. The recent deterioration in American social capital has been sufficiently great that (if no other country changed its position in the meantime) another quarter-century of change at the same rate would bring the United States, roughly speaking, to the midpoint among all these countries, roughly equivalent to South Korea, Belgium, or Estonia today. Two generations' decline at the same rate would leave the United States at the level of today's Chile, Portugal, and Slovenia.

WHY IS U.S. SOCIAL CAPITAL ERODING?

As we have seen, something has happened in America in the last two or three decades to diminish civic engagement and social connectedness. What could that "something" be? Here are several possible explanations, along with some initial evidence on each.

The movement of women into the labor force. Over these same two or three decades, many millions of American women have moved out of the home into paid employment. This is the primary, though not the sole, reason why the weekly working hours of the average American have increased significantly during these years. It seems highly plausible that this social revolution should have reduced the time and energy available for building social capital. For certain organizations, such as the PTA, the League of Women Voters, the Federation of Women's Clubs, and the Red Cross, this is almost certainly an important part of the story. The sharpest decline in women's civic participation seems to have come in the 1970s; membership in such "women's" organizations as these has been virtually halved since the late 1960s. By contrast, most of the decline in participation in men's organizations occurred about ten years later; the total decline to date has been approximately 25 percent for the typical organization. On the other hand, the survey data imply that the aggregate declines for men are virtually as great as those for women. It is logically possible, of course, that the male declines might represent the knock-on effect of women's liberation, as dishwashing crowded out the lodge, but time-budget studies suggest that most husbands of working wives have assumed only a minor part of the housework. In short, something besides the women's revolution seems to lie behind the erosion of social capital.

Mobility: The "re-potting" hypothesis. Numerous studies of organizational involvement have shown that residential stability and such related phenomena as homeownership are clearly associated with greater civic engagement. Mobility, like frequent re-potting of plants, tends to disrupt root systems, and it takes time for an uprooted individual to put down new roots. It seems plausible that the automobile, suburbanization, and the movement to the Sun Belt have reduced the social rootedness of the average American, but one fundamental difficulty with this hypothesis is apparent: the best evidence shows that residential stability and homeownership in America have risen modestly since 1965, and are surely higher now than during the 1950s, when civic engagement and social connectedness by our measures was definitely higher.

Other demographic transformations. A range of additional changes have transformed the American family since the 1960s – fewer marriages, more divorces, fewer children, lower real wages, and so on. Each of these changes might account for some of the slackening of civic engagement, since married, middle-class parents are generally more socially involved than other people. Moreover, the changes in scale that have swept over the American economy in these years – illustrated by the replacement of the corner grocery by the supermarket and now perhaps of the supermarket by electronic shopping at home, or the replacement of community-based enterprises by outposts of distant multinational firms – may perhaps have undermined the material and even physical basis for civic engagement.

The technological transformation of leisure. There is reason to believe that deep-seated technological trends are radically "privatizing" or "individualizing" our use of leisure time and thus disrupting many opportunities for social-capital formation. The most obvious and probably the most powerful instrument of this revolution is television. Time-budget studies in the 1960s showed that the growth in time spent watching television dwarfed all other changes in the way Americans passed their days and nights. Television has made our communities (or, rather, what we experience as our communities) wider and shallower. In the language of economics, electronic technology enables individual tastes to be satisfied more fully, but at the cost of the positive social externalities associated with more primitive forms of entertainment. The same logic applies to the replacement of vaudeville by the movies and now of movies by the VCR. The new "virtual reality" helmets that we will soon don to be entertained in total isolation are merely the latest extension of this trend. Is technology thus driving a wedge between our individual interests and our collective interests? It is a question that seems worth exploring more systematically.

WHAT IS TO BE DONE?

The last refuge of a social-scientific scoundrel is to call for more research. Nevertheless, I cannot forbear from suggesting some further lines of inquiry.

- We must sort out the dimensions of social capital, which clearly is not a unidimensional concept, despite language (even in this essay) that implies the contrary. What types of organizations and networks most effectively embody – or generate – social capital, in the sense of mutual reciprocity, the resolution of dilemmas of collective action, and the broadening of social identities? In this essay I have emphasized the density of associational life. In earlier work I stressed the structure of networks, arguing that "horizontal" ties represented more productive social capital than vertical ties.
- Another set of important issues involves macrosociological crosscurrents that might intersect with the trends described here. What will be the impact, for example, of electronic networks on social capital? My hunch is that meeting in an electronic forum is not the equivalent of meeting in a bowling alley – or even in a saloon – but hard empirical research is needed. What about the development of social capital in the workplace? Is it growing in counterpoint to the decline of civic engagement, reflecting some social analogue of the first law of thermodynamics – social capital is neither created nor destroyed, merely redistributed? Or do the trends described in this essay represent a deadweight loss?
- A rounded assessment of changes in American social capital over the last quarter-century needs to count the costs as well as the benefits of community engagement. We must not romanticize small-town, middle-class civic life in the America of the 1950s. In addition to the deleterious trends emphasized in this essay, recent decades have witnessed a substantial decline in intolerance and probably also in overt discrimination, and those beneficent trends may be related in complex ways to the erosion of traditional social capital. Moreover, a balanced accounting of the social-capital books would need to reconcile the insights of this approach with the undoubted insights offered by Mancur

Olson and others who stress that closely knit social, economic, and political organizations are prone to inefficient cartelization and to what political economists term "rent seeking" and ordinary men and women call corruption.

- Finally, and perhaps most urgently, we need to explore creatively how public policy impinges on (or might impinge on) social-capital formation. In some well-known instances, public policy has destroyed highly effective social networks and norms. American slum-clearance policy of the 1950s and 1960s, for example, renovated physical capital, but at a very high cost to existing social capital. The consolidation of country post offices and small school districts has promised administrative and financial efficiencies, but full-cost accounting for the effects of these policies on social capital might produce a more negative verdict. On the other hand, such past initiatives as the county agricultural-agent system, community colleges, and tax deductions for charitable contributions illustrate that government can encourage social-capital formation. Even a recent proposal in San Luis Obispo, California, to require that all new houses have front porches illustrates the power of government to influence where and how networks are formed.

The concept of "civil society" has played a central role in the recent global debate about the preconditions for democracy and democratization. In the newer democracies this phrase has properly focused attention on the need to foster a vibrant civic life in soils traditionally inhospitable to self-government. In the established democracies, ironically, growing numbers of citizens are questioning the effectiveness of their public institutions at the very moment when liberal democracy has swept the battlefield, both ideologically and geopolitically. In America, at least, there is reason to suspect that this democratic disarray may be linked to a broad and continuing erosion of civic engagement that began a quarter-century ago. High on our scholarly agenda should be the question of whether a comparable erosion of social capital may be under way in other advanced democracies, perhaps in different institutional and behavioral guises. High on America's agenda should be the question of how to reverse these adverse trends in social connectedness, thus restoring civic engagement and civic trust.

PART TWO

Understanding Urban Growth in the Capitalist City

Plate 3 Manhattan, New York City, USA 1959 by Henri Cartier-Bresson. In the upper part of the city, around 103rd Street, some 61 blocks from Times Square, slums are being torn down with ruthless speed to make way for low cost housing projects such as these seen against the skyline. Reproduced by permission of Magnum Photos.

INTRODUCTION TO PART TWO

Intellectual perspectives on the nature of urban growth in the capitalist city have shifted considerably across the course of the last hundred years. The positivistic structural-functional perspective of the human ecology school prevailed in the early and middle decades of the twentieth century. Human ecologists described the tendency of competitive market forces in the city to stabilize into centralized structures of economic and political dominance and moral authority. Out of the Depression and wartime years, the Fordist industrial model and Keynesian commitments in government spending created the promise of an affluent society. The social conflicts of the 1960s and the oil crisis of the 1970s fueled growing conditions of economic uncertainty and social unrest in contemporary capitalist societies. These external shocks fed intellectual activism in urban sociology against the domination of the human ecology school, marking shifts to a "new urban sociology" that includes strands of Marxism, poststructuralism, and postmodernism.

These approaches provoke a more systematic interrogation of the nature of urban power and social inequality. There is a growing sense of recurring crisis and structural transformation as global capitalism knits greater interdependencies between the fates of nations. The prosperity created by Fordist-Keynesian capital/labor cooperation has given way to calls for neoliberal restructuring, government entitlement roll-backs, and a return to free-market privatism. Class struggle in the new millennium has stronger ideological dimensions in the debates over social regulation of family, sexuality, and privacy. There is a growing concern about problems of urban sprawl and urban policies to promote a more sustainable society as the ecological impacts of urbanization appear closer to reaching consumption limits.

American urban sociology was initially formulated around the analysis of the city of Chicago, the great urban crossroads of the early twentieth century. As the Midwestern agro-industrial capital, Chicago was a transportation and trade gateway for foreign immigrants into the American heartland. Its status as a thriving international metropolis was established with its World's Columbian Exposition of 1893. The city was a main node in the hub and spoke system of cities in the railroad era. An elevated train, the Loop, circled the downtown, boosting the significance of the urban center. The dominant center was significant to the theories of "human ecology" associated with the "Chicago School" of urban sociology, which Robert Park and his associates established at the University of Chicago beginning in 1913. Like London and New York, Chicago incited a literary and social work tradition that testified to the presence of economic Darwinism and social inequality. This included Upton Sinclair's *The Jungle* (1906), Theodore Dreiser's *Sister Carrie* (1900), and Jane Addams' Hull House for settlement work. The poet Carl Sandburg described Chicago as "hog butcher to the world."

The human ecologists applied ideas of Charles Darwin to the urban scene, justifying the presence of urban social inequality through comparison with the "struggle for existence" in the evolutionary life of plant and animal communities. Principles of "natural selection" explained the dominance of banks and corporations in the central business district (CBD), and dynamics of invasion-succession, now commonly identified as the gentrification and displacement process. Park and Burgess each discussed how the land market "sifted and sorted" a population engaged in competition and conflict, eventually leading to a new social equilibrium. Their notion of "equilibrium" drew from thermodynamics and economic theory.

Human ecology applied concepts of economics and market competition to the biological and social realm. Park had a more passive understanding of Darwinism than Herbert Spencer, who coined the phrase "survival of the fittest" to justify free-market liberalism, colonialism, and racial eugenics. Park drew from the evolutionary naturalism of John Dewey, an advocate for educational reform. He was an apologist for the status quo, rarely critical of social inequality, and he sought to devalue the reform tradition in early twentieth-century sociology in favor of scientific analysis incorporating active fieldwork.

Burgess presented Chicago as the prototype of the modern city, with a dominant CBD ringed by successive concentric zones of settlement. This is one of the more durable images in the annals of urban sociology, and has been critiqued and seen less relevant to the decentralized U.S. cities of the post-war era, which are often poly-nucleated and multi-nodal metropolitan networks of sprawling freeways. Burgess also drew from biology, relating the metabolism of the human body to the moving equilibrium of human ecology between zones of organization and disorganization. He suggested that the "pulse" of a city is related to the incidence of increasing movement and mobility of people. Mobility was not only geographic, but also cultural, with reference to changing cultural attitudes and unconventional social behaviors. He linked mobility to social disorganization in breeding greater demoralization, promiscuity, and vice. Burgess considered social disorganization not as pathological, but as normal. Burgess identified the "zone-in-transition" as an area of high mobility that featured transitional housing, immigrant colonies, vice, and criminal activity.

Writing in the aftermath of the early 1970s oil shock, British geographer David Harvey presents a Marxist perspective on the urban process under capitalism that illuminates our understanding of capitalist crisis and its relationship to contradiction, overaccumulation, and the mobility of different circuits of capital. He draws attention to the tendency for capitalists to make a priority of profitability over productivity and promote overaccumulation in the primary circuit of capital in production, which is resolved through capital switching into the secondary circuit (which encompasses finance capital and the built environment for production) as well as the tertiary circuit in science and technology. The same tendency to overaccumulation creates additional contradictions and shocks in the secondary and tertiary circuits, doing much to help explain recent experiences of crises in the banking and housing industries and the crisis of ecological sustainability. Labor struggles may provoke capitalist investment or switching into such sectors as housing or health, which boost worker productivity but also provide additional sectors for overaccumulation.

Class struggle entered new social arenas when the social movements of the 1960s and 1970s brought new attention to problems of racial segregation, immigrant exclusion, poverty, and social inequality in urban life. A concatenation of urban riots, protests, and community action movements arose to address issues such as racial inequality, pollution and toxic waste, transit and expressway expansion, and rights to housing, education, and welfare. The protestors defended the integrity of urban neighborhoods and their quality of life, advanced the civil rights and legal entitlements of urban citizens, and promoted the economic and social development of urban communities. The spirit of revolutionary change in urban society also stimulated a paradigm shift in urban sociology, promoting a host of writing that was critical of human ecology and addressed issues of elite control, political and class interests, and social inequality in cities.

Logan and Molotch offer a seminal American interpretation of the political-economy approach by applying Marxian concepts of use and exchange value to the analysis of urban land markets. They observe the idiosyncratic special use values associated with real estate, including the sentimental preciousness of place attachments people may associate with homes and small businesses. Another special use value is that urban property markets determine access to resources and successful "life chances" for urban denizens. They also recognize the special use values of real estate as limited resources that are under the monopolized control of place entrepreneurs. These place entrepreneurs self-aggrandize their property interests by forging pro-growth coalitions with local business elites and governmental units to effectively create an apparatus of interlocking interests that they call an urban "growth machine." They lobby with government to focus the development of urban infrastructure to their interests. Joe Feagin discusses the local growth elite known as the "Suite 8F" crowd in his 1988 book *Free Enterprise City*.

Gregory Squires is another new urban sociologist, who draws our attention to the dynamics of public/private partnership and the role of the state in U.S. urban history. He questions the legitimacy of the credo of "partnership" in U.S. urban development history and exposes an underlying historical and contemporary ideology of privatism. He contests the political autonomy of the state and questions its legitimate authority to truly act in the public interest. He discerns episodic waves in U.S. history when private capital procured governmental subsidies to aggrandize their interests, including the canal and railroad building era, the freeway construction and urban renewal era, and the postindustrial era of downtown redevelopment and growth of the service sector. He considers the prospects for alternative community-based partnerships focused upon more balanced growth and shared prosperity.

The postmodern perspective in new urban sociology is represented by Michael Dear, a principal voice in the "L.A. School" that extends the assault against human ecology that was launched by Marxism. A British geographer turned Angeleno, Michael Dear has written widely in codifying the L.A. School perspective. Los Angeles replaces Chicago as the urban prototype in the era of the postmodern metropolis. The decentralized freeway system has replaced the hub-and-spoke system and created a sprawling polycentric, polyglot, and polycultural metropolitan region where the hinterlands can command as much authority as the center. The waning economic and cultural dominance of the city center correlates with the onset of a variety of other economic and social trends, such as deindustrialization and globalization, the growth of ethnic suburbs, and the rise of gated communities in the exurbs. The metropolitan sprawl of Los Angeles resembles a giant keno game board of ethnoburbs, theme parks, and gated communities.

The growth of neoliberal doctrines in urban planning and politics is examined by Neil Brenner and Nik Theodore. They address how the trend towards open, competitive, deregulated markets in contemporary cities is an outgrowth of broader global free trade and structural adjustment programs instituted by global institutions like the International Monetary Fund, the World Bank, and the World Trade Organization. While neoliberal restructuring occurs at multiple geographic scales from local to global, they draw special attention to the urban scale. The city offers an illuminating window on the scaling back of Fordist-Keynesian systems of government spending and social regulation. The retrenchment and devolution of power from centralized bureaucracies to states and localities, in a climate of economic austerity and deficit reduction, has been abetted by the ascendance of neoconservative political and social movements that vilify subcultural groups and the poor and label them as "dangerous classes" that endanger the traditional family and public morality. The disenfranchised have responded with struggles of their own, marking the relevance of cities as significant flashpoints of contention over the rolling out of neoliberalism.

The problems of urban sprawl in worsening segregation of the inner city, socioeconomic divides between city and suburb, and environmental degradation are confronted by Peter Dreier, John Mollenkopf, and Todd Swanstrom in the excerpt from their book *Place Matters*. Like Logan and Molotch, they draw attention to the differential power of urban locations in determining access to resources and better life-chances. They argue that addressing inequality can greatly help the commonwealth and that cities still hold great promise as "engines of prosperity" in the postindustrial economy. They argue for "smart growth" initiatives built upon regionalist policy and planning initiatives that bridge city/suburban divides and promote economic efficiency, competitiveness, environmentalism, and social equity. They contribute to a "metropolitics" that achieves "the new regionalism" through concrete suggestions for building new political coalitions.

Finally, William Rees and Mathis Wackernagel address the environmental consequences of urbanization and transform our way of understanding human ecology, through their concept of ecological carrying capacity. They articulate a mathematical method for estimating urban and regional "ecological footprints" that measures human carrying capacity or "load." They multiply population size with a per capita estimation of a typical market basket of major consumption items, including units of energy, food, and forest products. They furthermore compute an estimation of ecological impact or "deficit" that measures the "overshoot" between what an urban population consumes in human load versus what can be produced on its available ecologically productive land area. They address how ecological deficit estimation has enlarged our comprehension of North–South inequalities in the capitalist world-system. They describe cities as entropic dissipative structures, black holes of energy and goods consumption and waste. They articulate several concrete policy recommendations to help make cities more sustainable.

"Human Ecology"

from *American Journal of Sociology* (1936)

Robert Ezra Park

Editors' Introduction

Robert Park was born in Harveyville, Pennsylvania while his father was serving in the Civil War. They later settled in Minnesota and his father became a prosperous grocer. Park entered the University of Michigan in 1882 and was particularly drawn to the philosophy courses of John Dewey. He acquired ideas of evolutionary naturalism from Dewey, coming to see society as set in the natural order, in a competitive arena, but also held together by cognitive and moral consensus.

Upon graduation he began a career as a newspaper reporter, moving from Minneapolis, to Detroit, to Denver, to New York, and finally to Chicago. He wrote on the corruption of urban political machines, the immigrant areas of the city, crime, and other urban affairs. Journalism, particularly in Manhattan, satisfied his thirst for adventure and multifarious experience, but a persisting interest in the grand questions of life led him to return to academia to study philosophy at Harvard University in 1898. He subsequently grew interested in social thought and thus was impelled to move to Germany and the University of Berlin, which was then seen by many to be the intellectual center of Europe. While in Berlin, he came under the influence of Georg Simmel, then a *Privatdozent* lecturing in sociology. He obtained a Ph.D. from the University of Heidelberg in 1904.

Park returned from Germany to Massachusetts in 1903 and became a teaching assistant at Harvard. Through a chance encounter with a missionary, however, he discovered the work of the Congo Reform Association, and soon accepted work as their secretary and chief publicity agent. Through his work lobbying Congress to take action on the state of brutality and exploitation in the Congo Free State, Park met Booker T. Washington, who in 1905 was at the height of his notoriety as an accommodationist spokesman for black causes among the political elites. He became Washington's stenographer/ghostwriter, counselor, and press agent for the next seven years, working mainly at the Tuskegee Institute in Macon, Georgia, with regular visits to New England and occasional tours to Europe with Washington. This migratory lifestyle did not suit his family, however, and he decided to return to academic life, at the invitation of W. I. Thomas, then a professor of sociology at the University of Chicago.

In the fall of 1913, at the age of 49, Robert Park began the quarter-century of teaching and research leadership during which the University of Chicago sociology department became a celebrated center of the discipline in America. Through the tremendous surge of field research that he supervised, he was instrumental in drawing sociology away from a normative and reform-oriented focus of the Progressive era to a more scientific analysis that still accounted for the social importance of knowledge. His seminal essay titled, "The City: Suggestions for the Investigation of Human Behavior in the City," published in 1915 in the *American Journal of Sociology*, became a kind of manifesto for the use of city as a research laboratory. In it, he called for the study of urban life using the same ethnographic methods used by anthropologists to study the Native Americans. With Ernest W. Burgess, Park wrote and edited a textbook, *Introduction to the Science of Sociology* (Chicago, IL: University of Chicago Press, 1921), which became the most influential reader in the early history of American sociology. Park served as President of the American Sociological Society (later changed to Association) in 1925.

R. D. McKenzie provided the first exposition on human ecology in an essay titled "The Ecological Approach to the Study of the Human Community," published in the *American Journal of Sociology* in 1924. Robert Park codified his beliefs in the same journal in 1936, in a paper titled, "Human Ecology." In this essay, Park applied the principles of Charles Darwin's "web of life" and "struggle for existence" in plant and animal communities to the study of human communities. Through his explanation of concepts such as dominance, invasion-succession, and natural areas, Park provided a justification for urban inequality and free market competition that is often associated with the beliefs of Social Darwinists, from whom he drew some of his thinking.

Social Darwinism was a body of late nineteenth-century philosophy and social thought expounded by the British Herbert Spencer and American William Sumner that applied Charles Darwin's principles of "natural selection" to the analysis of human social evolution. While Darwin held a passive sense of the interplay between variation and heredity, the Social Darwinists were more akin to Jean-Baptiste Lamarck, who had a more active conception of the inheritance of acquired characteristics. Spencer coined the concept "survival of the fittest" to express the concept that the rich and powerful are rewarded for their greater intelligence, talents, ambition, and industriousness, while the poor are doomed to failure for their lack of these characteristics. Free market liberalism was promoted in the economy, while charitable and state redistributional programs were opposed. Social Darwinism was eventually used to justify colonialism, racial eugenics, and policies of cultural assimilation.

Park was inspired by Charles Darwin, but ultimately diverges from Social Darwinism through his recognition that human societies participate in a social and moral order that has no counterpart on the nonhuman level. There is a dualism in human ecology in that there is competition as well as cooperation and symbiosis, especially at higher levels of the interactional pyramid. Park furthermore accounted for process, or social change, and was concerned that ecological equilibrium could commonly be disrupted by external changes.

Robert Park was driven by the philosophy of pragmatism that he learned from John Dewey, who exhorted American educators to school their students to engage in active learning through direct service in communities. He was influenced by the turn-of-the-century social reform and Progressive movements, as evidenced by his early passion for journalistic muckraking and devotion to anti-colonialist and black causes, distinguishing him from the conservative and racist Social Darwinists. Though liberal-minded, he did not buck the status quo, as attested by his association with the accommodationism of Booker T. Washington. Park died in 1944 in Nashville.

For further writing on the legacy of Robert Park, see Fred H. Matthews, *Quest for an American Sociology: Robert E. Park and the Chicago School* (Montreal: McGill-Queen's University Press, 1977) and Edward Shils, "Robert E. Park, 1864–1944," *The American Scholar* (Winter, 1991): 120–127. See also the section on Robert Park in Lewis A. Coser's *Masters of Sociological Thought: Ideas in Historical and Social Context* (2nd Edition) (Fort Worth, TX: Harcourt Brace Jovanovich, 1977).

THE WEB OF LIFE

Naturalists of the last century were greatly intrigued by their observation of the interrelations and co-ordinations, within the realm of animate nature, of the numerous, divergent, and widely scattered species. Their successors, the botanists, and zoologists of the present day, have turned their attention to more specific inquiries, and the "realm of nature," like the concept of evolution, has come to be for them a notion remote and speculative.

The "web of life," in which all living organisms, plants and animals alike, are bound together in a vast system of interlinked and interdependent lives, is nevertheless, as J. Arthur Thompson puts it, "one of the fundamental biological concepts" and is "as characteristically Darwinian as the struggle for existence."

Darwin's famous instance of the cats and the clover is the classic illustration of this interdependence. He found, he explains, that bumblebees were almost indispensable to the fertilization of the heartsease, since other bees do not visit this flower. The same thing is true with some kinds of clover. Bumblebees alone visit red clover, as other bees cannot reach the nectar. The inference is that if the bumblebees became extinct or very rare in England, the heartsease and red clover would become very rare, or wholly disappear. However, the number of bumblebees in any district depends in a great measure on the number of field mice, which destroy their combs and nests. It is estimated that more than two-thirds of them are thus destroyed all over England. Near villages and small towns the nests of bumblebees are more numerous than elsewhere and this is attributed to the number of cats that destroy the mice. Thus next year's crop of purple clover in certain parts of England depends on the number of bumblebees in the district; the number of bumblebees depends upon the number of field mice, the number of field mice upon the number and the enterprise of the cats, and the number of cats – as someone has added – depends on the number of old maids and others in neighboring villages who keep cats.

These large food chains, as they are called, each link of which eats the other, have as their logical prototype the familiar nursery rhyme, "The House that Jack Built." You recall:

The cow with the crumpled horn,
That tossed the dog,
That worried the cat,
That killed the rat,
That ate the malt
That lay in the house that Jack built.

Darwin and the naturalists of his day were particularly interested in observing and recording these curious illustrations of the mutual adaptation and correlation of plants and animals because they seemed to throw light on the origin of the species. Both the species and their mutual interdependence, within a common habitat, seem to be a product of the same Darwinian struggle for existence.

It is interesting to note that it was the application to organic life of a sociological principle – the principle, namely, of "competitive co-operation" – that gave Darwin the first clue to the formulation of his theory of evolution.

[. . .]

The active principle in the ordering and regulating of life within the realm of animate nature is, as Darwin described it, "the struggle for existence." By this means the numbers of living organisms are regulated, their distribution controlled, and the balance of nature maintained. Finally, it is by means of this elementary form of competition that the existing species, the survivors in the struggle, find their niches in the physical environment and in the existing correlation or division of labor between the different species.

[. . .]

These manifestations of a living, changing, but persistent order among competing organisms – organisms embodying "conflicting yet correlated interests" – seem to be the basis for the conception of a social order transcending the individual species, and of a society based on a biotic rather than a cultural basis, a conception later developed by the plant and animal ecologists.

In recent years the plant geographers have been the first to revive something of the earlier field naturalists' interest in the interrelations of species. Haeckel, in 1878, was the first to give to these studies a name, "ecology," and by so doing gave them the character of a distinct and separate science.

The interrelation and interdependence of the species are naturally more obvious and more intimate within the common habitat than elsewhere. Furthermore, as correlations have multiplied and competition has decreased, in consequence of mutual adaptations of the competing species, the habitat and habitants have tended to assume the character of a more or less completely closed system.

Within the limits of this system the individual units of the population are involved in a process of competitive co-operation, which has given to their interrelations the character of a natural economy. To such a habitat and its inhabitants – whether plant, animal, or human – the ecologists have applied the term "community."

The essential characteristics of a community, so conceived, are those of: (1) a population, territorially organized, (2) more or less completely rooted in the soil it occupies, (3) its individual units living in a relationship of mutual interdependence that is symbiotic rather than societal, in the sense in which that term applies to human beings.

These symbiotic societies are not merely unorganized assemblages of plants and animals which happen to live together in the same habitat. On the contrary, they are interrelated in the most complex manner. Every community has something of the character of an organic unit. It has a more or less definite structure and it has a life history in which juvenile, adult, and senile phases can be observed. If it is an organism, it is one of the organisms which are other organisms. It is, to use Spencer's phrase, a superorganism.

What more than anything else gives the symbiotic community the character of an organism is the fact that it possesses a mechanism (competition) for (1) regulating the numbers, and (2) preserving the balance between the competing species of which it is composed. It is by maintaining this biotic balance that the community preserves its identity and integrity as an individual unit through the changes and the vicissitudes to which it is subject in the course of its progress from the earlier to the later phases of its existence.

THE BALANCE OF NATURE

The balance of nature, as plant and animal ecologists have conceived it, seems to be largely a question of numbers. When the pressure of population upon the natural resources of the habitat reaches a certain degree of intensity, something invariably happens. In the one case the population may swarm and relieve the pressure of population by migration. In another, where the disequilibrium between population and natural resources is the result of some change, sudden or gradual, in the conditions of life, the pre-existing correlation of the species may be totally destroyed.

Change may be brought about by a famine, an epidemic, or an invasion of the habitat by some alien species. Such an invasion may result in a rapid increase of the invading population and a sudden decline in the numbers if not the destruction of the original population. Change of some sort is continuous, although the rate and pace of change sometimes vary greatly.

[. . .]

Under ordinary circumstances, such minor fluctuations in the biotic balance as occur are mediated and absorbed without profoundly disturbing the existing equilibrium and routine of life. When, on the other hand, some sudden and catastrophic change occurs – it may be a war, a famine, or pestilence – it upsets the biotic balance, breaks "the cake of custom," and releases energies up to that time held in check. A series of rapid and even violent changes may ensue which profoundly alter the existing organization of communal life and give a new direction to the future course of events.

The advent of the boll weevil in the southern cotton fields is a minor instance but illustrates the principle. The boll weevil crossed the Rio Grande at Brownsville in the summer of 1892. By 1894 the pest had spread to a dozen counties in Texas, bringing destruction to the cotton and great losses to the planters. From that point it advanced, with every recurring season, until by 1928 it had covered practically all the cotton producing area in the United States. Its progress took the form of a territorial succession. The consequences to agriculture were catastrophic but not wholly for the worse, since they served to give an impulse to changes in the organization of the industry long overdue. It also hastened the northward migration of the Negro tenant farmer.

The case of the boll weevil is typical. In this mobile modern world, where space and time have been measurably abolished, not men only but all the minor organisms (including the microbes) seem to be, as never before, in motion. Commerce, in progressively destroying the isolation upon which the ancient order of nature rested, has intensified the struggle for existence over an ever widening area of the habitable world. Out of this struggle a new equilibrium and a new system of animate nature, the new biotic basis of the new world society, is emerging.

[. . .]

The conditions which affect and control the movements and numbers of populations are more complex in human societies than in plant and animal communities, but they exhibit extraordinary similarities.

The boll weevil, moving out of its ancient habitat in the central Mexican plateau and into the virgin territory of the southern cotton plantations, incidentally multiplying its population to the limit of the territories and resources, is not unlike the Boers of Cape Colony, South Africa, trekking out into the high veldt of the central South African plateau and filling it, within a period of one hundred years, with a population of their own descendants.

Competition operates in the human (as it does in the plant and animal) community to bring about and restore the communal equilibrium, when, either by the advent of some intrusive factor from without or in the normal course of its life-history, that equilibrium is disturbed.

Thus every crisis that initiates a period of rapid change, during which competition is intensified, moves over finally into a period of more or less stable equilibrium and a new division of labor. In this manner competition brings about a condition in which competition is superseded by co-operation.

It is when, and to the extent that, competition declines that the kind of order which we call society may be said to exist. In short, society, from the ecological point of view, and in so far as it is a territorial unit, is just the area within which biotic competition has declined and the struggle for existence has assumed higher and more sublimated forms.

COMPETITION, DOMINANCE AND SUCCESSION

There are other and less obvious ways in which competition exercises control over the relations of individuals and species within the communal habitat. The two ecological principles, dominance and succession, which operate to establish and maintain such communal order as here described are functions of, and dependent upon, competition.

In every life-community there is always one or more dominant species. In a plant community this dominance is ordinarily the result of struggle among the different species for light. In a climate which supports a forest the dominant species will invariably be trees. On the prairie and steppes they will be grasses.

[. . .]

But the principle of dominance operates in the human as well as in the plant and animal communities. The so-called natural or functional areas of a metropolitan community – for example, the slum, the rooming-house area, the central shopping section and the banking center – each and all owe their existence directly to the factor of dominance, and indirectly to competition.

The struggle of industries and commercial institutions for a strategic location determines in the long run the main outlines of the urban community. The distribution of population, as well as the location and limits of the residential areas which they occupy, are determined by another similar but subordinate system of forces.

The area of dominance in any community is usually the area of highest land values. Ordinarily there are in every large city two such positions of highest land value – one in the central shopping district, the other in the central banking area. From these points land values decline at first precipitantly and then more gradually toward the periphery of the urban community. It is these land values that determine the location of social institutions and business enterprises. Both the one and the other are bound up in a kind of territorial complex within which they are at once competing and interdependent units.

As the metropolitan community expands into the suburbs the pressure of professions, business enterprises, and social institutions of various sorts destined to serve the whole metropolitan region steadily increases the demand for space at the center. Thus not merely the growth of the suburban area, but any change in the method of transportation which makes the central business area of the city more accessible, tends to increase the pressure at the center. From thence this pressure is transmitted and diffused, as the profile of land values discloses, to every other part of the city.

Thus the principle of dominance, operating within the limits imposed by the terrain and other natural features of the location, tends to determine the general ecological pattern of the city and the functional relation of each of the different areas of the city to all others.

Dominance is, furthermore, in so far as it tends to stabilize either the biotic or the cultural community, indirectly responsible for the phenomenon of succession.

The term "succession" is used by ecologists to describe and designate that orderly sequence of changes through which a biotic community passes in the course of its development from a primary and relatively unstable to a relatively permanent or climax stage. The main point is that not merely do the individual plants and animals within the communal habitat grow but the community itself, i.e., the system of relations between the species, is likewise involved in an orderly process of change and development.

The fact that, in the course of this development, the community moves through a series of more or less clearly defined stages is the fact that gives this development the serial character which the term "succession" suggests.

The explanation of the serial character of the changes involved in succession is the fact that at every stage in the process a more or less stable equilibrium is achieved, and as a result of progressive changes in life-conditions, possibly due to growth and decay, the equilibrium achieved in the earlier stages is eventually undermined. In this case the energies previously held in balance will be released, competition will be intensified, and change will continue at a relatively rapid rate until a new equilibrium is achieved.

The climax phase of community development corresponds with the adult phase of an individual's life.

[. . .]

The cultural community develops in comparable ways to that of the biotic, but the process is more complicated. Inventions, as well as sudden or catastrophic changes, seem to play a more important part in bringing about serial changes in the cultural than in the biotic community. But the principle involved seems to be substantially the same. In any case, all or most of the fundamental processes seem to be functionally related and dependent upon competition.

Competition, which on the biotic level functions to control and regulate the interrelations of organisms, tends to assume on the social level the form of conflict. The intimate relation between competition and conflict is indicated by the fact that wars frequently, if not always, have, or seem to have, their source and origin in economic competition which, in that case, assumes the more sublimated form of a struggle for power and prestige. The social function of war, on the other hand, seems to be to extend the area over which it is possible to maintain peace.

BIOLOGICAL ECONOMICS

If population pressure, on the one hand, co-operates with changes in local and environmental conditions to disturb at once the biotic balance and social equilibrium, it tends at the same time to intensify competition. In so doing it functions, indirectly, to bring about a new, more minute and, at the same time, territorially extensive division of labor.

Under the influence of an intensified competition, and the increased activity which competition involves, every individual and every species, each for itself, tends to discover the particular niche in the physical and living environment where it can survive and flourish with the greatest possible expansiveness consistent with its necessary dependence upon its neighbors.

It is in this way that a territorial organization and a biological division of labor, within the communal habitat, is established and maintained. This explains, in part at least, the fact that the biotic community has been conceived at one time as a kind of superorganism and at another as a kind of economic organization for the exploitation of the natural resources of its habitat.

In their interesting survey, *The Science of Life*, H. G. Wells and his collaborators, Julian Huxley and G. P. Wells, have described ecology as "biological economics," and as such very largely concerned with "the balances and mutual pressures of species living in the same habitat."

"Ecology," as they put it, is "an extension of Economics to the whole of life." On the other hand the science of economics as traditionally conceived, though it is a whole century older, is merely a branch of a more general science of ecology which includes man with all other living creatures. Under the circumstances what has been traditionally described as economics and conceived as restricted to human affairs, might very properly be described as Barrows some years ago described geography, namely as human ecology. It is in this sense that Wells and his collaborators would use the term.

Since human ecology cannot be at the same time both geography and economics, one may adopt, as a working hypothesis, the notion that it is neither one nor the other but something independent of both. Even so the motives for identifying ecology with geography on the one hand, and economics on the other, are fairly obvious.

From the point of view of geography, the plant, animal, and human population, including their habitations and other evidence of man's occupation of the soil, are merely part of the landscape, of which the geographer is seeking a detailed description and picture.

On the other hand ecology (biologic economics), even when it involves some sort of unconscious co-operation and a natural, spontaneous, and non-rational division of labor, is something different from the economics of commerce; something quite apart from the bargaining of the market place. Commerce, as Simmel somewhere remarks, is one of the latest and most complicated of all the social relationships into which human beings have entered. Man is the only animal that trades and traffics.

Ecology, and human ecology, if it is not identical with economics on the distinctively human and cultural level is, nevertheless, something more than and different from the static order which the human geographer discovers when he surveys the cultural landscape.

The community of the geographer is not, for one thing, like that of the ecologist, a closed system, and the web of communication which man has spread over the earth is something different from the "web of life" which binds living creatures all over the world in a vital nexus.

SYMBIOSIS AND SOCIETY

Human ecology, if it is neither economics on one hand nor geography on the other, but just ecology, differs, nevertheless, in important respects from plant and animal ecology. The interrelations of human beings and interactions of man and his habitat are comparable but not identical with interrelations of other forms of life that live together and carry on a kind of "biological economy" within the limits of a common habitat.

For one thing man is not so immediately dependent upon his physical environment as other animals. As a result of the existing world-wide division of labor, man's relation to his physical environment has been mediated through the intervention of other men. The exchange of goods and services have co-operated to emancipate him from dependence upon his local habitat.

Furthermore man has, by means of inventions and technical devices of the most diverse sorts, enormously increased his capacity for reacting upon and remaking, not only his habitat but his world. Finally, man has erected upon the basis of the biotic community an institutional structure rooted in custom and tradition.

Structure, where it exists, tends to resist change, at least change coming from without; while it possibly facilitates the cumulation of change within. In plant and animal communities structure is biologically determined, and so far as any division of labor exists at all it has a physiological and instinctive basis. The social insects afford a conspicuous example of this fact, and one interest in studying their habits is that they show the extent to which social organization can be developed on a purely physiological and instinctive basis, as is the case among human beings in the natural as distinguished from the institutional family.

In a society of human beings, however, this communal structure is reinforced by custom and assumes an institutional character. In human as contrasted with animal societies, competition and the freedom of the individual is limited on every level above the biotic by custom and consensus.

The incidence of this more or less arbitrary control which custom and consensus imposes upon the natural social order complicates the social process but does not fundamentally alter it – or, if it does, the effects of biotic competition will still be manifest in the succeeding social order and the subsequent course of events.

The fact seems to be, then, that human society, as distinguished from plant and animal society, is organized on two levels, the biotic and the cultural. There is a symbiotic society based on competition and a cultural society based on communication and consensus. As a matter of fact the two societies are merely different aspects of one society, which, in the vicissitudes and changes to which they are subject remain, nevertheless, in some sort of mutual dependence each upon the other. The cultural superstructure rests on the basis of the symbiotic substructure, and the emergent energies that manifest themselves on the biotic level in movements and actions reveal themselves on the higher social level in more subtle and sublimated forms.

However, the interrelations of human beings are more diverse and complicated than this dichotomy, symbiotic and cultural, indicates. This fact is attested by the divergent systems of human interrelations which have been the subject of the special social sciences. Thus human society, certainly in its mature and more rational expression, exhibits not merely an ecological, but an economic, a political, and a moral order. The social sciences include not merely human geography and ecology, but economics, political science, and cultural anthropology.

It is interesting also that these divergent social orders seem to arrange themselves in a kind of hierarchy. In fact they may be said to form a pyramid of which the ecological order constitutes the base and the moral order the apex. Upon each succeeding one of these levels, the ecological, economic, political, and moral, the individual finds himself more completely incorporated into and subordinated to the social order of which he is a part than upon the preceding.

Society is everywhere a control organization. Its function is to organize, integrate, and direct the energies resident in the individuals of which it is composed. One might, perhaps, say that the function of society was everywhere to restrict competition and by so doing bring about a more effective co-operation of the organic units of which society is composed.

Competition, on the biotic level, as we observe it in the plant and animal communities, seems to be relatively unrestricted. Society, so far as it exists, is anarchic and free. On the cultural level, this freedom of the individual to compete is restricted by conventions, understandings, and law. The individual is more free upon the economic level than upon the political, more free on the political than the moral.

As society matures control is extended and intensified and free commerce of individuals restricted, if not by law then by what Gilbert Murray refers to as "the normal expectation of mankind." The mores are merely what men, in a situation that is defined, have come to expect.

Human ecology, in so far as it is concerned with a social order that is based on competition rather than consensus, is identical, in principle at least, with plant and animal ecology. The problems with which plant and animal ecology have been traditionally concerned are fundamentally population problems. Society, as ecologists have conceived it, is a population settled and limited to its habitat. The ties that unite its individual units are those of a free and natural economy, based on a natural division of labor. Such a society is territorially organized and the ties which hold it together are physical and vital rather than customary and moral.

Human ecology has, however, to reckon with the fact that in human society competition is limited by custom and culture. The cultural superstructure imposes itself as an instrument of direction and control upon the biotic substructure.

Reduced to its elements the human community, so conceived, may be said to consist of a population and a culture, including in the term culture (1) a body of customs and beliefs and (2) a corresponding body of artifacts and technological devices.

To these three elements or factors – (1) population, (2) artifact (technological culture), (3) custom and beliefs (non-material culture) – into which the social complex resolves itself, one should, perhaps, add a fourth, namely, the natural resources of the habitat.

It is the interaction of these four factors – (1) population, (2) artifacts (technological culture), (3) custom and beliefs (non-material culture), and (4) the natural resources that maintain at once the biotic balance and the social equilibrium, when and where they exist.

The changes in which ecology is interested are the movements of population and of artifacts (commodities) and changes in location and occupation – any sort of change, in fact, which affects an existing division of labor or the relation of the population to the soil.

Human ecology is, fundamentally, an attempt to investigate the processes by which the biotic balance and the social equilibrium (1) are maintained once they are achieved and (2) the processes by which, when the biotic balance and the social equilibrium are disturbed, the transition is made from one relatively stable order to another.

"The Growth of the City: An Introduction to a Research Project"

from Robert Park *et al.* (eds), *The City* (1925)

Ernest W. Burgess

Editors' Introduction

Ernest W. Burgess (1886–1966), together with Robert Park, established a distinctive program of urban research in the sociology department at the University of Chicago in the early twentieth century. Ernest Watson Burgess was born on May 16, 1886 in Tilbury, Ontario, Canada. He obtained his Ph.D. from the University of Chicago in 1913 and served on the faculty there from 1916 to 1951. One of the important concepts he disseminated was succession, a term borrowed from plant ecology. Burgess was the originator of concentric zone theory, which predicted that cities would take the form of five concentric rings growing outwards, with a zone of deterioration immediately surrounding the city center, succeeding to increasingly prosperous residential zones moving out to the city's edge. Burgess understood the invasion–succession process as a "moving equilibrium" of the social order, a "process of distribution takes place which sifts and sorts and relocates individuals and groups by residence and occupation."

The human ecological research program also involved the extensive use of mapping to reveal the spatial distribution of social problems and to permit comparison between areas. Burgess was particularly interested in maps and used them extensively, requiring all his students to acquire proficiency in basic mapmaking techniques. Burgess and his students scoured the city of Chicago for data that could be used for maps, gleaning information from city agencies and making more extensive use of census data than any other social scientists of the time. This was one of the most important legacies of the urban ecology studies undertaken at the University of Chicago in the 1920s as mapmaking became part of the methodological toolkit of the developing disciplines of sociology, criminology, and public policy. Burgess was not a systematic theoretician but an eclectic promoter of theory and methodology. He sought to develop reliable tools for the prediction of social phenomena such as delinquency, parole violation, divorce, city growth, and adjustment in old age.

Human ecology drew criticism for its formalism in the postwar era, and newer sunbelt cities became more decentralized with the decline of railroad transportation and the onset of automobile travel and highways. In particular, Park and Burgess' search for "natural" or "organic" process was criticized as a superficial undertaking that neglected both the social and cultural dimensions of urban life and the political-economic impact of industrialization on urban geography. Overall, the urban ecology studies of the 1920s were largely oblivious to issues of class, race, gender, and ethnicity. However, the concentric rings model has become one of the more famous formulations in urban sociology and is still applied creatively to studies of urban processes. In *The Ecology of Fear* (1992), Mike Davis adapts the concentric rings model to describe the exclusionary fortress of Los Angeles, a metropolis with zones of "homeless containment" and drug enforcement and neighborhood watches in the central city and near suburbs, with gated communities and prisons on the urban periphery.

Burgess drew from biology more than did Robert Park, through his references to concepts like "metabolism" and "pulse." Through the organic metaphor, the city can be compared to a body, with metabolic and circulatory processes. He thought of mobility as the "pulse of the community." He connected the growth of cities with increasing mobility and movement of people and cultures. Geographic and cultural mobility put people in contact with increasingly diverse and unconventional behaviors. Mobility could thus breed social disorganization in the form of crime, deviance, and promiscuous behavior. Burgess viewed mobility and social disorganization not as pathological but as normal. Unusual behaviors were particularly focused in the central city and "zone-in-transition" encompassing both the "bright light" and "red light" areas of the city. As displayed in Times Square in New York and Hollywood in Los Angeles, the respectable theater district of a city attracts a critical mass of nocturnal pedestrians that can also support more licentious activities or cheap amusements. Like a body, the city can be seen as having a "heart," erogenous zones, and something like a hormonal metabolism. The city is a site of excitement, adventure, and thrills. The bright light and red light areas of the city are crucial components of the metropolitan mosaic of social worlds. They give a city its cultural and subcultural identity and are the focal points for the emergence of artistic and bohemian communities. While Burgess mainly fretted about the negative effects of mobility on social disorganization, we may see mobility can also produce positive effects on cultural life and community.

Ernest W. Burgess served as the twenty-fourth President of the American Sociological Association. His Presidential Address "Social Planning and the Mores" was delivered at the organization's annual meeting in Chicago in December 1934. His editing roles were extensive, and he was Editor of the *American Journal of Sociology* from 1936 to 1940. He founded the Family Study Center at the University of Chicago, and was involved in a number of professional associations.

The outstanding fact of modern society is the growth of great cities. Nowhere else have the enormous changes which the machine industry has made in our social life registered themselves with such obviousness as in the cities. In the United States the transition from a rural to an urban civilization, though beginning later than in Europe, has taken place, if not more rapidly and completely, at any rate more logically in its most characteristic forms.

All the manifestations of modern life which are peculiarly urban – the skyscraper, the subway, the department store, the daily newspaper, and social work – are characteristically American. The more subtle changes in our social life, which in their cruder manifestations are termed "social problems," problems that alarm and bewilder us, as divorce, delinquency, and social unrest, are to be found in their most acute forms in our largest American cities. The profound and "subversive" forces which have wrought these changes are measured in the physical growth and expansion of cities. That is the significance of the comparative statistics of Weber, Bücher, and other students.

These statistical studies, although dealing mainly with the effects of urban growth, brought out into clear relief certain distinctive characteristics of urban as compared with rural populations. The larger proportion of women to men in the cities than in the open country, the greater percentage of youth and middle-aged, the higher ratio of the foreign-born, the increased heterogeneity of occupation increase with the growth of the city and profoundly alter its social structure. These variations in the composition of population are indicative of all the changes going on in the social organization of the community. In fact, these changes are part of the growth of the city and suggest the nature of the processes of growth.

The only aspect of growth adequately described by Bücher and other students of Weber was the rather obvious process of the *aggregation* of urban population. Almost as overt a process, that of *expansion*, has been investigated from a different and very practical point of view by groups interested in city planning, zoning, and regional surveys. Even more significant than the increasing density of urban population is its correlative tendency to overflow, and so to extend over wider areas, and to incorporate these areas into a larger communal life. This paper, therefore, will treat first of all the expansion of the city, and then of the less-known processes of urban metabolism and mobility which are closely related to expansion.

EXPANSION AS PHYSICAL GROWTH

The expansion of the city from the standpoint of the city plan, zoning, and regional surveys is thought of almost wholly in terms of its physical growth. Traction studies have dealt with the development of transportation in its relation to the distribution of population throughout the city. The surveys made by the Bell Telephone Company and other public utilities have attempted to forecast the direction and the rate of growth of the city in order to anticipate the future demands for the extension of their services. In the city plan the location of parks and boulevards, the widening of traffic streets, the provision for a civic center, are all in the interest of the future control of the physical development of the city.

This expansion in area of our largest cities is now being brought forcibly to our attention by the Plan for the Study of New York and Its Environs, and by the formation of the Chicago Regional Planning Association, which extends the metropolitan district of the city to a radius of 50 miles, embracing 4,000 square miles of territory. Both are attempting to measure expansion in order to deal with the changes that accompany city growth. In England, where more than one-half of the inhabitants live in cities having a population of 100,000 and over, the lively appreciation of the bearing of urban expansion on social organization is thus expressed by C. B. Fawcett:

> One of the most important and striking developments in the growth of the urban populations of the more advanced peoples of the world during the last few decades has been the appearance of a number of vast urban aggregates, or conurbations, far larger and more numerous than the great cities of any preceding age. These have usually been formed by the simultaneous expansion of a number of neighboring towns, which have grown out toward each other until they have reached a practical coalescence in one continuous urban area. Each such conurbation still has within it many nuclei of denser town growth, most of which represent the central areas of the various towns from which it has grown, and these nuclear patches are connected by the less densely urbanized areas which began as suburbs of these towns. The latter are still usually rather less continuously occupied by buildings, and often have many open spaces.

These great aggregates of town dwellers are a new feature in the distribution of man over the earth. At the present day there are from thirty to forty of them, each containing more than a million people, whereas only a hundred years ago there were, outside the great centers of population on the waterways of China, not more than two or three. Such aggregations of people are phenomena of great geographical and social importance; they give rise to new problems in the organization of the life and well-being of their inhabitants and in their varied activities. Few of them have yet developed a social consciousness at all proportionate to their magnitude, or fully realized themselves as definite groupings of people with many common interests, emotions and thoughts.

In Europe and America the tendency of the great city to expand has been recognized in the term "the metropolitan area of the city," which far overruns its political limits, and in the case of New York and Chicago, even state lines. The metropolitan area may be taken to include urban territory that is physically contiguous, but it is coming to be defined by that facility of transportation that enables a business man to live in a suburb of Chicago and to work in the loop, and his wife to shop at Marshall Field's and attend grand opera in the Auditorium.

EXPANSION AS A PROCESS

No study of expansion as a process has yet been made, although the materials for such a study and intimations of different aspects of the process are contained in city planning, zoning, and regional surveys. The typical processes of the expansion of the city can best be illustrated, perhaps, by a series of concentric circles, which may be numbered to designate both the successive zones of urban extension and the types of areas differentiated in the process of expansion.

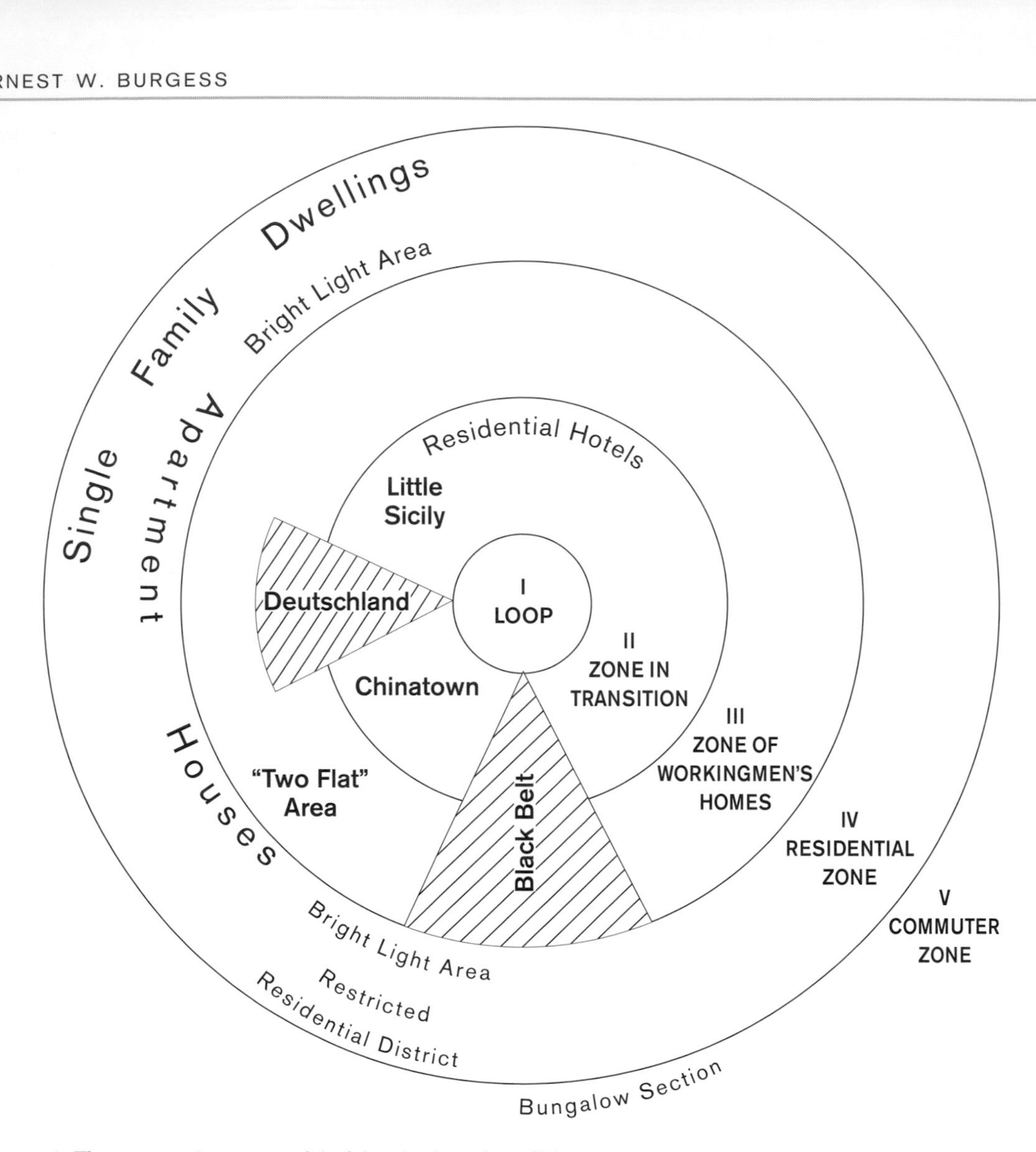

Figure 1 The concentric zone model of the city, based on Chicago

This chart represents an ideal construction of the tendencies of any town or city to expand radially from its central business district – on the map "The Loop" (I). Encircling the downtown area there is normally an area in transition, which is being invaded by business and light manufacture (II). A third area (III) is inhabited by the workers in industries who have escaped from the area of deterioration (II) but who desire to live within easy access of their work. Beyond this zone is the "residential area" (IV) of high-class apartment buildings or of exclusive "restricted" districts of single family dwellings. Still farther, out beyond the city limits, is the commuters' zone – suburban areas, or satellite cities – within a thirty- to sixty-minute ride of the central business district.

This chart brings out clearly the main fact of expansion, namely, the tendency of each inner zone to extend its area by the invasion of the next outer zone. This aspect of expansion may be called *succession*, a process which has been studied in detail in plant ecology. If this chart is applied to Chicago, all four of these zones were in its early history included in the circumference of the inner zone, the present business district. The present boundaries of the area of deterioration were not many years ago those of the zone now inhabited by independent wage-earners, and within the memories of thousands of Chicagoans contained the residences of the "best families." It hardly needs to be added that neither Chicago nor any other city fits perfectly into this ideal scheme. Complications are introduced by the lake front, the Chicago River, railroad lines, historical factors in the location of industry, the relative degree of the resistance of communities to invasion, etc.

Besides extension and succession, the general process of expansion in urban growth involves the antagonistic and yet complementary processes of concentration and decentralization. In all cities there is the natural tendency for local and outside transportation to converge in the central business district. In the downtown section of every large city we expect to find the department stores, the skyscraper office buildings, the railroad stations, the great hotels, the theaters, the art museum, and the city hall. Quite naturally, almost inevitably, the economic, cultural, and political life centers here. The relation of centralization to the other processes of city life may be roughly gauged by the fact that over half a million people daily enter and leave Chicago's "loop." More recently sub-business centers have grown up in outlying zones. These "satellite loops" do not, it seems, represent the "hoped for" revival of the neighborhood, but rather a telescoping of several local communities into a larger economic unity. The Chicago of yesterday, an agglomeration of country towns and immigrant colonies, is undergoing a process of reorganization into a centralized decentralized system of local communities coalescing into sub-business areas visibly or invisibly dominated by the central business district. The actual processes of what may be called centralized decentralization are now being studied in the development of the chain store, which is only one illustration of the change in the basis of the urban organization.

Expansion, as we have seen, deals with the physical growth of the city, and with the extension of the technical services that have made city life not only livable, but comfortable, even luxurious. Certain of these basic necessities of urban life are possible only through a tremendous development of communal existence. Three millions of people in Chicago are dependent upon one unified water system, one giant gas company, and one huge electric light plant. Yet, like most of the other aspects of our communal urban life, this economic co-operation is an example of co-operation without a shred of what the "spirit of co-operation" is commonly thought to signify. The great public utilities are a part of the mechanization of life in great cities, and have little or no other meaning for social organization.

Yet the processes of expansion, and especially the rate of expansion, may be studied not only in the physical growth and business development, but also in the consequent changes in the social organization and in personality types. How far is the growth of the city, in its physical and technical aspects, matched by a natural but adequate re-adjustment in the social organization. What, for a city, is a normal rate of expansion, a rate of expansion with which controlled changes in the social organization might successfully keep pace?

SOCIAL ORGANIZATION AND DISORGANIZATION AS PROCESSES OF METABOLISM

These questions may best be answered, perhaps, by thinking of urban growth as a resultant of organization and disorganization analogous to the anabolic and katabolic processes of metabolism in the body. In what way are individuals incorporated into the life of a city? By what process does a person become an organic part of his society? The natural process of acquiring culture is by birth. A person is born into a family already adjusted to a social environment – in this case the modern city. The natural rate of increase of population most favorable for assimilation may then be taken as the excess of the birth-rate over the death-rate, but is this the normal rate of city growth? Certainly, modern cities have increased and are increasing in population at a far higher rate. However, the natural rate of growth may he used to measure the disturbances of metabolism caused by any excessive increase, as those which followed the great influx of southern Negroes into northern cities since the war. In a similar way all cities show deviations in composition by age and sex from a standard population such as that of Sweden, unaffected in recent years by any great emigration or immigration. Here again, marked variations, as any great excess of males over females, or of females over males, or in the proportion of children, or of grown men or women, are symptomatic of abnormalities in social metabolism.

Normally the processes of disorganization and organization may be thought of as in reciprocal relationship to each other, and as co-operating in a moving equilibrium of social order toward an end vaguely or definitely regarded as progressive. So far as disorganization points to reorganization and makes for more efficient adjustment, disorganization must be conceived not as pathological, but as normal. Disorganization as preliminary to reorganization of attitudes and conduct is almost invariably the lot of the newcomer to the city, and the discarding of the habitual, and often of what has been to him the moral, is not infrequently accompanied by sharp mental conflict and sense of personal loss. Oftener, perhaps, the change gives sooner or later a feeling of emancipation and an urge toward new goals.

In the expansion of the city a process of distribution takes place which sifts and sorts and relocates individuals and groups by residence and occupation. The resulting differentiation of the cosmopolitan American city into areas is typically all from one pattern, with only interesting minor modifications. Within the central business district or on an adjoining street is the "main stem" of "hobohemia," the teeming Rialto of the homeless migratory man of the Middle West.[1] In the zone of deterioration encircling the central business section are always to be found the so-called "slums" and "bad lands," with their submerged regions of poverty, degradation, and disease, and their underworlds of crime and vice. Within a deteriorating area are rooming-house districts, the purgatory of "lost souls." Near by is the Latin Quarter, where creative and rebellious spirits resort. The slums are also crowded to overflowing with immigrant colonies – the Ghetto, Little Sicily, Greektown, Chinatown – fascinatingly combining old world heritages and American adaptations. Wedging out from here is the Black Belt, with its free and disorderly life. The area of deterioration, while essentially one of decay, of stationary or declining population, is also one of regeneration, as witness the mission, the settlement, the artists' colony, radical centers – all obsessed with the vision of a new and better world.

The next zone is also inhabited predominately by factory and shop workers, but skilled and thrifty. This is an area of second immigrant settlement, generally of the second generation. It is the region of escape from the slum, the *Deutschland* of the aspiring Ghetto family. For *Deutschland* (literally "Germany") is the name given, half in envy, half in derision, to that region beyond the Ghetto where successful neighbors appear to be imitating German Jewish standards of living. But the inhabitant of this area in turn looks to the "Promised Land" beyond, to its residential hotels, its apartment-house region, its "satellite loops," and its "bright light" areas.

This differentiation into natural economic and cultural groupings gives form and character to the city. For segregation offers the group, and thereby the individuals who compose the group, a place and a role in the total organization of city life. Segregation limits development in certain directions, but releases it in others. These areas tend to accentuate certain traits, to attract and develop their kind of individuals, and so to become further differentiated.

The division of labor in the city likewise illustrates disorganization, reorganization, and increasing differentiation. The immigrant from rural communities in Europe and America seldom brings with him economic skill of any great value in our industrial, commercial, or professional life. Yet interesting occupational selection has taken place by nationality, explainable more by racial temperament or circumstance than by old-world economic background, as Irish policemen, Greek ice-cream parlors, Chinese laundries, Negro porters, Belgian janitors, etc.

The facts that in Chicago one million (996,589) individuals gainfully employed reported 509 occupations, and that over 1,000 men and women in *Who's Who* gave 116 different vocations, give some notion of how in the city the minute differentiation of occupation "analyzes and sifts the population, separating and classifying the diverse elements." These figures also afford some intimation of the complexity and complication of the modern industrial mechanism and the intricate segregation and isolation of divergent economic groups. Interrelated with this economic division of labor is a corresponding division into social classes and into cultural and recreational groups. From this multiplicity of groups, with their different patterns of life, the person finds his congenial social world and – what is not feasible in the narrow confines of a village – may move and live in widely separated, and perchance conflicting, worlds. Personal disorganization may be but the failure to harmonize the canons of conduct of two divergent groups.

If the phenomena of expansion and metabolism indicate that a moderate degree of disorganization may and does facilitate social organization, they indicate as well that rapid urban expansion is accompanied by excessive increases in disease, crime, disorder, vice, insanity, and suicide, rough indexes of social disorganization. But what are the indexes of the causes, rather than of the effects, of the disordered social metabolism of the city? The excess of the actual over the natural increase of population has already been suggested as a criterion. The significance of this increase consists in the immigration into a metropolitan city like New York and Chicago of tens of thousands of persons annually. Their invasion of the city has the effect of a tidal wave inundating first the immigrant colonies, the ports of first entry, dislodging thousands of inhabitants who overflow into the next zone, and so on and on until the momentum of the wave has spent its force on the last urban zone. The whole effect is to speed up expansion, to speed up industry, to speed up the "junking" process in the area of deterioration (II). These internal movements of the population become the more significant for study. What movement is going on in the city, and how may this movement be measured? It is easier, of course, to classify movement within the city than to measure it. There is the movement from residence to residence, change of occupation, labor turnover, movement to and from work, movement for recreation and adventure. This leads to the question: What is the significant aspect of movement for the study of the changes in city life? The answer to this question leads directly to the important distinction between movement and mobility.

MOBILITY AS THE PULSE OF THE COMMUNITY

Movement, per se, is not an evidence of change or of growth. In fact, movement may be a fixed and unchanging order of motion, designed to control a constant situation, as in routine movement. Movement that is significant for growth implies a change of movement in response to a new stimulus or situation. Change of movement of this type is called *mobility*. Movement of the nature of routine finds its typical expression in work. Change of movement, or mobility, is characteristically expressed in adventure. The great city, with its "bright lights," its emporiums of novelties and bargains, its palaces of amusement, its underworld of vice and crime, its risks of life and property from accident, robbery, and homicide, has become the region of the most intense degree of adventure and danger, excitement and thrill.

Mobility, it is evident, involves change, new experience, stimulation. Stimulation induces a response of the person to those objects in his environment which afford expression for his wishes. For the person, as for the physical organism, stimulation is essential to growth. Response to stimulation is wholesome so long as it is a correlated *integral* reaction of the entire personality. When the reaction is *segmental*, that is, detached from, and uncontrolled by, the organization of personality, it tends to become disorganizing or pathological. That is why stimulation for the sake of stimulation, as in the restless pursuit of pleasure, partakes of the nature of vice.

The mobility of city life, with its increase in the number and intensity of stimulations, tends inevitably to confuse and to demoralize the person. For an essential element in the mores and in personal morality is consistency, consistency of the type that is natural in the social control of the primary group. Where mobility is the greatest, and where in consequence primary controls break down completely, as in the zone of deterioration in the modern city, there develop areas of demoralization, of promiscuity, and of vice.

In our studies of the city it is found that areas of mobility are also the regions in which are found juvenile delinquency, boys' gangs, crime, poverty, wife desertion, divorce, abandoned infants, vice.

These concrete situations show why mobility is perhaps the best index of the state of metabolism of the city. Mobility may be thought of in more than a fanciful sense, as the "pulse of the community." Like the pulse of the human body, it is a process which reflects and is indicative of all the changes that are taking place in the community, and which is susceptible of analysis into elements which may be stated numerically.

The elements entering into mobility may be classified under two main heads: (1) the state of mutability of the person, and (2) the number and kind of contacts or stimulations in his environment. The mutability of city populations varies with sex and age composition, the degree of detachment of the person from the family and from other groups. All these factors may be expressed numerically. The new stimulations to which a population responds can be measured in terms of change of movement or of increasing contacts. Statistics on the movement of urban population may only measure routine, but an increase at a higher ratio than the increase of population measures mobility. In 1860 the horse-car lines of New York City carried about 50,000,000 passengers; in 1890 the trolley cars (and a few surviving horse-cars) transported about 500,000,000; in 1921, the elevated, subway, surface, and electric and steam suburban lines carried a total of more than 2,500,000,000 passengers. In Chicago the total annual rides per capita on the surface and elevated lines were 164 in 1890; 215 in 1900; 320 in 1910; and 338 in 1921. In addition, the rides per capita on steam and electric suburban lines almost doubled between 1916 (23) and 1921 (41), and the increasing use of the automobile must not be overlooked. For example, the number of automobiles in Illinois increased from 131,140 in 1915 to 833,920 in 1923.

Mobility may be measured not only by these changes of movement, but also by increase of contacts. While the increase of population of Chicago in 1912–22 was less than 25 per cent (23.6 per cent), the increase of letters delivered to Chicagoans was double that (49.6 per cent) – (from 693,084,196 to 1,038,007,854). In 1912 New York had 8.8 telephones; in 1922, 16.9 per 100 inhabitants. Boston had, in 1912, 10.1 telephones; ten years later, 19.5 telephones per 100 inhabitants. In the same decade the figures for Chicago increased from 12.3 to 21.6 per 100 population. But increase of the use of the telephone is probably more significant than increase in the number of telephones. The number of telephone calls in Chicago increased from 606,131,928 in 1914 to 944,010,586 in 1922, an increase of 55.7 per cent, while the population increased only 13.4 per cent.

Land values, since they reflect movement, afford one of the most sensitive indexes of mobility. The highest land values in Chicago are at the point of greatest mobility in the city, at the corner of State and Madison Streets, in the Loop. A traffic count showed that at the rush period 31,000 people an hour, or 210,000 men and women in sixteen and one-half hours, passed the southwest corner. For over ten years land values in the Loop have been stationary, but in the same time they have doubled, quadrupled, and even sextupled in the strategic corners of the "satellite loops," an accurate index of the changes which have occurred. Our investigations so far seem to indicate that variations in land values, especially where correlated with differences in rents, offer perhaps the best single measure of mobility, and so of all the changes taking place in the expansion and growth of the city.

In general outline, I have attempted to present the point of view and methods of investigation which the department of sociology is employing in its studies in the growth of the city, namely, to describe urban expansion in terms of extension, succession, and concentration; to determine how expansion disturbs metabolism when disorganization is in excess of organization; and, finally, to define mobility and to propose it as a measure both of expansion and metabolism, susceptible to precise quantitative formulation, so that it may be regarded almost literally as the pulse of the community. In a way, this statement might serve as an introduction to any one of five or six research projects under way in the department: The project, however, in which I am directly engaged is an attempt to apply these methods of investigation to a cross-section of the city – to put this area, as it were, under the microscope, and so to study in more detail and with greater control and precision the processes which have been described here in the large. For this purpose the West Side Jewish community has been selected. This community includes the so-called "Ghetto," or area of first settlement, and Lawndale, the so-called "Deutschland," or area of second settlement. This area has certain obvious advantages for this study, from the standpoint of expansion, metabolism, and mobility. It exemplifies the tendency to expansion radially from the business center of the city. It is now relatively a homogeneous cultural group. Lawndale is itself an area in flux, with the tide of migrants still flowing in from the Ghetto and a constant egress to more desirable regions of the residential zone. In this area, too, it is also possible to study how the expected outcome of this high rate of mobility in social and personal disorganization is counteracted in large measure by the efficient communal organization of the Jewish community.

NOTE

1 For a study of this cultural area of city life see Nels Anderson (1923) *The Hobo*. Chicago: University of Chicago Press.

"The Urban Process under Capitalism: A Framework for Analysis"

from *International Journal of Urban and Regional Research* (1978)

David Harvey

Editors' Introduction

In this selection, British geographer David Harvey applies ideas from Karl Marx's *Capital*, volumes two and three, to the analysis of the urban process under capitalism. In the wake of the global oil shock of 1973, he opens up our understanding of recurring conditions of contradiction and crisis in capitalism, and their relation to tendencies of overaccumulation and capital switching into various circuits of investment. He explains how the expropriation of surplus value and the subsuming of productivity to profitability under capitalism eventually results in two kinds of social contradiction. One is the conflict and overproduction that results from the competition between capitalists, and the second is the conflict and violence that are the outcome of capitalist exploitation of labor.

Harvey further expands on how the exploitation of labor and overproduction of commodities contribute to a falling rate of profit. The problem of overaccumulation in the *primary circuit* of capital in production of manufactured goods can be resolved through capital reinvestment in other outlets. This surplus capital can be reinvested in fixed capital machinery applied to the production process, or switched into the *secondary circuit* of the built environment for production such as infrastructure. The secondary circuit also includes commodities for worker consumption, as well as the built environment for consumption (which includes worker housing and infrastructure necessary for the reproduction of labor). The money and banking system, or *fictional capital* sector that is controlled by financial and state institutions, functions as a "nerve center" that mediates and regulates capital flows between the circuits. Capital can also be switched into the *tertiary circuit* of investments in science and technology.

The tendency towards overaccumulation also besets secondary and tertiary circuits of capital, thereby creating additional opportunities for shocks and contradictions to occur. While state planning can help to ensure that investments in other sectors are potentially productive, there is a pervasive tendency towards chronic overinvestment in the secondary and tertiary sectors, reflecting again that capitalist investment is guided by the needs of capital rather than by the real needs of people. A variety of crises can result, including financial, credit, and mortgage crises, the fiscal crisis of the state, etc. Crises can furthermore be classified into: a) partial crises affecting subregions or subsectors, b) geographical switching crises such as from developed to developing world, and c) global crises, such as the 1973 oil shock and the current global economic crisis that stemmed in large part from the collapse of the speculative subprime mortgage market.

Harvey also discusses how labor struggles may provoke capitalist investment or switching into sectors such as housing or other consumption amenities. Collective action by labor in the early twentieth century, for instance, created pressure for state investments in sectors such as mortgage financing and infrastructure for the workforce. At the same time, circuits of secondary investment such as housing are subject to new tendencies towards overproduction and crisis.

Kevin Fox Gotham has recently applied these same Marxian concepts, regarding capitalist crises and their relation to the primary and secondary circuits of capital, to an interpretation of the subprime mortgage crisis. See, e.g., "The Secondary Circuit of Capital Reconsidered: Globalization and the U.S. Real Estate Sector," *American Journal of Sociology* 112, 1 (July 2006): 231–275, and "Creating Liquidity out of Spatial Fixity: The Secondary Circuit of Capital and the Subprime Mortgage Crisis," *International Journal of Urban and Regional Research* 33, 2 (June 2009): 355–371.

David Harvey taught at Johns Hopkins University in Baltimore, Maryland, for many years and currently is Distinguished Professor of Geography and Anthropology at the Graduate Center of the City of New York. He has published widely in the area of urban geography. Some representative book titles include *Social Justice and the City* (London: Edward Arnold, 1973), *The Condition of Postmodernity* (Oxford: Blackwell, 1989), and *Spaces of Global Capitalism: Towards a Theory of Uneven Geographical Development* (London: Verso, 2006).

My objective is to understand the urban process under capitalism. I confine myself to the capitalist forms of urbanization because I accept the idea that the "urban" has a specific meaning under the capitalist mode of production which cannot be carried over without a radical transformation of meaning (and of reality) into other social contexts.

Within the framework of capitalism, I hang my interpretation of the urban process on the twin themes of *accumulation* and *class struggle*. The two themes are integral to each other and have to be regarded as different sides of the same coin – different windows from which to view the totality of capitalist activity. The class character of capitalist society means the domination of labor by capital. Put more concretely, a class of capitalists is in command of the work process and organizes that process for the purposes of producing profit. The laborer, on the other hand, has command only over his or her labor power, which must be sold as a commodity on the market. The domination arises because the laborer must yield the capitalist a profit (surplus value) in return for a living wage. All of this is extremely simplistic, of course, and actual class relations (and relations between factions of classes) within an actual system of production (comprising production, services, necessary costs of circulation, distribution, exchange, etc.) are highly complex. The essential Marxian insight, however, is that profit arises out of the domination of labor by capital and that the capitalists as a class must, if they are to reproduce themselves, continuously expand the basis for profit. We thus arrive at a conception of a society founded on the principle of "accumulation for accumulation's sake, production for production's sake." The theory of accumulation which Marx constructs in *Capital* amounts to a careful enquiry into the dynamics of accumulation and an exploration of its contradictory character. This may sound rather "economistic" as a framework for analysis, but we have to recall that accumulation is the means whereby the capitalist class reproduces both itself and its domination over labor. Accumulation cannot, therefore, be isolated from class struggle.

THE CONTRADICTIONS OF CAPITALISM

We can spin a whole web of arguments concerning the urban process out of an analysis of the contradictions of capitalism. Let me set out the principal forms these contradictions take.

Consider, first, the contradiction which lies within the capitalist class itself. In the realm of exchange each capitalist operates in a world of individualism, freedom and equality and can and must act spontaneously and creatively. Through competition, however, the inherent laws of capitalist production are asserted as "external coercive laws having power over every individual capitalist." A world of individuality and freedom on the surface conceals a world of conformity and coercion underneath. But the translation from individual action to behavior according to class norms is neither complete nor perfect – it never can be because the *process* of exchange under capitalist rules always presumes individuality while the law of value always asserts itself in social terms. As a consequence, individual capitalists, each acting in their own immediate self-interest, can produce an aggregative result which is wholly antagonistic to their collective class interest. To take a rather dramatic example, competition may force each capitalist to so lengthen and intensify the work process that the capacity of the labor force to produce surplus value is seriously impaired. The collective effects of individual entrepreneurial activity can seriously endanger the social basis for future accumulation.

Consider, secondly, the implications of accumulation for the laborers. We know from the theory of surplus value that the exploitation of labor power is the source of capitalist profit. The capitalist form of accumulation therefore rests upon a certain violence which the capitalist class inflicts upon labor. Marx showed, however, that this appropriation could be worked out in such a way that it did not offend the rules of equality, individuality and freedom as they must prevail in the realms of exchange. Laborers, like the capitalists, "freely" trade the commodity they have for sale in the market place. But laborers are also in competition with each other for employment while the work process is under the command of the capitalist. Under conditions of unbridled competition, the capitalists are forced willy-nilly into inflicting greater and greater violence upon those whom they employ. The individual laborer is powerless to resist this onslaught. The only solution is for the laborers to constitute themselves as a class and find collective means to resist the depredations of capital. The capitalist form of accumulation consequently calls into being overt and explicit class struggle between labor and capital. This contradiction between the classes explains much of the dynamic of capitalist history and is in many respects quite fundamental to understanding the accumulation process.

The two forms of contradiction are integral to each other. They express an underlying unity and are to be construed as different aspects of the same reality. Yet we can usefully separate them in certain respects. The internal contradiction within the capitalist class is rather different from the class confrontation between capital and labor, no matter how closely the two may be linked. In what follows I will focus on the accumulation process in the absence of any overt response on the part of the working class to the violence which the capitalist class must necessarily inflict upon it. I will then broaden the perspective and consider how the organization of the working class and its capacity to mount an overt class response affects the urban process under capitalism.

Various other forms of contradiction could enter in to supplement the analysis. For example, the capitalist production system often exists in an antagonistic relationship to non- or pre-capitalist sectors which may exist within (the domestic economy, peasant and artisan production sectors, etc.) or without it (pre-capitalist societies, socialist countries, etc.). We should also note the contradiction with "nature" which inevitably arises out of the relation between the dynamics of accumulation and the "natural" resource base as it is defined in capitalist terms. Lack of space precludes any examination of these matters here. But they would obviously have to be taken into account in any analysis of the history of urbanization under capitalism.

THE LAWS OF ACCUMULATION

[. . .]

The primary circuit of capital

In volume one of *Capital*, Marx presents an analysis of the capitalist production process. The drive to create surplus value rests either on an increase in the length of the working day (absolute surplus value) or on the gains to be made from continuous revolutions in the "productive forces" through reorganizations of the work process which raise the productivity of labor power (relative surplus value). The capitalist captures relative surplus value from the organization of cooperation and division of labor within the work process or by the application of fixed capital (machinery). The motor for these continuous revolutions in the work process, for the rising productivity of labor, lies in capitalist competition as each capitalist seeks an excess profit by adopting a superior production technique to the social average.

The implications of all of this for labor are explored in a chapter entitled "The general law of capitalist accumulation." Marx here examines alterations in the rate of exploitation and in the temporal rhythm of changes in the work process in relation to the supply conditions of labor power (in particular, the formation of an industrial reserve army), assuming, all the while, that a positive rate of accumulation must be sustained if the capitalist class is to reproduce itself. The analysis proceeds around a strictly circumscribed set of interactions with all other problems assumed away or held constant . . . The second volume of *Capital* closes with a "model" of accumulation on an expanded scale. The problems of proportionality involved in the aggregative production of means of production and means of consumption are examined with all other problems held constant (including technological change, investment in fixed capital, etc.). The objective here is to show the potential for crises of disproportionality within the production process. But Marx has now broadened the structure of relationships put under the microscope. We note, however, that in both cases Marx assumes, tacitly, that all commodities are produced and consumed within one time period. The structure of relations can be characterized as the *primary circuit of capital*.

Much of the analysis of the falling rate of profit and its countervailing tendencies in volume three similarly presupposes production and consumption within one time period although there is some evidence that Marx intended to broaden the scope of this if he had lived to complete the work. But it is useful to consider the volume three analysis as a synthesis of the arguments presented in the first two volumes and as at the very least a cogent statement of the internal contradictions which exist within the primary circuit. Here we can clearly see the contradictions which arise out of the tendency for individual capitalists to act in a way which, when aggregated, runs counter to their own class interest. This contradiction produces a tendency to *overaccumulation* – too much capital is produced in aggregate relative to the opportunities to employ that capital. This tendency is manifest in a variety of guises. We have:

1 Overproduction of commodities – a glut on the market.
2 Falling rates of profit (in pricing terms, to be distinguished from the falling rate of profit in value terms which is a theoretical construct).
3 Surplus capital which can be manifest either as idle productive capacity or as money capital lacking opportunities for profitable employment.
4 Surplus labor and/or rising rate of exploitation of labor power. One or a combination of these manifestations may be present at the same time. We have here a preliminary framework for the analysis of capitalist crises.

The secondary circuit of capital

We now drop the tacit assumption of production and consumption within one time period and consider the problems posed by production and use of commodities requiring different working periods, circulation periods, and the like. This is an extraordinarily complex problem which Marx addresses to some degree in volume two of *Capital* and in the *Grundrisse*. I cannot do justice to it here so I will confine myself to some remarks regarding the formation of *fixed capital* and the *consumption fund*. Fixed capital, Marx argues, requires special analysis because of certain peculiarities which attach to its mode of production and realization. These peculiarities arise because fixed capital items can be produced in the normal course of capitalist commodity production but they are used as aids to the production process rather than as direct raw material inputs. They are also used over a relatively long time period. We can also usefully distinguish between fixed capital enclosed within the production process and fixed capital which functions as a physical framework for production. The latter I will call the *built environment for production*.

On the consumption side, we have a parallel structure. A *consumption fund* is formed out of commodities which function as aids rather than as direct inputs to consumption. Some items are directly enclosed within the consumption process (consumer durables, such as cookers, washing machines, etc.) while others act as a physical framework for consumption (houses, sidewalks, etc.) – the latter I will call the *built environment for consumption*. We should note that some items in the built environment function jointly for both production and consumption – the transport network, for example – and that items can be transferred from one category to another by changes in use. Also, fixed capital in the built environment is immobile in space in the sense that the value incorporated in it cannot be moved without being destroyed. Investment in the built environment therefore entails the creation of a whole physical landscape for purposes of production, circulation, exchange and consumption.

We will call the capital flows into fixed asset and consumption fund formation the *secondary circuit of capital*. Consider, now, the manner in which such flows can occur. There must obviously be a "surplus" of both capital and labor in relation to current production and consumption needs in order to facilitate the movement of capital into the formation of long-term assets, particularly those comprising the built environment. The tendency towards overaccumulation produces such conditions within the primary circuit on a periodic basis. One feasible if *temporary* solution to this overaccumulation problem would therefore be to switch capital flows into the secondary circuit.

Individual capitalists will often find it difficult to bring about such a switch in flows for a variety of reasons. The barriers to individual switching of capital are particularly acute with respect to the built environment where investments tend to be large-scale and long-lasting, often difficult to price in the ordinary way and in many cases open to collective use by all individual capitalists. Indeed, individual capitalists left to themselves will tend to under-supply their own collective needs for production precisely because of such barriers. Individual capitalists tend to overaccumulate in the primary circuit and to under-invest in the secondary circuit; they have considerable difficulty in organizing a balanced flow of capital between the primary and secondary circuits.

A general condition for the flow of capital into the secondary circuit is, therefore, the existence of a functioning capital market and, perhaps, a state willing to finance and guarantee long-term, large-scale projects with respect to the creation of the built environment. At times of overaccumulation, a switch of flows from the primary to the secondary circuit can be accomplished only if the various manifestations of overaccumulation can be transformed into money capital which can move freely and unhindered into these forms of investment. This switch of resources cannot be accomplished without a money supply and credit system which creates "fictional capital" *in advance* of actual production and consumption. This applies as much to the consumption fund (hence the importance of consumer credit, housing mortgages, municipal debt) as it does to fixed capital. Since the production of money and credit are relatively autonomous processes, we have to conceive of the financial and state institutions controlling them as a kind of collective nerve centre governing and *mediating* the relations between the primary and secondary circuits of capital. The nature and form of these financial and state institutions and the policies they adopt can play an important role in checking or enhancing flows of capital into the secondary circuit of capital or into certain specific aspects of it (such as transportation, housing, public facilities, and so on). An alteration in these mediating structures can therefore affect both the volume and direction of the capital flows by constricting movement down some channels and opening up new conduits elsewhere.

The tertiary circuit of capital

In order to complete the picture of the circulation of capital in general, we have to conceive of a *tertiary circuit of capital* which comprises, first, investment in science and technology (the purpose of which is to harness science to production and thereby to contribute to the processes which continuously revolutionize the productive forces in society) and second, a wide range of social expenditures which relate primarily to the processes of reproduction of labor power. The latter can usefully be divided into investments directed towards the qualitative improvement of labor power from the standpoint of capital (investment in education and health by means of which the capacity of the laborers to engage in the work process will be enhanced) and investment in cooptation, integration and repression of the labor force by ideological, military and other means. Individual capitalists find it hard to make such investments as individuals, no matter how desirable they may regard them. Once again, the capitalists are forced to some degree to constitute themselves as a class usually through the agency of the state – and thereby to find ways to channel investment into research and development and into the quantitative and qualitative improvement of labor power. We should recognize that capitalists often *need* to make such investments in order to fashion an adequate social basis for further accumulation. But with regard to social expenditures, the investment flows are very strongly affected by the state of class struggle. The amount of investment in repression and in ideological control is directly related to the threat of organized working-class resistance to the depredations of capital. And the need to coopt labor arises only when the working class has accumulated sufficient power to require cooptation. Since the state can become a field of active class struggle, the mediations which are accomplished by no means fit exactly with the requirements of the capitalist class. The role of the state requires careful theoretical and historical elaboration in relation to the organization of capital flows into the tertiary circuit.

THE CIRCULATION OF CAPITAL AS A WHOLE AND ITS CONTRADICTIONS

[. . .]

We have already seen how the contradictions internal to the capitalist class generate a tendency towards overaccumulation within the primary circuit of capital. And we have argued that this tendency can be overcome temporarily at least by switching capital into the secondary or tertiary circuits. Capital has, therefore, a variety of investment options open to it – fixed capital or consumption fund formation, investment in science and technology, investment in "human capital" or outright repression. At particular historical conjunctures capitalists may not be capable of taking up all of these options with equal vigor, depending upon the degree of their own organization, the institutions which they have created and the objective possibilities dictated by the state of production and the state of class struggle. I shall assume away such problems for the moment in order to concentrate on how the tendency towards overaccumulation, which we have identified so far only with respect to the primary circuit, manifests itself within the overall structure of circulation of capital. To do this we first need to specify a concept of productivity of investment.

On the productivity of investments in the secondary and tertiary circuits

I choose the concept of "productivity" rather than "profitability" for a variety of reasons. First of all, the rate of profit as Marx treats of it in volume three of *Capital* is measured in value rather than pricing terms and takes no account of the distribution of the surplus value into its component parts of interest on money capital, profit on productive capital, rent on land, profit on merchant's capital, etc. The rate of profit is regarded as a social average earned by individual capitalists in all sectors and it is assumed that competition effectively ensures its equalization. This is hardly a suitable conception for examining the flows between the three circuits of capital. To begin with, the formation of fixed capital in the built environment – particularly the collective means of production – cannot be understood without understanding the formation of a capital market and the distribution of part of the surplus in the form of interest. Second, many of the commodities produced in relation to the secondary and tertiary circuits cannot be priced in the ordinary way, while collective action by way of the state cannot be examined in terms of the normal criteria of profitability. Third, the rate of profit which holds is perfectly appropriate for understanding the behaviors of individual capitalists in competition, but cannot be translated into a concept suitable for examining the behavior of capitalists as a class without some major assumptions (treating the total profit as equal to the total surplus value, for example).

The concept of productivity helps to by-pass some of these problems if we specify it carefully enough. For the fact is that capitalists as a class – often through the agency of the state – do invest in the production of conditions which they hope will be favorable to accumulation, to their own reproduction as a class and to their continuing domination over labor. This leads us immediately to a definition of a productive investment as one which directly or indirectly expands the basis for the production of surplus value. Plainly, investments in the secondary and tertiary circuits have the *potential* under certain conditions to do this. The problem – which besets the capitalists as much as it confuses us – is to identify the conditions and means which will allow this potential to be realized.

Investment in new machinery is the easiest case to consider. The new machinery is directly productive if it expands the basis for producing surplus value and unproductive if these benefits fail to materialize. Similarly, investment in science and technology may or may not produce new forms of scientific knowledge which can be applied to expand accumulation. But what of investment in roads, housing, health care and education, police forces and the military, and so on? If workers are being recalcitrant in the work place, then judicious investment by the capitalist class in a police force, to intimidate the workers and to break their collective power, may indeed be productive indirectly of surplus value for the capitalists. If, on the other hand, the police are employed to protect the bourgeoisie in the conspicuous consumption of their revenues in callous disregard of the poverty and misery which surrounds them, then the police are not acting to facilitate accumulation. The distinction may be fine but it demonstrates the dilemma. How can the capitalist class identify, with reasonable precision, the opportunities for indirectly and directly productive investment in the secondary and tertiary circuits of capital?

The main thrust of the modern commitment to planning (whether at the state or corporate level) rests on the idea that certain forms of investment in the secondary and tertiary circuits are potentially productive. The whole apparatus of cost–benefit analysis and of programming and budgeting, of analysis of social benefits, as well as notions regarding investment in human capital, express this commitment and testify to the complexity of the problem. And at the back of all of this is the difficulty of determining an appropriate basis for decision-making in the absence of clear and unequivocal profit signals. Yet the cost of bad investment decisions – investments which do not contribute directly or indirectly to accumulation of capital – must emerge somewhere. They must, as Marx would put it, come to the surface and thereby indicate the errors which lie beneath. We can begin to grapple with this question by considering the origins of crises within the capitalist mode of production.

On the forms of crisis under capitalism

Crises are the real manifestation of the underlying contradictions within the capitalist process of accumulation. The argument which Marx puts forward throughout much of *Capital* is that there is always the potential within capitalism to achieve "balanced growth" but that this potentiality can never be realized because of the structure of the social relations prevailing in a capitalist society. This structure leads individual capitalists to produce results collectively which are antagonistic to their own class interest and leads them also to inflict an insupportable violence upon the working class which is bound to elicit its own response in the field of overt class struggle. We have already seen how the capitalists tend to generate states of overaccumulation within the primary circuit of capital and considered the various manifestations which result. As the pressure builds, either the accumulation process grinds to a halt or new investment opportunities are found as capital flows down various channels into the secondary and tertiary circuits. This movement may start as a trickle and become a flood as the potential for expanding the production of surplus value by such means becomes apparent. But the tendency towards overaccumulation is not eliminated. It is transformed, rather, into a pervasive tendency towards over-investment in the secondary and tertiary circuits. This over-investment, we should stress, is in relation solely to the needs of capital and has nothing to do with the real needs of people, which inevitably remain unfulfilled. Manifestations of crisis thus appear in both the secondary and tertiary circuits of capital.

As regards fixed capital and the consumption fund, the crisis takes the form of a crisis in the valuation of assets. Chronic overproduction results in the devaluation of fixed capital and consumption fund items – a process which affects the built environment as well as producer and consumer durables . . . In each case the crisis occurs because the potentiality for productive investment within each of these spheres is exhausted. Further flows of capital do not expand the basis for the production of surplus value. We should also note that a crisis of any magnitude in any of these spheres is automatically registered as a crisis within the financial and state structures while the latter, because of the relative autonomy which attaches to them, can be an independent source of crisis (we can thus speak of financial, credit and monetary crises, the fiscal crises of the state, and so on).

Crises are the "irrational rationalizers" within the capitalist mode of production. They are indicators of imbalance and force a rationalization (which may be painful for certain sectors of the capitalist class as well as for labor) of the processes of production, exchange, distribution and consumption. They may also force a rationalization of institutional structures (financial and state institutions in particular). From the standpoint of the total structure of relationships we have portrayed, we can distinguish different kinds of crises:

a) *Partial crises* which affect a particular sector, geographical region or set of mediating institutions. These can arise for any number of reasons but are potentially capable of being resolved within that sector, region or set of institutions. We can witness autonomously forming monetary crises, for example, which can be resolved by institutional reforms, crises in the formation of the built environment which can be resolved by reorganization of production for that sector, etc.
b) *Switching crises* which involve a major reorganization and restructuring of capital flows and/or a major restructuring of mediating institutions in order to open up new channels for productive investments. It is useful to distinguish between two kinds of switching crises:
   1 *Sectoral switching crises* which entail switching the allocation of capital from one sphere (e.g. fixed capital formation) to another (e.g. education);
   2 *Geographical switching crises* which involve switching the flows of capital from one place to another. We note here that this form of crisis is particularly important in relation to investment in the built environment because the latter is immobile in space and requires interregional or international flows of money capital to facilitate its production.
c) *Global crises* which affect, to greater or lesser degree, all sectors, spheres and regions within the capitalist production system. We will thus see devaluations of fixed capital and the consumption fund, a crisis in science and technology, a fiscal crisis in state expenditures, a crisis in the productivity of labor, all manifest at the same time across all or most regions within the capitalist system. I note, in passing, that there have been only two global crises within the totality of the capitalist system – the first during the 1930s and its Second World War aftermath; the second, that which became most evident after 1973 but which had been steadily building throughout the 1960s.

A complete theory of capitalist crises should show how these various forms and manifestations relate in both space and time. Such a task is beyond the scope of a short article, but we can shed some light by returning to our fundamental theme – that of understanding the urban process under capitalism.

CLASS STRUGGLE, ACCUMULATION AND THE URBAN PROCESS UNDER CAPITALISM

[. . .]

The strategies of dispersal, community improvement and community competition, arising as they do out of the bourgeois response to class antagonisms, are fundamental to understanding the material history of the urban process under capitalism. And they are not without their implications for the circulation of capital either. The direct victories and concessions won by the working class have their impacts. But at this point we come back to the principles of accumulation, because if the capitalist class is to reproduce itself and its domination over labor it must effectively render whatever concessions labor wins from it consistent with the rules governing the productivity of investments under capitalist accumulation. Investments may switch from one sphere to another in response to class struggle to the degree that the rules for the accumulation of capital are observed. Investment in working-class housing or in a national health service can thus be transformed into a vehicle for accumulation via commodity production for these sectors. Class struggle can, then, provoke "switching crises," the outcome of which can change the structure of investment flows to the advantage of the working class. But those demands which lie within the economic possibilities of accumulation as a whole can in the end be conceded by the capitalist class without loss. Only when class struggle pushes the system beyond its own internal potentialities is the accumulation of capital and the reproduction of the capitalist class called into question. How the bourgeoisie responds to such a situation depends on the possibilities open to it. For example, if capital can switch geographically to pastures where the working class is more compliant, then it may seek to escape the consequences of heightened class struggle in this way. Otherwise it must invest in economic, political and physical repression or simply fall before the working-class onslaught.

"The City as a Growth Machine"

from *Urban Fortunes: The Political Economy of Place* (1987)

John Logan and Harvey Molotch

Editors' Introduction

The concept of the "growth machine" was initially formulated by Harvey Molotch as an outgrowth of his denunciation of the environmentally destructive effects of a massive 1969 oil tanker spill off the beautiful coastline of Santa Barbara, California. The spill was considered by many to be a watershed for the national environmental movement, and an even more irrevocable turning point for California. Molotch contributed to the national debate with a 1969 article in *Ramparts* magazine titled "Oil in the Velvet Playground." In this and other articles he expressed the outrage of local businesses and residents, who perceived that they gained little wealth or tax benefits from the oil companies working in offshore federal waters. The water and air pollution from the drilling operations furthermore hampered the tourism industry, another major component of the regional economic base. In the ensuing years, Molotch would continue to reflect on the damaging environmental and social consequences of American capitalism and urbanization, and move towards a more generalized urban political economy that understands cities as growth machines that serve elite interests, promote social inequality, and harm the environment. He eventually collaborated with John Logan and they published *Urban Fortunes: The Political Economy of Place* in 1987.

Logan and Molotch reject the human ecology view that places and land markets are the natural outcome of Darwinian and market processes in favor of a Marxist-influenced political economy perspective. They comprehend homes not just as places that are "lived," but also as commodities within real estate markets that can be bought and exchanged, generating use and exchange values for producers and consumers. They see prices and markets, importantly, as *social phenomena*, governed not by natural laws of competition, supply, and demand, but by inequalities of wealth, ownership, and power. Places organize and distribute life chances in a class stratification system. As commodities, places also acquire special use and exchange values. In terms of use value, homes, neighborhoods, local businesses, localities, and other places obtain a special preciousness for people characterized by intense sentiment, commitment, or attachment. In terms of exchange values, places are idiosyncratic and not substitutable in the fashion of other commodities such as cars, clothes, and food, because real estate is a more limited and finite resource. Because property markets are structured by access to infrastructure such as jobs, housing, transportation, schools, hospitals, and other resources, places determine life chances and are key components of the American class stratification system.

Place entrepreneurs such as landlords, businessmen, developers, transportation and utility companies, banks, and corporations gain profit from their control of land and from the proceeds of economic growth. There are serendipitous, active, and structural entrepreneurs, with varying levels of access to capital as well as place attachment. Place entrepreneurs form pro-growth coalitions with governmental units and other economic interests to focus infrastructure and urban development in areas that intensify the profitability of their own interests. They promote a good business climate and an ideology of growth as a public good through their influence on politicians and the media. They foster a booster spirit through partnerships with schools and civic organizations with essay contests, public celebrations and spectacles such as dedications, soapbox derbies, parade floats, and beauty contests. Growth machines have evolved from small groups of local power brokers to include a more multifaceted matrix of auxiliary interests and institutions that include universities, museums, convention centers, sports franchises, entertainment conglomerates, and tourism interests such as theme parks. The growth machine works to suppress or deflect public consciousness from the negative social and environmental consequences of urban development. Citizens' movements have sprung up to contest the negative externalities of growth through protests, lobbying, public hearings, and environmental impact reviews.

The political scientist John Mollenkopf offers a similar concept of "growth coalitions" in his 1983 book, *The Contested City*. Joe Feagin offers an excellent historical and empirical case study of urban political economy in his 1988 book, *Free Enterprise City*. He charts the succession of the Houston growth elite from the days of the "Suite 8F crowd" to the global corporate interests of the late twentieth century. Andrew Kirby and A. Karen Lynch examine the negative social consequences of rapid growth and urban sprawl in Houston in "A Ghost in the Growth Machine: The Aftermath of Rapid Population Growth in Houston," *Urban Studies* 24 (1987): 587–596. Mark Gottdiener and Joe Feagin "The Paradigm Shift in Urban Sociology," *Urban Affairs Quarterly* 24, 2 (1987): 163–187, offer another useful articulation of the new urban sociology. John Walton published a useful historical perspective on urban political economy in "Urban Sociology: The Contribution and Limits of Political Economy," *Annual Review of Sociology* 19 (1993): 301–320.

Harvey L. Molotch obtained his Ph.D. in Sociology from the University of Chicago in 1968. He was on the faculty of the University of California at Santa Barbara from 1967 to 2003, and also served as Centennial Professor at the London School of Economics in 1998–99. He is now Professor of Metropolitan Studies and Sociology at New York University. In 2003, Molotch won the Robert and Helen Lynd Award for lifetime career contribution from the Community and Urban Section of the American Sociological Association. His latest book is *Where Stuff Comes From: How Toasters, Toilets, Cars, Computers and Many Other Things Come to Be as They Are* (New York and London: Routledge, 2003).

John Logan received his Ph.D. from the University of California, Berkeley, in 1974. He was on the faculty of the State University of New York, Stony Brook from 1972 to 1980; then he moved to the State University of New York, Albany, where he is now Distinguished Professor in the Department of Sociology and Department of Public Administration and Policy. He is also Director of the Lewis Mumford Center for Comparative Urban and Regional Research. He served as the Chair of the Community and Urban Sociology Section of the American Sociological Association in 1993–94. Logan has published hundreds of articles in the areas of urban sociology, race and ethnicity, political sociology, immigration, family, aging/gerontology, and social movements.

Urban Fortunes won the 1988 Robert Park Award for best book on community and urban sociology from the American Sociological Association (ASA). It also won the 1990 Distinguished Scholarly Publication Award of the ASA. *Urban Fortunes: The Political Economy of Place* is one of the best articulations of urban political economy, sometimes described as the new urban sociology.

THE SOCIAL CONSTRUCTION OF CITIES

The earth below, the roof above, and the walls around make up a special sort of commodity: a place to be bought and sold, rented and leased, as well as used for making a life. At least in the United States, this is the standing of place in legal statutes and in ordinary people's imaginations. Places can (and should) be the basis not only for carrying on a life but also for exchange in a market. We consider this commodification of place fundamental to urban life and necessary in any urban analysis of market societies.

Yet in contrast to the way neoclassical economists (and their followers in sociology) have undertaken the task of understanding the property commodity, we focus on how markets work as social phenomena. Markets are not mere meetings between producers and consumers, whose relations are ordered by the impersonal "laws" of supply and demand. For us, the fundamental attributes of all commodities, but particularly of land and buildings, are the social contexts through which they are used and exchanged. Any given piece of real estate has both a use value and an exchange value.

[. . .]

PLACES AS COMMODITIES

For us, as for many of our intellectual predecessors, the market in land and buildings orders urban phenomena and determines what city life can be. This means we must show how real estate markets actually work and how their operations fail to meet the neoclassical economists' assumptions. In short, we will find the substance of urban phenomena in the actual operations of markets. Our goal is to identify the specific processes, the sociological processes, through which the pursuit of use and exchange values fixes property prices, responds to prices, and in so doing determines land uses and the distribution of fortunes. Since economic sociology is still without a clear analytical foundation, we must begin our work in this chapter by laying a conceptual basis for the empirical descriptions that will be presented later.

Special use values

People use place in ways contrary to the neoclassical assumptions of how commodities are purchased and consumed. We do not dispose of place after it has been bought and used. Places have a certain preciousness for their users that is not part of the conventional concept of a commodity. A crucial initial difference is that place is indispensable; all human activity must occur somewhere. Individuals cannot do without place by substituting another product. They can, of course, do with less place and less desirable place, but they cannot do without place altogether.

Even when compared to other indispensable commodities – food, for example – place is still idiosyncratic. The use of a particular place creates and sustains access to additional use values. One's home in a particular place, for example, provides access to school, friends, work place, and shops. Changing homes disrupts connections to these other places and their related values as well. Place is thus not a discrete element, like a toy or even food; the precise conditions of its use determine how other elements, including other commodities, will be used. Any individual residential location connects people to a range of complementary persons, organizations, and physical resources.

The stakes involved in the relationship to place can be high, reflecting all manner of material, spiritual, and psychological connections to land and buildings. Numerous scholars have shown that given places achieve significance beyond the more casual relations people have to other commodities. The connection to place can vary in intensity for different class, age, gender, and ethnic groups, individual relationships to place are often characterized by intense feelings and commitments appropriate to long-term and multifaceted social and material attachments.

This special intensity creates an asymmetrical market relation between buyers and sellers. People pay what the landlord demands, not because the housing unit is worth it, but because the property is held to have idiosyncratic locational benefits. Access to resources like friends, jobs, and schools is so important that residents (as continuous consumers–buyers) are willing to resort to all sorts of "extramarket" mechanisms to fight for their right to keep locational relations intact. They organize, protest, use violence, and seek political regulation. They strive not just for tenure in a given home but for stability in the surrounding neighborhood as well.

Location establishes a special collective interest among individuals. People who have "bought" into the same neighborhood share a quality of public services (garbage pickup, police behavior); residents have a common stake in the area's future. Residents also share the same fate when natural disasters such as floods and hurricanes threaten and when institutions alter the local landscape by creating highways, parks, or toxic dumps. Individuals are not only mutually dependent on what goes on inside a neighborhood (including "compositional effects"); they are affected by what goes on outside it as well. The standing of a neighborhood vis-à-vis other neighborhoods creates conditions that its residents experience in common. Each place has a particular political or economic standing vis-à-vis other places that affect the quality of life and opportunities available to those who live within its boundaries. A neighborhood with a critical voting bloc (for example, Chicago's Irish wards in the 1930s) may generate high levels of public services or large numbers of patronage jobs for its working-class residents, thereby aiding their well being. A rich neighborhood can protect its residents' life styles from external threats (sewer plants, public housing) in a way that transcends personal resources, even those typically associated with the affluent. The community in itself can be a local force.

Neighborhoods organize life chances in the same sense as do the more familiar dimensions of class and caste. . . . Like class and status groupings, and even more than many other associations, places create communities of fate. Thus we must consider the stratification of places along with the stratification of individuals in order to understand the distribution of life chances. People's sense of these dynamics, perceived as the relative "standing" of their neighborhood, gives them some of their spiritual or sentimental stake in place – thus further distinguishing home from other, less life-significant, commodities.

Contrary to much academic debate on the subject, we hold that the material use of place cannot be separated from psychological use; the daily round that makes physical survival possible takes on emotional meanings through that very capacity to fulfill life's crucial goals. The material and psychic rewards thus combine to create feelings of "community." Much of residents' striving as members of community organizations or just as responsible neighbors represents an effort to preserve and enhance their networks of sustenance. Appreciation of neighborhood resources, so varied and diffusely experienced, gives rise to "sentiment." Sentiment is the inadequately articulated sense that a particular place uniquely fulfills a complex set of needs. When we speak of residents' use values, we imply fulfillment of all these needs, material and non-material.

Homeownership gives some residents exchange value interests along with use value goals. Their houses are the basis of a lifetime wealth strategy. For those who pay rent to landlords, use values are the only values at issue. Owners and tenants can thus sometimes have divergent interests. When rising property values portend neighborhood transformation, tenants and owners may adopt different community roles; but ordinarily, the exchange interests of owners are not sufficiently significant to divide them from other residents.

[. . .]

Special exchange values

Exchange values from place appear as "rent." We use the term broadly to include outright purchase expenditures as well as payments that homebuyers or tenants make to landlords, realtors, mortgage lenders, real estate lawyers, title companies, and so forth. As with use values, people pursue exchange values in ways that differ from the manner in which they create other commodities. Suppliers cannot "produce" places in the usual sense of the term. All places consist, at least in part, of land, which "is only another name for nature, which is not produced by man" (Karl Polanyi, *The Great Transformation*. Boston: Beacon Press, 1944, p. 72) and obviously not produced for sale in a market. The quantity is fixed. It is not, says David Harvey (*The Limits to Capital*. Chicago: Chicago University Press, 1982, p. 357), "the product of labor." This makes the commodity description of land, in Marx's word, "fictitious." Michael Storper and Richard Walker ("The Theory of Labor and the Theory of Location," *International Journal of Urban and Regional Research* 7, 1 (1983): 43) describe land, like labor, as a "pseudocommodity."

Place as monopoly

Perhaps the fundamental "curiosity" is that land markets are inherently monopolistic, providing owners, as a class, with complete control over the total commodity supply. There can be no additional entrepreneurs or any new product. The individual owner also has a monopoly over a subsection of the marketplace. Every parcel of land is unique in the idiosyncratic access it provides to other parcels and uses, and this quality underscores the specialness of property as a commodity. Unlike widgets or Ford Pintos, more of the same product cannot be added as market demand grows. Instead the owner of a particular parcel controls all access to it and its given set of spatial relations. In setting prices and other conditions of use, the owner operates with this constraint on competition in mind.

Property prices do go down as well as up, but less because of what entrepreneurs do with their own holdings than because of the changing relations among properties. This dynamic accounts for much of the energy of the urban system as place entrepreneurs strive to increase their rent by revamping the spatial organization of the city. Rent levels are based on the location of a property vis-à-vis other places, on its particularity. In Marxian conceptual terms, entrepreneurs establish the rent according to the "differential" locational advantage of one site over another. Gaining "differential rent" necessarily depends on the fate of other parcels and those who own and use them. In economists' language, each property use "spills over" to other parcels and, as part of these "externality effects," crucially determines what every other property will be. The "web of externalities" affects an entrepreneur's particular holding. When a favorable relationship can be made permanent (for example, by freezing out competitors through restrictive zoning), spatial monopolies that yield even higher rents – "monopoly rents" in the Marxian lexicon – are created. But all property tends to have a monopolistic character. . . .

Nevertheless, property owners can and do inventively alter the content of their holdings. Sometimes they build higher and more densely, increasing the supply of dwellings, stores, or offices on their land. According to neoclassical thinking, this manner of increase should balance supply and demand, thus making property respond to market pressures as other commodities supposedly do. But new construction has less bearing on market dynamics than such reasoning would imply. New units on the same land can never duplicate previous products; condominiums stacked in a high-rise building are not the same as split-levels surrounded by lawn. Office space on the top roof of a skyscraper is more desirable than the same square footage just one floor lower. Conversely, the advantages of street-level retail space cannot be duplicated on a floor above. Each product, old or new, is different and unique, and each therefore reinforces the monopoly character property and the resulting price system.

Another curious aspect of the real estate market is its essentially second-hand nature. Buildings and land parcels are sold and resold, rented and rerented. In a typical area, no more than 3 percent of the product for sale or rent consists of new construction. Not only land, but even the structures on any piece of land can have infinite (for all practical purposes) lives; neither utility nor market price need decrease through continuous use. . . . Moreover, since the amount of "new" property on the market at any given moment is ordinarily only a small part of the total that is for sale, entrepreneurs' decisions to add to this supply by building additional structures will have a much more limited impact on price than would the same decisions with other types of commodities. Indeed, recent studies indicate that U.S. cities with more rapid rates of housing construction have higher, not lower, housing costs, even when demand factors are statistically controlled. Similarly, relatively high vacancy rates are not associated with lower rent levels, which suggests that new construction "leads" local markets to a new, higher pricing structure rather than equilibrating a previous one. Given the fixed supply of land and the monopolies over relational advantages, more money entering an area's real estate market not only results in more structures being built but also increases the price of land and, quite plausibly, the rents on previously existing "comparable" buildings. Thus higher investment levels can push the entire price structure upward.

[. . .]

GROWTH MACHINES

Those seeking exchange value often share interests with others who control property in the same block, city, or region. Like residents, entrepreneurs in similar situations also make up communities of fate, and they often get together to help fate along a remunerative path.

Whether the geographical unit of their interest is as small as a neighborhood shopping district or as large as a national region, place entrepreneurs attempt, through collective action and often in alliance with other business people, to create conditions that will intensify future land use in an area. There is an unrelenting search, even in already successful places, for more and more. An apparatus of interlocking progrowth associations and governmental units makes up . . . the "growth machine." Growth machine activists are largely free from concern for what goes on within production processes (for example, occupational safety), for the actual use value of the products made locally (for example, cigarettes), or for spillover consequences in the lives of residents (for example, pollution). They tend to oppose any intervention that might regulate development on behalf of use values. They may quarrel among themselves over exactly how rents will be distributed among parcels, over how, that is, they will share the spoils of aggregate growth. But virtually all place entrepreneurs and their growth machine associates, regardless of geographical or social location, easily agree on the issue of growth itself.

They unite behind a doctrine of value-free development – the notion that free markets alone should determine land use. In the entrepreneur's view, land-use regulation endangers both society at large and the specific localities favored as production sites.

[. . .]

Growth machines in U.S. history

The role of the growth machine as a driving force in U.S. urban development has long been a factor in U.S. history, and is nowhere more clearly documented than in the histories of eighteenth- and nineteenth-century American cities. Indeed, although historians have chronicled many types of mass opposition to capitalist organization (for example, labor unions and the Wobblie movement), there is precious little evidence of resistance to the dynamics of value-free city building characteristic of the American past. . . . The creators of towns and the builders of cities strained to use all the resources at their disposal, including crude political clout, to make great fortunes out of place. . . . Sometimes, the "communities" were merely subdivided parcels with town names on them, on whose behalf governmental actions could nonetheless be taken. The competition among them was primarily among growth elites.

These communities competed to attract federal land offices, colleges and academies, or installations such as arsenals and prisons as a means of stimulating development. . . . The other important arena of competition was also dependent on government decision making and funding: the development of a transportation infrastructure that would give a locality better access to raw materials and markets. First came the myriad efforts to attract state and federal funds to link towns to waterways through canals. Then came efforts to subsidize and direct the paths of railroads. Town leaders used their governmental authority to determine routes and subsidies, motivated by their private interest in rents.

The people who engaged in this city building have often been celebrated for their inspired vision and "absolute faith." . . . But more important than their personalities, these urban founders were in the business of manipulating place for its exchange values. Their occupations most often were real estate or banking. Even those who initially practiced law, medicine, or pharmacy were rentiers in the making.

[. . .]

The city-building activities of these growth entrepreneurs in frontier towns became the springboard for the much celebrated taming of the American wilderness. The upstart western cities functioned as market, finance, and administrative outposts that made rural pioneering possible. This conquering of the West, accomplished through the machinations of "the urban frontier," was critically bound up with a coordinated effort to gain rents. . . .

Perhaps the most spectacular case of urban ingenuity was the Chicago of William Ogden. When Ogden came to Chicago in 1835, its population was under four thousand. He succeeded in becoming its mayor, its great railway developer, and the owner of much of its best real estate. As the organizer and first president of the Union Pacific (among other railroads) and in combination with his other business and civic roles, he was able to make Chicago (as a "public duty") the crossroads of America, and hence the dominant metropolis of the Midwest. Chicago became a crossroads not only because it was "central" (other places were also in the "middle") but because a small group of people (led by Ogden) had the power to literally have the roads cross in the spot they chose....

This tendency to use land and government activity to make money was not invented in nineteenth-century America, nor did it end then. The development of the American Midwest was only one particularly noticed (and celebrated) moment in the total process. One of the more fascinating instances, farther to the West and later in history, was the rapid development of Los Angeles, an anomaly to many because it had none of the "natural" features that are thought to support urban growth: no centrality, no harbor, no transportation crossroads, not even a water supply. Indeed, the rise of Los Angeles as the preeminent city of the West, eclipsing its rivals San Diego and San Francisco, can only be explained as a remarkable victory of human cunning over the so-called limits of nature. Much of the development of western cities hinged on access to a railroad; the termination of the first continental railroad at San Francisco, therefore, secured that city's early lead over other western towns. The railroad was thus crucial to the fortunes of the barons with extensive real estate and commercial interests in San Francisco – Stanford, Crocker, Huntington, and Hopkins. These men feared the coming of a second cross-country railroad (the southern route), for its urban terminus might threaten the San Francisco investments. San Diego, with its natural port, could become a rival to San Francisco, but Los Angeles, which had no comparable advantage, would remain forever in its shadow. Hence the San Francisco elites used their economic and political power to keep San Diego from becoming the terminus of the southern route. ... Of course, Los Angeles won in the end, but here again the wiles of boosters were crucial: the Los Angeles interests managed to secure millions in federal funds to construct a port, today the world's largest artificial harbor – as well as federal backing to gain water.

The same dynamic accounts for the other great harbor in the Southwest. Houston beat out Galveston as the major port of Texas (ranked third in the country in 1979) only when Congressman Tom Ball of Houston successfully won, at the beginning of this century, a million-dollar federal appropriation to construct a canal linking landlocked Houston to the Gulf of Mexico. That was the crucial event that, capitalizing on Galveston's susceptibility to hurricanes, put Houston permanently in the lead.

In more recent times, the mammoth federal interstate highway system ... has similarly made and unmade urban fortunes. To use one clear case, Colorado's leaders made Denver a highway crossroads by convincing President Eisenhower in 1956 to add three hundred miles to the system to link Denver to Salt Lake City by an expensive mountain route. A presidential stroke of the pen removed the prospects of Cheyenne, Wyoming, of replacing Denver as a major western transportation center. In a case reminiscent of the nineteenth-century canal era, the Tennessee-Tolnbigbee Waterway opened in 1985, dramatically altering the shipping distances to the Gulf of Mexico for many inland cities. The largest project ever built by the U.S. Corps of Engineers, the $2 billion project was questioned as a boondoggle in Baltimore, which will lose port business because of it, but praised in Decatur, Alabama, and Knoxville, Tennessee, which expect to profit from it. The opening of the canal cut by four-fifths the distance from Chattanooga, Tennessee, to the Gulf, but did almost nothing for places like Minneapolis and Pittsburgh, which were previously about the same nautical distance from the Gulf as Chattanooga.

Despite the general hometown hoopla of boosters who have won infrastructural victories, not everyone gains when the structural speculators of a city defeat their competition. Given the stakes, the rentier elites would obviously become engulfed by the "booster spirit." ... Researchers have made little effort to question the linkage between public betterment and growth, even when they could see that specific social groups were being hurt. Zunz reports that in industrializing Detroit, city authorities extended utility service into uninhabitated areas to help development rather than into existing residential zones, whose working-class residents went without service even as they bore the costs (through taxes) of the new installations.

[...]

The modern-day good business climate

The jockeying for canals, railroads, and arsenals of the previous century has given way in this one to more complex and subtle efforts to manipulate space and redistribute rents. The fusing of public duty and private gain has become much less acceptable (both in public opinion and in the criminal courts); the replacing of frontiers by complex cities has given important roles to mass media, urban professionals, and skilled political entrepreneurs. The growth machine is less personalized, with fewer local heroes, and has become instead a multifaceted matrix of important social institutions pressing along complementary lines.

With a transportation and communication grid already in place, modern cities typically seek growth in basic economic functions, particularly job intensive ones. Economic growth sets in motion the migration of labor and a demand for ancillary production services, housing, retailing, and wholesaling ("multiplier effects"). Contemporary places differ in the type of economic base they strive to build (for example, manufacturing, research and development, information processing, or tourism). But any one of the rainbows leads to the same pot of gold: more intense land use and thus higher rent collections, with associated professional fees and locally based profits.

Cities are in a position to affect the factors of production that are widely believed to channel the capital investments that drive local growth. They can, for example, lower access costs of raw materials and markets through the creation of shipping ports and airfields (either by using local subsidies or by facilitating state and federal support). Localities can decrease corporate overhead costs through sympathetic policies on pollution abatement, employee health standards, and taxes. Labor costs can be indirectly lowered by pushing welfare recipients into low-paying jobs and through the use of police to constrain union organizing. Moral laws can be changed; for example, drinking alcohol can be legalized (as in Ann Arbor, Mich., and Evanston, Ill.) or gambling can be promoted (as in Atlantic City, N.J.) to build tourism and convention business. Increased utility costs caused by new development can be borne, as they usually are, by the public at large rather than by those responsible for the "excess" demand they generate. Federally financed programs can be harnessed to provide cheap water supplies; state agencies can be manipulated to subsidize insurance rates; local political units can forgive business property taxes. Government installations of various sorts (universities, military bases) can be used to leverage additional development by guaranteeing the presence of skilled labor, retailing customers, or proximate markets for subcontractors. For some analytical purposes, it doesn't even matter that a number of these factors have little bearing on corporate locational decisions (some certainly do; others are debated); just the possibility that they might matter invigorates local growth activism and dominates policy agendas.

Following the lead of St. Petersburg, Florida, the first city to hire a press agent (in 1918) to boost growth, virtually all major urban areas now use experts to attract outside investment. One city, Dixon, Illinois, has gone so far as to systematically contact former residents who might be in a position to help (as many as twenty thousand people) and offer them a finder's fee up to $10,000 for directing corporate investment toward their old home town. More pervasively, each city tries to create a "good business climate." The ingredients are well known in city-building circles and have even been codified and turned into "official" lists for each regional area. The much-used Fantus rankings of business climates are based on factors like taxation, labor legislation, unemployment compensation, scale of government, and public indebtedness (Fantus ranks Texas as number one and New York as number forty-eight). In 1975, the Industrial Development Research Council, made up of corporate executives responsible for site selection decisions, conducted a survey of its members. In that survey, states were rated more simply as "cooperative," "indifferent," or "antigrowth"; the results closely paralleled the Fantus rankings of the same year.

Any issue of a major business magazine is replete with advertisements from localities of all types (including whole countries) striving to portray themselves in a manner attractive to business. Consider these claims culled from one issue of *Business Week* (February 12, 1979):

> New York City is open for business. No other city in America offers more financial incentives to expand or relocate. . . .

The state of Louisiana advertises

> Nature made it perfect. We made it profitable.

On another page we find the claim that "Northern Ireland works" and has a work force with "positive attitudes toward company loyalty, productivity and labor relations." Georgia asserts, "Government should strive to improve business conditions, not hinder them." Atlanta headlines that as "A City Without Limits" it "has ways of getting people like you out of town" and then details its transportation advantages to business. Some places describe attributes that would enhance the life style of executives and professional employees (not a dimension of Fantus rankings); thus a number of cities push an image of artistic refinement. No advertisements in this issue (or in any other, we suspect) show city workers living in nice homes or influencing their working conditions.

While a good opera or ballet company may subtly enhance the growth potential of some cities, other cultural ingredients are crucial for a good business climate. There should be no violent class or ethnic conflict. Racial violence in South Africa is finally leading to the disinvestment that reformers could not bring about through moral suasion. In the good business climate, the work force should be sufficiently quiescent and healthy to be productive; this was the rationale originally behind many programs in work place relations and public health. Labor must, in other words, be "reproduced," but only under conditions that least interfere with local growth trajectories.

Perhaps most important of all, local publics should favor growth and support the ideology of value-free development. This public attitude reassures investors that the concrete enticements of a locality will be upheld by future politicians. The challenge is to connect civic pride to the growth goal, tying the presumed economic and social benefits of growth in general to growth in the local area. Probably only partly aware of this, elites generate and sustain the place patriotism of the masses. . . . In the nineteenth-century cities, the great rivalries over canal and railway installations were the political spectacles of the day, with attention devoted to their public, not private, benefits. With the drama of the new railway technology, ordinary people were swept into the competition among places, rooting for their own town to become the new "crossroads" or at least a way station.

The celebration of local growth continues to be a theme in the culture of localities. Schoolchildren are taught to view local history as a series of breakthroughs in the expansion of the economic base of their city and region, celebrating its numerical leadership in one sort of production or another; more generally, increases in population tend to be equated with local progress. Civic organizations sponsor essay contests on the topic of local greatness. They encourage public celebrations and spectacles in which the locality name can be proudly advanced for the benefit of both locals and outsiders. They subsidize soapbox derbies, parade floats, and beauty contests to "spread around" the locality's name in the media and at distant competitive sites.

One case can illustrate the link between growth goals and cultural institutions. In the Los Angeles area, St. Patrick's Day parades are held at four different locales, because the city's Irish leaders can't agree on the venue for a joint celebration. The source of the difficulty (and much acrimony) is that these parades march down the main business streets in each locale, thereby making them a symbol of the life of the city. Business groups associated with each of the strips want to claim the parade as exclusively their own, leading to charges by still a fifth parade organization that the other groups are only out to make money. The countercharge, vehemently denied, was that the leader of the challenging business street was not even Irish. Thus even an ethnic celebration can receive its special form from the machinations of growth interests and the competitions among them.

The growth machine avidly supports whatever cultural institutions can play a role in building locality. Always ready to oppose cultural and political developments contrary to their interests (for example, black nationalism and communal cults), rentiers and their associates encourage activities that will connect feelings of community . . . to the goal of local growth. The overall ideological thrust is to deemphasize the connection between growth and exchange values and to reinforce the link between growth goals and better lives for the majority. We do not mean to suggest that the only source of civic pride is the desire to collect rents; certainly the cultural pride of tribal groups predates growth machines. Nevertheless, the growth machine coalition mobilizes these cultural motivations, legitimizes them, and channels them into activites that are consistent with growth goals.

"Partnership and the Pursuit of the Private City"

from Mark Gottdiener and Chris Pickvance (eds), *Urban Life in Transition* (1991)

Gregory Squires

Editors' Introduction

Gregory Squires makes an important contribution to new urban sociology that further interrogates the vested interests behind urban planning and development that Logan and Molotch describe as urban growth machines. He interrogates the credo of "public-private partnership" that has become a rallying cry for urban planners and officials in the contemporary era of fiscal austerity and economic decline in America. He exposes "partnership" as a foil for an underlying agenda of capitalist accumulation and privatism to undermine government spending and entitlements while providing subsidies and inducements to the private sector. At the same time he questions the real political autonomy of the state and its authority to act in the interests of the public. The privatism he describes in the climate of global restructuring has more recently been dubbed neoliberalism by urban scholars making comparisons with Europe and the developing and postcommunist world (see Brenner and Theodore).

Squires describes recurring phases of work of public-private partnership and privatism in U.S. cities, leading to dramatic structural, spatial, and social consequences. In the nineteenth century came public subsidization of canal and railroad building that benefited many studies on the expanding frontier. Urban renewal and the freeway system were the great public spending programs of the mid-twentieth century, stimulating an exodus of jobs and people to the suburbs and devastating many inner cities. Urban renewal also sparked the emergence of the first postwar downtown redevelopment coalitions, who profited from governmental land-grabs of valuable central city neighborhoods through strategies of condemnation and slum clearance.

The 1980s were banner years for public-private partnership formation in U.S. cities as industrial decline and global competitiveness, and a "supply-side" revolution in the federal government under the Reagan administration, spurred state abandonment of civil rights initiatives and labor and environmental regulations in favor of business and urban entrepreneurial strategies. Landmark downtown redevelopmental projects proliferated in U.S. cities as urban politicians and planners put their money on a postindustrial future built on the sectoral shift from manufacturing to administrative functions and the service sector. While creating short-term profitability, he asserts these policies led to a deepening of social inequality and created social and mental health strains for many urban families.

Squires also highlights pioneering alternative, community-based partnerships that promote sustainable urban quality of life, social equity, and shared prosperity over short-term capitalist gain. He offers a number of demonstration projects, including: a) a public participation campaign under Mayor Harold Washington in Chicago, b) a linkage program under Mayor Raymond Flynn in Boston, and c) a metropolitan-wide fair lending initiative program in Milwaukee. He says that socially progressive partnerships will continue to be relevant in the years to come as an antidote to social contradictions and ill consequences created by laissez-faire privatism in urban governance and planning. His prognosis still has resonance in the new millennium as cities contend with the current fiscal and social crisis sparked by the collapse of the deregulated banking system and speculative subprime home mortgage market.

Gregory D. Squires is a Professor of Sociology and Public Policy and Public Administration at George Washington University. He has written extensively and edited books on important related topics such as racial discrimination in housing, insurance redlining, residential segregation, the urban effects of Hurricane Katrina, and the subprime mortgage lending crisis. He serves on a number of non-profit and educational boards and has been a consultant and expert witness for fair housing groups and civil rights organizations around the country, including HUD, the National Fair Housing Alliance, and the National Community Reinvestment Coalition. He also served a three-year term as a member of the Consumer Advisory Council of the Federal Reserve Board.

Public-private partnerships have become the rallying cry for economic development professionals throughout the United States. As federal revenues for economic development, social service, and other urban programs diminish such partnerships are increasingly looked to as the key for urban revitalization. These partnerships take many forms. Formal organizations of executives from leading businesses have been established that work directly with public officials. In some cases public officials as well as representatives from various community organizations are also members. Some partnerships have persisted for decades working on an array of issues while others are ad hoc arrangements that focus on a particular time-limited project. Direct subsidies from public agencies to private firms have been described as public-private partnerships. If economic development has emerged as a major function of local government, public-private partnerships are increasingly viewed as the critical tool.

The concept of partnership is widely perceived to be an innovative approach that is timely in an age of austerity. In fact, "public-private partnership" is little more than a new label for a long-standing relationship between the public and private sectors. Growth has been the constant, central objective of that relationship, though in recent years subsidization of dramatic economic restructuring has become a complementary concern. While that relationship has evolved throughout U.S. history, it has long been shaped by an ideology of privatism that has dominated urban redevelopment from colonial America through the so-called postindustrial era. The central tenet of privatism is the belief in the supremacy of the private sector and market forces in nurturing development, with the public sector as a junior partner whose principal obligation is to facilitate private capital accumulation. Individual material acquisitiveness is explicitly avowed, but that selfishness is justified by the public benefits that are assumed to flow from the dynamics of such relations.

One need look no further than the roadways, canals, and railroads of the eighteenth and nineteenth centuries to see early concrete manifestations of large-scale public subsidization of private economic activity and the hierarchical relationship between the public and private sectors. These relationships crystallized in the urban renewal days of the 1950s and 1960s and the widely celebrated partnerships of the 1980s. Structural changes in the political economy of cities, regions, and nations altered the configuration of specific public-private partnerships, but not the fundamental relationship between the public and private sectors. These structural changes have, however, influenced the spatial development of cities and exacerbated the social problems of urban America.

The continuity reflected by public-private partnerships, despite some new formulations in recent years, is revealed by the persistence in the corporate sector's efforts to utilize government to protect private wealth, and primarily on its terms. Demands on the state to subsidize painful restructuring process have placed added strains on public-private relations. The glue that holds these efforts together, despite these tensions, is the commitment to privatism.

Focusing on the postwar years, this chapter examines the ideology of privatism, its influence on the evolution of public-private partnerships, and their combined effects on the structural, spatial, and social development of cities in the United States, and the lives of people residing in the nation's urban neighborhoods. Perhaps the most striking feature of the evolution of American cities, to be explored in the following pages, is the uneven nature of urban development. To many, such uneven development simply reflects the "creative destruction" that Schumpeter (1942) asserted was essential for further economic progress in a capitalist economy. To others, however, the unevenness generated by unrestrained market-based private capital accumulation constitutes the core of the nation's urban problems.

[...]

PRIVATISM

The American tradition of privatism was firmly established by the time of the Revolution in the 1700s. According to this tradition individual and community happiness are to be achieved through the search for personal wealth. Individual loyalties are to the family first, and the primary obligation of political authorities is to keep the peace among individual money-makers.

Public policy, from this perspective, should serve private interests. Government has an important role, but one that should focus on the facilitation of private capital accumulation via the free market. (Privatism should not be confused with privatization. The former refers to a broader ideological view of the world generally and relationships between the public and private sectors in particular. The latter constitutes a specific policy of transferring ownership of particular industries or services from government agencies to private entrepreneurs.) While urban policy must acknowledge the well-known problems of big cities, it can do so best by encouraging private economic growth. A critical assumption is that the city constitutes a unitary interest and all citizens benefit from policies that enhance aggregate private economic growth.

The ideology of privatism has been tested in recent years by regional shifts in investment and globalization of the economy in general that have devastated entire communities. Advocates of privatism attribute such developments primarily to technological innovation and growing international competition. They claim the appropriate response is to accommodate changes in the national and international economy. Given that redevelopment is presumed to be principally a technical rather than political process, cities must work more closely with private industry to facilitate such restructuring in order to establish more effectively their comparative advantages and market themselves in an increasingly competitive economic climate. Such partnership, it is assumed, will bring society's best and brightest resources (such reside in the private sector) to bear on its most severe public problems.

[...]

Concretely, the policies of privatism consist of financial incentives to private economic actors that are intended to reduce factor costs of production and encourage private capital accumulation, thus stimulating investment, ultimately serving both private and public interests. The search for new manufacturing sites, retooling of obsolete facilities, and restructuring from manufacturing to services have all been facilitated by such subsidization. During the postwar years cities have been dramatically affected by the focus on downtown development that has generally taken the form of office towers, luxury hotels, convention centers, recreational facilities, and other paeans to the postindustrial society. Real estate investment itself is frequently viewed as part of the antidote to deindustrialization. All of this is justified, however, by the assumption that a revitalized economy generally and a reinvigorated downtown in particular will lead to regeneration throughout the city. As more jobs are created and space is more intensively utilized, more money is earned and spent by local residents, new property and income tax dollars bolster local treasuries, and new wealth trickles down throughout the metropolitan area. Among the specific policy tools are tax abatements, low-interest loans, land cost writedowns, tax increment finance districts (TIFs), enterprise zones, urban development action grants (UDAGs), industrial revenue bonds (IRBs), redevelopment authorities, eminent domain, and other public-private activities through which private investment is publicly subsidized. The object of such incentives, again, is the enhancement of aggregate private economic growth by which it is assumed the public needs of the city can be most effectively and efficiently met.

[...]

STRUCTURAL, SPATIAL, AND SOCIAL DEVELOPMENT

Urban renewal and the prosperous postwar years

The United States emerged from World War II as a growing and internationally dominant economic power. Given its privileged structural position at that time, the end of ideology was declared and optimism for future growth and prosperity was widespread.

Yet blighted conditions within the nation's central cities posed problems for residents trapped in poverty and for local businesses threatened by conditions within and immediately surrounding the downtown business center. Recognizing the "higher uses" (i.e., more profitable for developers and related businesses) for which such land could be utilized, a policy of urban renewal evolved that brought together local business and government entities in working partnerships with the support of the federal government. At the same time, federal housing policy and highway construction stimulated homeownership and opened up the suburbs, while reinforcing the racial exclusivity of neighborhoods.

As Mollenkopf has observed, urban renewal and related federal programs reflected a political coalition of disparate groups. Local entrepreneurial Democratic politicians, along with their counterparts at the federal level, created large-scale downtown construction projects that benefited key local contractors and unions, machine politicians and reformers, and white ethnic groups along with at least some racial minorities. These emerging political alliances were clearly, though not always explicitly, committed to economic growth (particularly downtown) with the private sector as the primary engine for, and beneficiary of, that development.

Although urban renewal was launched and initially justified as an effort to improve the housing conditions of low-income urban residents, it quickly became a massive public subsidy for private business development, particularly downtown commercial real estate interests. Shopping malls, office buildings, and convention centers rather than housing became the focus of urban renewal programs. Following the lead of the Allegheny Conference on Community Development formed in Pittsburgh in 1943, coalitions of local business leaders were organized in most large cities to encourage public subsidization of downtown development. Examples include the Greater Milwaukee Committee, Central Atlanta Progress, Inc., Greater Philadelphia Movement, Cleveland Development Foundation, Detroit Renaissance, the Vault (Boston), the Blyth-Zellerbeck Committee (San Francisco), Greater Baltimore Committee, and Chicago Central Area Committee. Using their powers of eminent domain, city officials generally would assemble land parcels and provide land cost writedowns for private developers. In the process local business associations frequently operated as private governments as they designed and implemented plans that had dramatic public consequences but did so with little public accountability.

If such developments were justified rhetorically as meeting important public needs, indeed urban renewal took sides. Not all sides were represented in the planning process and the impact of urban renewal reflected such unequal participation. Some people were forcefully relocated so that others could benefit . . .

At the same time that the public sector was subsidizing downtown commercial development, it was also subsidizing homeownership and highway construction programs to stimulate suburban development. Through Federal Housing Administration (FHA), Veterans Administration (VA), and related federally subsidized and insured mortgage programs launched around the war years, long-term mortgages requiring relatively low down payments made homeownership possible for many families who previously could not afford to buy. With the federal insurance, lenders were far more willing to make such loans. An equally if not more compelling factor leading to the creation of these programs was the financial assistance they provided to real estate agents, contractors, financial institutions, and other housing related industries. Since half the FHA and VA loans made during the 1950s and 1960s financed suburban housing, the federal government began, perhaps unwittingly, to subsidize the exodus from central cities to suburban rings that characterized metropolitan development during these decades. The Interstate Highway Act of 1956, launching construction of the nation's high-speed roadway system, further subsidized and encouraged that exodus.

A significant feature of these developments was the racial exclusivity that was solidified in part because the federal government encouraged it . . . If redlining practices originated within the nation's financial institutions, the federal government sanctioned and reinforced such discriminatory practices at a critical time in the history of suburban development. The official stance of the federal government has changed in subsequent decades, but the patterns established by these policies have proven to be difficult to alter.

During the prosperous postwar years of the 1950s and 1960s urban redevelopment strategies were shaped by public-private partnership. But the private partner dominated as the public sector's role consisted principally of "preparing the ground for capital." Spatially, the focus was on downtown and the suburbs. Socially, the dominant feature was the creation and reinforcement of racially discriminatory dual housing markets and homogenous urban and suburban communities. These basic patterns have persisted in subsequent years when the national economy was not so favorable.

Partnerships in an age of decline

The celebrated partnerships of the 1980s reflect an emerging effort to undermine the public sector, particularly the social safety net it has provided, and to reaffirm the "privileged position of business" (Lindblom 1977) in the face of declining profitability brought on by globalization of the U.S. economy and its declining position in that changing marketplace. Government has a role, but again it is a subordinate one.

[. . .]

Global domination by the U.S. economy peaked roughly 25 years following the conclusion of World War II. After more than two decades of substantial economic growth subsequent to the war, international competition, particularly from Japan and West Germany but also from several Third World countries, began to challenge the U.S. position as productivity and profitability at home began to decline. As both a cause and effect of the general decline beginning in the late 1960s and early 1970s the U.S. economy experienced significant shifts out of manufacturing and into service industries. Perhaps even more important than the overall trajectory of decline has been the response to these developments on the part of corporate America and its partners in government . . .

Rather than directing investment into manufacturing plants and equipment or research and development to improve the productivity of U.S. industry, corporate America pursued what Robert B. Reich labeled "paper entrepreneurialism." That is, capital was expended on mergers and acquisitions, speculative real estate ventures, and other investments in which "some money will change hands, and no new wealth will be created" (Reich 1983). Rather than strategic planning for long-term productivity growth, the pursuit of short-term gain has been the objective.

Reducing labor costs has constituted a second component of an overall strategy aimed at short-term profitability. A number of tactics have been utilized to reduce the wage bill including decentralizing and globalizing production, expanding part-time work at the expense of full-time positions, contracting out work from union to non-union shops, aggressively fighting union organizing campaigns, implementing two-tiered wage scales and outright demands for wage concessions. Rather than viewing human capital as a resource in which to invest to secure productivity in the long run, labor has increasingly been viewed as a cost of production to be minimized in the interests of short-term profitability.

[. . .]

True to the spirit of privatism, government has nurtured these developments through various forms of assistance to the private sector. Federal tax laws encourage investment in new facilities, particularly overseas rather than reinvestment in older but still usable equipment, thus exacerbating the velocity of capital mobility. State and local governments have offered their own inducements to encourage the pirating of employers of all industries ranging from heavy manufacturing to religious organizations. Further inducements have been offered to the private sector through reductions in various regulatory functions of government. Civil rights, labor law, occupational health and safety rules, and environmental protection were enforced less aggressively in the 1980s than had been the case in the immediately preceding decades. If the expansion of such financial incentives and reductions in regulatory activity were initially justified in terms of the public benefits that would accrue from a revitalized private sector, in recent years unbridled competition and minimal government have become their own justification and not simply means to some other end.

The impact of these structural developments is clearly visible on the spatial development of American cities. Accommodating these national and international trends, local partnerships have nurtured downtown development to service the growing service economy. If steel is no longer produced in Pittsburgh, the Golden Triangle has risen as the city's major employers now include financial, educational, and health care institutions. If auto workers have lost jobs by the thousands in Detroit, the Renaissance Center, a major medical center, and the Joe Louis Sports Arena have been built downtown. Most major breweries have left Milwaukee, but the Grand Avenue Shopping Mall, several office buildings for legal, financial, and insurance companies, a new Performing Arts Center, and the Bradley Center housing the professional basketball Milwaukee Bucks are growing up in the central business district. With the U.S. economy deindustrializing and corporations consolidating administrative functions, downtown development to accommodate these changes is booming. These initiatives are more ambitious than urban renewal efforts that focused on rescuing downtown real estate, but many of the actors are the same and the fundamental relationships between the public and private entities prevail. In city after city such developments are initiated by the private side of local partnerships, usually with substantial public economic development assistance in the forms of UDAGs, IRBs, and other subsidies.

As cities increasingly become centers of administration, they experience an influx of relatively high-paid professional workers, the majority of whom are suburban residents. Despite some pockets of gentrification, most of the increasing demand for housing for such workers has been in the suburbs. Retail and commercial businesses have expanded into the suburbs to service that growing population. To the extent that metropolitan areas have experienced an expansion of existing manufacturing facilities or have attracted new facilities, this growth has also disproportionately gone to the suburbs. Extending a trend that goes back before the war years, suburban communities have continued to grow.

[. . .]

Throughout urban America, the rise of service industry jobs has fueled downtown and suburban development while the loss of manufacturing jobs has devastated blue-collar urban communities. Such uneven development is not simply the logical or natural outcome of impersonal market forces. The "supply-side" revolution at the federal level with the concomitant paper entrepreneurialism in private industry, the array of subsidies offered by state and local governments, and other forms of public intervention into the workings of the economy and the spatial development of cities, reveal the centrality of politics. As Mollenkopf concluded in reference to the postindustrial transformation of the largest central cities in the United States, "while its origins may be found in economic forces, federal urban development programs and the local progrowth coalitions which implemented them have magnified and channeled those economic forces" (1983). Uneven development therefore reflects conscious decisions made in both the public and private sectors in accordance with the logic of privatism, to further certain interests at the expense of others. Ideology has remained very much alive. Consequently, serious social costs have been paid.

Many of the social costs of both sudden economic decline and dramatic growth have been fully documented. As indicated above they include a range of economic and social strains for families, mental and physical health difficulties for current and former employees, fiscal crises for cities, and a range of environmental and community development problems. Among the more intangible yet clearly most consequential costs have been a reduction in the income of the average family and increasing inequality among wage earners and their families. Uneven economic and spatial development of cities has yielded unequal access to income and wealth and city residents.

[. . .]

ALTERNATIVES TO THE PURSUIT OF THE PRIVATE CITY

Privatism and the policies that flow logically from that ideology have benefited those shaping redevelopment policy, including members of most public-private partnerships. But these policies have not stimulated redevelopment of cities generally. Structural, spatial, and social imbalances remain and are reinforced by the dynamics of privatism. To address the well-known social problems of urban America successfully, policies must be responsive to the structural and spatial forces impinging on cities. At least fragmented challenges to privatism have emerged in local redevelopment struggles in recent years. Alternative conceptions of development, the nature of city life, and human relations in general have been articulated and have had some impact on redevelopment efforts.

In several cities community groups have organized, and in some cases captured, the mayor's office, in efforts to pursue more balanced redevelopment policies. Explicitly viewing the city in terms of its use value rather than as a profit center for the local growth machine, initiatives have been launched to democratize the redevelopment process and to assure more equitable outcomes of redevelopment policy. Among the specific ingredients of this somewhat inchoate challenge to privatism are programs to retain and attract diverse industries including manufacturing, targeting of initiatives to those neighborhoods and population groups most in need, human capital development, and other public investments in the infrastructure of cities. A critical dimension of many of these programs is a conscious effort to bring neighborhood groups and residents, long victimized by uneven development, into the planning and implementation process as integral parts of urban partnerships.

When Harold Washington was elected mayor of Chicago in 1983, he launched a redevelopment plan that incorporated several of these components. The planning actually began during the campaign when people from various racial groups, economic classes, and geographic areas were brought together to identify goals and policies to achieve them under a Washington administration. Shortly after the election Washington released *Chicago Works Together: Chicago Development Plan 1984*, which reflected that involvement. Explicitly advocating a strategic approach to pursuing development with equity, the plan articulated five major goals: increased job opportunities for Chicagoans; balanced growth; neighborhood development via partnerships and coordinated investment; enhanced public participation in decision making; and pursuit of a regional, state, and national legislative agenda. As development initiatives proceeded under Washington, strategic plans were implemented that involved industrial and geographic sector-specific approaches to retain manufacturing and regenerate older neighborhoods, affirmative action plans to bring more minorities and women into city government as employees and as city contractors, provision of business incentives that were conditioned on locational choices and other public needs, and a planning process that involved community groups, public officials, and private industry . . .

In 1983 Boston also held a significant mayoral election. At the height of the Massachusetts miracle the city's economy was prospering and Raymond L. Flynn was elected with a mandate to "share the prosperity." Several policies have been implemented in order to do so.

Boston's strong real estate market in the early 1980s led to a shortage of low- and middle-income housing. Flynn played a central role in the implementation of a linkage program that took effect one month before he was elected. Under the linkage program a fee was levied on downtown development projects to assist construction of housing for the city's low- and middle-income residents. Shortly after taking office, the Flynn administration negotiated inclusionary zoning agreements with individual housing developers to provide below-market rate units in their housing developments or to pay an "in lieu of" fee into the linkage fund. To further alleviate the housing shortage, in 1983 the Boston Housing Partnership was formed to assist community development corporations in rehabilitating and managing housing units in their neighborhoods. The partnership's board includes executives from leading banks, utility companies, and insurance firms; city and state housing officials; and directors of local community development corporations.

Boston also established a residents job policy under which developers and employers are required to target city residents, minorities, and women for construction jobs and in the permanent jobs created by these developments. These commitments hold for publicly subsidized developments and, in an agreement reached by the mayor's office, the Greater Boston Real Estate Board, the Buildings Trade Council, and leaders of the city's minority community, for private developments as well.

The Boston Compact represents another creative partnership in that city. Under this program the public schools agreed to make commitments to improve the school's performance in return for the business community's agreement to give hiring preferences to their graduates. Schools have designed programs to encourage students to stay in school, develop their academic abilities, and learn job readiness skills. Several local employers, including members of the Vault, have agreed to provide jobs paying more than the minimum wage and financial assistance for college tuition to students who succeed in the public schools.

As in Chicago, the Flynn administration in Boston has consciously pursued balanced development and efforts to bring previously disenfranchised groups into the development process. The specific focus has been in housing and jobs, but the broader objective has been to share the benefits of development generally throughout the city.

The Community Reinvestment Act (CRA) passed by Congress in 1977 has led to partnerships for urban reinvestment in cities across the nation. The CRA requires federally regulated banks and savings and loans to assess and be responsive to the credit needs of their service areas. Failure to do so can result in lenders being denied charters, new branches, or other corporate changes they intend to make. Neighborhood groups can challenge lenders' applications for such business operations with federal regulators, thus providing lenders with incentives to meet their CRA obligations . . .

A unique lending partnership was created in Milwaukee in 1989. In response to a 1989 study finding Milwaukee to have the nation's highest racial disparity in mortgage loan rejection rates, the city's Democratic Mayor and Republican Governor created a committee to find ways to increase lending in the city's minority community. The Fair Lending Action Committee (FLAC) (1989) included lenders, lending regulators, real estate agents, community organizers, civil rights leaders, a city alderman, and others. An ambitious set of recommendations was unanimously agreed to in its report *Equal Access to Mortgage Lending: Milwaukee Plan*. The key recommendation in the report was that area lenders would direct 13% of all residential, commercial real estate, and business loans to racial minorities by 1991. (After much debate the 13% figure was agreed upon because that was the current minority representation in the population of the four-county Milwaukee metropolitan area.) Several low-interest loan programs were proposed to be financed and administered by lenders, city officials, and neighborhood groups. Fair housing training programs were recommended for all segments of the housing industry including lenders, real estate agents, insurers, and appraisers. The lending community was advised to provide $75,000 to support housing counseling centers that assist first-time home buyers. The city, county, and state were called upon to consider a linked deposit program to assure that public funds would go to those lenders in response to the credit needs of the entire community. Specific recommendations were made to increase minority employment in the housing industry. And a permanent FLAC was called for to monitor progress in implementing the report's recommendations.

[. . .]

These diverse initiatives are illustrative of experiments being launched in small towns and large cities in all regions of the United States. While they constitute an array of programs addressing a variety of problems, there are important underlying commonalities. They are responsive to the structural and spatial underpinnings of critical urban social problems. They are premised on a commitment to growth with equity; the notion that economic productivity and social justice can be mutually reinforcing. And the objective is to make cities more livable, not just more profitable. A more progressive city is certainly not inevitable, but these efforts are vivid reminders that the major impediments have as much to do with politics as markets.

BEYOND LAISSEZ-FAIRE?

The trajectory of future redevelopment activity is blurred. The ideology of privatism is being challenged. Experiments with more progressive policies have occurred. But no linear path in the overall direction of public-private partnerships in particular or urban redevelopment in general has emerged. Harold Washington was soon followed by Daley in Chicago. Boston's economy in the early 1990s does not look as promising as it did in the early 1980s and the demand for more incentives to the business community is getting louder in the wake of the Massachusetts miracle. Milwaukee's mayor frequently expresses concern about the local business climate as civil rights groups challenge him to respond to the city's racial problems. Redevelopment remains a highly contentious political matter.

The grip of privatism has waned since the height of the Reagan years, HUD abuses, the savings and loan bailout, insider trading scandals, and other manifestations of the excesses of the pursuit of personal wealth serve as reminders of the importance of a public sector role beyond subsidization of private capital accumulation. Experiments in strategic planning to achieve balanced growth in Chicago, to share the prosperity in Boston, and to expand memberships in partnerships in Milwaukee and elsewhere demonstrate the capacity to conceive a different image of the city and the ability to implement programs in hopes of realizing that image. Yet as Warner concluded, "The quality which above all else characterizes our urban inheritance is privatism" (1987). For better or worse, that remains the bedrock on which future plans will be built.

"Los Angeles and the Chicago School: Invitation to a Debate"

from *City and Community* (2002)

Michael Dear

Editors' Introduction

Michael Dear discusses the emergence of the "L.A. School" of urban studies, a concatenation of geographers, sociologists, planners, and architects that considers Los Angeles as the prototype of a new kind of urbanism. The L.A. School is linked with urban political economy but has a closer engagement with postmodern social theory and cultural studies. Along the way, it has forcefully indicted Chicago School human ecology theory for being a narrative of modernist hegemony that served to justify the dominant political, economic and cultural interests in favor of a paradigm of Los Angeles as the quintessential postmodern metropolis, a polycentric, polyglot, and polycultural pastiche that suggests a different paradigm of urban development in the new millennium.

Dear attacks the classic concentric zone formulation of urban development promulgated by Ernest Burgess of the Chicago School. Chicago was the prototype of the world metropolis during the early twentieth century, during which railroad infrastructure concentrated the business district in a central location and transportation lines radiated outwards in a hub-and-spoke fashion into the suburbs. Racial/ethnic and subcultural minorities were subsumed, marginalized, or assimilated both socially and spatially into urban life, while the elite interests ruled the city center, controlling politics, land markets, and the cultural values of hegemonic interests. Since World War II, with the emergence of cities like Los Angeles built around a freeway infrastructure, the business district has become poly-nucleated or multi-polar, and the center has relinquished authority to the hinterlands. The forces of cultural assimilation have also become more fragmented with the arrival of new immigrants and overseas capital investment from Asia, Latin America, and other regions, and the ascendance of Los Angeles as a major command and manufacturing center in the new global economy. Immigrant colonies do not disappear so much anymore with the process of invasion–succession, while ethnic suburbs are emerging on the urban periphery. These "ethnoburbs" are emerging to become new poles of economic and cultural activity, acting as transaction nodes to other world trading regions.

Los Angeles is also a center of postmodern architecture, a showplace for architects such as Frank Gehry (the Disney Music Hall), John Portman (the Bonaventure Hotel), and Charles Moore. There is an "L.A. School" of architecture, design, arts, and even filmmaking. The L.A. School invites a cultural interpretation of Los Angeles as harbinger of twenty-first-century urbanism, just as Walter Benjamin described Paris as the capital of the nineteenth century. Ed Soja, in *Thirdspace: Journeys to Los Angeles and Other Real-and-Imagined Places* (Oxford: Blackwell, 1996) and *Postmetropolis: Critical Studies of Cities and Regions* (Oxford: Blackwell, 2000) discusses Los Angeles from the standpoint of identity politics, globalization, and theme parks. Mike Davis pontificates on the experience of Los Angeles as a fortress or citadel for transnational capitalism, in which the underclass and the homeless are marginalized in *City of Quartz* (New York: Verso, 1990) and on the environmental disaster of Los Angeles urbanism in *Ecology of Fear* (New York: Metropolitan Books, 1998).

The German expatriate turned Angeleno, Roger Keil, also contributed to the L.A. School with his book, *Los Angeles: Globalization, Urbanization and Social Struggles* (New York: John Wiley and Sons, 1998). Allen Scott joined with Ed Soja to edit the important reader, *The City: Los Angeles and Urban Theory at the End of the Twentieth Century* (Berkeley, CA: University of California Press, 1996). As Michael Dear discusses, the decentralized mosaic of greater Los Angeles resembles a game board of "keno capitalism." Many white and middle-class residents have abandoned the insecurity and social unrest found in the city center for gated communities in the exurbs.

Michael Dear, along with Allen Scott, published an earlier edited text, *Urbanization and Urban Planning in Capitalist Society* (New York: Methuen, 1981), a cross-national reader on urban political economy that was effectively an early precursor to the new urban sociology in the United States. He also published an edited reader, with H. Eric Schockman and Greg Hise, called *Rethinking Los Angeles* (Thousand Oaks, CA: Sage Publications, 1996) and, more recently, the book *The Postmodern Urban Condition* (Oxford: Blackwell, 2000). He published an edited reader on the L.A. School, with J. Dallas Dishman, *From Chicago to L.A.: Making Sense of Urban Theory* (Thousand Oaks, CA: Sage Publications, 2002). The selection presented in this reader was published initially in the inaugural issue of *City and Community*. There are five response essays engaging Michael Dear in lively debate.

Michael Dear is Professor of Geography at the University of Southern California, and Director of the Southern California Studies Center. He received his higher education in Regional Science, Town Planning, and Geography in both the United Kingdom and United States. Dear conducts research on Los Angeles, postmodern urbanism, and political and social geography. He is often cited as an authority in geography, and is the author or editor of ten books and over 100 journal articles and reports. He was a Fellow at the Center for Advanced Study in the Behavioral Sciences at Stanford in 1995–96, and held a Guggenheim Fellowship in 1989. He received Honors from the Association of American Geographers in 1995 and, in the same year, received the University of Southern California's Associates Award for highest honors for creativity in research, teaching and service.

More than 75 years ago, the University of Chicago Press published a book of essays entitled *The City: Suggestions for Investigation of Human Behavior in the Urban Environment*. The book is still in print. Six of its 10 essays are by Robert E. Park, then Chair of the University's Sociology Department. There are also two essays by Ernest W. Burgess, and one each from Roderick D. McKenzie and Louis Wirth. In essence, the book announced the arrival of the "Chicago School" of urban sociology, defining an agenda for urban studies that persists to this day. Shrugging off challenges from competing visions, the School has maintained a remarkable longevity that is a tribute to its model's beguiling simplicity, to the tenacity of its adherents who subsequently constructed a formidable literature, and to the fact that the model "worked" in its application to so many different cities over such a long period of time.

The present essay begins the task of defining an alternative agenda for urban studies, based on the precepts of what I shall refer to as the "Los Angeles School." Quite evidently, adherents of the Los Angeles School take many cues from the Los Angeles metropolitan region, or (more generally) from Southern California – a five-county region encompassing Los Angeles, Orange, Riverside, San Bernardino, and Ventura Counties. This exceptionally complex, fast-growing megalopolis is already home to more than 16 million people. It is likely soon to overtake New York as the nation's premier urban region. Yet, for most of its history it has been regarded as an exception to the rules governing American urban development, an aberrant outlier on the continent's western edge.

All this is changing. During the past two decades, Southern California has attracted increasing attention from scholars, the media, and other social commentators. The region has become not the exception to but rather a prototype of our urban future. For many current observers, L.A. is simply confirming what contemporaries knew throughout its history: that the city posited a set of different rules for understanding urban growth. An alternative urban metric is now overdue, since as Joel Garreau (1991, p. 3) observed in his study of edge cities: "Every American city that is growing, is growing in the fashion of Los Angeles."

Just as the Chicago School emerged at a time when that city was reaching new national prominence, Los Angeles is now making its impression on the minds of urbanists across the world. Few argue that the city is unique, or necessarily a harbinger of the future, even though both viewpoints are at some level demonstrable – true. However, at a very minimum, they all assert that Southern California is an unusual amalgam – a polycentric, polyglot, polycultural pastiche that is deeply involved in rewriting American urbanism. Moreover, their theoretical inquiries do not end with Southern California, but are also focused on more general questions concerning broader urban socio-spatial processes. The variety, volume, and pace of contemporary urban change requires the development of alternative analytical frameworks; one can no longer make an unchallenged appeal to a single model for the myriad global and local trends that surround us.

[. . .]

The particular conditions that have led now to the emergence of a Los Angeles School may be almost coincidental: (1) that an especially powerful intersection of empirical and theoretical research projects have come together in this particular place at this particular time; (2) that these trends are occurring in what has historically been the most understudied major city in the United States; (3) that these projects have attracted the attention of an assemblage of increasingly self-conscious scholars and practitioners; and (4) that the world is facing the prospect of a Pacific century, in which Southern California is likely to become a global capital. The vitality and potential of the Los Angeles School derive from the intersection of these events, and the promise they hold for a renaissance of urban theory.

[. . .]

THE LOS ANGELES SCHOOL EMERGES

[. . .]

It was during the 1980s that a group of loosely associated scholars, professionals, and advocates based in Southern California became convinced that what was happening in the region was somehow symptomatic of a broader socio-geographic transformation taking place within the United States as a whole. Their common, but then unarticulated, project was based on certain shared theoretical assumptions, as well as on the view that L.A. was emblematic of a more general urban dynamic. One of the earliest expressions of the emergent "Los Angeles School" came with the appearance in 1986 of a special issue of the journal *Society and Space*, devoted entirely to understanding Los Angeles. In their prefatory remarks to that issue, Allen Scott and Edward Soja (1986, p. 249) referred to L.A. as the "capital of the twentieth century," deliberately invoking Walter Benjamin's designation of Paris as capital of the 19th. They predicted that the volume of scholarly work on Los Angeles would quickly overtake that on Chicago, the dominant model of the American industrial metropolis.

Ed Soja's celebrated tour of Los Angeles (which first appeared in the 1986 *Society and Space* issue, and was later incorporated into his 1989 *Postmodern Geographies*) most effectively achieved the conversion of L.A. from the exception to the rule – the prototype of late 20th-century postmodern geographies:

> What better place can there be to illustrate and synthesize the dynamics of capitalist spatialization? In so many ways, Los Angeles is the place where "it all comes together". . . one might call the sprawling urban region . . . a prototopos, a paradigmatic place; or a mesocosm, an ordered world in which the micro and the macro, the idiographic and the nomothetic, the concrete and the abstract, can be seen simultaneously in an articulated and interactive combination.
>
> (Soja, 1989, p. 191)

Soja went on to assert that L.A. "insistently presents itself as one of the most informative palimpsests and paradigms of twentieth-century urban development and popular consciousness," comparable to Borges's *Aleph*: "the only place on earth where all places are seen from every angle, each standing clear, without any confusion or blending" (Soja, 1989, p. 248).

As ever, Charles Jencks (1993, p. 132) quickly picked up on the trend toward an L.A.-based urbanism, taking care to distinguish its practitioners from the L.A. school of architecture:

> The L.A. School of geographers and planners had quite a separate and independent formulation in the 1980s, which stemmed from the analysis of the city as a new post-modern urban type. Its themes vary from L.A. as the post-Fordist, post-modern city of many fragments in search of a unity, to the nightmare city of social inequities.

This same group of geographers and planners (accompanied by a few dissidents from other disciplines) gathered at Lake Arrowhead in the San Bernardino Mountains on October 11–12, 1987, to discuss the wisdom of engaging in a Los Angeles School. The participants included, if memory serves, Dana Cuff, Mike Davis, Michael Dear, Margaret FitzSimmons, Rebecca Morales, Allen Scott, Ed Soja, Michael Storper, and Jennifer Wolch. Mike Davis (1989, p. 9) later provided the first description of the putative school:

> I am incautious enough to describe the "Los Angeles School." In a categorical sense, the twenty or so researchers I include within this signatory are a new wave of Marxist geographers – or, as one of my friends put it, "political economists with their space suits on" – although a few of us are also errant urban sociologists, or, in my case, a fallen labor historian. The "School," of course, is based in Los Angeles, at UCLA and USC, but it includes members in Riverside, San Bernardino, Santa Barbara, and even Frankfurt, West Germany.

[. . .]

Mike Davis was, to the best of my knowledge, the first to mention a specific L.A. school of urbanism, and he repeated the claim in his popular contemporary history of Los Angeles, *City of Quartz* (1990).

[. . .]

FROM CHICAGO TO L.A.

The basic primer of the Chicago school was *The City*. Originally published in 1925, the book retains a tremendous vitality far beyond its interest as a historical document. I regard the book as emblematic of a modernist analytical paradigm that remained popular for most of the 20th century: Its assumptions included:

- a "modernist" view of the city as a unified whole, i.e., a coherent regional system in which the center organizes its hinterland;
- an individual-centered understanding of the urban condition; urban process in *The City* is typically grounded in the individual subjectivities of urbanites, their personal choices ultimately explaining the overall urban condition, including spatial structure, crime, poverty, and racism; and
- a linear evolutionist paradigm, in which processes lead from tradition to modernity, from primitive to advanced, from community to society, and so on.

There may be other important assumptions of the Chicago School, as represented in *The City*, that are not listed here. Finding them and identifying what is right or wrong about them is one of the tasks at hand, rather than excoriating the book's contributors for not accurately foreseeing some distant future.

The most enduring of the Chicago School models was the zonal or *concentric ring theory*, an account of the evolution of differentiated urban social areas by E. W. Burgess (1925). Based on assumptions that included a uniform land surface, universal access to a single-centered city, free competition for space, and the notion that development would take place outward from a central core, Burgess concluded that the city would tend to form a series of concentric zones.

[. . .]

Other urbanists subsequently noted the tendency for cities to grow in star-shaped rather than concentric form, along highways that radiate from a center with contrasting land uses in the interstices. This observation gave rise to a *sector theory* of urban structure, an idea advanced in the late 1930s by Homer Hoyt (1933, 1939), who observed that once variations arose in land uses near the city center, they tended to persist as the city expanded. Distinctive sectors thus grew out from the CBD, often organized along major highways.

Hoyt emphasized that "non-rational" factors could alter urban form, as when skillful promotion influenced the direction of speculative development. He also understood that the age of the buildings could still reflect a concentric ring structure, and that sectors may not be internally homogeneous at one point in time.

The complexities of real-world urbanism were further taken up in the multiple nuclei theory of C. D. Harris and E. Ullman (1945). They proposed that cities have a cellular structure in which land-uses develop around multiple growth-nuclei within the metropolis – a consequence of accessibility-induced variations in the land-rent surface and agglomeration (dis)economics. Harris and Ullman also allow that real-world urban structure is determined by broader social and economic forces, the influence of history, and international influences. But whatever the precise reasons for their origin, once nuclei have been established, general growth forces reinforce their pre-existing patterns.

Much of the urban research agenda of the 20th century has been predicted on the precepts of the concentric zone, sector, and multiple-nuclei theories of urban structure. Their influences can be seen directly in factorial ecologies of intra-urban structure, land-rent models, studies of urban economies and diseconomies of scale, and designs for ideal cities and neighborhoods. The specific and persistent popularity of the Chicago concentric ring model is harder to explain, however, given the proliferation of evidence in support of alternative theories. The most likely reasons for its endurance (as I have mentioned) are related to its beguiling simplicity and the enormous volume of publications produced by adherents of the Chicago School. . . .

In the final chapter of *The City*, the same Louis Wirth (1925) had already provided a magisterial review of the field of urban sociology, entitled (with deceptive simplicity and astonishing self-effacement) "A Bibliography of the Urban Community." But what Wirth does in this chapter, in a remarkably prescient way, is to summarize the fundamental premises of the Chicago School and to isolate two fundamental features of the urban condition that was to rise to prominence at the beginning of the 21st century. Specifically, Wirth establishes that the city lies at the center of, and provides the organizational logic for, a complex regional hinterland based on trade. But he also notes that the development of "satellite cities" is characteristic of the "latest phases" of city growth and that the location of such satellites can exert a "determining influence" on the direction of growth (1925, p. 185). He further observes that modern communications have transformed the world into a "single mechanism," where the global and the local intersect decisively and continuously (Wirth, 1925, p. 186).

And there, in a sense, you have it. In a few short paragraphs, Wirth anticipates the pivotal moments that characterize Chicago-style urbanism, those primitives that eventually will separate it from an L.A.-style urbanism. He effectively foreshadowed *avant la lettre* the shift from what I term a "modernist" to a "postmodern" city, and, in so doing, the necessity of the transition from the Chicago to the Los Angeles School. *For it is no longer the center that organizes the urban hinterlands, but the hinterlands that determine what remains of the center.* The imperatives toward decentralization (including suburbanization) have become the principal dynamic in contemporary cities; and the 21st century's emerging world cities (including L.A.) are ground-zero loci in a communications-driven globalizing political economy. From a few, relatively humble first steps, we gaze out over the abyss – the yawning gap of an intellectual fault line separating Chicago from Los Angeles.

CONTEMPORARY URBANISMS IN SOUTHERN CALIFORNIA

I turn now to review the empirical evidence of recent urban developments in Southern California. In this task, I take my lead from what exists rather than what may be considered as a normative taxonomy of urban research. From this, I move quickly to a synthesis that is prefigurative of a proto-postmodern urbanism that serves as a basis for a distinctive L.A. school of urbanism.

Edge cities

Joel Garreau noted the central significance of Los Angeles in understanding contemporary metropolitan growth in the United States. He refers to L.A. as the "great-granddaddy" of edge cities, claiming there are 26 of them within a five-county area in Southern California (Garreau, 1991, p. 9). For Garreau, edge cities represent the crucible of America's urban future. . . . One essential feature of the edge city is that politics is not yet established there. Into the political vacuum moves a "shadow government" – a privatized protogovernment that is essentially a plutocratic alternative to normal politics. Shadow governments can tax, legislate for, and police their communities, but they are rarely accountable, are responsive primarily to wealth (as opposed to numbers of voters), and subject to few constitutional constraints.

Privatopia

Privatopia, perhaps the quintessential edge city residential form, is a private housing development based in common-interest developments (CIDs) and administered by homeowner associations. There were fewer than 500 such associations in 1964; by 1992, there were 130,000 associations privately governing approximately 32 million Americans. . . . In her futuristic novel of L.A. wars between walled-community dwellers and those beyond the walls, Octavia Butler (1993) envisioned a dystopian privatopian future. It includes a balkanized nation of defended neighborhoods at odds with one another, where entire communities are wiped out for a handful of fresh lemons or a few cups of potable water; where torture and murder of one's enemies is common; and where company-town slavery is attractive to those who are fortunate enough to sell their services to the hyper-defended enclaves of the very rich.

Cultures of heteropolis

One of the most prominent sociocultural tendencies in contemporary Southern California is the rise of minority populations. Provoked to comprehend the causes and implications of the 1992 civil disturbances in Los Angeles, Charles Jencks zeroes in on the city's diversity as the key to L.A.'s emergent urbanism: "Los Angeles is a combination of enclaves with high identity, and multienclaves with mixed identity, and, taken as a whole, it is perhaps the most heterogenenous city in the world" (Jencks, 1993, p. 32).

City as theme park

California in general, and Los Angeles in particular, have often been promoted as places where the American (suburban) Dream is most easily realized. Its oft-noted qualities of optimism and tolerance coupled with a balmy climate have given rise to an architecture and society fostered by a spirit of experimentation, risk-taking, and hope. Many writers have used the "theme park" metaphor to describe the emergence of such variegated cityscapes. . . . Disneyland is the archetype, described by Sorkin (1992, p. 227) as a place of "Taylorized fun," the "Holy See of Creative Geography." What is missing in this new cybernetic suburbia is not a particular building or place, but the spaces between, i.e., the connections that make sense of forms. What is missing, then, is connectivity and community. . . .

Fortified city

The downside of the Southern Californian dream has, of course, been the subject of countless dystopian visions in histories, movies, and novels. In one powerful account, Mike Davis (1992a) noted how Southern Californians' obsession with security has transformed the region into a fortress. This shift is accurately manifested in the physical form of the city, which is divided into fortified cells of affluence and places of terror where police battle the criminalized poor. These urban phenomena, according to Davis (1992a, p. 155), have placed Los Angeles "on the hard edge of postmodernity." The dynamics of fortification involve the omnipresent application of high-tech policing methods to protect the security of gated residential developments and panopticon malls. It extends to space policing, including a proposed satellite observation capacity that would create an invisible Haussmannization of Los Angeles. In the consequent carceral city, the working poor and destitute are spatially sequestered on the mean streets, and excluded from the affluent forbidden cities through security by design.

Interdictory spaces

Elaborating upon Davis' fortress urbanism, Steven Flusty (1994) observed how various types of fortification have extended a canopy of suppression and surveillance across the entire city. His taxonomy identifies how spaces are designed to exclude by a combination of their function and cognitive sensibilities. . . . One consequence of the socio-spatial differentiation described by Davis and Flusty is an acute fragmentation of the urban landscape. Commentators who remark upon the strict division of residential neighborhoods along race and class lines miss the fact that L.A's microgeography is incredibly volatile and varied. In many neighborhoods, simply turning a street corner will lead the pedestrian/driver into totally different social and physical configurations. . . .

Historical geographies of restructuring

. . . In his history of Los Angeles between 1965 and 1992, Soja (1996) attempts to link the emergent patterns of urban form with underlying social processes. He identified six kinds of restructuring, which together define the region's contemporary urban process. In addition to *Exopolis* (noted earlier), Soja lists: *Flexcities*, associated with the transition to post-Fordism, especially deindustrialization and the rise of the information economy; and *Cosmopolis*, referring to the globalization of Los Angeles both in terms of its emergent world city status and its internal multicultural diversification. According to Soja, peripheralization, post-Fordism, and globalization together define the experience of urban restructuring in Los Angeles. Three specific geographies are consequent upon these dynamics: *Splintered Labyrinth*, which describes the extreme forms of social, economic, and political polarization characteristic of the postmodern city; *Carceral City*, referring to the new "incendiary urban geography" brought about by the amalgam of violence and police surveillance; and *Simcities*, the term Soja uses to describe the new ways of seeing the city that are emerging from the study of Los Angeles – a kind of epistemological restructuring that foregrounds a postmodern perspective.

Fordist versus Post-Fordist regimes of accumulation and regulation

. . . In a series of important books, Allen Scott (1988a, 1988b, 1993, 2000) has portrayed the burgeoning urbanism of Southern California as a consequence of this deep-seated structural change in the capitalist political economy. Scott's basic argument is that there have been two major phases of urbanization in the United States. The first related to an era of Fordist mass production, during which the paradigmatic cities of industrial capitalism (Detroit, Chicago, Pittsburgh, etc.) coalesced around industries that were themselves based on ideas of mass production. The second phase is associated with the decline of the Fordist era and the rise of a post-Fordist "flexible production" (what some refer to as "flexible accumulation"). This is a form of industrial activity based on small-size, small-batch units of (typically sub-contracted) production that are nevertheless integrated into clusters of economic activity. Such clusters have been observed in two manifestations: labor-intensive craft forms (in Los Angeles, typically garments and jewelry); and high technology (especially the defense and aerospace industries). According to Scott, these so-called "technopoles" until recently constituted the principal geographical loci of contemporary (sub)urbanization in Southern California. An equally important facet of post-Fordism is the significant informal sector that mirrors the gloss of the high-tech sectors. Post-Fordist regimes of accumulation are associated with analogous regimes of regulation, or social control. . . .

Globalization

Needless to say, any consideration of the changing nature of industrial production sooner or later must encompass the globalization question. In his reference to the global context of L.A.'s localisms, Mike Davis (1992b) claims that if L.A. is in any sense paradigmatic, it is because the city condenses the intended and unintended spatial consequences of a global post-Fordism. He insists that there is no simple master-logic of restructuring, focusing instead on two key localized macroprocesses: the overaccumulation in Southern California of bank and real estate capital principally from the East Asian trade surplus of the 1980s; and the reflux of low-wage manufacturing and labor-intensive service industries following upon immigration from Mexico and Central America. For instance, Davis (1992b, p. 26) noted how the City of Los Angeles used tax dollars gleaned from international capital investments to subsidize its downtown (Bunker Hill) urban renewal, a process he refers to as "municipalized land speculation." Through such connections, what happens today in Asia and Central America will tomorrow have an effect in Los Angeles. This global/local dialectic has already become an important (if somewhat imprecise) leitmotif of contemporary urban theory, most especially via notions of "world cities" and global "city-regions" (Scott, 1998, 2001).

Politics of nature

The natural environment of Southern California has been under constant assault since the first colonial settlements. Human habitation on a metropolitan scale has only been possible through a widespread manipulation of nature, especially the control of water resources in the American West. On the one hand, Southern Californians tend to hold a grudging respect for nature, living as they do adjacent to one of the earth's major geological hazards, and in a desert environment that is prone to flood, landslide, and fire. On the other hand, its inhabitants have been energetically, ceaselessly, and often carelessly unrolling the carpet of urbanization over the natural landscape for more than a century. This uninhibited occupation has engendered its own range of environmental problems, most notoriously air pollution, but also issues related to habitat loss and encounters between humans and other animals. . . .

LOS ANGELES AS POSTMODERN URBANISM

If all these observers of the Southern California scene could talk with each other, how might they synthesize their visions? At the risk of misrepresenting their work, I can suggest a synthesis that outlines a "proto-postmodern" urban process. It is driven by a global restructuring that is permeated and balkanized by a series of interdictory networks; whose populations are socially and culturally heterogeneous, but politically and economically polarized; whose residents are educated and persuaded to the consumption of dreamscapes even as the poorest are consigned to carceral cities; whose built environment, reflective of these processes consists of edge cities, privatopias, and the like; and whose natural environment is being erased to the point of unlivability while at the same time providing a focus for political action.

[. . .]

The Los Angeles School is distinguishable from the Chicago precepts (as noted above) by the following counter-propositions:

- Traditional concepts of urban form imagine the city organized around a central core; in a revised theory, the urban peripheries are organizing what remains of the center.
- A global, corporate-dominated connectivity is balancing, even offsetting, individual-centered agency in urban processes.
- A linear evolutionist urban paradigm has been usurped by a nonlinear, chaotic process that includes pathological forms such as transnational criminal organizations, common-interest developments (CIDs), and life-threatening environmental degradation (e.g., global warming).

[. . .]

"Keno capitalism" is the synoptic term that Steven Flusty and I have adopted to describe the spatial manifestations that are consequent upon the (postmodern) urban condition implied by these assumptions (see Figure 1). Urbanization is occurring on a quasi-random field of opportunities, in which each space is (in principle) equally available through its connection with the information superhighway (Dear and Flusty, 1998). Capital touches down as if by chance on a parcel of land, ignoring the opportunities on intervening lots, thus sparking the development process. The relationship between development of one parcel and non-development of another is a disjointed, seemingly unrelated affair. While not truly a random process, it is evident that the traditional, center-driven agglomeration economies that have guided urban

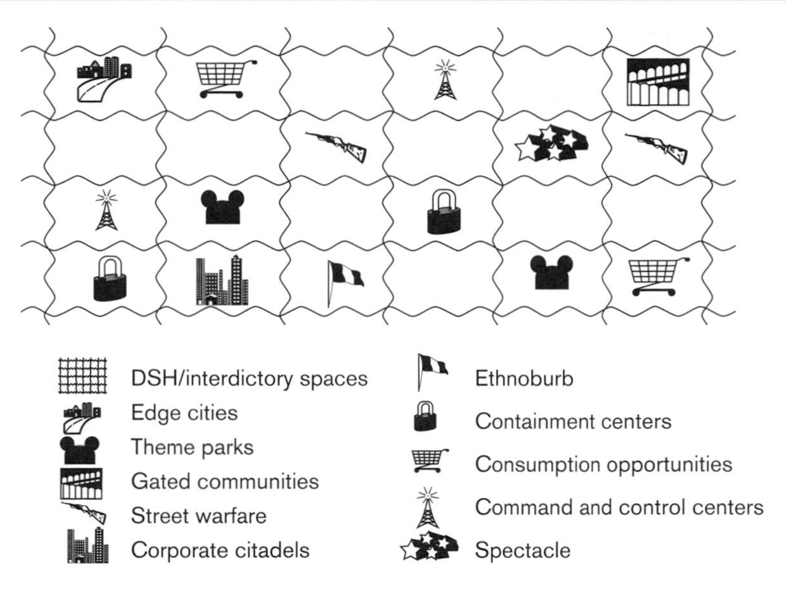

Figure 1 The "keno gaming board" model of postmodern Los Angeles

development in the past no longer generally apply. Conventional city form, Chicago-style, is sacrificed in favor of a non-contiguous collage of parcelized, consumption-oriented landscapes devoid of conventional centers yet wired into electronic propinquity and nominally unified by the mythologies of the (dis)information superhighway.

[. . .]

INVITATION TO A DEBATE

In these postmodern times, the gesture to a Los Angeles School might appear to be a deeply contradictory intellectual strategy: A "school" has semantic overtones of codification and hegemony; it has structure and authority. Modernists and postmodernists alike might shudder at the irony implied by these associations. And yet, ultimately, I am comfortable in proclaiming the existence of an L.A. school of urbanism for two reasons. First, the Los Angeles School exists as a body of literature, as this essay attests. It exhibits an evolution through history, beginning with analysis of Los Angeles as an aberrant curiosity distinct from other forms of urbanism. The tone of that history has shifted gradually to the point that the city is now commonly represented as indicative of a new form of urbanism supplanting the older forms against which Los Angeles was once judged deviant. Second, the Los Angeles School exists as a discursive strategy demarcating a space both for the exploration of new realities and for resistance to old hegemonies. It is proving to be far more successful than its detractors at explaining the form and function of the urban.

[. . .]

The fragmented and globally oriented nature of the Los Angeles School counters any potential for a new hegemony. The avowal of a Los Angeles School can become a decolonizing, postcolonial impulse, even as it alerts us to new colonialisms lurking along the historical path. Those who worry about the hegemonic intent of a Los Angeles School may rest assured that its adherents are in fact pathologically anti-leadership. Nor will everyone who writes on L.A. readily identify as a member of the Los Angeles School; some adamantly reject such a notion (e.g., Ethington and Meeker, 2001). The programmatic intent of the Los Angeles School remains fractured, incoherent, and idiosyncratic even to its constituent scholars, who most often perceive themselves as occupying a place on the periphery rather than at the center. The Los Angeles School promotes inclusiveness by inviting as members all those who take Los Angeles as a worthy object of study and a source of insight into the nature of contemporary urbanism. Such a school evades dogma by including divergent empirical and theoretical approaches rooted in philosophies, both modern and postmodern, ranging from Marxist to Libertarian. Admittedly, such a school will be a fragmentary and loosely connected entity, always on the verge of disintegration – but, then again, so is Los Angeles itself.

A unified, consensual description of Los Angeles is equally unlikely, since it would necessitate excluding a plethora of valuable readings on the region. For instance, numerous discursive battles have been fought in L.A. since the events of April 1992 to decide what term best describes them or, more cynically, which term most effectively recasts them as a weapon adaptable to a particular rhetorical arsenal. Those who read the events as a spontaneous, visceral, opportunistic reaction to the acquittal of Rodney King employ the term *riot*. For those who read the events within the context of economic evisceration and social polarization, the term *uprising* is preferred. And those who see in them a more conscious political intentionality apply the term *rebellion*. For its part, civic authority skirts these issues by relying on the supposedly depoliticized term, *civil unrest*. But those concerned with the perspective of Korean participants, literally caught in the middle of the turmoil itself as well as the subsequent rhetoric wars, deploy the Korean tradition of naming an occurrence by its principal date and so make use of the term, *Sa-I-Gu*. Which name is definitive? The polyvocality of the Los Angeles School permits us to replace the question, "Which is it?" with, "Which is it, at which stage of events, at which location in the region, and from whose perspective?" Such an approach may well entail a loss of clarity and certitude, but in exchange it offers a richness of description and interpretation that would otherwise be forfeited in the name of achieving an "official" narrative.

Finally, the temptation to adopt L.A. as a world city template is avoidable because the urban landscapes of Los Angeles are not necessarily original to L.A. The luxury compound atop a matrix of impoverished misery, and self-contained communities of fortified homes can also he found in places like Manila and São Paulo. Indeed, Anthony King has suggested that all things ascribed to postmodern urbanism can be seen decades earlier in the principal cities of the colonial world. The Los Angeles School justifies a presentation of L.A. not as *the* model of contemporary urbanism, nor as the privileged locale whence a cabal of regal theoreticians issue proclamations about the way things really are, but as one of a number of space-time geographical prisms through which current processes of urban (re)formation may be advantageously viewed. Hence, the literature of the Los Angeles School largely (although not exclusively) shows itself to be less concerned with looking to L.A. for models of the urban, and more about looking for contemporary expressions of the urban in L.A. Thus, the school and its concepts of contemporary Angeleno urbanism do not represent an emerging vision of contemporary urbanism in total; instead they are but one component in a new comparative urban studies working out of Los Angeles but inviting the participation of (and placing equal importance upon) the on-going experiences and voices of Tijuana, São Paulo, Hong Kong, and the like (cf. Sassen, 1991).

[. . .]

REFERENCES

Burgess, E. W. (1925) "The Growth of the City," in R. E. Park, E. W. Burgess, and R. McKenzie, *The City: Suggestions of Investigation of Human Behavior in the Urban Environment*, pp. 47–62. Chicago, IL: University of Chicago Press.

Butler, O. E. (1993) *Parable of the Sower*. New York: Four Walls Eight Windows.

Davis, M. (1989) "Homeowners and Homeboys: Urban Restructuring in LA," *Enclitic*, Summer, 9–16.

Davis, M. (1990) *City of Quartz: Excavating the Future in Los Angeles*. New York: Verso.

Davis, M. (1992a) "Fortress Los Angeles: The Militarization of Urban Space," in M. Sorkin (ed.), *Variations on a Theme Park*, p. 155. New York: Noonday Press.

Davis, M. (1992b) "*Chinatown* Revisited? The 'Internationalization' of Downtown Los Angeles," in Reid, D. (ed.), *Sex, God and Death in L.A.*, pp. 19–53. New York: Pantheon Books.

Dear, M. and Flusty, S. (1998) "Postmodern Urbanism," *Annals, Association of American Geographers*, 88 (1), 50–72.

Ethington, P. and Meeker, M. (2001) "'*Saber y Conocer*': The Metropolis of Urban Inquiry," in M. Dear (ed.), *From Chicago to L.A.: Making Sense of Urban Theory*, pp. 403–420. Thousand Oaks, CA: Sage Publications.

Flusty, S. (1994) *Building Paranoia. The Proliferation of Interdictory Space and the Erosion of Spatial Justice.* West Hollywood, CA: Los Angeles Forum for Architecture and Urban Design.

Garreau, J. (1991) *Edge City: Life on the New Frontier.* New York: Anchor Books.

Hoyt, H. (1933) *One Hundred Years of Land Values in Chicago.* Chicago, IL: University of Chicago Press.

Hoyt, H. (1939) *The Structure and Growth of Residential Neighborhoods in American Cities.* Washington, DC: U.S. Federal Housing Administration.

Jencks, C. (1993) *Heteropolis: Los Angeles, the Riots and the Strange Beauty of Hetero-Architecture.* New York: St. Martin's Press.

Sassen, S. (1991) *The Global City*, Princeton, NJ: Princeton University Press.

Scott, A. J. (1988a) *New Industrial Spaces: Flexible Production Organization and Regional Development in North America and Western Europe.* London: Pion.

Scott, A. J. (1988b) *Metropolis: From the Division of Labor to Urban Form.* Berkeley, CA: University of California Press.

Scott, A. J. (1993) *Technopolis: High-Technology Industry and Regional Development in Southern California.* Berkeley, CA: University of California Press.

Scott, A. J. (1998) *Regions and the World Economy: The Coming Shape of Global Production, Competition, and Political Order.* Oxford: Oxford University Press.

Scott, A. J. (2000) *The Cultural Economy of Cities.* London: Sage Publications.

Scott, A. J. (ed.) (2001) *Global City-Regions: Trends, Theory, Policy.* Oxford: Oxford University Press.

Scott, A. J. and Soja, E. W. (1986) "Los Angeles: Capital of the Late 20th Century," *Society and Space*, 4, 249–254.

Soja, E. W. (1989) *Postmodern Geographies: The Reassertion of Space in Critical Social Theory.* New York: Verso.

Soja, E. (1996) "Los Angeles 1965–1992: The Six Geographies of Urban Restructuring," in A. J. Scott and E. Soja (eds), *The City: Los Angeles and Urban Theory at the End of the Twentieth Century*, pp. 426–462. Los Angeles, CA: University of California Press.

Sorkin, M. (ed.) (1992) *Variations on a Theme Park. The New American City and the End of Public Space.* New York: Hill and Wang.

Wirth, L. (1925) "A Bibliography of the Urban Community" in R. E. Park, E. W. Burgess, and R. McKenzie, *The City: Suggestions of Investigation of Human. Behavior in the Urban Environment*, pp. 161–228. Chicago, IL: University of Chicago Press.

TWO

"Cities and the Geographies of 'Actually Existing Neoliberalism'"

from *Antipode* (2002)

Neil Brenner and Nik Theodore

Editors' Introduction

Neil Brenner and Nik Theodore cut through the smoke and mirrors of neoliberal doctrine as espoused by international organizations like the World Trade Organization and World Bank that purport to "unleash" optimal economic development through the mechanisms of open markets and financial deregulation. They unmask the operations of "actually existing neoliberalism," as contextually embedded practices of economic restructuring that occur on multiple scales from local, national, to global. This restructuring involves the rolling back of inherited government spending programs and social regulatory frameworks that were entrenched during the Fordist-Keynesian period of capitalist development. They apply the Marxian concept of creative destruction, which recognizes that new capitalist development arises out of the destruction of the prior economic order.

They believe that cities and city-regions are focal points of neoliberal transition that have experienced particularly intense problems and social conflicts caused by the dismantling of financial and governmental regulatory mechanisms, and the recent tendency towards overproduction and periodic economic crises. They see cities as institutional laboratories for the rolling out of neoliberalism, as climates of economic austerity and deficit reduction have led to municipal cost-cutting and retreat from urban services. Similarly to the experience in the developing world, neoliberal cities in the developed world have shifted increasingly to free trade and entrepreneurial strategies, enterprise zones, and privatization schemes. The ascendance of neoconservative political and social movements has also led to demonizing of subcultural groups, the poor, and immigrants as "dangerous classes" that threaten public morality and family life. But the disenfranchised have responded with struggles in response, marking cities as flashpoints of contention in the neoliberal period.

They distinguish a more negotiated "roll-out" period as following the initial contested "roll-back" period of neoliberal transition. This is analogous to the transition from the radical, antistatist neoliberalisms of Reagan and Thatcher in the 1980s to the more socially moderate neoliberalisms of Clinton, Blair, and Schröder in the 1990s (see Mele, this volume). The roll-out period includes more benign institutional realignments such as deployment of community-based programs and new forms of inter-organizational networking. There will be trial-and-error and social contestation, they point out, along with recurring conflict. Rather than a linear or predictable process that can help generate coherent and sustainable solutions, they describe neoliberalism as a deeply contradictory set of restructuring strategies that can generate further instability.

Neil Brenner is a Professor of Urban Theory at the Harvard Graduate School of Design. Nik Theodore is an Associate Professor of Urban Planning and Policy at the University of Illinois-Chicago.

INTRODUCTION

The linchpin of neoliberal ideology is the belief that open, competitive, and unregulated markets, liberated from all forms of state interference, represent the optimal mechanism for economic development. Although the intellectual roots of this "utopia of unlimited exploitation" (Bourdieu 1998) can be traced to the postwar writings of Friedrich Hayek and Milton Friedman, neoliberalism first gained widespread prominence during the late 1970s and early 1980s as a strategic political response to the sustained global recession of the preceding decade.

Faced with the declining profitability of traditional mass-production industries and the crisis of Keynesian welfare policies, national and local states throughout the older industrialized world began, if hesitantly at first, to dismantle the basic institutional components of the postwar settlement and to mobilize a range of policies intended to extend market discipline, competition, and commodification throughout all sectors of society. In this context, neoliberal doctrines were deployed to justify, among other projects, the deregulation of state control over major industries, assaults on organized labor, the reduction of corporate taxes, the shrinking and/or privatization of public services, the dismantling of welfare programs, the enhancement of international capital mobility, the intensification of interlocality competition, and the criminalization of the urban poor.

If Thatcherism and Reaganism represented particularly aggressive programs of neoliberal restructuring during the 1980s, more moderate forms of a neoliberal politics were also mobilized during this same period in traditionally social democratic or social Christian democratic states such as Canada, New Zealand, Germany, the Netherlands, France, Italy, and even Sweden. Following the debt crisis of the early 1980s, neoliberal programs of restructuring were extended globally through the efforts of the USA and other G-7 states to subject peripheral and semiperipheral states to the discipline of capital markets. Bretton Woods institutions such as the General Agreement on Tariffs and Trade (GATT), World Trade Organization (WTO), the World Bank, and the International Monetary Fund (IMF) were subsequently transformed into the agents of a transnational neoliberalism and were mobilized to institutionalize this extension of market forces and commodification in the Third World through various structural adjustment and fiscal austerity programs. By the mid-1980s, in the wake of this dramatic U-turn of policy agendas throughout the world, neoliberalism had become the dominant political and ideological form of capitalist globalization.

The global imposition of neoliberalism has, of course, been highly uneven, both socially and geographically, and its institutional forms and sociopolitical consequences have varied significantly across spatial scales and among each of the major supraregional zones of the world economy. While recognizing the polycentric and multiscalar character of neoliberalism as a geopolitical and geoeconomic project, the goal of this collection is to explore the role of neoliberalism in ongoing processes of urban restructuring. The supranational and national parameters of neoliberalism have been widely recognized in the literatures on geopolitical economy. However, the contention that neoliberalism has also generated powerful impacts at subnational scales – within cities and city-regions – deserves to be elaborated more systematically.

This essay provides a "first cut" towards theorizing and exploring the complex institutional, geographical, and social interfaces between neoliberalism and urban restructuring. We begin by presenting the methodological foundations for an approach to the geographies of what we term "actually existing neoliberalism." In contrast to neoliberal ideology, in which market forces are assumed to operate according to immutable laws no matter where they are "unleashed," we emphasize the contextual embeddedness of neoliberal restructuring projects insofar as they have been produced within national, regional, and local contexts defined by the legacies of inherited institutional frameworks, policy regimes, regulatory practices, and political struggles. An understanding of actually existing neoliberalism must therefore explore the path-dependent, contextually specific interactions between inherited regulatory landscapes and emergent neoliberal, market-oriented restructuring projects at a broad range of geographical scales. These considerations lead to a conceptualization of contemporary neoliberalization processes as catalysts and expressions of an ongoing creative destruction of political-economic space at multiple geographical scales. While the neoliberal restructuring projects of the last two decades have failed to establish a coherent basis for sustainable capitalist growth, they have nonetheless profoundly reworked the institutional infrastructures upon which Fordist-Keynesian capitalism was grounded. The concept of creative destruction is presented to describe the geographically uneven, socially regressive, and politically volatile trajectories of institutional/spatial change that have been crystallizing under these conditions. The essay concludes by discussing the role of urban spaces within the contradictory and chronically unstable geographies of actually existing neoliberalism. Throughout the advanced capitalist world, we suggest, cities have become strategically crucial geographical arenas in which a variety of neoliberal initiatives – along with closely intertwined strategies of crisis displacement and crisis management – have been articulated.

TOWARDS A POLITICAL ECONOMY OF ACTUALLY EXISTING NEOLBERALISM

The 1990s was a decade in which the term "neoliberalism" became a major rallying point for a wide range of anticapitalist popular struggles, from the Zapatista rebellion in Chiapas, the subsequent series of Gatherings for Humanity and Against Neoliberalism, and the December 1995 mass strikes in France to the mass protests against the WTO, the IMF, the World Bank, and the World Economic Forum in locations such as Davos, Genoa, London, Melbourne, Mumbai, Nice, Prague, Seattle, Sydney, Washington DC, and Zürich, among many others. As such struggles continue to proliferate in the new millennium, anticapitalist forces throughout the world have come to identify neoliberalism as a major target for oppositional mobilization. Among activists and radical academics alike, there is considerable agreement regarding the basic elements of neoliberalism as an ideological project.

[. . .]

While neoliberalism aspires to create a "utopia" of free markets liberated from all forms of state interference, it has in practice entailed a dramatic intensification of coercive, disciplinary forms of state intervention in order to impose market rule upon all aspects of social life. Whereas neoliberal ideology implies that self-regulating markets will generate an optimal allocation of investments and resources, neoliberal political practice has generated pervasive market failures, new forms of social polarization, and a dramatic intensification of uneven development at all spatial scales . . . During the last two decades, the dysfunctional effects of neoliberal approaches to capitalist restructuring have been manifested in diverse institutional arenas and at a range of spatial scales. As such studies have indicated, the disjuncture between the ideology of self-regulating markets and the everyday reality of persistent economic stagnation – intensifying inequality, destructive interplace competition, and generalized social insecurity – has been particularly blatant in precisely those political-economic contexts in which neoliberal doctrines have been imposed most extensively.

Crucially, the manifold disjunctures that have accompanied the worldwide imposition of neoliberalism – between ideology and practice; doctrine and reality; vision and consequence – are not merely accidental side effects of this disciplinary project of imposing a new "market civilization." Rather, they are among its most essential features. We would argue a purely definitional approach to the political economy of neoliberal restructuring contains significant analytical limitations . . . The somewhat elusive phenomenon that needs definition must be construed as a historically specific, ongoing, and internally contradictory process of market-driven sociospatial transformation, rather than as a fully actualized policy regime, ideological form, or regulatory framework. An adequate understanding of contemporary neoliberalization processes requires not only a grasp of their politico-ideological foundations but also, just as importantly, a systematic inquiry into their multifarious institutional forms, their developmental tendencies, their diverse sociopolitical effects, and their multiple contradictions.

We shall describe these ongoing neoliberalization processes through the concept of actually existing neoliberalism. This concept is intended not only to underscore the contradictory, destructive character of neoliberal policies, but also to highlight the ways in which neoliberal ideology systematically misrepresents the real effects of such policies upon the macroinstitutional structures and evolutionary trajectories of capitalism. In this context, two issues deserve particular attention. First, neoliberal doctrine represents states and markets as if they were diametrically opposed principles of social organization, rather than recognizing the politically constructed character of all economic relations. Second, neoliberal doctrine is premised upon a "one size fits all" model of policy implementation that assumes that identical results will follow the imposition of market-oriented reforms, rather than recognizing the extraordinary variations that arise as neoliberal reform initiatives are imposed within contextually specific institutional landscapes and policy environments.

[...]

First and foremost, the preceding considerations suggest that an analysis of actually existing neoliberalism must begin by exploring the entrenched landscapes of capitalist regulation, derived from the Fordist-Keynesian period of capitalist development, within which neoliberal programs were first mobilized following the geoeconomic crises of the early 1970s. From this perspective, the impacts of neoliberal restructuring strategies cannot be understood adequately through abstract or decontextualized debates regarding the relative merits of market-based reform initiatives or the purported limits of particular forms of state policy.

As numerous scholars in the regulationist tradition have indicated, the Fordist-Keynesian configuration of capitalist development was grounded upon a historically specific set of regulatory arrangements and political compromises that provisionally stabilized the conflicts and contradictions that are endemic to capitalism. Although the sources of this unprecedented "golden age" of capitalist expansion remain a matter of considerable academic dispute, numerous scholars have emphasized the key role of the *national* scale as the pre-eminent geographical basis for accumulation and for the regulation of political-economic life during this period. Of course, the exact configuration of regulatory arrangements and political compromises varied considerably according to the specific model of capitalism that was adopted in each national context. Nonetheless, a number of broad generalizations can be articulated regarding the basic regulatory institutional architecture that underpinned North Atlantic Fordism.

- *Wage relation.* Collective bargaining occurred at the national scale, often through corporatist accommodations between capital, labor, and the state; wage labor was extended and standardized with the spread of mass-production systems throughout national social formations; and wages were tied to productivity growth and tendentially increased in order to underwrite mass consumption.
- *Form of intercapitalist competition.* Monopolistic forms of regulation enabled corporate concentration and centralization within major national industrial sectors; competition between large firms was mediated through strategies to rationalize mass-production technologies; and national states mobilized various forms of industrial policy in order to bolster the world-market positions of their largest firms as national champions.
- *Monetary and financial regulation.* The money supply was regulated at a national scale through the US-dominated Bretton Woods system of fixed exchange rates; national central banks oversaw the distribution of credit to corporations and consumers; and long-term investment decisions by capital were enabled by a stabilized pattern of macroeconomic growth.
- *The state and other forms of governance.* National states became extensively engaged in managing aggregate demand, containing swings in the business cycle, generalizing mass consumption, redistributing the social product through welfare programs, and mediating social unrest.
- *International configuration.* The world economy was parcelized among relatively autocentric national economies and policed by the US global hegemon. Meanwhile, as the Fordist accumulation regime matured, global interdependencies among national economic spaces intensified due to enhanced competition among transnational corporations, the expansion of trade relations, and the ascendancy of the US dollar as world currency.
- *The regulation of uneven spatial development.* National states introduced a range of compensatory regional policies and spatial planning initiatives intended to alleviate intranational sociospatial polarization by spreading industry and population across the surface of the national territory. Entrenched world-scale patterns of

uneven development were nonetheless maintained under the rubric of US global hegemony and Cold War geopolitics.

During the early 1970s, however, the key link between (national) mass production and (national) mass consumption was shattered due to a range of interconnected trends and developments, including: the declining profitability of Fordist sectors; the intensification of international competition; the spread of deindustrialization and mass unemployment; and the abandonment of the Bretton Woods system of national currencies. Subsequently, the Fordist system was subjected to a variety of pressures and crisis-tendencies, leading to a profound shaking-up and reworking of the forms of territorial organization that had underpinned the "golden age" of postwar economic prosperity. The global political-economic transformations of the post-1970s period radically destabilized the Fordist accumulation regime, decentered the entrenched role of the national scale as the predominant locus for state regulation, and undermined the coherence of the national economy as a target of state policies. This reshuffling of the hierarchy of spaces has arguably been the most far-reaching geographical consequence of the crisis of North Atlantic Fordism in the early 1970s.

SPACES OF NEOLIBERALIZATION: CITIES

The preceding discussion underscored the ways in which the worldwide ascendancy of neoliberalism during the early 1980s was closely intertwined with a pervasive rescaling of capital–labor relations, intercapitalist competition, financial and monetary regulation, state power, the international configuration, and uneven development throughout the world economy. As the taken-for-granted primacy of the national scale has been undermined in each of these arenas, inherited formations of urban governance have likewise been reconfigured quite systematically throughout the older industrialized world. While the processes of institutional creative destruction associated with actually existing neoliberalism are clearly transpiring at all spatial scales, it can be argued that they are occurring with particular intensity at the urban scale, within major cities and city-regions. On the one hand, cities today are embedded within a highly uncertain geoeconomic environment characterized by monetary chaos, speculative movements of financial capital, global location strategies by major transnational corporations, and rapidly intensifying interlocality competition. In the context of this deepening "global–local disorder," most local governments have been constrained – to some degree, independently of their political orientation and national context – to adjust to heightened levels of economic uncertainty by engaging in short-termist forms of interspatial competition, place-marketing, and regulatory undercutting in order to attract investments and jobs. Meanwhile, the retrenchment of national welfare state regimes and national intergovernmental systems has likewise imposed powerful new fiscal constraints upon cities, leading to major budgetary cuts during a period in which local social problems and conflicts have intensified in conjunction with rapid economic restructuring.

On the other hand, in many cases, neoliberal programs have also been directly "interiorized" into urban policy regimes, as newly formed territorial alliances attempt to rejuvenate local economies through a shock treatment of deregulation, privatization, liberalization, and enhanced fiscal austerity. In this context, cities – including their suburban peripheries – have become increasingly important geographical targets and institutional laboratories for a variety of neoliberal policy experiments, from place-marketing, enterprise and empowerment zones, local tax abatements, urban development corporations, public-private partnerships, and new forms of local boosterism to workfare policies, property-redevelopment schemes, business-incubator projects, new strategies of social control, policing, and surveillance, and a host of other institutional modifications within the local and regional state apparatus. As the contributions to this volume indicate in detail, the overarching goal of such neoliberal urban policy experiments is to mobilize city space as an arena both for market-oriented economic growth and for elite consumption practices. Table 1 schematically illustrates some of the many politico-institutional mechanisms through which neoliberal projects have been localized within North American and Western European cities during the past two

Table 1 Destructive and creative moments of neoliberal localization

Mechanisms of Neoliberal Localization	*Moment of Destruction*	*Moment of Creation*
Recalibration of intergovernmental relations	■ Dismantling of earlier systems of central government support for municipal activities	■ Devolution of new tasks, burdens, and responsibilities to municipalities; creation of new incentive structures to reward local entrepreneurialism and to catalyze "endogenous growth"
Retrenchment of public finance	■ Imposition of fiscal austerity measures upon municipal governments	■ Creation of new revenue-collection districts and increased reliance of municipalities upon local sources of revenue, user fees, and other instruments of private finance
Restructuring the welfare state	■ Local relays of national welfare service-provision are retrenched; assault on managerial-welfarist local state apparatuses	■ Expansion of community based sectors and private approaches to social service provision ■ Imposition of mandatory work requirements on urban welfare recipients; new (local) forms of workfare experimentation
Reconfiguring the institutional structure of the local state	■ Dismantling of bureaucratized, hierarchical forms of local public administration ■ Devolution of erstwhile state tasks to voluntary community networks ■ Assault on traditional relays of local democratic accountability	■ "Rolling forward" of new networked forms of local governance based upon public-private partnerships, "quangos," and the "new public management" ■ Establishment of new institutional relays through which elite business interests can directly influence major local development decisions
Privatization of the municipal public sector and collective infrastructures	■ Elimination of public monopolies for the provision of standardized municipal services (utilities, sanitation, public safety, mass transit, etc.)	■ Privatization and competitive contracting of municipal services ■ Creation of new markets for service delivery and infrastructure maintenance ■ Creation of privatized, customized, and networked urban infrastructures intended to (re)position cities within supranational capital flows
Restructuring urban housing markets	■ Razing public housing and other forms of low-rent accommodation ■ Elimination of rent controls and project-based construction subsidies	■ Creation of new opportunities for speculative investment in central-city real estate markets ■ Emergency shelters become "warehouses" for the homeless ■ Introduction of market rents and tenant-based vouchers in low-rent niches of urban housing markets

Table 1 *Continued*

Mechanisms of Neoliberal Localization	*Moment of Destruction*	*Moment of Creation*
Reworking labor market regulation	■ Dismantling of traditional, publicly funded education, skills training, and apprenticeship programs for youth, displaced workers, and the unemployed	■ Creation of a new regulatory environment in which temporary staffing agencies, unregulated "labor corners," and other forms of contingent work can proliferate ■ Implementation of workreadiness programs aimed at the conscription of workers into low-wage jobs ■ Expansion of informal economies
Restructuring strategies of territorial development	■ Dismantling of autocentric national models of capitalist growth ■ Destruction of traditional compensatory regional policies ■ Increasing exposure of local and regional economies to global competitive forces ■ Fragmentation of national space-economies into discrete urban and regional industrial systems	■ Creation of free trade zones, enterprise zones, and other deregulated spaces within major urban regions ■ Creation of new development areas, technopoles, and other new industrial spaces at subnational scales ■ Mobilization of new "glocal" strategies intended to rechannel economic capacities and infrastructure investments into "globally connected" local/regional agglomerations
Transformations of the built environment and urban form	■ Elimination and/or intensified surveillance of urban public spaces ■ Destruction of traditional working-class neighborhoods in order to make way for speculative redevelopment ■ Retreat from community oriented planning initiatives	■ Creation of new privatized spaces of elite/corporate consumption ■ Construction of large-scale megaprojects intended to attract corporate investment and reconfigure local land-use patterns ■ Creation of gated communities, urban enclaves, and other "purified" spaces of social reproduction ■ "Rolling forward" of the gentrification frontier and the intensification of sociospatial polarization ■ Adoption of the principle of "highest and best use" as the basis for major land-use planning decisions
Interlocal policy transfer	■ Erosion of contextually sensitive approaches to local policymaking ■ Marginalization of "home-grown" solutions to localized market failures and governance failures	■ Diffusion of generic, prototypical approaches to "modernizing" reform among policymakers in search of quick fixes for local social problems (e.g. welfare-to-work programs, place-marketing strategies, zero-tolerance crime policies, etc.) ■ Imposition of decontextualized "best practice" models upon local policy environments

Table 1 *Continued*

Mechanisms of Neoliberal Localization	*Moment of Destruction*	*Moment of Creation*
Re-regulation of urban civil society	■ Destruction of the "liberal city" in which all inhabitants are entitled to basic civil liberties, social services, and political rights	■ Mobilization of zero tolerance crime policies and "broken windows" policing ■ Introduction of new discriminatory forms of surveillance and social control ■ Introduction of new policies to combat social exclusion by reinserting individuals into the labor market
Re-representing the city	■ Postwar image of the industrial, working-class city is recast through a (re-)emphasis on urban disorder, "dangerous classes," and economic decline	■ Mobilization of entrepreneurial discourses and representations focused on the need for revitalization, reinvestment, and rejuvenation within major metropolitan areas

decades, distinguishing in turn their constituent (partially) destructive and (tendentially) creative moments. Table 1 is intended to provide a broad overview of the manifold ways in which contemporary processes of neoliberalization have affected the institutional geographies of cities throughout North America and Western Europe. For present purposes, two additional aspects of the processes of creative destruction depicted in the table deserve explication.

First, it is important to underscore that the processes of neoliberal localization outlined in the table necessarily unfold in place-specific forms and combinations within particular local and national contexts. Indeed, building upon the conceptualization of actually existing neoliberalism developed above, we would argue that patterns of neoliberal localization in any national or local context can be understood adequately only through an exploration of their complex, contested interactions with inherited national and local regulatory landscapes. The contributions to this volume provide abundant evidence for this proposition with reference to diverse pathways of neoliberal localization. Moreover, as these essays demonstrate, the different pathways of neoliberal urban restructuring that have crystallized throughout the older industrialized world reflect not only the diversity of neoliberal political projects but also the contextually specific interactions of such projects with inherited frameworks of urban political-economic regulation. An examination of the diverse pathways through which neoliberal political agendas have been imposed upon and reproduced within cities is therefore central to any comprehensive inquiry into the geographies of actually existing neoliberalism.

A second, equally important issue concerns the evolution and/or reconstitution of neoliberal forms of urban policy since their initial deployment in North American and Western European cities during the late 1970s and early 1980s. We have already alluded above to the various mutations that neoliberalization processes have undergone since the late 1970s. The essential point at this juncture of our discussion is that these mutations of neoliberalism have unfolded in particularly pronounced forms within major cities and city-regions. Indeed, we would argue that each of the broader phases of neoliberalization has been anchored and fought out within strategic urban spaces.

- During the initial phase of "proto-neoliberalism," cities became flashpoints both for major economic dislocations and for various forms of sociopolitical struggle, particularly in the sphere of social reproduction. Indeed, the problematic of collective consumption acquired such political prominence during this period that Castells (1972) interpreted it as the sociological essence of the urban phenomenon itself under capitalism. In this context, cities became battlegrounds in which preservationist and modernizing alliances struggled to influence the form and trajectory of economic restructuring during a period in which the postwar growth regime was being systematically undermined throughout the older industrialized world. Consequently, local economic initiatives were adopted in many older industrial cities in order to promote renewed growth from below while maintaining established sociopolitical settlements and redistributive arrangements.

- During the era of "roll-back" neoliberalism in the 1980s, the dominant form of neoliberal urban policy shifted significantly. In this era of lean government, municipalities were increasingly constrained to introduce various kinds of cost-cutting measures – including tax abatements, land grants, cutbacks in public services, the privatization of infrastructural facilities, and so forth – in order to lower the costs of state administration, capitalist production, and social reproduction within their jurisdictions, and thereby to accelerate external investment. Traditional Fordist-Keynesian forms of localized collective consumption were retrenched, in this context, as fiscal austerity measures were imposed upon local governments by neoliberalizing national state apparatuses. Under these conditions, enhanced administrative efficiency and direct and indirect state subsidies to large corporations and an increasing privatization of social reproduction functions were widely viewed as the "best practices" for promoting a good business climate within major cities. The contradictions of this zero-sum, cost-cutting form of urban entrepreneurialism are now evident throughout North America and Western Europe. In addition to its highly polarizing consequences for major segments of local, regional, and national populations, the effectiveness of such strategies for promoting economic rejuvenation has been shown to decline quite precipitously as they are diffused throughout urban systems.
- The subsequent consolidation of "roll-out" neoliberalism in the early 1990s may be viewed as an evolutionary reconstitution of the neoliberal project in response to its own immanent contradictions and crisis tendencies. Throughout this decade, a marked reconstitution of neoliberal strategies occurred at the urban scale as well. On the one hand, the basic neoliberal imperative of mobilizing economic space – in this case, city space – as a purified arena for capitalist growth, commodification, and market discipline remained the dominant political project for municipal governments throughout the world economy . . . The institutionally destructive neoliberalisms of the 1980s were thus apparently superseded by qualitatively new forms of neoliberal localization that actively addressed the problem of establishing nonmarket forms of coordination and cooperation through which to sustain the accumulation process.

  Under these circumstances, the neoliberal project of institutional creation is no longer oriented simply towards the promotion of market-driven capitalist growth; it is also oriented towards the establishment of new flanking mechanisms and modes of crisis displacement through which to insulate powerful economic actors from the manifold failures of the market, the state, and governance that are persistently generated within a neoliberal political framework. Just as crucially, these mutations have also entailed a number of significant institutional realignments at the urban scale, including: (a) the establishment of cooperative business-led networks in local politics; (b) the mobilization of new forms of local economic development policy that foster interfirm cooperation and industrial clustering; (c) the deployment of community-based programs to alleviate social exclusion; (d) the promotion of new forms of coordination and interorganizational networking among previously distinct spheres of local state intervention; and (e) the creation of new regional institutions to promote metropolitan-wide place-marketing and intergovernmental coordination.

Clearly, then, as this schematic discussion indicates, the creative destruction of institutional space at the urban scale does not entail a linear transition from a generic model of the "welfare city" towards a new model of the "neoliberal city." Rather, these multifaceted processes of local institutional change involve a contested, trial-and-error searching process in which neoliberal strategies are being mobilized in place-specific forms and combinations in order to confront some of the many regulatory problems that have afflicted advanced capitalist cities during the post-1970s period. Even in the contemporary "roll-out" phase, neoliberal strategies of localization severely exacerbate many of the regulatory problems they ostensibly aspire to resolve – such as economic stagnation, unemployment, sociospatial polarization, and uneven development – leading in turn to unpredictable mutations of those very strategies and the institutional spaces in which they are deployed. Consequently, the manifold forms and pathways of neoliberal localization must be viewed, not as coherent, sustainable solutions to the regulatory problems of post-1970s capitalism, but rather as deeply contradictory restructuring strategies that are significantly destabilizing inherited landscapes of urban governance and socioeconomic regulation throughout the older industrialized world.

CONCLUSION: FROM NEOLIBERALIZED CITIES TO THE URBANIZATION OF NEOLIBERALISM

It would appear, then, that cities are not merely localized arenas in which broader global or national projects of neoliberal restructuring unfold. On the contrary, cities have become increasingly central to the reproduction, mutation, and continual reconstitution of neoliberalism itself during the last two decades. Indeed, it might be argued that a marked urbanization of neoliberalism has been occurring during this period, as cities have become strategic targets for an increasingly broad range of neoliberal policy experiments, institutional innovations, and politico-ideological projects. Under these conditions, cities have become the incubators for many of the major political and ideological strategies through which the dominance of neoliberalism is being maintained.

[. . .]

At the present time, it remains to be seen whether the powerful contradictions inherent within the current urbanized formation of roll-out neoliberalism will provide openings for more progressive, radical democratic reappropriations of city space, or whether, by contrast, neoliberal agendas will be entrenched still further within the underlying institutional structures of urban governance. Should this latter outcome occur, we have every reason to anticipate the crystallization of still leaner and meaner urban geographies in which cities engage aggressively in mutually destructive place-marketing policies, in which transnational capital is permitted to opt out from supporting local social reproduction, and in which the power of urban citizens to influence the basic conditions of their everyday lives is increasingly undermined.

"Metropolitics for the Twenty-First Century"

from *Place Matters* (2001)

Peter Dreier, John Mollenkopf, and Todd Swanstrom

Editors' Introduction

This team-authored selection by Peter Dreier, John Mollenkopf, and Todd Swanstrom, is an excerpt from their book *Place Matters: Metropolitics for the Twenty-First Century*. They draw urgent attention to the distressed inner city and, like Logan and Molotch, draw attention to the power of "place" or geographic location, in determining future life chances for urban residents through differential access to quality jobs, resources, and public services. Greater inequality in contemporary U.S. society has deepened the importance of place, trumping the benefits brought by new communication technologies and the advent of the Internet. They cite the Nobel Prize-winning economist Amarya Sen, who calls for the abolition of inequality as a means of achieving the full potential of all of humanity.

They foreground the national problem of urban sprawl, which has deepened the socioeconomic and political divide between city and suburb, worsened economic distress and segregation in the inner city, and further degraded the environment. The exodus to the suburbs appealed to urban expatriates searching for middle-class privacy and exclusiveness, who were increasingly divested from public commitments to rental and subsidized housing, parks, schools, transit, and other urban services. Meanwhile a new environment of competitiveness among central cities means public officials are favoring new gentrified neighborhoods attractive to the professional-managerial classes amidst long-term decline in the manufacturing workforce. Dreier and his co-authors trumpet the ongoing relevance of cities as "engines of prosperity," evident in the rise of Sun Belt cities, urban technology and innovation centers, and command centers of the new global economy.

The authors argue for "smart growth" and regional planning policies to combat the political fragmentation that is rife in the suburbs, to help overcome central city–suburban divides, and to improve overall metropolitan quality of life. They observe that the "new regionalism" of the contemporary era has its early twentieth-century precursors in the "garden city" movement of Ebenezer Howard in the UK, and such initiatives as the Regional Planning Association led by Lewis Mumford and others in the New York City metropolitan region. They present several arguments addressing how the new regionalism can promote economic efficiency, competitiveness, environmentalism, and social equity. They contribute to a "metropolitics" through practical suggestions for building new political coalitions.

A sociologist by training, Peter Dreier is the E. P. Clapp Distinguished Professor of Politics, and Chair of the Urban and Environmental Policy Department at Occidental College. John Mollenkopf is Distinguished Professor of Political Science and Sociology at the Graduate Center of the City University of New York and Director of their Center for Urban Research. Todd Swanstrom is the E. Desmond Lee Endowed Professor in Community Collaboration and Public Policy at the University of Missouri-St. Louis.

PLACE STILL MATTERS

Place matters. Where we live makes a big difference in the quality of our lives, and how the places in which we live function has a big impact on the quality of our society. The evidence shows that places are becoming more unequal. Economic classes are becoming more spatially separate from each other, with the rich increasingly living with other rich people and the poor with other poor. The latter are concentrated in central cities and distressed inner suburbs, and the former are in exclusive central-city neighborhoods and more distant suburbs.

This rising economic segregation has many negative consequences, ranging from reinforcing disadvantage in central city neighborhoods to heightening the cost of suburban sprawl as families flee deteriorating central cities and inner suburbs. This trend in the spatial organization of American metropolitan areas is not the simple result of individuals making choices in free markets. Rather, federal and state policies have biased metropolitan development in favor of economic segregation, concentrated urban poverty, and suburban sprawl. We need new policies for metropolitan governance that will level the playing field and stop the drift toward greater spatial inequality. We also need a political strategy for uniting residents of central cities and suburbs in a new coalition that will support these policies.

It may seem odd to argue that place matters when technology appears to be conquering space and Americans are so mobile. Cars and planes have made it possible for us to move about more quickly than ever before. More importantly, cable television, telephones, faxes, computers, and, above all, the Internet enable us to access many of society's benefits without leaving our homes. With a satellite dish or cable service, we can choose from a menu of entertainment options ranging from tractor pulls to Tolstoy, from rap to Rachmaninoff. Distance learning is growing rapidly, with "virtual" universities enabling students to pursue college degrees from their homes. With e-commerce, we hardly even need to drive to the mall anymore, and more people are working at home instead of commuting to the office every day. Where you live, in short, seems to have less and less of an effect on the type of person you are and what you do. Technology has eclipsed the traditional reasons people gathered together in cities: to be close to jobs, culture, and shopping. Cities, it seems, are becoming obsolete.

In fact, this idea is nonsense. As places of intense personal interaction, cities are as important as ever. If technology were truly abolishing the importance of space and place, real estate values would flatten out. Soaring house prices in Silicon Valley, Boston, and New York City are proof positive that people will pay dearly to live in certain places. Indeed, the vast majority of Americans, over 80 percent, have chosen to live within metropolitan areas and have not spread themselves across the countryside.

It is true that mass ownership of automobiles has made it possible for people to live farther away from where they work within these metropolitan areas than they could fifty years ago. If they can afford it, Americans generally prefer to live in low-density suburbs. But people still care about where they live, perhaps more than ever. Higher-income professionals have geographically dispersed networks that transcend their neighborhoods and cities, sometimes extending to the entire globe. They use these "weak" ties, or dispersed networks for gathering information, seeking opportunities, and finding jobs. But where they choose to live still affects how much they pay in taxes, where their children go to school, and who their friends are.

Place becomes more important as one moves down the economic ladder. On the wrong side of the "digital divide," poor and working-class families are less likely to own a computer, have Internet access, or send and receive e-mail. They rely more on local networks to find out about jobs and other opportunities. Often lacking a car (and adequate mass transit), they must live close to where they work. Unable to send their children to private schools, they must rely on local public schools. Unable to afford day care, lower-income families must rely on informal day care provided by nearby relatives and friends.

In short, whether we are highly skilled professionals or minimum-wage workers, where we live matters. Place affects our access to jobs and public services (especially education), our access to shopping and culture, our level of personal security, the availability of medical services, and even the air we breathe. People still care deeply about where they live. The adage still holds true: the three most important factors in real estate are location, location, location.

[. . .]

Place-based inequalities

Missing from the debate about rising inequality has been an understanding of the critical role of place. The Nobel Prize-winning economist Amarya Sen provides a broader way of understanding inequality. He argues that we should understand inequality not simply in terms of income or wealth but in terms of our ability to achieve the good life, by which he means being active members of society and realizing our full potential as human beings. According to Sen, "relevant functionings can vary from such elementary things as being adequately nourished, being in good health, avoiding escapable morbidity and premature mortality, etc., to more complex achievements such as being happy, having self-respect, taking part in the life of the community, and so on." Sen adds that we must also be concerned about "capabilities," or our ability to choose different activities or functionings. For example, a starving person is very different from one who has chosen to go on a hunger strike.

Other things being equal, people are better off if they have real choices in life.

Sen would be the first to admit that having money, or access to jobs, services, and credit, is essential to free choice and a high quality of life. But he argues that equality of income cannot be equated to true equality. A focus on income or wealth confuses the means to the good life with the good life itself. People's ability to convert income into the good life, Sen observes, varies tremendously. A person who suffers from severe kidney disease, for example, cannot enjoy the same quality of life as a perfectly healthy person with the same income, because of the daily monetary and emotional costs of dialysis. Thus, Sen argues, we cannot look at inequality simply in terms of income; we must take into account the actual situations and activities of people. Health, age, gender, race, education, and many other conditions besides income affect our ability to function effectively.

The thesis of this book is that where we live has a powerful effect on the choices we have and our capacity to achieve a higher quality of life. Following Sen, we examine inequality in light of how place shapes and constrains our opportunities not only to acquire income and convert it into quality of life but also to become fully functioning members of the economy, society, and polity. As one example, the increasing devolution of public functions from the federal government to state and local governments means that geographical location has become more important in determining what we pay in taxes and what public goods and services we enjoy. The segregation of income groups into different local governments means that supposedly equal citizens have unequal access to public goods such as schools, parks, and clean air.

[. . .]

Cities as engines of prosperity

Because we emphasize the concentration of poverty in cities, readers may get the impression that cities are basket cases – like sick people with so few resources that they only serve to burden society. Nothing could be further from the truth. In fact, cities are economic dynamos that provide extraordinary benefits to society as a whole. Cities are both reservations for the poor (with all the burdens that entails) and centers of economic productivity and innovation. The contradictory nature of American cities is reflected in the fact that most cities have daytime working populations that are significantly higher than their nighttime residential populations. They export income to the suburbs. Ultimately, we argue, the residential concentration of poverty at the core undermines the entire regional economy's economic efficiency and ability to innovate. Greater regional cooperation, aimed at less economic segregation and sprawl, would benefit the entire society.

Over the last half century, suburbanization, deindustrialization, and the rise of new cities in the South and West have dramatically transformed the older cities of the Northeast and Midwest. In the process, metropolitan areas became far larger, encompassing four-fifths of the nation's population, but central cities became less dominant within them. A few old cities lost half their populations, and others remained roughly the same size but lost population relative to the surrounding suburbs. Some new cities such as Phoenix and San Diego, mainly in the Sun Belt, grew dramatically. Even in these cities, however, much of the growth took place on the periphery, leading some observers to claim that cities are now "obsolete." In their eyes, the "old" industrial economy required dense cities with many factories and much face-to-face contact, whereas the "new" information economy is more comfortable in the suburbs, where computers, the Internet, and cell phones have made the dense face-to-face interactions of the older cities unnecessary.

This view profoundly misreads the key functions that most central cities continue to play in our national economy. Regional economies are integrated wholes, with different parts of the metropolitan area specializing in different economic functions. For routine goods production and distribution activities and even many corporate headquarters, suburban locations may be economically preferable; however, older central cities continue to provide large pools of private assets, accumulated knowledge, sophisticated skills, cultural resources, and social networks. Cities house most of the leading global, national, and regional corporate services firms, such as banks, law firms, and management consultants. They are still centers of innovation, skill, fashion, and market exchange.

Urban density enhances economic efficiency and innovation. What economists call "agglomeration economies" are still important in the global economy. The density of employment in cities reduces the costs of transportation and increases each business's access to skilled and specialized labor. The geographical clustering of industries in certain cities further enhances productivity. In many industries, understanding ambiguous information is the key to innovation. It cannot be communicated in an e-mail message or even a phone call; it requires the kind of face-to-face interaction that cities are good at fostering. The cultural production of these cities has been just as important as their economic role.

[. . .]

THE PROBLEM

The competition among metropolitan jurisdictions to attract higher-income residents and exclude the less well-off has been a powerful factor promoting the concentration of poor people in central cities. In the typical metropolitan area, dozens, even hundreds, of suburban towns seek to establish their places within the metropolitan pecking order. A "favored quarter" houses upper-income people and businesses that pay taxes but do not demand many services and do not lessen the quality of life. These places use zoning regulations, high prices, and even racial prejudice to keep out the unwanted, or at least the less privileged. Elsewhere, less exclusive residential suburbs, suburban commercial and industrial areas, and aging inner-ring suburbs seek to carve out their own places in the hierarchy of municipalities. Even those aging, economically stagnating working-class suburbs on the city's edge often try to keep out inner-city residents. For those with limited means, living in the central city may be their only choice.

A similar but less keen competition takes place among central-city neighborhoods. Here, neighborhoods do not have their own formal authority to exclude some residents and attract others, but factors such as housing quality and price, neighborhood amenities (particularly a good neighborhood primary school), and city zoning and land-use regulations can serve the same purpose. Unlike exclusive suburban towns, however, even well-off central-city residents pay into the city's common budget. As a result, higher-income central-city residents who do not want to pay for services for the less fortunate, or who do not use the central-city services for which they are paying (such as schools, public transit, and parks), have a strong incentive to move to a more exclusive suburban jurisdiction.

Federal policies heightened this competition and made the suburbs attractive by building freeways, fostering suburban home ownership, and encouraging central cities to specialize in social services for needy constituents. Many suburbs reinforce this arrangement by regulating land uses to maximize property values and tax revenues, by customizing their services for middle-class professionals, and by declining to build subsidized housing or sometimes even any kind of rental housing. Central-city public officials have also contributed to this state of affairs by relying on the growth of social services to enhance their budgets, provide jobs for constituents, and build political support. They are no more willing to give up responsibility for, and control over, these activities than exclusive suburbs are to embrace them. Defenders of this system draw on widely held beliefs that localism, private property, and homogeneous communities are sacred parts of the American way of metropolitan living . . .

Although this deeply embedded system may seem rational to suburban residents and public officials, it has produced dysfunctional consequences for the larger society. Metropolitan political fragmentation has encouraged unplanned, costly sprawl on the urban fringe. It has imposed longer journeys to work on commuters, allowing them less time for family life. It has undermined the quality of life in older suburbs, hardened conflicts between suburbs and their central cities, hampered financing for regional public facilities such as mass transit, and encouraged disinvestment from central cities. Countries with strong national land-use regulations and regional governments have avoided many of these problems. Indeed, the United States could have avoided them if we had chosen a more intelligent path for metropolitan growth over the last fifty years.

As these problems became increasingly evident, they drew criticism from scholars, planners, and good-government groups. These critics have focused on how metropolitan political fragmentation undermines administrative efficiency, environmental quality, economic competitiveness, and social equity. As early as the 1930s, administrative experts promoted regional solutions in ways to address the overlap, duplication, lack of coordination, and waste in the provision of public services. Concern today is becoming widespread. Even longtime suburban residents have expressed concern over the environmental costs of sprawl as they see their countryside being gobbled up by new development and find themselves in traffic jams even while doing their Saturday morning shopping. They have made "smart growth" a hot-button issue across the country. Executives of large firms, transportation planners, and economic development officials most often express concern that fragmented metropolitan areas undermine the economic competitiveness of urban regions.

Finally, those who crusade for civil rights and racial desegregation, who care about the plight of the inner-city poor, and who champion greater civic participation have criticized the ways in which metropolitan fragmentation has hurt the social and economic fabric of our communities. They argue in favor of fair housing and housing mobility programs, tax-base sharing, and metropolitan school districts. Indeed, metropolitan fragmentation, sprawl, and inequality are major causes of the decline in community and civic participation . . .

ORIGINS OF THE NEW REGIONALISM

The current debate over regionalism echoes earlier concerns. At the end of the nineteenth century, New York, Chicago, and many other cities moved to annex or consolidate most of the adjacent territory likely to be developed in the next fifty years. The formation of the Greater New York from New York City, Brooklyn, the hamlets of Queens County, the Bronx, and Staten Island in 1898 was a grand and highly successful experiment in metropolitan government. Indeed, much of the current vitality of cities such as Chicago and New York stems from these early actions. By the 1920s, however, middle-class urbanites, dismayed by the rapidly increasing density and inequality of their industrial cities, and frustrated by their inability to continue to dominate their politics, increasingly fled to the suburbs. Once they established themselves as suburbanites, they persuaded state legislatures to pass laws hindering cities from further annexation. Metropolitan growth would henceforth take place largely outside the jurisdiction of the central city.

Responding to such trends, an early group of regionalists foresaw, as early as the 1920s, the need for new forms of metropolitan planning and cooperation. Echoing the "garden city" idea first promulgated by Ebenezer Howard in England, Lewis Mumford and his colleagues in the Regional Planning Association of America, formed in 1923, hoped that coherent regions would gradually emerge to dissolve the problems of the industrial city. "The hope of the city," Mumford wrote in 1925, "lies outside itself." Other radical (though less utopian) thinkers called for regional land-use planning as an antidote to fragmentation. In 1927, the Regional Plan Association's magisterial plan for metropolitan New York called for knitting the region together with a comprehensive system of highways and rail transit that would concentrate economic growth in Manhattan and in a few suburban centers. In 1937, the New Deal's National Resources Committee called for federal efforts to foster regional planning, including the establishment of multistate metropolitan planning agencies. In the face of resistance from those who saw such measures as abrogating private property and local democracy, however, none of these visionary blueprints had much impact on the post-World War II evolution of American cities. As a result, many of the problems they anticipated did indeed come to pass.

Efficiency arguments for regionalism

The negative consequences of unplanned metropolitan growth triggered new strains of regionalist thinking and new political constituencies that favored the creation of new regional institutions. Public administrators, city planners, and municipal reformers viewed regional planning as the best way to promote regional economic efficiency and maintain a sound environment. Writing for the National Municipal League in 1930, Paul Studentski criticized metropolitan fragmentation and called for a framework that would support "real, democratic, comprehensive, and permanent organization of the metropolitan community." Prominent academics and planners such as Charles Merriam, Victor Jones, and Luther Gulick elaborated on these themes in the postwar period. Robert Wood's 1961 classic *1400 Governments* argued that postwar suburbanization in metropolitan New York was irrational, inefficient, and unaccountable. The federal Advisory Commission on Intergovernmental Relations, created by Congress in 1947 and eliminated in 1996, recommended ways to broaden the urban tax base and improve the regional distribution of services.

A common theme was that fragmented metropolitan governments promoted wasteful duplication, uneven standards of public service, and wasteful competition between local governments. These trenchant critiques were not more effective than the work of earlier metropolitan visionaries had been in restraining the construction of freeways, suburban shopping malls, and tract housing in the 1950s and 1960s. It nevertheless remained a key doctrine of public administration that regions required some level of metropolitan planning in order to function well. The 1960s saw the creation of many regional councils of government (often called COGs) and single-purpose regional agencies such as water and sewer systems, garbage disposal, and transportation. City–county consolidation took place in Miami, Nashville, Jacksonville, and Indianapolis, but voters in many other areas rejected proposals to merge city and suburban governments into regional governments. Bucking this trend, in the November 2000 election, voters in Louisville, Kentucky, and its suburbs approved (by a 54 to 46 percent margin) the consolidation of the city of Louisville with Richmond County. This is the first major city–county merger to win approval in thirty years. Fast-growing cities of the South and Southwest, such as Phoenix and Albuquerque, also annexed surrounding territory long after that practice ended elsewhere in the country. Finally, voters and public officials sought to streamline and modernize the governance of suburban areas in the postwar period by enhancing county government and consolidating school districts. All these efforts, however, fell short of the goals espoused by the first and second generations of regionalists.

Environmental arguments for regionalism

Concern for administrative efficiency motivated the early advocates of metropolitan governance. Beginning in the 1960s, a new generation of regionalists emerged, concerned about environmental protection, sustainable development, and "smart growth." Development rapidly swallowed the metropolitan countryside after World War II, and the freeway construction of the 1960s and 1970s increased traffic congestion to new, more disturbing levels. Confronting these realities, suburban residents and those who represented them became less enthusiastic about unbridled metropolitan growth.

In 1974, the Council on Environmental Quality issued a report titled *The Costs of Sprawl*, calling for greater regulation of suburban development. Today, several national environmental organizations, including the Sierra Club, are campaigning against sprawl, Oregon and Maryland have adopted state legislation for "smart growth," and many other states are actively discussing similar measures. Some approaches call for the establishment of regional growth boundaries monitored by the state; others merely provide incentives to channel new investment toward already developed areas while attempting to preserve agricultural land uses or otherwise protect undeveloped areas. Regional groups are active in the San Francisco Bay Area, Washington, D.C., Pittsburgh, and Suffolk County, New York. A movement for "new urbanism" has emerged among architects and city planners who favor denser, more pedestrian- and transit-oriented forms of neighborhood development. Despite naysayers who see an adverse impact on housing affordability and consumer choice, a growing consensus has emerged in many places on behalf of rethinking older growth policies. Residents of Cook County, Illinois (an old urban area), and Santa Clara County, California (the high-tech Silicon Valley), both showed overwhelming support for regional approaches to solving urban problems. Voters across the country have approved ballot measures to limit suburban sprawl and preserve open spaces. Responding to such sentiments, Vice President Al Gore made smart growth a central theme in the Clinton–Gore administration's "livability agenda," which focused on preserving open spaces, redeveloping brownfield areas, mitigating congestion, and improving urban air quality.

[. . .]

Economic competitiveness arguments for regionalism

Contemporary regionalists have also argued that metropolitan areas divided against themselves cannot compete successfully in the new global economy. In particular, business leaders and regional planning organizations have recognized that regionally oriented planning and development policies could make metropolitan economies more competitive. Although business-supported regional planning groups have existed at least since the Regional Plan Association was established in New York in 1923, groups like San Francisco's Bay Area Council and Pittsburgh's Allegheny Council for Community Development became more common after World War II. The radical changes in technology, business organization, and global competition since the 1980s gave this perspective new force. The high-technology companies of Silicon Valley took the lead in supporting regional approaches to the South Bay's housing, transportation, and development issues. One scholar, Annalee Saxenian, found Silicon Valley entrepreneurs' ability to collaborate in this way made that region more successful, over the long haul, than the similar technology complex along Route 128 around Boston.

Social scientists who examine the interrelationships between central-city and suburban economies find a high correlation between the two, though in some cases, substantial central-city decline has not prevented the surrounding suburbs from prospering. Nevertheless, there are obvious linkages in the economic conditions and income growth rates of central cities and their suburbs. Central cities continue to perform functions and provide services that are critical to regional growth. The evidence suggests that cooperative regions are more likely to prosper than are more competitive, divided regions.

These realities have given new impetus to certain efforts to form regional public–private partnerships to promote regional growth. Syndicated columnist Neal Peirce, who has given visibility to all forms of new regionalism, and his colleague Curtis Johnson have been particularly active in inspiring and advising a new generation of such organizations. They have not promoted any particular organizational forms, stressing instead the general need for collaboration, trust, dialogue, and leadership. These collaborations have produced many Web sites and a journal, *The Regionalist.*

Closely associated with this perspective is the growing focus on industrial clusters as a basis for urban and regional economic development. Advanced by Michael Porter's *The Competitive Advantage of Nations*, this perspective argues that a region's competitiveness is based on the quality of networks and interactions among related and often physically close firms. Even though they may compete against one another in some ways, they share technical knowledge, a skilled labor pool, and support services and spur one another to rapid technological innovation. This has led some policy makers to promote clusters as a way to address such regional needs as better-paid jobs, more rapid innovation, and a better quality of life for industrial workers. This perspective was given national prominence in the 1996 report *America's New Economy and the Challenge of the Cities.*

Many state and local agencies hoping to emulate Silicon Valley's success have embraced this way of thinking about economic development, such as the Regional Technology Alliance organized by the San Diego Association of Governments. San Diego has also surveyed employers in these clusters to understand their workforce needs and has developed programs to address them.

Equity arguments for regionalism

The spatial concentration of urban poverty is another concern motivating many new regionalists. As Michael Schill observed, "although segregating themselves in the suburbs may serve the interests of large numbers of Americans today, the long term costs of doing nothing to alleviate concentrated ghetto poverty are likely to be tremendous." Distinguished public intellectuals, including Anthony Downs of the Brookings Institution, former Albuquerque mayor David Rusk, Minnesota state legislator Myron Orfield, and Harvard professor Gary Orfield, among others, have concluded that regional approaches are the only way that the problem of inner-city poverty can be solved. Increasingly, they have been joined by officials representing the inner-ring suburbs who are facing the growth of "urban" problems that they cannot solve within the limits of their own jurisdictions. With foundation funding, Orfield and his associates have launched initiatives to document and remedy "growing social and economic polarization" in twenty-two metropolitan areas.

Increasingly, planners are arguing that there need not be a trade-off between equality and efficiency, or growth. Indeed, regional strategies that lift up the central city can make regions more competitive. The Center on Wisconsin Strategy at the University of Wisconsin has advocated that workforce developers and regional employers cooperate to train central-city residents for high-wage jobs in technologically innovative firms. This "high road" approach to regional economic development would simultaneously enhance wages, upward mobility, and employer competitiveness.

[. . .]

THE POLITICS OF REGIONALISM IN THE NEW MILLENNIUM

Historically, many forces have worked against a regional perspective in urban governance. Suburbs have been happy to benefit from being located in a large metropolitan area while excluding the less well-off and avoiding the payment of taxes to support services required by the urban poor. Central cities, for their part, have made a virtue of necessity by increasing spending on social services as a way of expanding the employment of central-city constituents. This spending has become an increasingly substantial part of municipal budgets and an important form of "new patronage" in city politics. As federal benefits have increasingly flowed to needy people, as opposed to needy places, it has helped to expand these functions. Legislators elected to represent areas where the minority poor are concentrated develop a stake in this state of affairs.

This dynamic is gradually but steadily shifting. The first wave of inner, working-class suburbs has long since been built out, their populations have aged, and their residents' incomes have stagnated since the early 1970s. These suburbs have developed increasingly "urban" problems that they cannot solve on their own. Black and Hispanic central-city residents have increasingly moved to the suburbs. Although minority suburbs have significantly better conditions than inner-city minority neighborhoods, they still have higher rates of poverty and disadvantage than white suburbs and may face some of the same forces of decline that operate on inner cities. Even as metropolitan economic segregation has increased, metropolitan racial segregation has declined and suburban diversity has increased. As Orfield has pointed out, these inner suburbs are coming to realize that they share significant interests with their central cities. And as Juliet Gainsborough has shown, voters residing in more diverse suburbs are substantially more likely to vote like their urban neighbors.

[. . .]

It has thus become clear that many powerful players, including corporations, foundations, unions, political leaders, and community organizations, think that regional collaboration offers the surest route to competitive advantage in the new global economy. Regions divided against themselves are least likely to be able to undertake the necessary investments in physical, human, and social capital. Emerging metropolitan governance institutions should carry out functions that make the most sense from economic, environmental, and equity viewpoints.

These include regional capital investments, transportation, land-use planning, economic development, job training, education, and tax-base sharing. This form of cooperation can generate more widely distributed benefits if they follow a high-productivity, high-wage "high road" instead of engaging in a "race to the bottom" that competes for low-wage, low-value-added jobs in highly mobile industries.

To be truly effective, metropolitan cooperation must develop a broad, democratic base and the organizational capacity to articulate the common good, not merely to sum up the aims of the individual parts of the metropolis. Cooperation of the region's constituent elements must be secured through consent, not through unwanted mandates imposed on resistant local jurisdictions. To achieve this consent, the new regional form must provide tangible benefits to its constituent jurisdictions, and its actions cannot be subject to the veto of an exclusive "favored quarter."

"Urban Ecological Footprints: Why Cities Cannot Be Sustainable – And Why They Are a Key to Sustainability"

from *Urban Ecology* (2008) [1996]

William Rees and Mathis Wackernagel

Editors' Introduction

William Rees and Mathis Wackernagel are innovators in the development of ecological footprint analysis, through a mathematical method for estimating the huge demand of urban settlements on our earth ecosystems. Their measurement of human "carrying capacity" or "load" multiplies total population of any given urban or regional settlement by a per capita estimation of a typical market basket of major consumption items, including units of energy, food, and forest products. They also estimate the deficit or "overshoot" between what a given urban population consumes as human load, versus what can be produced on its available ecologically productive land.

Rees and Wackernagel describe cities as entropic structures that consume and dissipate energy, "black holes" of fossil fuels, foods, and goods consumption that generate enormous waste and pollute the atmosphere. This view of entropy and dissipation is a marked contrast from the "equilibrium" dynamics described by human ecologists of the last century. The estimation method of Rees and Wackernagel could be criticized for paying insufficient attention to the savings brought from collective consumption of goods and services in urbanized areas. Yet their essential proposition underscores the urgency of a conceptual shift that recognizes the world has reached an historic turning point. They have also opened up new perspectives on the nature of North–South inequality and the global overshoot in resource flows between the developing and developed world.

While drawing attention to the crisis of overconsumption besetting urban settlements, Rees and Wackernagel also recognize that cities offer key collective infrastructures for achieving sustainability on a global scale. They point out that cities offer economies of scale for collectivizing waste disposal and recycling, providing public transit and housing, and achieving higher population density, which relieves per capita demand for occupied land.

William Rees is a Professor at the University of British Columbia (UBC) and former Director of the School of Community and Regional Planning (SCARP) at UBC. Mathis Wackernagel received his Ph.D. at UBC, and is currently President of Global Footprint Network, an international sustainability think tank that works to develop and promote metrics for sustainability. Wackernagel developed the estimation model while working on his dissertation under the supervision of Rees. This selection is derived from a book published by Wackernagel and Rees in 1996, *Our Ecological Footprint: Reducing Human Impact on the Earth*, by New Society Publishers.

INTRODUCTION: TRANSFORMING HUMAN ECOLOGY

It is sometimes said that the industrial revolution stimulated the greatest human migration in history. This migration swept first through Australia, Europe, and North America and is still in the process of transforming Asia and the rest of the world. We refer, or course, to the mass movement of people from farms and rural villages to cities everywhere. The seeming abandonment of the countryside is creating an urban world – 75% or more of the people in the so-called industrialized countries now live in towns and cities, and half of humanity will be city dwellers by the end of the century.

Although usually seen as an economic or demographic phenomenon, urbanization also represents a human ecological transformation. Understanding the dramatic shift in human spatial and material relationships with the rest of nature is a key to sustainability. Our primary purpose, therefore, is to describe a novel approach to assessing the ecological role of cities and to estimate the scale of the impact they are having on the ecosphere. The analysis shows that, as nodes of energy and material consumption, cities are causally linked to accelerating global ecological decline and are not by themselves sustainable. At the same time, cities and their inhabitants can play a major role in helping to achieve global sustainability.

STARTING PREMISE

[. . .]

Carrying capacity as maximum human "load"

The notion that humanity may be up against a new kind of limit has rekindled the Malthusian debate about human carrying capacity. Carrying capacity is usually defined as the maximum population of a given species that can be supported indefinitely in a defined habitat without permanently impairing the productivity of that habitat. However, because we humans seem to be capable of continuously increasing the human carrying capacity of Earth by eliminating competing species, by importing locally scare resources, and through technology, conventional economists and planners generally reject the concept as applied to people . . .

We argue that the economy is an inextricably embedded subsystem of the ecosphere. Despite our technological and economic achievements, humankind remains in a state of "obligate dependence" on the productivity and life support services of the ecosphere. The trappings of technology and culture aside, human beings remain biophysical entities. From a trophic-dynamic perspective, the relationship of humankind to the rest of the ecosphere is similar to those of thousands of other consumer species with which we share the planet. We depend for both basic needs and the production of cultural artifacts on energy and material resources extracted from nature, and *all* this energy/matter is eventually returned in degraded form to the ecosphere as waste. The major material difference between humans and other species is that, in addition to our biological metabolism, the human enterprise is characterized by an industrial metabolism . . .

Because of continuing functional dependence on ecological processes, some analysts have stopped thinking of natural resources as mere "free goods of nature." Ecological economists now regard the species, ecosystems, and other biophysical entities that produce required resource flows as forms of "natural capital" and the flows themselves as types of essential "natural income." This capital theory approach provides a valuable insight into the meaning of sustainability – no development path is sustainable if it depends on the continuous depletion of productive capital. From this perspective, society can be said to be economically sustainable only if it passes on an undiminished per capita stock of essential capital from one generation to the next.

In the present context, the most relevant interpretation of this "constant natural stocks" criterion is as follows:

> Each generation should inherit an adequate per capita stock of natural capital assets no less than the stock of such assets inherited by the previous generations.

[. . .]

The fundamental question for ecological sustainability is whether remaining *natural* capital stocks (including other species populations and ecosystems) are adequate to provide the resources consumed and assimilate the wastes produced by the anticipated human population into the next century, while simultaneously maintaining the general life support functions of the ecosphere. In short, is there adequate human carrying capacity? At present, of course, both the human population and average consumption are increasing, whereas the total area of productive land and stocks of natural capital are fixed or in decline. In this light, we argue that shrinking carrying capacity may soon become the single most important issue confronting humanity.

The issue becomes clearer if we define human carrying capacity not as a maximum population but rather as the maximum (entropic) "load" that can safely be imposed on the environment by people. Human load is clearly a function not only of population but also of average per capita consumption. Significantly, the latter is increasing even more rapidly than the former due (ironically) to expanding trade, advancing technology, and rising incomes . . . For example, in 1790 the estimated average daily energy consumption by Americans was 11,000 kcal per capita. In 1980, this had increased almost 20-fold to 210,000 kcal/day. As a result of such trends, *load* pressure relative to carrying capacity is rising much faster than is implied by mere population increases.

Ecological footprints measuring human "load"

By inverting the standard carrying capacity ratio and extending the concept of load, we have developed a powerful tool for assessing human carrying capacity. Rather than asking what population a particular region can support sustainably, the critical question becomes: How large an area of productive land is needed to sustain a defined population indefinitely, *wherever on Earth that land is located?* Most importantly, this approach overcomes any objection to the concept of human carrying capacity based on trade and technological factors. In the language of the previous section, we ask how much of the Earth's surface is appropriated to support the "load" imposed by a referent population, whatever its dependence on trade or its level of technological sophistication.

Since most forms of natural income (resource and service flows) are produced by terrestrial ecosystems and associated aquatic ones, it should be possible to estimate the area of land/water required to produce sustainably the quantity of any resource or ecological service used by a defined population or economy at a given level of technology. The sum of such calculations for all significant categories of consumption would provide a conservative area-based estimate of the natural capital requirements for that population or economy. We call this area the population's true "ecological footprint."

A simple two-step mental experiment serves to illustrate the ecological principles behind this approach. First, imagine what would happen to any modern city as defined by its political boundaries if it were enclosed in a glass or plastic hemisphere completely closed to material flows. This means that the human system so contained would be able to depend only on whatever remnant ecosystems were initially trapped within the hemisphere.

It is obvious to most people that the city would cease to function, and its inhabitants would perish in a few days. The population and economy contained by the capsule would have been cut off from both vital resources and essential waste sinks leaving it to starve and suffocate at the same time. In other words, the ecosystems contained within our imaginary human terrarium – and any real world city – would have insufficient carrying capacity to service the ecological load imposed by the contained population.

The second step pushes us to contemplate urban ecological reality in more concrete terms. Let's assume that our experimental city is surrounded by a diverse landscape in which cropland and pasture, forests, and watersheds – all the different ecologically productive land-types – are represented in proportion to their actual abundance on the Earth and that adequate fossil energy is available to support current levels of consumption using prevailing technology. Let's also assume our imaginary glass enclosure is elastically expandable. The question now becomes: How large would the hemisphere have to grow before the city at its center could sustain itself indefinitely and exclusively on the land and water ecosystems and the energy resources contained within the capsule? In other words, what is the total area of different ecosystem types needed continuously to supply the material demands of the people of our city as they go about their daily activities?

Answering this question would provide an estimate of the de facto ecological footprint of the city. Formally defined, the ecological footprint (EF) is the total area of productive land and water required continuously to produce all the resources consumed and to assimilate all the wastes produced, by a defined population, wherever on Earth that land is located. As noted, the ecological footprint is a land-based surrogate measure of the population's demands on natural capital.

METHOD IN BRIEF

The basic calculations for ecological footprint estimates are conceptually simple. First we estimate the annual per capita consumption of major consumption items from aggregate regional or national data by dividing total consumption by population size. Much of the data needed for preliminary assessments is readily available from national statistical tables on, for example, energy, food, or forest products production and consumption. For many categories, national statistics provide both production and trade figures from which trade-corrected consumption can be assessed:

Trade-corrected consumption = production + imports – exports

The next step is to estimate the land area appropriated per capita for the production of each consumption item by dividing average annual consumption of that item by its average annual productivity or yield . . .

[. . .]

We then compile the total average per capita ecological footprint (ef) by summing all the ecosystem areas appropriated by an individual to fill his/her annual shopping basket of consumption goods and services.

Finally we get the ecological footprint (EF_P) of the study population by multiplying the average per capita footprint by population size (N): Thus, $EF_P = N \times ef$.

Our EF equation is structurally similar to the more familiar representation of human environmental impact (I) as a product of population (P), affluence (A), and technology (T), I = PAT. The ecological footprint is, in fact, a measure of population impact expressed in terms of appropriated land area. The size of the per capita footprint will, of course, reflect the affluence (material consumption) and technological sophistication of the subject population.

So far our EF calculations are based on five major categories of consumption – food, housing, transportation, consumer goods, and services – and on eight major land-use categories. However, we have examined only one class of waste flow in detail. We account for carbon dioxide emissions from fossil energy consumption by estimating the area of average carbon-sink forest that would be required to sequester them [carbon emissions/capital]/[carbon assimilation/hectare], on the assumption that atmospheric stability is a prerequisite of sustainability . . .

STRENGTHS AND LIMITATIONS OF FOOTPRINT ANALYSIS

[. . .]

Footprint analysis is not dynamic modeling and has no predictive capability. However, prediction was never our intent. Ecological footprinting acts, in effect, as an ecological camera – each analysis provides a snapshot of our current demands on nature, a portrait of how things stand *right now* under prevailing technology and social values. We believe that this in itself is an important contribution. We show that humanity has exceeded carrying capacity and that some people contribute significantly more to this ecological "overshoot" than do others. Ecological footprinting also estimates how much we have to reduce our consumption, improve our technology, or change our behavior to achieve sustainability.

Moreover, if used in a time-series study (repeated analytic "snap-shots" over years or decades) ecological footprinting can help monitor progress toward closing the sustainability gap as new technologies are introduced and consumer behavior changes . . .

[. . .]

Ecological footprinting is precisely that – it provides an index of biophysical impacts. It therefore tells us little about the sociopolitical dimensions of the global change crisis. Of equal relevance to achieving sustainability are considerations of political and economic power, the responsiveness of the political process to the ecological imperative, and chronic distributional inequity which is actually worsening (both within rich countries and between North and South), as the market economy becomes an increasingly global affair . . . As use of the concept spreads, however, the term "footprint" is increasingly being used to encompass the *overall* impacts of high-income economies on the developing world (or of cities on the countryside).

None of these limitations detracts from the fundamental message of ecological footprint analysis – that whatever the distribution of power or wealth, society will ultimately have to deal with the growing global ecological debt. Our original objective in advancing the ecological footprint concept was to bolster our critique of the prevailing development paradigm and to force the international development debate beyond its focus on GDP growth to include ecological reality.

[. . .]

Ecological footprints of modern cities and "developed" regions

Canada is one of the world's wealthiest countries. Its citizens enjoy very high material standards by any measure. Indeed, ecological footprint analysis shows that the total land required to support present consumption levels by the average Canadian is at least 4.3 hectares, including 2.3 hectares for carbon dioxide assimilation alone. Thus, the per capita ecological footprint of Canadians (their average "personal planetoid") is almost three times their "fair Earthshare" of 1.5 hectares.

Let's apply this result to a densely populated high-income region, the Lower Fraser Basin in the Province of British Columbia. Within this area, the city of Vancouver had a 1991 population of 472,000 and an area of 114 km^2 (11,400 hectares). Assuming a per capita land consumption rate of 4.3 hectares, the 472,000 people living in Vancouver require, conservatively, 2 million hectares of land for their exclusive use to maintain their current consumption patterns (assuming such land is being managed sustainably). This means that the city population appropriates the productive output of a land area *nearly 180 times larger than its political area* to support its present consumer lifestyle.

[. . .]

Extending our Canadian example to the Lower Fraser Basin (population = 1.78 million) reveals that even though only 18% of the region is dominated by urban land use (i.e., most of the area is rural agricultural or forested land), consumption by its human population "appropriates" through trade and biogeochemical flows the ecological output and services of a land area about 14 times larger than the home region of 5,550 square kilometers. In other words, the people of the Lower Fraser Basin, in enjoying their consumer lifestyles, have "overshot" the terrestrial carrying capacity of their geographic home territory by a factor of 14. Put another way, analysis of the ecological load imposed by the regional population shows that at prevailing material standards, *at least* 90% of the ecosystem area needed to support the Lower Fraser Basin actually lies outside the region itself. These results are summarized in Table 1.

It seems that the "sustainability" of the Lower Fraser Basin of British Columbia depends on imports of ecologically significant goods and services whose production requires an area elsewhere on Earth vastly larger than the internal area of the region itself. In effect, however healthy the region's economy appears to be in monetary terms, the Lower Fraser Basin is running a massive "ecological deficit" with the rest of Canada and the world.

Table 1 Estimated ecological footprints of Vancouver and the Lower Fraser Basin (terrestrial component only)

Geographic Unit	*Population*	*Land Area (ha)*	*Ecol. Footp (ha)*	*Overshoot Factor*
Vancouver City	472,000	11,400	2,029,600	178.0
L. Fraser Basin	1,780,000	555,000	7,654,000	13.8

GLOBAL CONTEXT

This situation is typical of high-income regions and even for some entire countries. Most highly urbanized industrial countries run an ecological deficit about an order of magnitude larger than the sustainable natural income generated by the ecologically productive land within their political territories (Table 2). The last two columns of Table 2 represent low estimates of these per capita deficits.

These data throw new light on current world development models. For example, Japan and the Netherlands both boast positive trade and current account balances measured in monetary terms, and their populations are among the most prosperous on earth. Densely populated yet relatively resource (natural capital) poor, these countries are regarded as stellar economic successes and held up as models for emulation by the developing world. At the same time, we estimate that Japan has 2.5 hectare/capita, and the Netherlands a 3.3 hectare/capita ecological footprint which gives these countries national ecological footprints about eight and 15 times larger than their total domestic territories respectively. The marked contrast between the physical and monetary accounts of such economic success stories raises difficult developmental questions in a world whose principal strategy for sustainability is economic growth. Global sustainability cannot be (ecological) deficit-financed; simple physics dictates that *not all countries or regions can be net importers of biophysical capacity.*

It is worth noting in this context that Canada is one of the few high-income countries that consumes less than its natural income domestically. Low in population and rich in natural resources, this country has yet to exceed domestic carrying capacity. However, Canada's natural capital stocks are being depleted by exports of energy, forest, fish, agricultural products, etc., to the rest of the world. In short, the apparent surpluses in Canada are being incorporated by trade into the ecological footprints of other countries, particularly that of the United States (although the entire Canadian surplus would be insufficient to satisfy the US deficit!). How should such biophysical realities be reflected in local and global strategies for ecologically sustainable socioeconomic development?

Discussion and conclusions: cities and sustainability

Ecological footprint analysis illustrates the fact that as a result of the enormous increase in per capita energy and material consumption made possible by (and required by) technology, and universally increasing dependencies on trade, *the ecological locations of high-density human settlements no longer coincide with their geographic locations.* Twentieth-century cities and industrial regions for survival and growth depend on a vast and increasingly global hinterland of ecologically productive landscapes. Cities necessarily "appropriate" the

Table 2 Ecological deficits of urban-industrial countries

Country	*Ecologically Productive Land (in Hectares)* *a*	*Population (1995)* *b*	*Ecol. Prod. Land Per Capita (in Hectares)* *c = a/b*	*Natural Ecological Deficit (in Hectares)* *d = Fprint – c*	*Natural Ecological Deficit (in % Available)* *e = d/c*
Japan	30,417,000	125,000,000	0.24	2.26	940%
Belgium	1,987,00	10,000,000	0.20	2.80	1400%
Britain	20,360,000	58,000,000	0.35	2.65	760%
France	45,385,000	57,800,000	0.78	2.22	280%
Germany	27,734,000	81,300,000	0.34	2.66	780%
Netherlands	2,300,000	15,500,000	0.15	2.85	1900%
Switzerland	3,073,000	7,000,000	0.44	2.56	580%
Canada	434,477,000	28,500,000	15.24	(10.94)	(250%)
United States	725,643,000	258,000,000	2.81	2.29	80%

ecological output and life support functions of distant regions all over the world through commercial trade and natural biogeochemical cycles. Perhaps the most important insight from this result is that *no city or urban region can achieve sustainability on its own*. Regardless of local land use and environmental policies, a prerequisite for sustainable cities is sustainable use of the global hinterland.

The other side of this dependency coin is the impact urban populations and cities have on rural environments and the ecosphere generally. Combined with rising material standards and the spread of consumerism, the mass migration of humans to the cities in this century has turned urban industrial regions into nodes of intense consumption. The wealthier the city and the more connected to the rest of the world, the greater the load it is able to impose on the ecosphere through trade and other forms of economic leverage. Seen in this light and contrary to popular wisdom, the seeming depopulation of many rural areas does not mean they are being abandoned in any ecofunctional sense. Whereas most of the people may have moved elsewhere, rural lands and ecosystem functions are being exploited more intensely than ever in the service of newly urbanized human populations.

CITIES AND THE ENTROPY LAW

The population of "advanced" high-income countries are 75% or more urban and estimates suggest that over 50% of the entire human population will be living in urban areas by the end of the century. If we accept the Brundtland Commission's estimate that the wealthy quarter of the world's population consume over three-quarters of the world's resources (and therefore produce at least 75% of the wastes), then the populations of wealthy cities are responsible for about 60% of the current levels of resource depletion and pollution. The global total contribution from cities is probably 70% or more.

In effect, cities have become entropic black holes drawing in energy and matter from all over the ecosphere (and returning all of it in degraded form back to the ecosphere). This relationship is an inevitable expression of the Second Law of Thermodynamics. The second law normally states that the entropy of any isolated system increases. That is, the available energy spontaneously dissipates, gradients disappear, and the system becomes more increasingly unstructured and disordered in an inexorable slide towards thermodynamic equilibrium. This is a state in which "nothing happens or can happen."

What is often forgotten is that all systems, whether isolated or not, are subject to the same forces of entropic decay. In other words, any complex differentiated system has a natural tendency to erode, dissipate, and unravel. The reason open, self-organizing systems such as modern cities do not run down in this way is that they are able to import available energy and material (essergy) from their host environments which they use to maintain their internal integrity. Such systems also export the resultant entropy (waste and disorder) into their hosts. The second law therefore also suggests that all highly-ordered systems can grow and develop (increase their internal order) only at the expense of increasing disorder at higher levels in the systems hierarchy. Because such systems continuously degrade and dissipate available energy and matter, they are called "dissipative structures."

Clearly, cities are prime examples of highly-ordered dissipative structures. At the same time, these nodes of intense economic activity are open sub-systems of the materially closed, nongrowing ecosphere. Thus, to grow, or simply to maintain their internal order and structure, cities necessarily appropriate large quantities of useful energy and matter from the ecosphere and "dissipate" an equivalent stream of degraded waste back into it.

This means that in the aggregate, cities (or the human economy) can operate sustainably only within the thermodynamic load-bearing capacity of the ecosphere. Beyond a certain point, the cost of material economic growth will be measured by increasing entropy or disorder in the "environment." We would expect this point (at which consumption by humans exceeds available natural income) to be revealed through the continuous depletion of natural capital – reduced biodiversity, fisheries collapse, air/water/land pollution, deforestation, desertification, etc. Such trends are the stuff of headlines today . . . It seems we have reached the entropic limits to growth.

This brings us back to our starting premise, that with the onset of global ecological change, the world has reached an historic turning point that requires a conceptual shift from empty-world to full-world economics (and ecology). Ecological footprint analysis underscores the urgency of making this shift. As noted, the productive land "available" to each person on Earth has decreased increasingly rapidly with the explosion of human population in this century. Today, there are only about 1.5 hectares of such land for each person, including wilderness areas that probably shouldn't be used for any other purpose.

At the same time, the land area appropriated by residents of richer countries has steadily increased. The present per capita ecological footprints of North Americans (4–5 ha) represents three times their fair share of the Earth's bounty. By extrapolation, if everyone on Earth lived like the average North American, the total land requirement would exceed 26 billion hectares. However, there are fewer than 9 billion hectares of such land on Earth. This means that we would need three such planets to support just the *present* human family. In fact, we estimate that resource consumption and waste disposal by the wealthy quarter of world's population alone exceeds global carrying capacity and that total global overshoot is as much as 30%.

[. . .]

TOWARD URBAN SUSTAINABILITY

Ecological footprint analysis not only measures the sustainability gap, it also provides insight into strategies for sustainable urban development. To begin, it is important to recognize that cities are themselves vulnerable to the negative consequences of overconsumption and global ecological mismanagement. How economically stable and socially secure can a city of 10 million be if distant sources of food, water, energy, or other vital resource flows are threatened by accelerating ecospheric change, increasing competition, and dwindling supplies? Does the present pattern of global development, one that increases interregional dependence on vital natural income flows that may be in jeopardy, make ecological or geopolitical sense? If the answer is "no," or even a cautious "possibly not," circumstances may already warrant a restoration of balance away from the present emphasis on global economic integration and interregional dependency toward enhanced ecological independence and greater intraregional self-reliance. (If all regions were in ecological steady-state, the aggregate effect would be global stability.)

To reduce their dependence on external flows, urban regions and whole countries may choose to develop explicit policies to invest in rehabilitating their own natural capital stocks and to promote the use of local fisheries, forests, agricultural land, etc. This would increase regional independence, thus creating a hedge against rising international demand, global ecological change, and potentially reduced productivity elsewhere.

Although greater regional self-reliance is a desirable goal on several grounds, we are not arguing for regional closure. In any event, self-sufficiency is not in the cards for most modern urban regions. The more important issue before us is to assure urban security and define an appropriate role of cities in achieving global sustainability . . .

Ecological pros and cons of cities

A major conclusion of ecological footprint analysis and similar studies is that urban policy should strive to minimize the disruption of ecosystems processes and massively reduce the energy and material consumption associated with cities . . .

Addressing these issues shows that cities present both unique problems and opportunities. First, the fact that cities concentrate both human populations and resource consumptions results in a variety of ecological impacts that would not occur, or would be less severe, with a more dispersed settlement pattern. For example, cities produce locally dangerous levels of various pollutants that might otherwise safely be dissipated, diluted, and assimilated over a much larger area.

More importantly from the perspective of ecosystems integrity, cities also significantly alter natural biogeochemical cycles of vital nutrients and other chemical resources. Removing people and livestock far from the land that supports them prevents the economic recycling of phosphorus, nitrogen, and other nutrients, and organic matter back onto farms and forests. As a consequence of urbanization, local, cyclically integrated ecological production systems have become global, horizontally disintegrated, throughput systems. For example, instead of being returned to the land, Vancouver's daily appropriation of Saskatchewan mineral nutrients goes straight out to sea. As a result, agricultural soils are degraded (half the natural nutrients and organic matter from much of Canada's once-rich prairie soils have been lost in a century of mechanized export agriculture), and we are forced to substitute nonrenewable artificial fertilizer for the once renewable real thing. All of this calls for much improved accounting for the hidden costs of cities, of transportation, and of mechanized agriculture, and a redefinition of economic efficiency to include biophysical factors.

While urban regions certainly disrupt the ecosystems of which they are a part, the sheer concentration of population and consumption also gives cities enormous leverage in the quest for global sustainability. Some of the advantages of urban settlements are as follows.

- Lower costs per capita of providing piped treated water, sewer systems, waste collection, and most other forms of infrastructure and public amenities;
- Greater possibilities for, and a greater range of options for, material recycling, re-use, remanufacturing and the specialized skills and enterprises needed to make these things happen;
- High population density, which reduces the per capita demand for occupied land;
- Great potential through economies of scale, co-generation, and the use of waste process heat from industry or power plants, to reduce the per capita use of fossil fuel for space-heating;
- Great potential for reducing (mostly fossil) energy consumption by motor vehicles through walking, cycling, and public transit.

For a fuller appreciation of urban leverage, let us examine this last point in more detail. It is commonplace to argue that the private automobile must give way to public transportation in our cities and just as commonplace to reject the idea (at least in North America) as political unfeasible. However, political feasibility depends greatly on public support. The popularity of the private car for urban transportation is in large part due to underpriced fossil fuel and numerous other hidden subsidies (up to $2500 per year per vehicle). Suppose we gradually move toward full cost pricing of urban auto use and reallocate a significant proportion of the considerable auto subsidy to public transit. This could make public transportation faster, more convenient, and more comfortable than at present, and vastly cheaper than private cars. Whither political feasibility? People would demand improved public transit with the same passion they presently reserve for increased road capacity for cars.

[. . .]

Epilogue

Cities are among the brightest stars in the constellation of human achievement. At the same time, ecological footprint analysis shows that they act as entropic black holes, sweeping up the output of whole regions of the ecosphere vastly larger than themselves. Given the causal linkage between global ecological change and concentrated local consumption, nations and provincial/state governments should assess what powers might be devolved to, or shared with, the municipal level to enable cities better to cope with the inherently urban dimensions of sustainability.

At the same time, international agencies and national powers must recognize that policies for local, provincial, or national sustainability have little meaning without firm international commitment to the protection and enhancements of remaining common-pool natural capital and global life support services. There can be no ecological sustainability without international agreement of the nature of the sustainability crisis and difficult solutions that may be necessary at all spatial scales.

PART THREE

Racial and Social Inequality

Plate 4 East 100th Street, Spanish Harlem, New York City, USA, 1966, by Bruce Davidson. Reproduced by permission of Magnum Photos.

Plate 5 Homeless, São Paulo, Brazil, 2003 by Ferdinando Scianna. Reproduced by permission of Magnum Photos.

INTRODUCTION TO PART THREE

The city is a landscape of racial and social inequality. Our urban society is divided by differences of socioeconomic class, race/ethnicity, and residential status. The dominant elite members of our society occupy the prime spaces of the central business district and affluent districts and neighborhoods such as Chicago's Gold Coast, New York's Upper East Side "silk stocking district," and the Hollywood hills of Los Angeles. The poor and the homeless inhabit the marginal spaces of the ghetto, the barrio, and skid row in places such as South Central Los Angeles, East Los Angeles, and New York's Bowery. These marginal spaces proliferate in the "inner city" and the "zone in transition" in the interstitial spaces surrounding the dominant spaces of the central city. The slums, skid rows, and other urban "badlands" of the cities both constitute and reproduce social inequality. There is a dialectic interaction between space and society. Segregation of the poor in deteriorated places disadvantages these people from access to good jobs, housing, schools, and hospitals. The poor and the homeless do not have opportunities for better life-chances in the marginal spaces of the city. This social isolation reproduces poverty through the generations. Spatial immobility acts as a barrier to social mobility.

The first essay, by W. E. B. Du Bois, addresses the deplorable housing conditions of African Americans segregated into the Seventh Ward of Philadelphia at the turn of the twentieth century. The overcrowding was compounded by the erection of residential tenement buildings in rear lots and back yards, with inadequate fresh water and plumbing. Vice and criminality proliferated in these back spaces. Black residents in the Seventh Ward endured relatively high rents and rent-to-income ratios because the neighborhood gave good access to the homes, hotels, and businesses of the white community. Subleasing was common and some subletters endured predatory rents in some rooming houses. Racial discrimination reinforced the segregation of African Americans in the Seventh Ward, but they were also drawn by the congregational life of historic churches.

Du Bois finds significant social stratification in Black Philadelphia, including a criminal underclass, a well-intentioned poor that includes the sick, widows, and orphans, a respectable working class, and an aristocracy. He wishes for greater public visibility for the respectable black elite, but regrets that they don't provide more leadership to unify the community. He blames the criminal underclass for perpetuating negative public impressions of African Americans and fears that the children of the worthy poor may fall under their sway. He calls for all races to unify and realize the promise of the intellectual and political Enlightenment. He addresses the duty of blacks to work diligently for social improvement. He cites the duty of whites to overcome prejudice and ignorance and give proper recognition to black promise. His study mixes social science with social reformism, as it was funded by the University of Pennsylvania and the College Settlement Association.

Loïc Wacquant and William Julius Wilson, nearly a century later, address the challenges faced by urban black America in an era of inner-city disinvestment and expansion of underclass poverty. Civil rights legislation has abated racial segregation and permitted the growth of black suburbs for affluent and middle-class African Americans. But the loss of stable manufacturing jobs in the urban core has eroded the status of working-class black families, who are increasingly challenged by chronic unemployment or underemployment, poor access to housing, education, and health care, and tendencies towards street crime, substance abuse, welfare dependency, out-of-wedlock births, and female-headed families. Attention has focused on how the spatial isolation of the underclass fosters their economic marginalization and severely reduces the access of residents to good health, education, jobs, and overall life-chances.

In their co-written chapter, Wacquant and Wilson draw central attention to the phenomenon of "hyperghettoization" that confronts Black America as a "crisis" of racial and class exclusion. They emphasize that massive job losses have beset the inner city through branch plant relocations and urban disinvestment and reinvestment in suburban locations, the Sun Belt, and overseas. This deindustrialization process in the metropolitan economy has involved a shift from traditional manufacturing industries to service employment or knowledge-intensive industries. Inner-city residents can no longer rely on relatively high-wage factory employment that does not require higher education, in which they acquire on-the-job training. They are relegated to low-paying service sector employment in such arenas as fast-food restaurants and building security. They suffer a jobs–skills mismatch in the postindustrial economy. They cannot acquire financial capital and homeownership has declined. The loss of social capital in the hyperghetto leaves them in social isolation, without access to good educational resources or job networks. The loss of social capital correlates with the decline of local businesses, good schools, and stable neighborhoods of working families.

The persistence of the ghetto contradicts some basic predictions of human ecology theory, namely that racial-ethnic enclaves and colonies of the inner city would eventually dissipate with the upward social mobility and cultural assimilation of minorities and immigrants into American cities and society. Douglas Massey later articulates this phenomenon as the "spatial assimilation" thesis in writings such as "Ethnic Residential Segregation: A Theoretical and Empirical Review," *Sociology and Social Research* 69 (1985): 315–350. Upward social mobility into better jobs and social status is connected with spatial mobility into better homes and neighborhoods. The continuing spatial entrapment of minorities in the inner city is a main factor in their social immobility.

Douglas Massey and Nancy Denton argue that the black ghetto was a conscious creation of white people seeking to isolate and contain poor African Americans during the Jim Crow era of the early twentieth century. Despite the efforts to ameliorate residential segregation and integrate blacks, a continuing more serious form of "hypersegregation" that fuses race and class factors has emerged in American cities. The prosecution of racial discrimination in housing and banking through civil rights legislation is not adequate to put a stop to racial steering by realtors, redlining practices, and predatory lending by banks and mortgage companies. The Fair Housing Acts address individual but not the structural roots of discrimination in our financial and housing markets. They find that, especially in older Midwestern and northern cities, underclass communities grew in the 1970s and 1980s and segregation increased. White flight has increasingly alienated white suburban voters from the concerns of the inner cities, and they oppose public spending that would assist the urban underclass. Segregation has undermined the potential for interracial coalitions, and America is fragmented into "white suburbs and chocolate cities." In the era of hypersegregation, the ghetto is politically isolated.

The social and spatial isolation intensifies the creation of an alternative "oppositional culture" or ghetto social world. Residents reject mainstream values from a system that marginalizes them and they embrace unconventional attitudes and operate in informal and underground economic activities. The alternative culture is communicated through Ebonics or Black English Vernacular, rich cultural phenomena that also impede mainstreaming of underclass children into our society. There is a correlation between spatial and cultural/linguistic isolation.

Immigrant colonies also exhibit characteristics of racial/ethnic residential segregation but with more positive functions. They have grown increasingly since the 1960s, especially among Asian and Latino immigrants, and are an outgrowth of changes in immigration law and the globalization of the U.S. economy. These immigrant enclaves include Chinatown, Koreatown, Little Tokyo, Little Saigon, Little Havana, and Little Haiti. Ethnic enclaves emerge as safety zones in the face of prejudice and discrimination by the mainstream host society, but also assist in immigrant adaptation. They offer an alternative sub-economy separate from the segmented mainstream economy, which is split into a primary sector of jobs with good mobility ladders, and a secondary sector of dead-end jobs in which minorities predominate. These ethnic enclaves offer a protected sector for immigrants newly arrived without English language skills, good education, or official papers. Co-ethnic bosses profit through their ability to self-exploit co-ethnics in arduous sweatshop conditions with marginal benefits and labor rights. Immigrant enclaves create jobs and revenue through the phenomenon of the economic multiplier, where every dollar that enters the enclave is re-circulated through interindustry and consumption linkages. While co-ethnic exploitation is common to the immigrant enclave, the enclave economy offers more opportunity for long-run upward mobility and income generation than the secondary sector of the mainstream economy.

Some immigrant entrepreneurs, however, focus on certain occupational niches as economic and social middlemen between dominant white groups and poor minorities. Chinese, Korean, and Indian immigrants commonly operate small business groceries, liquor stores, and motels. They fulfill a function undesired by white elites. These middlemen minorities act as social buffers between the dominant and oppressed groups of society and, in situations of crisis, may bear the brunt of underclass anger, as seen in the black attacks on Korean merchants that occurred during the Rodney King disturbances of 1992 in Los Angeles. Chinese merchants in Southeast Asia and Jewish merchants in Europe were also classic historical middlemen minorities.

Jan Lin and Paul Robinson probe deeper into issues of spatial and cultural assimilation by investigating the dynamics of ethnic suburbs, or "ethnoburbs," such as the Chinese ethnoburb of the San Gabriel Valley of Los Angeles. The appearance of immigrant enclaves in the suburbs challenges traditional assumptions, going back to the human ecology school, that immigrants would assimilate with spatial and social mobility into the suburbs. The authors find evidence of a core–fringe pattern, with a lower-class Chinese population in the core, and two middle-class settlement districts in higher-rent suburban cities. They examined measures of limited English-speaking ability and linguistic isolation and found broad evidence of ethnic persistence in the core and one fringe district, but the other fringe district exhibited more assimilation to English. The spread of the Chinese ethnoburb has taken place across a broader historical pattern of white flight into exurban residential communities of Los Angeles County or adjoining Orange County. The transition from white residential communities and commercial districts to Chinese settlements and business districts incited a movement for English-only signage and slow growth candidates in local government in the 1980s, but there was a transition to managed growth and emergence of Chinese political leaders by the 1990s. The racial/ethnic succession from white to Chinese residents and businesses in San Gabriel Valley communities and public school systems has been mirrored in cities such as Cupertino in the Silicon Valley area of northern California's Santa Clara County. These cases shed new light on the racial/ethnic dynamics of white flight beyond the traditional case of white–black succession.

Homeless people are the focus of the following article by James S. Duncan. The homeless and other social minorities face a condition of spatial and social exclusion very similar to that experienced by the underclass. Their presence is a threat to established conventions of behavior and social order, and people fear them as a health or crime hazard. Our vagrancy laws restrict their movement because an address and property ownership are prerequisites to the rights and entitlements of full citizenship. Thus they are segregated into marginal spaces such as alleyways and back roads and jurisdictional voids commonly known as "skid rows." Their presence may be tolerated in some prime public spaces if they operate under cover or with acceptable etiquette. The segregation of the homeless and related minorities such as hustlers, drug addicts, and runaways in the no man's lands of the city permits their containment by charitable institutions and the state. The recent gentrification and revival of central cities threatens to displace the homeless from marginal spaces under threat of redevelopment.

James Elliott and Jeremy Pais consider how race and class variables affected people in New Orleans before and after the devastation of Hurricane Katrina, and how race and class affected the human response to the disaster. They discuss how race/ethnicity and class are intertwined in determining the scale of response to human disaster. They found that ethnic minorities were less likely to evacuate ahead of the disaster. They found that black and poor residents were more likely to leave after the storm hit rather than before, as did most white residents. They also determined that race had a significant effect on disaster stress levels. When addressing the differences in social support, the authors found black people used religion for support, while white people used friends and family for support.

"The Environment of the Negro"

from *The Philadelphia Negro* (1900)

W. E. B. Du Bois

Editors' Introduction

W. E. B. Du Bois was one of the great African American intellectuals of the early twentieth century, and a co-founder of the National Association for the Advancement of Colored People (NAACP). His study on *The Philadelphia Negro* was commissioned by the University of Pennsylvania and the College Settlement Association, a reform organization interested in documenting and improving the social and moral life of what was then the fourth largest black urban settlement in the nation. Du Bois spent 15 months doing an extensive field survey focused on the Seventh Ward, the largest and most diverse black ward in the city. The study was a landmark in modern American sociology for its scientific rigor, which included careful analysis of housing data, house-to-house interviews, and field observation.

These excerpts draw mainly from Chapter 15, which describes the social environment of the African American community, with special attention to their housing conditions and social class stratification. The analysis of their overcrowded and deplorable housing conditions was reminiscent of the depictions of New York City tenement housing by Jacob Riis in *How the Other Half Lives* (1890). Just as in New York, Philadelphia tenement buildings were often crowded into rear lots and back yards, and ill-equipped with fresh water and toilet facilities. Criminality and prostitution, it was observed, congregated in rear lot housing. The overcrowding was partly an outcome of the comparatively high rents, and high rent-to-income ratios endured by African Americans in the Seventh Ward, a neighborhood that gave them relative proximity to central-city employment opportunities in the homes, hotels, and businesses of the white community. There was considerable subletting of rooms; in some rooming houses the rents were quite predatory for the quality of housing and plumbing. Black segregation in the Seventh Ward was enforced by racial discrimination in banking and the wider housing market. African Americans also gravitated to the neighborhood because of the historic churches.

Decrying outsider impressions of social class homogeneity in the black community, Du Bois draws our attention to what he characterizes as a fourfold stratification system that includes: a) well-to-do families, b) respectable working-class families, c) the well-intentioned poor, and d) criminals, loafers, and prostitutes. He excoriates the career criminals and the loafers whom they attract. He distinguishes between prostitutes in brothels and streetwalkers. He regrets the unfortunate lot of the poor and casual laborers who struggle to obtain worthy employment. This group includes the sick, disabled, widows, and orphans. He warns that the children of the poor are easily influenced by the criminal class they live adjacent to. He regrets that the respectable working class lack the education to obtain more remunerative employment to raise their station in life. He calls for greater visibility for the well-to-do black aristocracy, who are better representatives of the possibilities for the African American community than the criminal and poor classes that are more frequently of concern to the white community. Yet he also chastises the black aristocracy for being relatively removed from their poor and working-class brethren. He remarks they could be better moral and social leaders for African American underclasses.

Du Bois closes with a moral and ethical call to bring the spirit of the philosophical and political Enlightenment to the resolution of the "Negro problem," which he believes will bring about greater social unity and the mutual advancement of all races. He addresses the duty of blacks to work diligently to advance their station and not lower their standards. He addresses the duty of whites to have faith in the ability of African Americans and not to hinder them with ignorance or racism. With the spirit of self-help and cooperation, he hopes there can be a future of equal opportunity for all.

CHAPTER 1: THE SCOPE OF THIS STUDY

General aim

This study seeks to present the results of an inquiry undertaken by the University of Pennsylvania into the condition of the forty thousand or more people of Negro blood now living in the city of Philadelphia. This inquiry extended over a period of fifteen months and sought to ascertain something of the geographical distribution of this race, their occupations and daily life, their homes, their organizations, and, above all, their relation to their million white fellow-citizens. The final design of the work is to lay before the public such a body of information as may be a safe guide for all efforts toward the solution of the many Negro problems of a great American city.

CHAPTER 15: THE ENVIRONMENT OF THE NEGRO

Houses and rent

[. . .]

We see that in 1848 the average Negro family rented by the month or quarter, and paid between four and five dollars per month rent. The highest average rent for any section was less than fifteen dollars a month. For such rents the poorest accommodations were afforded, and we know from descriptions that the mass of Negroes had small and unhealthful homes, usually on the back streets and alleys.

[. . .]

The lodging system so prevalent in the Seventh Ward makes some rents appear higher than the real facts warrant. This ward is in the centre of the city, near the places of employment for the mass of the people and near the centre of their social life; consequently people crowd here in great numbers. Young couples just married engage lodging in one or two rooms; families join together and hire one house; and numbers of families take in single lodgers.

[. . .]

The practice of sub-renting is found of course in all degrees: from the business of boarding-house keeper to the case of a family which rents out its spare bed-chamber. In the first case the rent is practically all repaid, and must in some cases be regarded as income; in the other cases a small fraction of the rent is repaid and the real rent and the size of the home reduced. Let us endeavor to determine what proportion of the rents of the Seventh Ward are repaid in sub-rents, omitting some boarding and lodging-houses where the sub-rent is really the income of the housewife. In most cases the room-rent of lodgers covers some return for the care of the room.

[. . .]

Of the 2441 families only 334 had access to bathrooms and water-closets, or 13.7 per cent. Even these 334 families have poor accommodations in most instances. Many share the use of one bathroom with one or more other families. The bath-tubs usually are not supplied with hot water and very often have no water-connection at all. This condition is largely owing to the fact that the Seventh Ward belongs to the older part of Philadelphia, built when vaults in the yards were used exclusively and bathrooms could not be given space in the small houses. This was not so unhealthful before the houses were thick and when there were large back yards. To-day, however, the back yards have been filled by tenement houses and the bad sanitary results are shown in the death rate of the ward.

Of the latter only sixteen families had water-closets. So that over 20 per cent and possibly 30 per cent of the Negro families of this ward lack some of the very elementary accommodations necessary to health and decency. And this too in spite of the fact that they are paying comparatively high rents. Here too there comes another consideration, and that is the lack of public urinals and water-closets in this ward and, in fact, throughout Philadelphia. The result is that the closets of tenements are used by the public.

[. . .]

When, however, certain districts like the Seventh Ward became crowded and given over to tenants, the thirst for money-getting led landlords in large numbers of cases to build up their back yards. This is the origin of numbers of the blind alleys and dark holes which make some parts of the Fifth, Seventh and Eighth Wards notorious. The closets in such cases are sometimes divided into compartments for different tenants, but in many cases not even this is done; and in all cases the alley closet becomes a public resort for pedestrians and loafers. The back tenements thus formed rent usually for from $7 to $9 a month, and sometimes for more. They consist of three rooms one above the other, small, poorly lighted and poorly ventilated. The inhabitants of the alley are at the mercy of its worst tenants; here policy shops abound, prostitutes ply their trade, and criminals hide. Most of these houses have to get their water at a hydrant in the alley, and must store their fuel in the house. These tenement abominations of Philadelphia are perhaps better than the vast tenement houses of New York, but they are bad enough, and cry for reform in housing.

The fairly comfortable working class live in houses of 3–6 rooms, with water in the house, but seldom with a bath. A three room house on a small street rents from $10 up; on Lombard street a 5–8 room house can be rented for from $18 to $30 according to location. The great mass of comfortably situated working people live in houses of 6–10 rooms, and sub-rent a part or take lodgers. A 5–7 room house on South Eighteenth street can be had for $20; on Florida street for $18; such houses have usually a parlor, dining room and kitchen on the first floor and two to four bedrooms, of which one or two are apt to be rented to a waiter or coachman for $4 a month, or to a married couple at $6–10 a month. The more elaborate houses are on Lombard street and its cross streets.

The rents paid by the Negroes are without doubt far above their means and often from one-fourth to three-fourths of the total income of a family goes in rent. This leads to much non-payment of rent both intentional and unintentional, to frequent shifting of homes, and above all to stinting the families in many necessities of life in order to live in respectable dwellings. Many a Negro family eats less than it ought for the sake of living in a decent house.

Some of this waste of money in rent is sheer ignorance and carelessness. The Negroes have an inherited distrust of banks and companies, and have long neglected to take part in Building and Loan Associations. Others are simply careless in the spending of their money and lack the shrewdness and business sense of differently trained peoples. Ignorance and carelessness, however, will not explain all or even the greater part of the problem of rent among Negroes. There are three causes of even greater importance: these are the limited localities where Negroes may rent, the peculiar connection of dwelling and occupation among Negroes and the social organization of the Negro. The undeniable fact that most Philadelphia white people prefer not to live near Negroes limits the Negro very seriously in his choice of a home and especially in the choice of a cheap home. Moreover, real estate agents knowing the limited supply usually raise the rent a dollar or two for Negro tenants, if they do not refuse them altogether. Again, the occupations which the Negro follows, and which at present he is compelled to follow, are of a sort that makes it necessary for him to live near the best portions of the city; the mass of Negroes are in the economic world purveyors to the rich – working in private houses, in hotels, large stores, etc. In order to keep this work they must live near by; the laundress cannot bring her Spruce street family's clothes from the Thirtieth Ward, nor can the waiter at the Continental Hotel lodge in Germantown. With the mass of white workmen this same necessity of living near work does not hinder them from getting cheap dwellings; the factory is surrounded by cheap cottages, the foundry by long rows of houses, and even the white clerk and shop girl can, on account of their hours of labor, afford to live further out in the suburbs than the black porter who opens the store. Thus it is clear that the nature of the Negro's work compels him to crowd into the centre of the city much more than is the case with the mass of white working people. At the same time this necessity is apt in some cases to be overestimated, and a few hours of sleep or convenience serve to persuade a good many families to endure poverty in the Seventh Ward when they might be comfortable in the Twenty-fourth Ward. Nevertheless much of the Negro problem in this city finds adequate explanation when we reflect that here is a people receiving a little lower wages than usual for less desirable work, and compelled, in order to do that work, to live in a little less pleasant quarters than most people, and pay for them somewhat higher rents.

The final reason of the concentration of Negroes in certain localities is a social one and one peculiarly strong: the life of the Negroes of the city has for years centered in the Seventh Ward; here are the old churches, St. Thomas', Bethel, Central, Shiloh and Wesley; here are the halls of the secret societies; here are the homesteads of old families. To a race socially ostracized it means far more to move to remote parts of a city, than to those who will in any part of the city easily form congenial acquaintances and new ties. The Negro who ventures away from the mass of his people and their organized life, finds himself alone, shunned and taunted, stared at and made uncomfortable; he can make few new friends, for his neighbors however well-disposed would shrink to add a Negro to their list of acquaintances. Thus he remains far from friends and the concentered social life of the church, and feels in all its bitterness what it means to be a social outcast. Consequently emigration from the ward has gone in groups and centered itself about some church, and individual initiative is thus checked. At the same time color prejudice makes it difficult for groups to find suitable places to move to – one Negro family would be tolerated where six would be objected to; thus we have here a very decisive hindrance to emigration to the suburbs.

It is not surprising that this situation leads to considerable crowding in the homes, *i.e.*, to the endeavor to get as many people into the space hired as possible. It is this crowding that gives the casual observer many false notions as to the size of Negro families, since he often forgets that every other house has its sub-renters and lodgers. It is, however, difficult to measure this crowding on account of this very lodging system which makes it very often uncertain as to just the number of rooms a given group of people occupy.

[...]

As said before, this is probably something under the real truth, although perhaps not greatly so. The figures show considerable overcrowding, but not nearly as much as is often the case in other cities. This is largely due to the character of Philadelphia houses, which are small and low, and will not admit many inmates. Five persons in one room of an ordinary tenement would be almost suffocating. The large number of one-room tenements with two persons should be noted. These 572 families are for the most part young or childless couples, sub-renting a bedroom and working in the city.

[...]

Social classes

There is always a strong tendency on the part of the community to consider the Negroes as composing one practically homogeneous mass. This view has of course a certain justification: the people of Negro descent in this land have had a common history, suffer to-day common disabilities, and contribute to one general set of social problems. And yet if the foregoing statistics have emphasized any one fact it is that wide variations in antecedents, wealth, intelligence and general efficiency have already been differentiated within this group. These differences are not, to be sure, so great or so patent as those among the whites of to-day, and yet they undoubtedly equal the difference among the masses of the people in certain sections of the land fifty or one hundred years ago; and there is no surer way of misunderstanding the Negro or being misunderstood by him than by ignoring manifest differences of condition and power in the 40,000 black people of Philadelphia.

[...]

Nothing more exasperates the better class of Negroes than this tendency to ignore utterly their existence. The law-abiding, hard-working inhabitants of the Thirtieth Ward are aroused to righteous indignation when they see that the word Negro carries most Philadelphians' minds to the alleys of the Fifth Ward or the police courts. Since so much misunderstanding or rather forgetfulness and carelessness on this point is common, let us endeavor to try and fix with some definiteness the different social classes which are clearly enough defined among Negroes to deserve attention. When the statistics of the families of the Seventh Ward were gathered, each family was put in one of four grades as follows:

Grade 1. Families of undoubted respectability earning sufficient income to live well; not engaged in menial service of any kind; the wife engaged in no occupation save that of housewife, except in a few cases where she had special employment at home. The children not compelled to be bread-winners, but found in school; the family living in a well-kept home.

Grade 2. The respectable working-class; in comfortable circumstances, with a good home, and having steady remunerative work. The younger children in school.

Grade 3. The poor; persons not earning enough to keep them at all times above want; honest, although not always energetic or thrifty, and with no touch of gross immorality or crime. Including the very poor, and the poor.

Grade 4. The lowest class of criminals, prostitutes and loafers; the "submerged tenth."

Thus we have in these four grades the criminals, the poor, the laborers, and the well-to-do. The last class represents the ordinary middle-class folk of most modern countries, and contains the germs of other social classes which the Negro has not yet clearly differentiated. Let us begin first with the fourth class.

[. . .]

The size of the more desperate class of criminals and their shrewd abettors is of course comparatively small, but it is large enough to characterize the slum districts. Around this central body lies a large crowd of satellites and feeders: young idlers attracted by excitement, shiftless and lazy ne'er-do-wells, who have sunk from better things, and a rough crowd of pleasure seekers and libertines. These are the fellows who figure in the police courts for larceny and fighting, and drift thus into graver crime or shrewder dissoluteness. They are usually far more ignorant than their leaders, and rapidly die out from disease and excess. Proper measures for rescue and reform might save many of this class. Usually they are not natives of the city, but immigrants who have wandered from the small towns of the South to Richmond and Washington and thence to Philadelphia. Their environment in this city makes it easier for them to live by crime or the results of crime than by work, and being without ambition – or perhaps having lost ambition and grown bitter with the world – they drift with the stream.

One large element of these slums, a class we have barely mentioned, are the prostitutes [. . .] Seven of these women had small children with them and had probably been betrayed, and had then turned to this sort of life. There were fourteen recognized bawdy houses in the ward; ten of them were private dwellings where prostitutes lived and were not especially fitted up, although male visitors frequented them. Four of the houses were regularly fitted up, with elaborate furniture, and in one or two cases had young and beautiful girls on exhibition. All of these latter were seven- or eight-room houses for which $26 to $30 a month was paid. They are pretty well-known resorts, but are not disturbed. In the slums the lowest class of street walkers abound and ply their trade among Negroes, Italians and Americans. One can see men following them into alleys in broad daylight. They usually have male associates whom they support and who join them in "badger" thieving. Most of them are grown women though a few cases of girls under sixteen have been seen on the street.

This fairly characterizes the lowest class of Negroes. According to the inquiry in the Seventh Ward at least 138 families were estimated as belonging to this class out of 2395 reported, or 5.8 per cent. This would include between five and six hundred individuals. Perhaps this number reaches 1000 if the facts were known, but the evidence at hand furnishes only the number stated. In the whole city the number may reach 3000, although there is little data for an estimate.

The next class are the poor and unfortunate and the casual laborers; most of these are of the class of Negroes who in the contact with the life of a great city have failed to find an assured place. They include immigrants who cannot get steady work; good-natured, but unreliable and shiftless persons who cannot keep work or spend their earnings thoughtfully; those who have suffered accident and misfortune; the maimed and defective classes, and the sick; many widows and orphans and deserted wives; all these form a large class and are here considered. It is of course very difficult to separate the lowest of this class from the one below, and probably many are included here who, if the truth were known, ought to be classed lower. In most cases, however, they have been given the benefit of the doubt. The lowest ones of this class usually live in the slums and back streets, and next door or in the same house often, with criminals and lewd women. Ignorant and easily influenced, they readily go with the tide and now rise to industry and decency, now fall to crime. Others of this class get on fairly well in good times, but never get far ahead. They are the ones who earliest feel the weight of hard times and their latest blight. Some correspond to the "worthy poor" of most charitable organizations, and some fall a little below that class. The children of this class are the feeders of the criminal classes. Often in the same family one can find respectable and striving parents weighed down by idle, impudent sons and wayward daughters. This is partly because of poverty, more because of the poor home life. In the Seventh Ward 30½ per cent of the families or 728 may be put into this class, including the very poor, the poor and those who manage just to make ends meet in good times. In the whole city perhaps ten to twelve thousand Negroes fall in this third social grade.

Above these come the representative Negroes; the mass of the servant class, the porters and waiters, and the best of the laborers. They are hard-working people, proverbially good-natured; lacking a little in foresight and forehandedness, and in "push." They are honest and faithful, of fair and improving morals, and beginning to accumulate property. The great drawback to this class is lack of congenial occupation especially among the young men and women, and the consequent widespread dissatisfaction and complaint. As a class these persons are ambitious; the majority can read and write, many have a common school training, and all are anxious to rise in the world. Their wages are low compared with corresponding classes of white workmen, their rents are high, and the field of advancement opened to them is very limited. The best expression of the life of this group is the Negro church, where their social life centers, and where they discuss their situation and prospects.

[. . .]

Finally we come to the 277 families, 11.5 per cent of those of the Seventh Ward, and including perhaps 3,000 Negroes in the city, who form the aristocracy of the Negro population in education, wealth and general social efficiency. In many respects it is right and proper to judge a people by its best classes rather than by its worst classes or middle ranks. The highest class of any group represents its possibilities rather than its exceptions, as is so often assumed in regard to the Negro. The colored people are seldom judged by their best classes, and often the very existence of classes among them is ignored. This is partly due in the North to the anomalous position of those who compose this class: they are not the leaders or the ideal-makers of their own group in thought, work, or morals. They teach the masses to a very small extent, mingle with them but little, do not largely hire their labor. Instead then of social classes held together by strong ties of mutual interest we have in the case of the Negroes, classes who have much to keep them apart, and only community of blood and color prejudice to bind them together. If the Negroes were by themselves either a strong aristocratic system or a dictatorship would for the present prevail. With, however, democracy thus prematurely thrust upon them, the first impulse of the best, the wisest and richest is to segregate themselves from the mass. This action, however, causes more of dislike and jealousy on the part of the masses than usual, because those masses look to the whites for ideals and largely for leadership. It is natural therefore that even to-day the mass of Negroes should look upon the worshipers at St. Thomas' and Central as feeling themselves above them, and should dislike them for it. On the other hand it is just as natural for the well-educated and well-to-do Negroes to feel themselves far above the criminals and prostitutes of Seventh and Lombard streets, and even above the servant girls and porters of the middle class of workers. So far they are justified; but they make their mistake in failing to recognize that, however laudable an ambition to rise may be, the first duty of an upper class is to serve the lowest classes. The aristocracies of all peoples have been slow in learning this and perhaps the Negro is no slower than the rest, but his peculiar situation demands that in his case this lesson be learned sooner. Naturally the uncertain economic status even of this picked class makes it difficult for them to spare much time and energy in social reform; compared with their fellows they are rich, but compared with white Americans they are poor, and they can hardly fulfill their duty as the leaders of the Negroes until they are captains of industry over their people as well as richer and wiser. To-day the professional class among them is, compared with other callings, rather over-represented, and all have a struggle to maintain the position they have won.

This class is itself an answer to the question of the ability of the Negro to assimilate American culture. It is a class small in numbers and not sharply differentiated from other classes, although sufficiently so to be easily recognized. Its members are not to be met with in the ordinary assemblages of the Negroes, nor in their usual promenading places. They are largely Philadelphia born, and being descended from the house-servant class, contain many mulattoes. In their assemblies there are evidences of good breeding and taste, so that a foreigner would hardly think of ex-slaves. They are not to be sure people of wide culture and their mental horizon is as limited as that of the first families in a country town. Here and there may be noted, too, some faint trace of careless moral training. On the whole they strike one as sensible, good folks. Their conversation turns on the gossip of similar circles among the Negroes of Washington, Boston and New York; on questions of the day, and, less willingly, on the situation of the Negro. Strangers secure entrance to this circle with difficulty and only by introduction. For an ordinary white person it would be almost impossible to secure introduction even by a friend. Once in a while some well-known citizen meets a company of this class, but it is hard for the average white American to lay aside his patronizing way toward a Negro, and to talk of aught to him but the Negro question; the lack, therefore, of common ground even for conversation makes such meetings rather stiff and not often repeated. Fifty-two of these families keep servants regularly; they live in well-appointed homes, which give evidence of taste and even luxury.

[. . .]

CHAPTER 18: A FINAL WORD

The meaning of all this

Two sorts of answers are usually returned to the bewildered American who asks seriously: What is the Negro problem? The one is straightforward and clear: it is simply this, or simply that, and one simple remedy long enough applied will in time cause it to disappear. The other answer is apt to be hopelessly involved and complex – to indicate no simple panacea, and to end in a somewhat hopeless – There it is; what can we do? Both of these sorts of answers have something of truth in them: the Negro problem looked at in one way is but the old world questions of ignorance, poverty, crime, and the dislike of the stranger. On the other hand it is a mistake to think that attacking each of these questions single-handed without reference to the others will settle the matter: a combination of social problems is far more than a matter of mere addition, – the combination itself is a problem. Nevertheless the Negro problems are not more hopelessly complex than many others have been. Their elements despite their bewildering complication can be kept clearly in view: they are after all the same difficulties over which the world has grown gray: the question as to how far human intelligence can be trusted and trained; as to whether we must always have the poor with us; as to whether it is possible for the mass of men to attain righteousness on earth; and then to this is added that question of questions: after all who are Men? Is every featherless biped to be counted a man and brother? Are all races and types to be joint heirs of the new earth that men have striven to raise in thirty centuries and more? Shall we not swamp civilization in barbarism and drown genius in indulgence if we seek a mythical Humanity which shall shadow all men? The answer of the early centuries to this puzzle was clear: those of any nation who can be called Men and endowed with rights are few: they are the privileged classes – the well-born and the accidents of low birth called up by the King. The rest, the mass of the nation, the *pöbel*, the mob, are fit to follow, to obey, to dig and delve, but not to think or rule or play the gentleman. We who were born to another philosophy hardly realize how deep-seated and plausible this view of human capabilities and powers once was; how utterly incomprehensible this republic would have been to Charlemagne or Charles V or Charles I. We rather hasten to forget that once the courtiers of English kings looked upon the ancestors of most Americans with far greater contempt than these Americans look upon Negroes – and perhaps, indeed, had more cause. We forget that once French peasants were the "Niggers" of France, and that German princelings once discussed with doubt the brains and humanity of the *bauer*.

Much of this – or at least some of it – has passed and the world has glided by blood and iron into a wider humanity, a wider respect for simple manhood unadorned by ancestors or privilege. Not that we have discovered, as some hoped and some feared, that all men were created free and equal, but rather that the differences in men are not so vast as we had assumed. We still yield the well-born the advantages of birth, we still see that each nation has its dangerous flock of fools and rascals; but we also find most men have brains to be cultivated and souls to be saved.

[. . .]

1 The Negro is here to stay.
2 It is to the advantage of all, both black and white, that every Negro should make the best of himself.
3 It is the duty of the Negro to raise himself by every effort to the standards of modern civilization and not to lower those standards in any degree.
4 It is the duty of the white people to guard their civilization against debauchment by themselves or others; but in order to do this it is not necessary to hinder and retard the efforts of an earnest people to rise, simply because they lack faith in the ability of that people.
5 With these duties in mind and with a spirit of self-help, mutual aid and co-operation, the two races should strive side by side to realize the ideals of the republic and make this truly a land of equal opportunity for all men.

"The Cost of Racial and Class Exclusion in the Inner City"

from *Annals of the American Academy of Political and Social Science* (1989)

Loïc J. D. Wacquant and William Julius Wilson

Editors' Introduction

William Julius Wilson is one of the leading black sociologists and one of the most influential thinkers on issues of urban poverty, race, and social policy in America. His first major contribution to the national debate on the status of African Americans in the U.S. was *The Declining Significance of Race* (University of Chicago Press, 1976) in which he argued that socioeconomic issues were superseding racial issues as the main problems confronting black urban America. He applied his ideas more specifically to the conditions of the urban black poor with his second book, *The Truly Disadvantaged: The Inner City, the Underclass, and Public Policy* (University of Chicago Press, 1987).

In *The Truly Disadvantaged*, Wilson argues that the black ghetto has become a much more dangerous, deprived, and socially disorganized place across the course of the twentieth century. He begins with a discussion of the problem of labeling; the term "underclass" like the phrase "culture of poverty," has been used by political conservatives since the 1980s to blame the victims of urban poverty for their own plight. Wilson repudiates the arguments of political conservatives while challenging liberals to reestablish control of public discourse concerning the underclass. He analyzes the effect of structural economic change and the suburbanization of the black middle class in concentrating the problems of the black poor in the inner cities. He asserts that the urban black poor suffer from a "tangle of pathologies" and live in "social isolation" from the mainstream of social life in America. He also discusses the merits of social policies of universalism versus targeted income-tested or race-based programs to address the urban underclass.

In their co-written selection, Loïc Wacquant and William Wilson reiterate and reformulate some of the issues that Wilson initially addressed. They emphasize the dual importance of both class and racial dynamics in the exclusion of blacks in Chicago as a case study of national trends. The mass exodus of jobs and working families from the inner city, coupled with the growth of neoliberal policies of government privatization and reduction of public spending has triggered a process of "hyperghettoization," concentrating blacks in a crisis of joblessness and extreme poverty. They draw attention also to deindustrialization or structural shift in the economy, notably the decentralization of manufacturing employment from the inner city to the suburbs, Sunbelt states, and offshore locations in developing nations.

The decline of institutional structures in the ghetto, what Wilson in *The Truly Disadvantaged* called "social buffers," is described in this selection as the loss of the "pulpit and the press." The loss of the black leadership (such as teachers, clergy, journalists, lawyers, and businessmen) into the suburbs has left the inner city bereft of stable working families and resources for upward social mobility. Wacquant and Wilson describe the loss of educational resources in the hyperghetto, a situation that is all the more stark because of the loss of manufacturing employment from the inner city. These factory jobs were often available for the previous generation without formal education, as work skills could often be acquired on the job. They also paid a living wage, unlike the service sector jobs that have replaced factory jobs, with the "runaway plant," and deindustrialization process in American cities. Contemporary residents of the hyperghetto are also poorly suited for employment in the new information and technology-based sectors of the postindustrial economy. John Kasarda has described this problem as "jobs–skills mismatch" in a variety of writings, including a chapter titled "Urban Industrial Transition and the Underclass," in *The Ghetto Underclass: Social Science Perspectives*, edited by William Wilson (Newbury Park, CA: Sage Publications, 1993).

Wacquant and Wilson also consider the growing feminization of poverty in the hyperghetto, as poor households are increasingly headed by single women. They note the continuing erosion of financial resources for ghetto households, and the decline in homeownership. They note that the households left in the hyperghetto are bereft of links to solidarity groups, networks, and organizations, what the French sociologist Pierre Bourdieu calls "social capital" ("The Forms of Capital," in *Handbook of Theory and Research for the Sociology of Education*, edited by J. G. Richardson (New York: Greenwood Press, 1986). The political scientist Robert Putnam has recently received national attention for his writings on the general decline of social capital and community networks as a general process in postwar U.S. society (*Bowling Alone: The Collapse and Revival of American Community* (New York: Simon and Schuster, 2000).

William Wilson's *The Declining Significance of Race* was winner of the American Sociological Association's Sydney Spivack Award. *The Truly Disadvantaged* was selected by the editors of the *New York Times Book Review* as one of the 16 best books of 1987, and it also received the *Washington Monthly* Annual Book Award and the Society for the Study of Social Problems C. Wright Mills Award. *When Work Disappears: The World of the New Urban Poor* (New York: Alfred A. Knopf) was chosen as one of the notable books of 1996 by the editors of the *New York Times Book Review* and received the Sidney Hillman Foundation Award. He published *The Bridge over the Racial Divide: Rising Inequality and Coalition Politics* in 1999 (Berkeley, CA: University of California Press).

William Julius Wilson received his Ph.D. from Washington State University in 1996. He taught at the University of Massachusetts at Amherst before joining the University of Chicago faculty in 1972. In 1990 he became the Director of the Center for the Study of Urban Inequality at the University of Chicago. In 1996, he moved to become the Lewis P. and Linda L. Geyser University Professor at Harvard University. Wilson is a past President of the American Sociological Association. He was a MacArthur Prize fellow from 1987 to 1992 and has been elected to the National Academy of Sciences, the American Academy of Arts and Sciences, the National Academy of Education, and the American Philosophical Society. In June 1996 he was selected by *Time* magazine as one of America's 25 Most Influential People. In 1998, he received the National Medal of Science, the highest scientific honor in the United States.

Loïc Wacquant was a doctoral student at the University of Chicago, working as a research assistant at the Urban Poverty and Family Structure Project, when he and Wilson began the collaboration that led to this selection. Further biographical background on Wacquant is provided in the introduction to his selection on "Urban Outcasts: Stigma and Division in the Black American Ghetto and the French Urban Periphery."

After a long eclipse, the ghetto has made a stunning comeback into the collective consciousness of America. Not since the riots of the hot summers of 1966–68 have the black poor received so much attention in academic, activist, and policymaking quarters alike. Persistent and rising poverty, especially among children, mounting social disruptions, the continuing degradation of public housing and public schools, concern over the eroding tax base of cities plagued by large ghettos and by the dilemmas of gentrification, the disillusions of liberals over welfare have all combined to put the black inner-city poor back in the spotlight. Owing in large part to the pervasive and ascendant influence of conservative ideology in the United States, however, recent discussions of the plight of ghetto blacks have typically been cast in individualistic and moralistic terms. The poor are presented as a mere aggregation of personal cases, each with its own logic and self-contained causes. Severed from the struggles and structural changes in the society, economy, and polity that in fact determine them, inner-city dislocations are then portrayed as a self-imposed, self-sustaining phenomenon. This vision of poverty has found perhaps its most vivid expression in the lurid descriptions of ghetto residents that have flourished in the pages of popular magazines and on televised programs devoted to the emerging underclass. Descriptions and explanations of the current predicament of inner-city blacks put the emphasis on individual attributes and the alleged grip of the so-called culture of poverty.

This chapter, in sharp contrast, draws attention to the specific features of the proximate social structure in which ghetto residents evolve and strive, against formidable odds, to survive and, whenever they can, escape its poverty and degradation. We provide this different perspective by profiling blacks who live in Chicago's inner city, contrasting the situation of those who dwell in low-poverty areas with residents of the city's ghetto neighborhoods. Beyond its sociographic focus, the central argument running through this article is that the interrelated set of phenomena captured by the term "underclass" is primarily social-structural and that the ghetto is experiencing a "crisis" not because a "welfare ethos" has mysteriously taken over its residents but because joblessness and economic exclusion, having reached dramatic proportions, have triggered a process of hyperghettoization.

Indeed, the urban black poor of today differ both from their counterparts of earlier years and from the white poor in that they are becoming increasingly concentrated in dilapidated territorial enclaves that epitomize acute social and economic marginalization.

[. . .]

This growing social and spatial concentration of poverty creates a formidable and unprecedented set of obstacles for ghetto blacks. As we shall see, the social structure of today's inner city has been radically altered by the mass exodus of jobs and working families and by the rapid deterioration of housing, schools, businesses, recreational facilities, and other community organizations, further exacerbated by government policies of industrial and urban laissez-faire that have channeled a disproportionate share of federal, state, and municipal resources to the more affluent. The economic and social buffer provided by a stable black working class and a visible, if small, black middle class that cushioned the impact of downswings in the economy and tied ghetto residents to the world of work has all but disappeared. Moreover, the social networks of parents, friends, and associates, as well as the nexus of local institutions, have seen their resources for economic stability progressively depleted. In sum, today's ghetto residents face a closed opportunity structure.

[. . .]

DEINDUSTRIALIZATION AND HYPERGHETTOIZATION

Social conditions in the ghettos of Northern metropolises have never been enviable, but today they are scaling new heights in deprivation, oppression, and hardship. The situation of Chicago's black inner city is emblematic of the social changes that have sown despair and exclusion in these communities. An unprecedented tangle of social woes is gripping the black communities of the city's South Side and West Side. These racial enclaves have experienced rapid increases in the number and percentage of poor families, extensive out-migration of working- and middle-class households, stagnation – if not real regression – of income, and record levels of unemployment. . . .

The single largest force behind this increasing social and economic marginalization of large numbers of inner-city blacks has been a set of mutually reinforcing spatial and industrial changes in the country's urban political economy that have converged to undermine the material foundations of the traditional ghetto. Among these structural shifts are the decentralization of industrial plants, which commenced at the time of World War I but accelerated sharply after 1950, and the flight of manufacturing jobs abroad, to the Sunbelt states, or to the suburbs and exurbs at a time when blacks were continuing to migrate en masse to Rustbelt central cities; the general deconcentration of metropolitan economies and the turn toward service industries and occupations, promoted by the growing separation of banks and industry; and the emergence of post-Taylorist, so-called flexible forms of organizations and generalized corporate attacks on unions – expressed by, among other things, wage cutbacks and the spread of two-tier wage systems and labor contracting – which has intensified job competition and triggered an explosion of low-pay, part-time work. This means that even mild forms of racial discrimination – mild by historical standards – have a bigger impact on those at the bottom of the American class order. In the labor-surplus environment of the 1970s, the weakness of unions and the retrenchment of civil rights enforcement aggravated the structuring of unskilled labor markets along racial lines, marking large numbers of inner-city blacks with the stamp of economic redundancy.

In 1954, Chicago was still near the height of its industrial power. Over 10,000 manufacturing establishments operated within the city limits, employing a total of 616,000, including nearly half a million production workers. By 1982, the number of plants had been cut by half, providing a mere 277,000 jobs for fewer than 162,000 blue-collar employees – a loss of 63 percent, in sharp contrast with the overall growth of manufacturing employment in the country, which added almost 1 million production jobs in the quarter century starting in 1958. This crumbling of the city's industrial base was accompanied by substantial cuts in trade employment, with over 120,000 jobs lost in retail and wholesale from 1963 to 1982. The mild growth of services – which created an additional 57,000 jobs during the same period, excluding health, financial, and social services – came nowhere near to compensating for this collapse of Chicago's low-skilled employment pool. Because, traditionally, blacks have relied heavily on manufacturing and blue-collar employment for economic sustenance, the upshot of these structural economic changes for the inhabitants of the inner city has been a steep and accelerating rise in labor market exclusion. In the 1950s, ghetto blacks had roughly the same rate of employment as the average Chicagoan, with some 6 adults in 10 working. While this ratio has not changed citywide over the ensuing three decades, nowadays most residents of the Black Belt cannot find gainful employment and must resort to welfare, to participation in the second economy, or to illegal activities in order to survive. . . .

As the metropolitan economy moved away from smokestack industries and expanded outside of Chicago, emptying the Black Belt of most of its manufacturing jobs and employed residents, the gap between the ghetto and the rest of the city, not to mention its suburbs, widened dramatically. By 1980, median family income on the South and West sides had dropped to around one-third and one-half of the city average, respectively, compared with two-thirds and near parity thirty years earlier. Meanwhile, some of the city's white bourgeois neighborhoods and upper-class suburbs had reached over twice the citywide figure. Thus in 1980, half of the families of Oakland had to make do with less than $5,500 a year, while half of the families of Highland Park incurred incomes in excess of $43,000.

A recent ethnographic account by Arne Duncan on changes in North Kenwood, one of the poorest black sections on the city's South Side, vividly encapsulates the accelerated physical and social decay of the ghetto and is worth quoting at some length:

> In the 1960's, 47th Street was still the social hub of the South Side black community. Sue's eyes light up when she describes how the street used to be filled with stores, theaters and nightclubs in which one could listen to jazz bands well into the evening. Sue remembers the street as "soulful." Today the street might be better characterized as soulless. Some stores, currency exchanges, bars and liquor stores continue to exist on 47th. Yet, as one walks down the street, one is struck more by the death of the street than by its life. Quite literally, the destruction of human life occurs frequently on 47th. In terms of physical structures, many stores are boarded up and abandoned. A few buildings have bars across the front and are closed to the public, but they are not empty. They are used, not so secretly, by people involved in illegal activities. Other stretches of the street are

> simply barren, empty lots. Whatever buildings once stood on the lots are long gone. Nothing gets built on 47th. . . . Over the years one apartment building after another has been condemned by the city and torn down. Today many blocks have the bombed-out look of Berlin after World War II. There are huge, barren areas of Kenwood, covered by weeds, bricks, and broken bottles.

Duncan reports how this disappearance of businesses and loss of housing have stimulated the influx of drugs and criminal activities to undermine the strong sense of solidarity that once permeated the community. With no activities or organizations left to bring them together or to represent them as a collectivity, with half the population gone in 15 years, the remaining residents, some of whom now refer to North Kenwood as the "Wild West," seem to be engaged in a perpetual *bellum omnium contra omnes* for sheer survival. One informant expresses this succinctly: "It's gotten worse. They tore down all the buildings, deterioratin' the neighborhood. All your friends have to leave. They are just spreading out your mellahs [close friends]. It's not no neighborhood anymore." With the ever present threat of gentrification – much of the area is prime lake-front property that would bring in huge profits if it could be turned over to upper-class condominiums and apartment complexes to cater to the needs of the higher-income clientele of Hyde Park, which lies just to the south – the future of the community appears gloomy. One resident explains: "They want to put all the blacks in the projects. They want to build buildings for the rich, and not us poor people. They are trying to move us all out. In four or five years we will all be gone."

Fundamental changes in the organization of America's advanced economy have thus unleashed irresistible centrifugal pressures that have broken down the previous structure of the ghetto and set off a process of hyperghettoization. By this, we mean that the ghetto has lost much of its organizational strength – the "pulpit and the press," for instance, have virtually collapsed as collective agencies – as it has become increasingly marginal economically; its activities are no longer structured around an internal and relatively autonomous social space that duplicates the institutional structure of the larger society and provides basic minimal resources for social mobility, if only within a truncated black class structure. And the social ills that have long been associated with segregated poverty – violent crime, drugs, housing deterioration, family disruption, commercial blight, and educational failure – have reached qualitatively different proportions and have become articulated into a new configuration that endows each with a more deadly impact than before.

If the "organized," or institutional, ghetto of forty years ago described so graphically by Drake and Cayton imposed an enormous cost on blacks collectively, the "disorganized" ghetto, or hyperghetto, of today carries an even larger price. For, now, not only are ghetto residents, as before, dependent on the will and decisions of outside forces that rule the field of power – the mostly white dominant class, corporations, realtors, politicians, and welfare agencies – they have no control over and are forced to rely on services and institutions that are massively inferior to those of the wider society. Today's ghetto inhabitants comprise almost exclusively the most marginal and oppressed sections of the black community. Having lost the economic underpinnings and much of the fine texture of organizations and patterned activities that allowed previous generations of urban blacks to sustain family, community, and collectivity even in the face of continued economic hardship and unflinching racial subordination, the inner-city now presents a picture of radical class and racial exclusion. It is to a sociographic assessment of the latter that we now turn.

THE COST OF LIVING IN THE GHETTO

Let us contrast the social structure of ghetto neighborhoods with that of low-poverty black areas of the city of Chicago. For purposes of this comparison, we have classified as low-poverty neighborhoods all those tracts with rates of poverty – as measured by the number of persons below the official poverty line between 20 and 30 percent as of the 1980 census. Given that the overall poverty rate among black families in the city is about one-third, these low-poverty areas can be considered as roughly representative of the average non-ghetto, non-middle-class, black neighborhood of Chicago. In point of fact, nearly all – 97 percent – of the respondents in this category reside outside traditional ghetto areas. Extreme-poverty neighborhoods comprise tracts with at least 40 percent of their residents in poverty in 1980. These tracts make up the historic heart of Chicago's black ghetto: over 82 percent of the respondents in this category inhabit the West and South sides of the city, in areas most of which have been all black for half a century and more, and an additional 13 percent live in immediately adjacent tracts. Thus when we counterpose extreme-poverty areas with low-poverty areas, we are in effect comparing ghetto neighborhoods with other black areas, most of which are moderately poor, that are not part of Chicago's traditional Black Belt. Even though this comparison involves a truncated spectrum of types of neighborhoods, the contrasts it reveals between low-poverty and ghetto tracts are quite pronounced.

It should be noted that this distinction between low-poverty and ghetto neighborhoods is not merely analytical but captures differences that are clearly perceived by social agents themselves. First, the folk category of ghetto does, in Chicago, refer to the South Side and West Side, not just to any black area of the city; mundane usages of the term entail a social-historical and spatial referent rather than simply a racial dimension. Furthermore, blacks who live in extreme-poverty areas have a noticeably more negative opinion of their neighborhood. Only 16 percent rate it as a "good" to "very good" place to live in, compared to 41 percent among inhabitants of low-poverty tracts; almost 1 in 4 find their neighborhood "bad or very bad" compared to fewer than 1 in 10 among the latter. In short, the contrast between ghetto and non-ghetto poor areas is one that is socially meaningful to their residents.

The black class structure in and out of the ghetto

The first major difference between low- and extreme-poverty areas has to do with their class structure. A sizable majority of blacks in low-poverty tracts are gainfully employed: two-thirds hold a job, including 11 percent with middle-class occupations and 55 percent with working-class jobs, while one-third do not work. These proportions are exactly opposite in the ghetto, where fully 61 percent of adult residents do not work, one-third have working-class jobs and a mere 6 percent enjoy middle-class status. For those who reside in the urban core, then, being without a job is by far the most likely occurrence, while being employed is the exception. Controlling for gender does not affect this contrast, though it does reveal the greater economic vulnerability of women, who are twice as likely as men to be jobless. Men in both types of neighborhoods have a more favorable class mix resulting from their better rates of employment: 78 percent in low-poverty areas and 66 percent in the ghetto. If women are much less frequently employed – 42 percent in low-poverty areas and 69 percent in the ghetto do not work – they have comparable, that is, severely limited, overall access to middle-class status: in both types of neighborhood, only about 10 percent hold credentialed salaried positions or better.

These data are hardly surprising. They stand as a brutal reminder that joblessness and poverty are two sides of the same coin. The poorer the neighborhood, the more prevalent joblessness and the lower the class recruitment of its residents. But these results also reveal that the degree of economic exclusion observed in ghetto neighborhoods during the period of sluggish economic growth of the late 1970s is still very much with us nearly a decade later, in the midst of the most rapid expansion in recent American economic history.

As we would expect, there is a close association between class and educational credentials. Virtually every member of the middle class has at least graduated from high school; nearly two-thirds of working-class blacks have also completed secondary education; but less than half – 44 percent – of the jobless have a high school diploma or more. Looked at from another angle, 15 percent of our educated respondents – that is, high school graduates or better – have made it into the salaried middle class, half have become white-collar or blue-collar wage earners, and 36 percent are without a job. By comparison, those without a high school education are distributed as follows: 1.6 percent in the middle class, 37.9 percent in the working class, and a substantial majority of 60.5 percent in the jobless category. In other words, a high school degree is *a conditio sine qua non* for blacks for entering the world of work, let alone that of the middle class. Not finishing secondary education is synonymous with economic redundancy.

Ghetto residents are, on the whole, less educated than the inhabitants of other black neighborhoods. This results in part from their lower class composition but also from the much more modest academic background of the jobless: fewer than 4 in 10 jobless persons on the city's South Side and West Side have graduated from high school, compared to nearly 6 in 10 in low-poverty areas. It should be pointed out that education is one of the few areas in which women do not fare worse than men: females are as likely to hold a high school diploma as males in the ghetto – 50 percent – and more likely to do so in low-poverty areas – 69 percent versus 62 percent.

Moreover, ghetto residents have lower class origins, if one judges from the economic assets of their family of orientation. Fewer than 4 ghetto dwellers in 10 come from a family that owned its home and 6 in 10 have parents who owned nothing, that is, no home, business, or land. In low-poverty areas, 55 percent of the inhabitants are from a home-owning family while only 40 percent had no assets at all a generation ago. Women, both in and out of the ghetto, are least likely to come from a family with a home or any other asset – 46 percent and 37 percent, respectively. This difference in class origins is also captured by differential rates of welfare receipt during childhood: the proportion of respondents whose parents were on public aid at some time when they were growing up is 30 percent in low-poverty tracts and 41 percent in the ghetto. Women in extreme-poverty areas are by far the most likely to come from a family with a welfare record.

Class, gender, and welfare trajectories in low- and extreme-poverty areas

If they are more likely to have been raised in a household that drew public assistance in the past, ghetto dwellers are also much more likely to have been or to be currently on welfare themselves. Differences in class, gender, and neighborhood cumulate at each juncture of the welfare trajectory to produce much higher levels of welfare attachments among the ghetto population.

In low-poverty areas, only one resident in four are currently on aid while almost half have never personally received assistance. In the ghetto, by contrast, over half the residents are current welfare recipients, and only one in five have never been on aid. These differences are consistent with what we know from censuses and other studies: in 1980, about half of the black population of most community areas on the South Side and West Side was officially receiving public assistance, while working- and middle-class black neighborhoods of the far South Side, such as South Shore, Chatham, or Roseland, had rates of welfare receipt ranging between one-fifth and one-fourth.

None of the middle-class respondents who live in low-poverty tracts were on welfare at the time they were interviewed, and only one in five had ever been on aid in their lives. Among working-class residents, a mere 7 percent were on welfare and just over one-half had never had any welfare experience. This same relationship between class and welfare receipt is found among residents of extreme-poverty tracts, but with significantly higher rates of welfare receipt at all class levels: there, 12 percent of working-class residents are presently on aid and 39 percent received welfare before; even a few middle-class blacks – 9 percent – are drawing public assistance and only one-third of them have never received any aid, instead of three-quarters in low-poverty tracts. But it is among the jobless that the difference between low- and extreme-poverty areas is the largest: fully 86 percent of those in ghetto tracts are currently on welfare and only 7 percent have never had recourse to public aid, compared with 62 percent and 20 percent, respectively, among those who live outside the ghetto.

Neighborhood differences in patterns of welfare receipt are robust across genders, with women exhibiting noticeably higher rates than men in both types of areas and at all class levels. The handful of black middle-class women who reside in the ghetto are much more likely to admit to having received aid in the past than their male counterparts: one-third versus one-tenth. Among working-class respondents, levels of current welfare receipt are similar for both sexes – 5.0 percent and 8.5 percent, respectively – while levels of past receipt again display the greater economic vulnerability of women: one in two received aid before as against one male in five. This gender differential is somewhat attenuated in extreme-poverty areas by the general prevalence of welfare receipt, with two-thirds of all jobless males and 9 in 10 jobless women presently receiving public assistance.

The high incidence and persistence of joblessness and welfare in ghetto neighborhoods, reflecting the paucity of viable options for stable employment, take a heavy toll on those who are on aid by significantly depressing their expectations of finding a route to economic self-sufficiency. While a slim majority of welfare recipients living in low-poverty tracts expect to be self-supportive within a year and only a small minority anticipate receiving aid for longer than five years, in ghetto neighborhoods, by contrast, fewer than 1 in 3 public-aid recipients expect to be welfare-free within a year and fully 1 in 5 anticipate needing assistance for more than five years. This difference of expectations increases among the jobless of both genders. For instance, unemployed women in the ghetto are twice as likely as unemployed women in low-poverty areas to think that they will remain on aid for more than five years and half as likely to anticipate getting off the rolls within a year.

Thus if the likelihood of being on welfare increases sharply as one crosses the line between the employed and the jobless, it remains that, at each level of the class structure, welfare receipt is notably more frequent in extreme-poverty neighborhoods, especially among the unemployed, and among women.

[...]

Differences in economic and financial capital

A quick survey of the economic and financial assets of the residents of Chicago's poor black neighborhoods reveals the appalling degree of economic hardship, insecurity, and deprivation that they must confront day in and day out. The picture in low-poverty areas is grim; that in the ghetto is one of near-total destitution.

In 1986, the median family income for blacks nationally was pegged at $18,000, compared to $31,000 for white families. Black households in Chicago's low-poverty areas have roughly equivalent incomes, with 52 percent declaring over $20,000 annually. Those living in Chicago's ghetto, by contrast, command but a fraction of this figure: half of all ghetto respondents live in households that dispose of less than $7500 annually, twice the rate among residents of low-poverty neighborhoods. Women assign their households to much lower income brackets in both areas, with fewer than 1 in 3 in low-poverty areas and 1 in 10 in extreme-poverty areas enjoying more than $25,000 annually. Even those who work report smaller incomes in the ghetto: the proportion of working-class and middle-class households falling under the $7500 mark on the South and West sides – 12.5 percent and 6.5 percent, respectively – is double that of other black neighborhoods, while fully one-half of jobless respondents in extreme-poverty tracts do not reach the $5000 line. It is not surprising that ghetto dwellers also less frequently report an improvement of the financial situation of their household, with women again in the least enviable position. This reflects sharp class differences: 42 percent of our middle-class respondents and 36 percent of working-class blacks register a financial amelioration as against 13 percent of the jobless.

Due to meager and irregular income, those financial and banking services that most members of the larger society take for granted are, to put it mildly, not of obvious access to the black poor. Barely one-third of the residents of low-poverty areas maintain a personal checking account; only one in nine manage to do so in the ghetto, where nearly three of every four persons report no financial asset whatsoever from a possible list of six and only 8 percent have at least three of those six assets. Here, again, class and neighborhood lines are sharply drawn: in low-poverty areas, 10 percent of the jobless and 48 percent of working-class blacks have a personal checking account compared to 3 percent and 37 percent, respectively, in the ghetto; the proportion for members of the middle class is similar – 63 percent – in both areas.

The American dream of owning one's home remains well out of reach for a large majority of our black respondents, especially those in the ghetto, where barely 1 person in 10 belong to a home-owning household, compared to over 4 in 10 in low-poverty areas, a difference that is just as pronounced within each gender. The considerably more modest dream of owning an automobile is likewise one that has yet to materialize for ghetto residents, of which only one-third live in households with a car that runs. Again, this is due to a cumulation of sharp class and neighborhood differences: 79 percent of middle-class respondents and 62 percent of working-class blacks have an automobile in their household, contrasted with merely 28 percent of the jobless. But, in ghetto tracts, only 18 percent of the jobless have domestic access to a car – 34 percent for men and 13 percent for women.

The social consequences of such a paucity of income and assets as suffered by ghetto blacks cannot be overemphasized. For just as the lack of financial resources or possession of a home represents a critical handicap when one can only find low-paying and casual employment or when one loses one's job, in that it literally forces one to go on the welfare rolls, not owning a car severely curtails one's chances of competing for available jobs that are not located nearby or that are not readily accessible by public transportation.

Social capital and poverty concentration

Among the resources that individuals can draw upon to implement strategies of social mobility are those potentially provided by their lovers, kin, and friends and by the contacts they develop within the formal associations to which they belong – in sum, the resources they have access to by virtue of being socially integrated into solidarity groups, networks, or organizations, what Bourdieu calls "social capital." Our data indicate that not only do residents of extreme-poverty areas have fewer social ties but also that they tend to have ties of lesser social worth, as measured by the social position of their partners, parents, siblings, and best friends, for instance. In short, they possess lower volumes of social capital.

Living in the ghetto means being more socially isolated: nearly half of the residents of extreme-poverty tracts have no current partner – defined here as a person they are married to, live with, or are dating steadily – and one in five admit to having no one who would qualify as a best friend, compared to 32 percent and 12 percent, respectively, in low-poverty areas. It also means that intact marriages are less frequent. Jobless men are much less likely than working males to have current partners in both types of neighborhoods: 62 percent in low-poverty neighborhoods and 44 percent in extreme-poverty areas. Black women have a slightly better chance of having a partner if they live in a low-poverty area, and this partner is also more likely to have completed high school and to work steadily; for ghetto residence further affects the labor-market standing of the latter. The partners of women living in extreme-poverty areas are less stably employed than those of female respondents from low-poverty neighborhoods: 62 percent in extreme-poverty areas work regularly as compared to 84 percent in low-poverty areas.

Friends often play a crucial role in life in that they provide emotional and material support, help construct one's identity, and often open up opportunities that one would not have without them – particularly in the area of jobs. We have seen that ghetto residents are more likely than other black Chicagoans to have no close friend. If they have a best friend, furthermore, he or she is less likely to work, is less educated, and twice as likely to be on aid. Because friendships tend to develop primarily within genders and women have much higher rates of economic exclusion, female respondents are much more likely than men to have a best friend who does not work and who receives welfare assistance. Both of these characteristics, in turn, tend to be more prevalent among ghetto females.

Such differences in social capital are also evidenced by different rates and patterns of organizational participation. While being part of a formal organization, such as a block club or a community organization, a political party, a school-related association, or a sports, fraternal, or other social group, is a rare occurrence as a rule – with the notable exception of middle-class blacks, two-thirds of whom belong to at least one such group – it is more common for ghetto residents – 64 percent, versus 50 percent in low-poverty tracts – especially females – 64 percent, versus 46 percent in low-poverty areas – to belong to no organization. As for church membership, the small minority who profess to be, in Weber's felicitous expression, "religiously unmusical" is twice as large in the ghetto as outside: 12 percent versus 5 percent. For those with a religion, ghetto residence tends to depress church attendance slightly – 29 percent of ghetto inhabitants attend service at least once a week compared to 37 percent of respondents from low-poverty tracts – even though women tend to attend more regularly than men in both types of areas. Finally, black women who inhabit the ghetto are also slightly less likely to know most of their neighbors than their counterparts from low-poverty areas. All in all, then, poverty concentration has the effect of devaluing the social capital of those who live in its midst.

CONCLUSION: THE SOCIAL STRUCTURING OF GHETTO POVERTY

The extraordinary levels of economic hardship plaguing Chicago's inner city in the 1970s have not abated, and the ghetto seems to have gone unaffected by the economic boom of the past five years. If anything, conditions have continued to worsen. This points to the asymmetric causality between the economy and ghetto poverty and to the urgent need to study the social and political structures that mediate their relationship. The significant differences we have uncovered between low-poverty and extreme-poverty areas in Chicago are essentially a reflection of their different class mix and of the prevalence of economic exclusion in the ghetto.

Our conclusion, then, is that social analysts must pay more attention to the extreme levels of economic deprivation and social marginalization as uncovered in this article before they further entertain and spread so-called theories about the potency of a ghetto culture of poverty that has yet to receive rigorous empirical elaboration. Those who have been pushing moral–cultural or individualistic –behavioral explanations of the social dislocations that have swept through the inner city in recent years have created a fictitious normative divide between urban blacks that, no matter its reality – which has yet to be ascertained – cannot but pale when compared to the objective structural cleavage that separates ghetto residents from the larger society and to the collective material constraints that bear on them. It is the cumulative structural entrapment and forcible socioeconomic marginalization resulting from the historically evolving interplay of class, racial, and gender domination, together with sea changes in the organization of American capitalism and failed urban and social policies, not a "welfare ethos," that explain the plight of today's ghetto blacks. Thus, if the concept of underclass is used, it must be a structural concept: it must denote a new sociospatial patterning of class and racial domination, recognizable by the unprecedented concentration of the most socially excluded and economically marginal members of the dominated racial and economic group. It should not be used as a label to designate a new breed of individuals molded freely by a mythical and all-powerful culture of poverty.

"Segregation and the Making of the Underclass"

from *American Apartheid* (1993)

Douglas S. Massey and Nancy A. Denton

Editors' Introduction

One hundred years after W. E. DuBois first decried that "the problem of the Twentieth Century is the problem of the color-line," Douglas Massey and Nancy Denton warn that racial segregation is still an imposing obstacle in American society. The "hypersegregation" of blacks and Latinos in urban ghettos at the turn of the new millennium is unlike the residential segregation experienced by the white ethnic minorities that preceded them. The "dark ghetto" has "invisible walls" that may be social, political, educational, or economic. Contrary to popular belief, segregation is not at its worst in the South, but in the North. In 1966, Martin Luther King came to Chicago and declared it "The most segregated city in America." Despite King's and all the civil rights leaders' efforts, little has changed. Decades after the passage of the last of the great civil rights acts of the 1960s, blacks remain almost as segregated in American cities as they were in 1968.

In *American Apartheid*, Massey and Denton argue that white people created the underclass ghetto during the first half of the twentieth century in order to isolate growing urban black populations. Despite the Fair Housing Act of 1968, segregation is perpetuated today through an interlocking set of individual actions, institutional practices, and governmental policies. In some urban areas the degree of black segregation is so intense and occurs in so many dimensions simultaneously that it amounts to "hypersegregation." The authors demonstrate that this systematic segregation of African Americans leads inexorably to the creation of underclass communities during periods of economic downturn. Under conditions of extreme segregation, any increase in the overall rate of black poverty yields a marked increase in the geographic concentration of indigence and the deterioration of social and economic conditions in black communities. As ghetto residents adapt to this increasingly harsh environment under a climate of racial isolation, they evolve attitudes, behaviors, and practices that further marginalize their neighborhoods and undermine their chances of success in mainstream American society. Their book is a sober challenge to those who argue that race is of declining significance in the United States today.

This represents perhaps the greatest failure in the national effort to equalize the condition of American blacks. Compared with the substantial changes in employment and political representation and education – the growth of the black middle class, the great increase in black college attendance, the surge in the number of black mayors, state legislators, members of Congress – the indices of residential segregation show almost no change. Despite the overturning of *de jure* segregation, *de facto* segregation continues in America, through practices such as bank and insurance redlining and prejudicial real estate steering. Despite the Mortgage Disclosure Act and the Community Reinvestment Act, banks still discriminate against Blacks in the home loan market. There is "Segregation with a smile," where realtors actively steer blacks away from white neighborhoods, a practice that is revealed by housing audit studies.

White flight and continuing segregation have isolated racial minorities in central cities, undermining political coalitions and fragmenting the political landscape and the tax base between "white suburbs" and "chocolate cities." White ethnic immigrants of the early twentieth century were able to create pan-ethnic coalitions in urban patronage machines to allocate spending in their neighborhoods and maintain the quality of their schools and infrastructure. But segregation has undermined the ability of blacks to advance their interests, form coalitions, and establish common interests with white voters. The spatial and political isolation of the ghetto makes it easier for racists to act on their own prejudices.

The emergence and persistence of the urban ghetto attests to the condition of spatial apartheid that confronts racial minorities at the turn of the millennium in American cities. Using the index of dissimilarity as a measure of residential segregation, they have collaborated in numerous exhaustive studies that confirm that while ethnics are becoming more spatially assimilated, blacks experience significant continuing segregation. Especially in older cities of the U.S. Northeast and Midwest, there was a growth of underclass communities in the 1970s and 1980s in cities experiencing greater residential segregation. Barriers to spatial mobility are also barriers to social mobility. They liken this condition of spatial and social immobility to a condition of racial apartheid.

Under conditions of extreme segregation, any increase in the overall rate of black poverty yields a marked increase in the geographic concentration of indigence and the deterioration of social and economic conditions in black communities. As ghetto residents adapt to this increasingly harsh environment under a climate of racial isolation, they evolve attitudes, behaviors, and practices that further marginalize their neighborhoods and undermine their chances of success in mainstream American society. Ghetto residents live in a very limited social world that intensifies the growth of Ebonics or Black English vernacular and reinforces the sense of an "oppositional culture" to the American cultural mainstream. Ebonics is a rich cultural phenomenon, but the oppositional culture creates peer pressure against school attendance and social mobility. There can be a climate of intimidation, violence, and fear among the youth of underclass black communities. Spatial isolation feeds cultural and linguistic isolation.

The authors suggest a more vigorous prosecution of realtors and bankers that discriminate against African Americans, stricter enforcement of the Fair Housing Law of 1968, and rental vouchers to African Americans in order to ease segregation. Segregation is a problem without any easy answers. If Massey and Denton's proposals were enacted, would that stop the problem of whites moving away? Perhaps the only real solution requires whites learning how to get along with and not fear African Americans.

Douglas Massey received his Ph.D. in 1978 from Princeton University and has served on the faculties of the University of Chicago and the University of Pennsylvania. His research focuses on international migration, race and housing, discrimination, education, urban poverty, and Latin America, especially Mexico. He is the co-author, most recently, of *Beyond Smoke and Mirrors: Mexican Immigration in an Age of Economic Integration* (New York: Russell Sage Foundation, 2002), and *Source of the River: The Social Origins of Freshmen at America's Selective Colleges and Universities* (Princeton, NJ: Princeton University Press, 2006). He is a member of the National Academy of Sciences and the American Academy of Arts and Sciences and past-President of the American Sociological Association and the Population Association of America.

Nancy Denton received her Ph.D. in Demography from the University of Pennsylvania in 1984. She is an Associate Professor of Sociology and Director of Graduate Studies at the State University of New York at Albany, where she is also a Research Associate at the Center for Social and Demographic Analysis. She recently edited, with Steward Tolnay, a reader titled *American Diversity: A Demographic Challenge for the 21st Century*. She was chair of the Community and Urban Section of the American Sociological Association from 2001 to 2002.

THE MISSING LINK

> It is quite simple. As soon as there is a group area then all your uncertainties are removed and that is, after all, the primary purpose of this Bill [requiring racial segregation in housing].
>
> (Minister of the Interior, Union of South Africa legislative debate on the Group Areas Act of 1950)

During the 1970s and 1980s a word disappeared from the American vocabulary. It was not in the speeches of politicians decrying the multiple ills besetting American cities. It was not spoken by government officials responsible for administering the nation's social programs. It was not mentioned by journalists reporting on the rising tide of homelessness, drugs, and violence in urban America. It was not discussed by foundation executives and think-tank experts proposing new programs for unemployed parents and unwed mothers. It was not articulated by civil rights leaders speaking out against the persistence of racial inequality; and it was nowhere to be found in the thousands of pages written by social scientists on the urban underclass. The word was segregation.

Most Americans vaguely realize that urban America is still a residentially segregated society, but few appreciate the depth of black segregation or the degree to which it is maintained by ongoing institutional arrangements and contemporary individual actions. They view segregation as an unfortunate holdover from a racist past, one that is fading progressively over time. If racial residential segregation persists, they reason, it is only because civil rights laws passed during the 1960s have not had enough time to work or because many blacks still prefer to live in black neighborhoods. The residential segregation of blacks is viewed charitably as a "natural" outcome of impersonal social and economic forces, the same forces that produced Italian and Polish neighborhoods in the past and that yield Mexican and Korean areas today.

But black segregation is not comparable to the limited and transient segregation experienced by other racial and ethnic groups, now or in the past. No group in the history of the United States has ever experienced the sustained high level of residential segregation that has been imposed on blacks in large American cities for the past fifty years. This extreme racial isolation did not just happen; it was manufactured by whites through a series of self-conscious actions and purposeful institutional arrangements that continue today. Not only is the depth of black segregation unprecedented and utterly unique compared with that of other groups, but it shows little sign of change with the passage of time or improvements in socioeconomic status.

If policymakers, scholars, and the public have been reluctant to acknowledge segregation's persistence, they have likewise been blind to its consequences for American blacks. Residential segregation is not a neutral fact; it systematically undermines the social and economic well-being of blacks in the United States. Because of racial segregation, a significant share of black America is condemned to experience a social environment where poverty and joblessness are the norm, where a majority of children are born out of wedlock, where most families are on welfare, where educational failure prevails, and where social and physical deterioration abound. Through prolonged exposure to such an environment, black chances for social and economic success are drastically reduced.

Deleterious neighborhood conditions are built into the structure of the black community. They occur because segregation concentrates poverty to build a set of mutually reinforcing and self-feeding spirals of decline into black neighborhoods. When economic dislocations deprive a segregated group of employment and increase its rate of poverty, socioeconomic deprivation inevitably becomes more concentrated in neighborhoods where that group lives. The damaging social consequences that follow from increased poverty are spatially concentrated as well, creating uniquely disadvantaged environments that become progressively isolated – geographically, socially, and economically – from the rest of society.

[. . .]

We trace the historical construction of the black ghetto during the nineteenth and twentieth centuries. We show that high levels of black–white segregation were not always characteristic of American urban areas. Until the end of the nineteenth century blacks and whites were relatively integrated in both northern and southern cities; as late as 1900, the typical black urbanite still lived in a neighborhood that was predominantly white. The evolution of segregated, all-black neighborhoods occurred later and was not the result of impersonal market forces. It did not reflect the desires of African Americans themselves. On the contrary, the black ghetto was constructed through a series of well-defined institutional practices, private behaviors, and public policies by which whites sought to contain growing urban black populations.

The manner in which blacks were residentially incorporated into American cities differed fundamentally from the path of spatial assimilation followed by other ethnic groups. Even at the height of immigration from Europe, most Italians, Poles, and Jews lived in neighborhoods where members of their own group did not predominate, and as their socioeconomic status and generations spent in the United States rose, each group was progressively integrated into American society. In contrast, after the construction of the black ghetto the vast majority of blacks were forced to live in neighborhoods that were all black, yielding an extreme level of social isolation.

We show that high levels of black–white segregation had become universal in American cities by 1970, and despite the passage of the Fair Housing Act in 1968, this situation had not changed much in the nation's largest black communities by 1980. In these large urban areas black–white segregation persisted at very high levels, and the extent of black suburbanization lagged far behind that of other groups. Even within suburbs, levels of racial segregation remained exceptionally high, and in many urban areas the degree of racial separation between blacks and whites was profound. Within sixteen large metropolitan areas – containing one-third of all blacks in the United States – the extent of racial segregation was so intense and occurred on so many dimensions simultaneously that we label the pattern "hypersegregation."

We examine why black segregation continues to be so extreme. One possibility that we rule out is that high levels of racial segregation reflect socioeconomic differences between blacks and whites. Segregation cannot be attributed to income differences, because blacks are equally highly segregated at all levels of income. Whereas segregation declines steadily for most minority groups as socioeconomic status rises, levels of black–white segregation do not vary significantly by social class. Because segregation reflects the effects of white prejudice rather than objective market forces, blacks are segregated no matter how much money they earn.

Although whites now accept open housing in principle, they remain prejudiced against black neighbors in practice. Despite whites' endorsement of the ideal that people should be able to live wherever they can afford to regardless of race, a majority still feel uncomfortable in any neighborhood that contains more than a few black residents; and as the percentage of blacks rises, the number of whites who say they would refuse to enter or would try to move out increases sharply.

These patterns of white prejudice fuel a pattern of neighborhood resegregation because racially mixed neighborhoods are strongly desired by blacks. As the percentage of blacks in a neighborhood rises, white demand for homes within it falls sharply while black demand rises. The surge in black demand and the withering of white demand yield a process of racial turnover. As a result, the only urban areas where significant desegregation occurred during the 1970s were those where the black population was so small that integration could take place without threatening white preferences for limited contact with blacks.

Prejudice alone cannot account for high levels of black segregation, however, because whites seeking to avoid contact with blacks must have somewhere to go. That is, some all-white neighborhoods must be perpetuated and maintained, which requires the erection of systematic barriers to black residential mobility. In most urban housing markets, therefore, the effects of white prejudice are typically reinforced by direct discrimination against black homeseekers. Housing audits carried out over the past two decades have documented the persistence of widespread discrimination against black renters and homebuyers, and a recent comprehensive study carried out by the U.S. Department of Housing and Urban Development suggests that prior work has understated both the incidence and the severity of this racial bias. Evidence also suggests that blacks can expect to experience significant discrimination in the allocation of home mortgages as well.

We demonstrate theoretically how segregation creates underclass communities and systematically builds deprivation into the residential structure of black communities. We show how any increase in the poverty rate of a residentially segregated group leads to an immediate and automatic increase in the geographic concentration of poverty. When the rate of minority poverty is increased under conditions of high segregation, all of the increase is absorbed by a small number of neighborhoods. When the same increase in poverty occurs in an integrated group, the added poverty is spread evenly throughout the urban area, and the neighborhood environment that group members face does not change much.

During the 1970s and 1980s, therefore, when urban economic restructuring and inflation drove up rates of black and Hispanic poverty in many urban areas, underclass communities were created only where increased minority poverty coincided with a high degree of segregation – principally in older metropolitan areas of the northeast and the Midwest. Among Hispanics, only Puerto Ricans developed underclass communities, because only they were highly segregated; and this high degree of segregation is directly attributable to the fact that a large proportion of Puerto Ricans are of African origin.

The interaction of intense segregation and high poverty leaves black neighborhoods extremely vulnerable to fluctuations in the urban economy, because any dislocation that causes an upward shift in black poverty rates will also produce a rapid change in the concentration of poverty and, hence, a dramatic shift in the social and economic composition of black neighborhoods. The concentration of poverty, for example, is associated with the wholesale withdrawal of commercial institutions and the deterioration or elimination of goods and services distributed through the market.

Neighborhoods, of course, are dynamic and constantly changing, and given the high rates of residential turnover characteristic of contemporary American cities, their well-being depends to a great extent on the characteristics and actions of their residents. Decisions taken by one actor affect the subsequent decisions of others in the neighborhood. In this way isolated actions affect the well-being of the community and alter the stability of the neighborhood.

Because of this feedback between individual and collective behavior, neighborhood stability is characterized by a series of thresholds, beyond which various self-perpetuating processes of decay take hold. Above these thresholds, each actor who makes a decision that undermines neighborhood well-being makes it increasingly likely that other actors will do the same. Each property owner who decides not to invest in upkeep and maintenance, for example, lowers the incentive for others to maintain their properties. Likewise, each new crime promotes psychological and physical withdrawal from public life, which reduces vigilance within the neighborhood and undermines the capacity for collective organization, making additional criminal activity more likely.

Segregation increases the susceptibility of neighborhoods to these spirals of decline. During periods of economic dislocation, a rising concentration of black poverty is associated with the simultaneous concentration of other negative social and economic conditions. Given the high levels of racial segregation characteristic of American urban areas, increases in black poverty such as those observed during the 1970s can only lead to a concentration of housing abandonment, crime, and social disorder, pushing poor black neighborhoods beyond the threshold of stability.

By building physical decay, crime, and social disorder into the residential structure of black communities, segregation creates a harsh and extremely disadvantaged environment to which ghetto blacks must adapt. In concentrating poverty, moreover, segregation also concentrates conditions such as drug use, joblessness, welfare dependency, teenage childbearing, and unwed parenthood, producing a social context where these conditions are not only common but the norm. We argue that in adapting to this social environment, ghetto dwellers evolve a set of behaviors, attitudes, and expectations that are sharply at variance with those common in the rest of American society.

As a direct result of the high degree of racial and class isolation created by segregation, for example, Black English has become progressively more distant from Standard American English, and its speakers are at a clear disadvantage in U.S. schools and labor markets. Moreover, the isolation and intense poverty of the ghetto provides a supportive structural niche for the emergence of an "oppositional culture" that inverts the values of middle-class society. Anthropologists have found that young people in the ghetto experience strong peer pressure not to succeed in school, which severely limits their prospects for social mobility in the larger society. Quantitative research shows that growing up in a ghetto neighborhood increases the likelihood of dropping out of high school, reduces the probability of attending college, lowers the likelihood of employment, reduces income earned as an adult, and increases the risk of teenage childbearing and unwed pregnancy.

[. . .]

THE PERPETUATION OF THE UNDERCLASS

> One notable difference appears between the immigrant and Negro populations. In the case of the former, there is the possibility of escape, with improvement in economic status in the second generation.
>
> (1931 report to President Herbert Hoover by the Committee on Negro Housing)

If the black ghetto was deliberately constructed by whites through a series of private decisions and institutional practices, if racial discrimination persists at remarkably high levels in U.S. housing markets, if intensive residential segregation continues to be imposed on blacks by virtue of their skin color, and if segregation concentrates poverty to build a self-perpetuating spiral of decay into black neighborhoods, then a variety of deleterious consequences automatically follow for individual African Americans. A racially segregated society cannot be a race-blind society; as long as U.S. cities remain segregated – indeed, hypersegregated – the United States cannot claim to have equalized opportunities for blacks and whites. In a segregated world, the deck is stacked against black socioeconomic progress, political empowerment, and full participation in the mainstream of American life.

In considering how individuals fare in the world, social scientists make a fundamental distinction between individual, family, and structural characteristics. To a great extent, of course, a person's success depends on individual traits such as motivation, intelligence, and especially, education. Other things equal, those who are more highly motivated, smarter, and better educated will be rewarded more highly in the labor market and will achieve greater socioeconomic success.

Other things generally are not equal, however, because individual traits such as motivation and education are strongly affected by family background. Parents who are themselves educated, motivated, and economically successful tend to pass these traits on to their children. Children who enter the middle and upper classes through the accident of birth are more likely than other, equally intelligent children from other classes to acquire the schooling, motivation, and cultural knowledge required for socioeconomic success in contemporary society. Other aspects of family background, moreover, such as wealth and social connections, open the doors of opportunity irrespective of education or motivation.

Yet even when one adjusts for family background, other things are still not equal, because the structural organization of society also plays a profound role in shaping the life chances of individuals. Structural variables are elements of social and economic organization that lie beyond individual control, that are built into the way society is organized. Structural characteristics affect the fate of large numbers of people and families who share common locations in the social order.

Among the most important structural variables are those that are geographically defined. Where one lives – especially, where one grows up – exerts a profound effect on one's life chances. Identical individuals with similar family backgrounds and personal characteristics will lead very different lives and achieve different rates of socioeconomic success depending on where they reside. Because racial segregation confines blacks to a circumscribed and disadvantaged niche in the urban spatial order, it has profound consequences for individual and family well-being.

Social and spatial mobility

In a market society such as the United States, opportunities, resources, and benefits are not distributed evenly across the urban landscape. Rather, certain residential areas have more prestige, greater affluence, higher home values, better services, and safer streets than others. Marketing consultants have grown rich by taking advantage of this "clustering of America" to target specific groups of consumers for wealthy corporate clients. The geographic differentiation of American cities by socioeconomic status does more than conveniently rank neighborhoods for the benefit of the demographer, however; it also creates a crucial connection between social and spatial mobility.

As people get ahead, they not only move up the economic ladder, they move up the residential ladder as well. As early as the 1920s, sociologists at the University of Chicago noted this close connection between social and spatial mobility, a link that has been verified many times since. As socioeconomic status improves, families relocate to take advantage of opportunities and resources that are available in greater abundance elsewhere. By drawing on benefits acquired through residential mobility, aspiring parents not only consolidate their own class position but enhance their and their children's prospects for additional social mobility.

In a very real way, therefore, barriers to spatial mobility are barriers to social mobility, and where one lives determines a variety of salient factors that affect individual well-being: the quality of schooling, the value of housing, exposure to crime, the quality of public services, and the character of children's peers. As a result, residential integration has been a crucial component in the broader process of socioeconomic advancement among immigrants and their children. By moving to successively better neighborhoods, other racial and ethnic groups have gradually become integrated into American society. Although rates of spatial assimilation have varied, levels of segregation have fallen for each immigrant group as socioeconomic status and generations in the United States have increased.

The residential integration of most ethnic groups has been achieved as a by-product of broader processes of socioeconomic attainment, not because group members sought to live among native whites per se. The desire for integration is only one of a larger set of motivations, and not necessarily the most important. Some minorities may even be antagonistic to the idea of integration, but for spatial assimilation to occur, they need only be willing to put up with integration in order to gain access to socioeconomic resources that are more abundant in areas in which white families predominate.

To the extent that white prejudice and discrimination restrict the residential mobility of blacks and confine them to areas with poor schools, low home values, inferior services, high crime, and low educational aspirations, segregation undermines their social and economic well-being. The persistence of racial segregation makes it difficult for aspiring black families to escape the concentrated poverty of the ghetto and puts them at a distinct disadvantage in the larger competition for education, jobs, wealth, and power. The central issue is not whether African Americans "prefer" to live near white people or whether integration is a desirable social goal, but how the restrictions on individual liberty implied by severe segregation undermine the social and economic well-being of individuals.

Extensive research demonstrates that blacks face strong barriers to spatial assimilation within American society. Compared with other minority groups, they are markedly less able to convert their socioeconomic attainments into residential contact with whites, and because of this fact they are unable to gain access to crucial resources and benefits that are distributed through housing markets. Dollar for dollar, blacks are able to buy fewer neighborhood amenities with their income than other groups.

Among all groups in the United States, only Puerto Ricans share blacks' relative inability to assimilate spatially; but this disadvantage stems from the fact that many are of African origin. Although white Puerto Ricans achieve rates of spatial assimilation that are comparable with those found among other ethnic groups, those of African or racially mixed origins experience markedly lower abilities to convert socioeconomic attainments into contact with whites. Once race is controlled, the "paradox of Puerto Rican segregation" disappears.

Given the close connection between social and spatial mobility, the persistence of racial barriers implies the systematic exclusion of blacks from benefits and resources that are distributed through housing markets. We illustrate the severity of this black disadvantage with data specially compiled for the city of Philadelphia in 1980. The data allow us to consider the socioeconomic character of neighborhoods that poor, middle-income, and affluent blacks and whites can be expected to inhabit, holding education and occupational status constant.

In Philadelphia, poor blacks and poor whites both experience very bleak neighborhood environments; both groups live in areas where about 40 percent of the births are to unwed mothers, where median home values are under $30,000, and where nearly 40 percent of high school students score under the 15th percentile on a standardized achievement test. Families in such an environment would be unlikely to build wealth through home equity, and children growing up in such an environment would be exposed to a peer environment where unwed parenthood was common and where educational performance and aspirations were low.

[. . .]

For blacks, in other words, high incomes do not buy entree to residential circumstances that can serve as springboards for future socioeconomic mobility; in particular, blacks are unable to achieve a school environment conducive to later academic success. In Philadelphia, children from an affluent black family are likely to attend a public school where the percentage of low-achieving students is three times greater than the percentage in schools attended by affluent white children. Small wonder, then, that controlling for income in no way erases the large racial gap in SAT scores. Because of segregation, the same income buys black and white families educational environments that are of vastly different quality.

Given these limitations on the ability of black families to gain access to neighborhood resources, it is hardly surprising that government surveys reveal blacks to be less satisfied with their residential circumstances than socioeconomically equivalent whites. This negative evaluation reflects an accurate appraisal of their circumstances rather than different values or ideals on the part of blacks. Both races want the same things in homes and neighborhoods; blacks are just less able to achieve then. Compared with whites, blacks are less likely to be homeowners, and the homes they do own are of poorer quality, in poorer neighborhoods, and of lower value. Moreover, given the close connection between home equity and family wealth, the net worth of blacks is a small fraction of that of whites, even though their incomes have converged over the years. Finally, blacks tend to occupy older, more crowded dwellings that are structurally inadequate compared to those inhabited by whites; and because these racial differentials stem from segregation rather than income, adjusting for socioeconomic status does not erase them.

[. . .]

THE FUTURE OF THE GHETTO

> The isolation of Negro from white communities is increasing rather than decreasing . . . Negro poverty is not white poverty. Many of its causes . . . are the same. But there are differences – deep, corrosive, obstinate differences – radiating painful roots into the community, the family, and the nature of the individual.
>
> (President Lyndon Johnson, address to Howard University, June 4, 1965)

After persisting for more than fifty years, the black ghetto will not be dismantled by passing a few amendments to existing laws or by implementing a smattering of bureaucratic reforms. The ghetto is part and parcel of modern American society; it was manufactured by whites earlier in the century to isolate and control growing urban black populations, and it is maintained today by a set of institutions, attitudes, and practices that are deeply embedded in the structure of American life. Indeed, as conditions in the ghetto have worsened and as poor blacks have adapted socially and culturally to this deteriorating environment, the ghetto has assumed even greater importance as an institutional tool for isolating the by-products of racial oppression: crime, drugs, violence, illiteracy, poverty, despair, and their growing social and economic costs.

For the walls of the ghetto to be breached at this point will require an unprecedented commitment by the public and a fundamental change in leadership at the highest levels. Residential segregation will only be eliminated from American society when federal authorities, backed by the American people, become directly involved in guaranteeing open housing markets and eliminating discrimination from public life. Rather than relying on private individuals to identify and prosecute those who break the law, the U.S. Department of Housing and Urban Development and the Office of the Attorney General must throw their full institutional weight into locating instances of housing discrimination and bringing those who violate the Fair Housing Act to justice; they must vigorously prosecute white racists who harass and intimidate blacks seeking to exercise their rights of residential freedom; and they must establish new bureaucratic mechanisms to counterbalance the forces that continue to sustain the residential color line.

Given the fact that black poverty is exacerbated, reinforced, and perpetuated by racial segregation, that black–white segregation has not moderated despite the federal policies tried so far, and that the social costs of segregation inevitably cannot be contained in the ghetto, we argue that the nation has no choice but to launch a bold new initiative to eradicate the ghetto and eliminate segregation from American life. To do otherwise is to condemn the United States and the American people to a future of economic stagnation, social fragmentation, and political paralysis.

Race, class, and public policy

In the United States today, public policy discussions regarding the urban underclass frequently devolve into debates on the importance of race versus class. However one defines the underclass, it is clear that African Americans are overrepresented within in it. People who trace their ancestry to Africa are at greater risk than others of falling into poverty, remaining there for a long time, and residing in very poor neighborhoods. On almost any measure of social and economic well-being, blacks and Puerto Ricans come out near the bottom.

The complex of social and economic problems that beset people of African origin has led many observers to emphasize race over class in developing remedies for the urban underclass. According to these theories, institutional racism is pervasive, denying blacks equal access to the resources and benefits of American society, notably in education and employment. Given this assessment, these observers urge the adoption of racial remedies to assist urban minorities; proposals include everything from special preference in education to affirmative action in employment.

Other observers emphasize class over race. The liberal variant of the class argument holds that blacks have been caught in a web of institutional and industrial change. Like other migrants, they arrived in cities to take low-skilled jobs in manufacturing, but they had the bad fortune to become established in this sector just as rising energy costs, changing technologies, and increased foreign competition brought a wave of plant closings and layoffs. The service economy that arose to replace manufacturing industries generated high-paying jobs for those with education, but poorly paid jobs for those without it.

Just as this transformation was undermining the economic foundations of the black working class, the class theorists argue, the civil rights revolution opened up new opportunities for educated minorities. After the passage of the 1964 Civil Rights Act, well-educated blacks were recruited into positions of responsibility in government, academia, and business, and thus provided the basis for a new black middle class. But civil rights laws could not provide high-paying jobs to poorly educated minorities when there were no jobs to give out. As a result, the class structure of the black community bifurcated into an affluent class whose fortunes were improving and a poverty class whose position was deteriorating.

The conservative variant of the class argument focuses on the deleterious consequences of government policies intended to improve the economic position of the poor. According to conservative reasoning, federal antipoverty programs implemented during the 1960s – notably the increases in Aid to Families with Dependent Children – altered the incentives governing the behavior of poor men and women. The accessibility and generosity of federal welfare programs reduced the attractiveness of marriage to poor women, increased the benefits of out-of-wedlock childbearing, and reduced the appeal of low-wage labor for poor men. As a result, female-headed families proliferated, rates of unwed childbearing rose, and male labor force participation rates fell. These trends drove poverty rates upward and created a population of persistently poor, welfare-dependent families.

Race- and class-based explanations for the underclass are frequently discussed as if they were mutually exclusive. Although liberal and conservative class theorists may differ with respect to the specific explanations they propose, both agree that white racism plays a minor role as a continuing cause of urban poverty; except for acknowledging the historical legacy of racism, their accounts are essentially race-neutral. Race theorists, in contrast, insist on the primacy of race in American society and emphasize its continuing role in perpetuating urban poverty; they view class-based explanations suspiciously, seeing them as self-serving ideologies that blame the victim.

By presenting the case for segregation's present role as a central cause of urban poverty, we seek to end the specious opposition of race and class. The issue is not whether race *or* class perpetuates the urban underclass, but how race *and* class *interact* to undermine the social and economic well-being of black Americans. We argue that race operates powerfully through urban housing markets, and that racial segregation interacts with black class structure to produce a uniquely disadvantaged neighborhood environment for African Americans.

If the decline of manufacturing, the suburbanization of employment, and the proliferation of unskilled service jobs brought rising rates of poverty and income inequality to blacks, the negative consequences of these trends were exacerbated and magnified by segregation. Segregation concentrated the deprivation created during the 1970s and 1980s to yield intense levels of social and economic isolation. As poverty was concentrated, moreover, so were all social traits associated with it, producing a structural niche within which welfare dependency and joblessness could flourish and become normative. The expectations of the urban poor were changed not so much by generous AFDC payments as by the spatial concentration of welfare recipients, a condition that was structurally built into the black experience by segregation.

If our viewpoint is correct, then public policies must address both race and class issues if they are to be successful. Race-conscious steps need to be taken to dismantle the institutional apparatus of segregation, and class-specific policies must be implemented to improve the socioeconomic status of minorities. By themselves, programs targeted to low-income minorities will fail because they will be swamped by powerful environmental influences arising from the disastrous neighborhood conditions that blacks experience because of segregation. Likewise, efforts to reduce segregation will falter unless blacks acquire the socioeconomic resources that enable them to take full advantage of urban housing markets and the benefits they distribute.

Although we focus in this chapter on how to end racial segregation in American cities, the policies we advocate cannot be pursued to the exclusion of broader efforts to raise the class standing of urban minorities. Programs to dismantle the ghetto must be accompanied by vigorous efforts to end discrimination in other spheres of American life and by class-specific policies designed to raise educational levels, improve the quality of public schools, create employment, reduce crime, and strengthen the family. Only a simultaneous attack along all fronts has any hope of breaking the cycle of poverty that has become deeply rooted.

"The Immigrant Enclave: Theory and Empirical Examples"

from Susan Olzak and Joane Nagel (eds), *Competitive Ethnic Relations* (1986)

Alejandro Portes and Robert D. Manning

Editors' Introduction

Alejandro Portes and Robert Manning offer an insightful comparative overview of the major schools of literature on immigrant enclaves in the United States today. Immigrant enclaves are important phenomena to consider since the 1960s in American cities, during which time the United States experienced the growth of an urban underclass and persisting race and class segregation despite the success of social mobility and spatial integration for some of the black middle class. We observe also the onset of globalization and neoliberal economic policies that stimulated the mobility of labor and capital. The Hart–Cellar Immigration Law of 1965 and subsequent immigration, banking, and free trade accords have led to the widespread emergence of Latino and Asian residential and business enclaves in U.S. cities. Chinatown, Koreatown and Little Havana offer an interesting comparison with immigrant enclaves in American cities of the early twentieth century, which were more commonly white ethnic (such as Little Italy and the Jewish enclave), and operated during a different phase of economic growth. Immigrant enclaves challenge us to look beyond the traditional black/white dynamics of American cities, and consider how new immigrants are being inserted into the existing system of race and class stratification.

Portes and Manning first make the point that the proliferation of immigrant enclaves challenges traditional precepts of assimilation theory, as the persistence of ethnic identity and ethnic communities is more permissible and commonplace since the 1960s. Whereas social mobility in American society was in the early twentieth century more predicated on the suppression of ethnic ancestry and the acquisition of American cultural values, we have seen that socioeconomic prosperity in contemporary America can be promoted along with a continuing commitment to ethnicity and ethnic enclaves. Immigrant enclaves are growing as often as they are disappearing, and furthermore, they arise sometimes in the suburbs. Immigrant enclaves are furthermore an alternative for economic incorporation beyond the existing dichotomy of a segmented labor market in which there is an upper tier of jobs that offers good mobility ladders, and a lower tier of dead-end jobs in which minorities predominate.

The contradiction is that immigrant enclaves offer opportunity for some ethnic people through the exploitation of co-ethnic others. Ethnic enclaves offer a kind of protection by hiring immigrants who may be undocumented or lack good English language skills. They commonly offer on-the-job training rather than requiring higher education. While being in this "protective" sector, they may also be subject to severe exploitation, working with low wages, poor benefits, no labor rights, and "sweatshop" conditions that may violate labor law. On the other hand, some co-ethnics benefit, notably bosses and immigrant business owners. Forward, backward, and consumption linkages within immigrant enclaves commonly result in the re-circulating of dollars in the ethnic economy via the phenomenon of the economic multiplier.

Immigrant enclaves commonly employ co-ethnic workers and serve co-ethnic customers. There are, however, a number of ethnic actors that act as "middleman minorities" between white elites and blacks and other minorities. Chinese and Korean immigrants, for instance, commonly operate small businesses in lower-class black communities, filling a niche considered socially undesirable or unprofitable by white businesses and corporations. Middleman minorities operate as a social buffer between the elite and oppressed and underclass groups of a given society. In times of political or economic crisis, the oppressed minorities may lash out with racial violence or anger against the middleman groups. Intergroup conflicts of this sort occurred in Miami in the early 1980s between blacks and Cubans, and in the 1990s in Los Angeles and New York between Asians and blacks.

For further reading on middleman minorities, see Edna Bonacich, "A Theory of Middleman Minorities," *American Sociological Review* 38 (October 1973): 583–594, who gives a historical perspective that highlights the experience of global Jewish traders and the Chinese in Southeast Asia. Pyong Gap Min considers Korean/black American relations in *Caught in the Middle: Korean Communities in New York and Los Angeles* (Berkeley, CA: University of California Press, 1996). Spike Lee offers a provocative interpretation of black/Korean relations in Brooklyn, New York, in his film, *Do the Right Thing*, especially in the scene where Radio Raheem confronts Korean-American shopkeepers with a request for "D" batteries. The confrontation is reiterated in the Ice Cube song, "Black Korea," in the album *Death Certificate* (Priority Records, 1991).

Alejandro Portes is the Howard Harrison and Gabrielle Snyder Beck Professor of Sociology and Director of the Center for Migration and Development at Princeton University. He formerly taught at Johns Hopkins University, where he held the John Dewey Chair in Arts and Sciences, Duke University, and the University of Texas-Austin. He served as President of the American Sociological Association in 1997–99. Born in Havana, Cuba, he came to the United States in 1960. He was educated at the University of Havana, Catholic University of Argentina, and Creighton University. He received his M.A. and Ph.D. from the University of Wisconsin-Madison.

Portes is the author of some 200 articles and chapters on national development, international migration, Latin American and Caribbean urbanization, and economic sociology. His books include *City on the Edge – the Transformation of Miami* (Berkeley, CA: University of California Press, 1993), co-authored with Alex Stepick and winner of the 1995 Robert Park Award for best book in community and urban sociology and the Anthony Leeds Award for best book in urban anthropology in 1995. His current research is on the adaptation process of the immigrant second generation and the rise of transnational immigrant communities in the United States. His most recent books, co-authored with Rubén G. Rumbaut, are *Legacies: The Story of the Immigrant Second Generation* and *Ethnicities: Children of Immigrants in America* (Berkeley, CA: University of California Press, 2001). *Legacies* was the winner of the 2002 Distinguished Scholarship Award from the American Sociological Association and of the 2002 W. I. Thomas and Florian Znaniecki Award for best book from the International Migration Section of ASA.

Robert Manning is currently the Caroline Werner Gannett Professor at the Rochester Institute of Technology. He held previous faculty appointments at American University and Georgetown University. He is a specialist in comparative economic relations, immigration, and minority relations in the United States. He received his Ph.D. from the Johns Hopkins University, his M.A. from Northern Illinois University, and his B.A. from Duke University. His recent book, *Credit Card Nation* (New York: Basic Books, 2000) focuses on the damaging social and political consequences of America's increasing reliance on credit cards.

INTRODUCTION

The purpose of this chapter is to review existing theories about the process of immigrant adaptation to a new society and to recapitulate the empirical findings that have led to an emerging perspective on the topic. This emerging view revolves around the concepts of different modes of structural incorporation and of the immigrant enclave as one of them. These concepts are set in explicit opposition to two previous viewpoints on the adaptation process, generally identified as assimilation theory and the segmented labor markets approach.

The study of immigrant groups in the United States has produced a copious historical and sociological literature, written mostly from the assimilation perspective. Although the experiences of particular groups varied, the common theme of these writings is the unrelenting efforts of immigrant minorities to surmount obstacles impeding their entry into the "mainstream" of American society. From this perspective, the adaptation process of particular immigrant groups followed a sequential path from initial economic hardship and discrimination to eventual socioeconomic mobility arising from increasing knowledge of American culture and acceptance by the host society. The focus on a "core" culture, the emphasis on consensus building, and the assumption of a basic patterned sequence of adaptation represent central elements of assimilation theory.

[...]

The second general perspective takes issue with this psychosocial and culturalist orientation as well as with the assumption of a single basic assimilation path. This alternative view begins by noting that immigrants and their descendants do not necessarily "melt" into the mainstream and that many groups seem not to want to do so, preferring instead to preserve their distinct ethnic identities. A number of writers have focused on the resilience of these communities and described their functions as sources of mutual support and collective political power. Others have gone beyond descriptive accounts and attempted to establish the causes of the persistence of ethnicity. Without exception, these writers have identified the roots of the phenomenon in the economic sphere and, more specifically, in the labor-market roles that immigrants have been called on to play.

Within this general perspective, several specific theoretical approaches exist. The first focuses on the situation of the so-called unmeltable ethnics – blacks, Chicanos, and American Indians – and finds the source of their plight in a history of internal colonialism during which these groups have been confined to specific areas and made to work under uniquely unfavorable conditions. In a sense, the role of colonized minorities has been to bypass the free labor market, yielding in the process distinct benefits both to direct employers of their labor and, indirectly, to other members of the dominant racial group. This continuation of colonialist practices to our day explains, according to this view, the spatial isolation and occupational disadvantages of these minorities.

A second approach attempts to explain the persistence of ethnic politics and ethnic mobilization on the basis of the organization of subordinate groups to combat a "cultural division of labor." The latter confined members of specific minorities to a quasi-permanent situation of exploitation and social inferiority. Unlike the first view, the second approach does not envision the persistence of ethnicity as a consequence of continuing exploitation, but rather as a "reactive formation" on the part of the minority to reaffirm its identity and its interests. For this reason, ethnic mobilizations are often most common among groups who have already abandoned the bottom of the social ladder and started to compete for positions of advantage with members of the majority.

A final variant focuses on the situation of contemporary immigrants to the United States. Drawing on the dual labor market literature, this approach views recent immigrants as the latest entrants into the lower tier of a segmented labor market where women and other minorities already predominate. Relative to the latter, immigrants possess the advantages of their lack of experience in the new country, their legal vulnerability, and their greater initial motivation. All of these traits translate into higher productivity and lower labor costs for the firms that employ them. Jobs in the secondary labor market are poorly paid, require few skills, and offer limited mobility opportunities. Hence, confinement of immigrants to this sector insures that those who do not return home are relegated to a quasi-permanent status as disadvantaged and discriminated minorities.

What these various structural theories have in common is the view of resilient ethnic communities formed as a result of a consistently disadvantageous economic position and the consequent absence of a smooth path of assimilation. These situations, ranging from slave labor to permanent confinement to the secondary labor market, are not altered easily. They have given rise, in time, either to hopeless communities of "unmeltable" ethnics or to militant minorities, conscious of a common identity and willing to support a collective strategy of self-defense rather than rely on individual assimilation.

These structural theories have provided an effective critique of the excessively benign image of the adaptation process presented by earlier writings. However, while undermining the former, the new structural perspective may have erred in the opposite direction. The basic hypothesis advanced in this chapter is that several identifiable modes of labor-market incorporation exist and that not all of them relegate newcomers to a permanent situation of exploitation and inferiority. Thus, while agreeing with the basic thrust of structural theories, we propose several modifications that are necessary for an adequate understanding of the different types of immigrant flows and their distinct processes of adaptation.

MODES OF INCORPORATION

In the four decades since the end of World War II, immigration to the United States has experienced a vigorous surge reaching levels comparable only to those at the beginning of the century. Even if one restricts attention to this movement, disregarding multiple other migrations elsewhere in the world, it is not the case that the inflow has been of a homogeneous character. Low-wage labor immigration itself has taken different forms, including temporary contract flows, undocumented entries, and legal immigration. More importantly, it is not the case that all immigrants have been directed to the secondary labor market. For example, since the promulgation of the Immigration Act of 1965, thousands of professionals, technicians, and craftsmen have come to the United States, availing themselves of the occupational preference categories of the law. This type of inflow, dubbed "brain drain" in the sending nations, encompasses today sizable contingents of immigrants from such countries as India, South Korea, the Philippines, and Taiwan, each an important contributor to U.S. annual immigration.

The characteristics of this type of migration have been described in detail elsewhere. Two such traits deserve mention, however. First, occupationally skilled immigrants – including doctors, nurses, engineers, technicians, and craftsmen – generally enter the primary labor market; they contribute to alleviate domestic shortages in specific occupations and gain access, after a period of time, to the mobility ladders available to native workers. Second, immigration of this type does not generally give rise to spatially concentrated communities; instead, immigrants are dispersed throughout many cities and regions, following different career paths.

Another sizable contingent of entrants whose occupational future is not easily characterized a priori are political refugees. Large groups of refugees, primarily from Communist-controlled countries, have come to the United States, first after the occupation of Eastern Europe by the Soviet Army, then after the advent of Fidel Castro to power in Cuba, and finally in the aftermath of the Vietnam War. Unlike purely "economic" immigrants, refugees have often received resettlement assistance from various governmental agencies. The economic adaptation process of one of these groups, the Cubans, will be discussed in detail in this chapter. For the moment, it suffices to note that all the available evidence runs contrary to the notion of a uniform entry of political refugees into low-wage secondary occupations; on the contrary, there are indications of their employment in many different lines of work.

A third mode of incorporation has gained the attention of a number of scholars in recent years. It consists of small groups of immigrants who are inserted or insert themselves as commercial intermediaries in a particular country or region. These "middleman minorities" are distinct in nationality, culture, and sometimes race from both the superordinate and subordinate groups to which they relate. They can be used by dominant elites as a buffer to deflect mass frustration and also as an instrument to conduct commercial activities in impoverished areas. Middlemen accept these risks in exchange for the opportunity to share in the commercial and financial benefits gained through such instruments as taxation, higher retail prices, and usury. Jews in feudal and early modern Europe represent the classic instance of a middleman minority. Other examples include Indian merchants in East Africa, and Chinese entrepreneurs in Southeast Asia and throughout the Pacific Basin. Contemporary examples in the United States include Jewish, Korean, and other Oriental merchants in inner-city ghetto areas and Cubans in Puerto Rico.

Primary labor immigration and middleman entrepreneurship represent two modes of incorporation that differ from the image of an homogeneous flow into low-wage employment. Political refugees, in turn, have followed a variety of paths, including both of the above as well as insertion into an ethnic enclave economy. The latter represents a fourth distinct mode. Although frequently confused with middleman minorities, the emergence and structure of an immigrant enclave possess distinct characteristics. The latter have significant theoretical and practical implications, for they set apart groups adopting this entry mode from those following alternative paths. We turn now to several historical and contemporary examples of immigrant enclaves to clarify their internal dynamics and causes of their emergence.

IMMIGRANT ENCLAVES

Immigration to the United States before World War I was, overwhelmingly, an unskilled labor movement. Impoverished peasants from southern Italy, Poland, and the eastern reaches of the Austro-Hungarian Empire settled in dilapidated and crowded areas, often immediately adjacent to their points of debarkation, and took any menial jobs available. From these harsh beginnings, immigrants commenced a slow and often painful process of acculturation and economic mobility. Theirs was the saga captured by innumerable subsequent volumes written from both the assimilation and the structural perspectives.

Two sizable immigrant groups did not follow this pattern, however. Their most apparent characteristic was the economic success of the first generation, even in the absence of extensive acculturation. On the contrary, both groups struggled fiercely to preserve their cultural identity and internal solidarity. Their approach to adaptation thus directly contradicted subsequent assimilation predictions concerning the causal priority of acculturation to economic mobility. Economic success and "clannishness" also earned for each minority the hostility of the surrounding population. These two immigrant groups did not have a language, religion, or even race in common and they never overlapped in significant numbers in any part of the United States. Yet, arriving at opposite ends of the continent, Jews and Japanese pursued patterns of economic adaptation that were quite similar both in content and in their eventual consequences.

Jews in Manhattan

The first major wave of Jewish immigration to the United States consisted of approximately 50,000 newcomers of German origin, arriving between 1840 and 1870. These immigrants went primarily into commerce and achieved, in the course of a few decades, remarkable success. By 1900, the average income of German-Jewish immigrants surpassed that of the American population. Many individuals who started as street peddlers and small merchants had become, by that time, heads of major industrial, retail, and financial enterprises.

The second wave of Jewish immigration exhibited quite different characteristics. Between 1870 and 1914, over two million Jews left the Pale of Settlement and other Russian-dominated regions, escaping Czarist persecution. Major pogroms occurred before and during this exodus. Thus, unlike most immigrants of the period, the migration of Russian and Eastern European Jews was politically motivated and their move was much more permanent. In contrast to German Jews, who were relatively well educated, the Yiddish-speaking newcomers came, for the most part, from modest origins and had only a rudimentary education. Although they viewed the new Russian wave with great apprehension, German Jews promptly realized that their future as an ethnic minority depended on the successful integration of the newcomers. Charitable societies were established to provide food, shelter, and other necessities, and private schools were set up to teach immigrants English, civics, and the customs of the new country.

Aside from its size and rapidity of arrival, turn-of-the-century Jewish immigration had two other distinct characteristics. First was its strong propensity toward commerce and self-employment in general in preference to wage labor; as German

Jews before them, many Russian immigrants moved directly into street peddling and other commercial activities of the most modest sort. Second was its concentration into a single, densely populated urban area – the lower East Side of Manhattan. Within this area, those who did not become storekeepers and peddlers from the start found employment in factories owned by German Jews, learning the necessary rudiments for future self-employment. . . .

The economic success of many of these ventures did not require and did not entail rapid acculturation. Immigrants learned English and those instrumental aspects of the new culture required for economic advancement. For the rest, they preferred to remain with their own and maintained, for the most part, close adherence to their original religion, language, and values. Jewish enclave capitalism depended, for its emergence and advancement, precisely on those resources made available by a solidaristic ethnic community: protected access to labor and markets, informal sources of credit, and business information. It was through these resources that upstart immigrant enterprises could survive and eventually compete effectively with better-established firms in the general economy.

The emergence of a Jewish enclave in East Manhattan helped this group bypass the conventional assimilation path and achieve significant economic mobility in the course of the first generation, well ahead of complete acculturation. Subsequent generations also pursued this path, but the resources accumulated through early immigrant entrepreneurship were dedicated primarily to furthering the education of children and their entry into the professions. It was at this point that outside hostility became most patent, as one university after another established quotas to prevent the onrush of Jewish students. The last of these quotas did not come to an end until after World War II.

Despite these and other obstacles, the movement of Jews into higher education continued. Building on the economic success of the first generation, subsequent ones achieved levels of education, occupation, and income that significantly exceeded the national average. The original enclave is now only a memory, but it provided in its time the necessary platform for furthering the rapid social and economic mobility of the minority. Jews did enter the mainstream of American society, but they did not do so starting uniformly at the bottom, as most immigrant groups had done; instead, they translated resources made available by early ethnic entrepreneurship into rapid access to positions of social prestige and economic advantage.

Japanese on the West Coast

The specific features of Japanese immigration differ significantly from the movement of European Jews, but their subsequent adaptation and mobility patterns are similar. Beginning in 1890 and ending with the enactment of the Gentlemen's Agreement of 1908, approximately 150,000 Japanese men immigrated to the West Coast. They were followed primarily by their spouses until the Immigration Act of 1924 banned any further Asiatic immigration. Although nearly 300,000 Japanese immigrants are documented in this period, less than half of this total remained in the United States. This is due, in contrast to the case of the Jews, to the sojourner character of Japanese immigrants: the intention of many was to accumulate sufficient capital for purchasing farm land or settling debts in Japan. Hence this population movement included commercial and other members of the Japanese middle class who, not incidentally, were explicitly sponsored by their national government.

[. . .]

Japanese immigrants were initially welcomed and recruited as a form of cheap agricultural labor. Their reputation as thrifty and diligent workers made them preferable to other labor sources. Nativist hostilities crystallized, however, when Japanese immigrants shifted from wage labor to independent ownership and smallscale farming. This action not only reduced the supply of laborers but it also increased competition for domestic growers in the fresh-produce market. In 1900, only about 40 Japanese farmers in the entire United States leased or owned a total of 5,000 acres of farmland. By 1909, the number of Japanese farmers had risen to 6000 and their collective holdings exceeded 210,000 acres. Faced with such "unfair" competition, California growers turned to the political means at their disposal. In 1913, the state legislature passed the first Alien Land Law, which restricted land ownership by foreigners. This legislation did not prove sufficient, however, and, in subsequent years, the ever-accommodating legislature passed a series of acts closing other legal loopholes to Japanese farming.

[. . .]

The ability of the first-generation Issei to escape the status of stoop labor in agriculture was based on the social cohesion of their community. Rotating credit associations offered scarce venture capital, while mutual-aid organizations provided assistance in operating farms and urban businesses. Capitalizations as high as $100,000 were financed through ethnic credit networks. Economic success was again accompanied by limited instrumental acculturation and by careful preservation of national identity and values. It was the availability of investment capital, cooperative business associations, and marketing practices (forward and backward economic linkages) within the ethnic enclave that enabled Japanese entrepreneurs to expand beyond its boundaries and compete effectively in the general economy. This is illustrated by the production and marketing of fresh produce. In 1920, the value of Japanese crops was about 10 percent of the total for California, when the Japanese comprised less than 1 percent of the state's population; many retail outlets traded exclusively with a non-Japanese clientele.

During the early 1940s, the Japanese ethnic economy was seriously disrupted but not eliminated by the property confiscations and camp internments accompanying World War II. After the war, economic prosperity and other factors combined to reduce local hostility toward the Japanese. Older Issei and many of their children returned to small business, while other secondgeneration Nisei, like their Jewish predecessors, pursued higher education and entered the whitecollar occupations en masse, This mobility path was completed by the third or Sansei generation, with 88 percent of their members attending college. Other third-generation Japanese have continued, however, the entrepreneurial tradition of their parents. Like Jews before them, Japanese-Americans have made use of the resources made available by early immigrant entrepreneurship to enter the mainstream of society in positions of relative advantage.

[. . .]

CONTEMPORARY EXAMPLES

As a mode of incorporation, the immigrant enclave is not only of historical interest since there are also several contemporary examples. Enclaves continue to be, however, the exception in the post-World War II period, standing in sharp contrast to the more typical pattern of secondary labor immigration. Furthermore, there is no guarantee that the emergence and development of contemporary ethnic enclaves will have the same consequences for their members that they had among turn-of-the-century immigrants.

Koreans in Los Angeles

The Korean community of Los Angeles is a recent product of liberalized U.S. immigration laws and strengthened political and economic ties between the two nations. Since 1965–1968, South Korean immigration to the United States has increased sixfold, swelling the Korean population of Los Angeles from less than 9000 in 1970 to over 65,000 in 1975. Approximately 60 percent of Korean immigrants settle in Los Angeles. In addition to increasing the size of this population flow, U.S. immigration law has altered its class composition. Korean immigrants come predominantly from the highly educated, Westernized, Christian strata of urban Korea.

[. . .]

Korean entrepreneurs, like Jewish and Japanese immigrants before them, are highly dependent on the social and economic resources of their ethnic community. Some immigrants managed to smuggle capital out of Korea, but most rely on individual thrift and ethnic credit systems. For instance, a Korean husband and wife may save their wages from several service and factory jobs until enough capital is accumulated to purchase a small business. This process usually takes 2 or 3 years. Rotating credit systems, which are based on mutual trust and honor, offer another common source of venture capital. This economic institution could not exist without a high degree of social solidarity within the ethnic community. There are more than 500 community social and business associations in Los Angeles, and nearly every Korean is an active member of one or more of them.

In addition, Korean businessmen have utilized public resources from the U.S. Small Business Administration as well as loans and training programs sponsored by the South Korean government.

The ability of the Korean community to generate a self-sustaining entrepreneurial class has had a profound impact on intraethnic labor relations and patterns of ethnic property transfers. For example, labor relations are enmeshed in extended kinship and friendship networks. In this context of "labor paternalism," working in the ethnic economy frequently entails the obligation of accepting low pay and long hours in exchange for on-the-job training and possible future assistance in establishing a small business. Hence, employment in the ethnic economy possesses a potential for advancement entirely absent from comparable low-wage labor in the secondary labor market.

Along the same lines, business practices are fundamentally influenced by cultural patterns. Koreans patronize coethnic businesses and frequently rely on referrals from members of their social networks. Korean-owned businesses, moreover, tend to remain in the community through intraethnic transactions. This is because economic mobility typically proceeds through the rapid turnover of immigrant-owned enterprises. A common pattern of succession, for instance, may begin with a business requiring a relatively small investment, such as a wig shop, and continue with the acquisition of enterprises requiring progressively larger capitalizations: grocery stores, restaurants, gas stations, liquor stores, and finally real estate. This circulation of businesses within the ethnic economy provides a continuous source of economic mobility for aspiring immigrant entrepreneurs.

[. . .]

Cubans in Miami

Over the past 20 years, nearly 900,000 Cubans or about 10 percent of the island's population have emigrated, mostly to the United States. The overwhelming proportion of the Cuban population in America, estimated at roughly 800,000, resides in the metropolitan areas of south Florida and New York. This movement of Cuban emigres has not been a continuous or socially homogeneous flow. Instead, it is more accurately described as a series of "waves," marked by abrupt shifts and sudden discontinuities. This pattern has supported the emergence of an enclave economy through such features as spatial concentration, the initial arrival of a moneyed, entrepreneurial class, and subsequent replenishments of the labor pool with refugees coming from more modest class origins.

In 1959, when Fidel Castro overthrew the regime of Fulgencio Batista, the Cuban community in the United States numbered probably less than 30,000. The political upheavals of the Revolution, however, precipitated a massive emigration from the island. Not surprisingly, the Cuban propertied class, including landowners, industrialists, and former Cuban managers of U.S.-owned corporations, were the first to leave, following close on the heels of leaders of the deposed regime. In the first year of the exodus, approximately 37,000 emigres settled in the United States; most were well-to-do and brought considerable assets with them. After the defeat of the exile force in the Bay of Pigs, in April 1961, Cuban emigration accelerated and its social base expanded to include the middle and urban working classes. By the end of 1962, the first phase of Cuban emigration had concluded and over 215,000 refugees had been admitted to the United States. The emerging Cuban community in south Florida, unlike earlier Japanese and contemporary Korean settlements, was thus fundamentally conditioned by political forces.

Political factors continued to shape the ups and downs of Cuban emigration as well as its reception by American society over the next two decades. In this period, three additional phases can be distinguished: November 1962 to November 1965, December 1965 to April 1973, and May 1973 to November 1980, including 74,000 in the second phase, 340,000 in the third, and 124,769 in the last. This massive influx of refugees to south Florida generated local complaints about the social and economic strains placed in the area. Accordingly, the policy of the Cuban Refugee Program, originally established by the Kennedy Administration, was oriented from the start to resettle Cubans away from Miami. Assistance to the refugees was often made contingent on their willingness to relocate. Although over 469,000 Cubans elected to move by 1978, many subsequently returned to metropolitan Miami. There is evidence that many of these "returnees" made use of their employment in relatively high-wage Northern areas to accumulate savings with which to start new business ventures in Miami. By 1980, the Cuban-born population of the city, composed to a large extent of returnees from the North, was six times greater than the second largest Cuban concentration in New Jersey.

Although a number of Cuban businesses appeared in Miami in the 1960s, they were mostly restaurants and ethnic shops catering to a small exile clientele. An enclave economy only emerged in the 1970s as a result of a combination of factors, including capital availability, access to low-wage labor provided by new refugee cohorts, and the increasingly tenuous hope of returning to Cuba. Cuban-owned firms in Dade County increased from 919 in 1967 to about 8000 in 1976 and approximately 12,000 in 1982.

[. . .]

CONCLUSION: A TYPOLOGY OF THE PROCESS OF INCORPORATION

Having reviewed several historical and contemporary examples, we can now attempt a summary description of the characteristics of immigrant enclaves and how they differ from other paths. The emergence of an ethnic enclave economy has three prerequisites: first, the presence of a substantial number of immigrants with business experience acquired in the sending country; second, the availability of sources of capital; and third, the availability of sources of labor. The latter two conditions are not too difficult to meet. The requisite labor can usually be drawn from family members and, more commonly, from recent arrivals. Surprisingly perhaps, capital is not a major impediment either since the sums initially required are usually small. When immigrants did not bring them from abroad, they could be accumulated through individual savings or pooled resources in the community. It is the first condition that appears critical. The presence of a number of immigrants skilled in the art of buying and selling is common to all four cases reviewed above. Such an entrepreneurial-commercial class among early immigrant cohorts can usually overcome other obstacles; conversely, its absence within an immigrant community will confine the community to wage employment even if sufficient resources of capital and labor are available.

Enclave businesses typically start small and cater exclusively to an ethnic clientele. Their expansion and entry into the broader market requires, as seen above, an effective mobilization of community resources. The social mechanism at work here seems to be a strong sense of reciprocity supported by collective solidarity that transcends the purely contractual character of business transactions. For example, receipt of a loan from a rotating credit association entails the duty of continuing to make contributions so that others can have access to the same source of capital. Although, in principle, it would make sense for the individual to withdraw once his loan is received, such action would cut him off from the very sources of community support on which his future business success depends.

Similarly, relations between enclave employers and employees generally transcend a contractual wage bond. It is understood by both parties that the wage paid is inferior to the value of labor contributed. This is willingly accepted by many immigrant workers because the wage is only one form of compensation. Use of their labor represents often the key advantage making poorly capitalized enclave firms competitive. In reciprocity, employers are expected to respond to emergency needs of their workers and to promote their advancement through such means as on-the-job training, advancement to supervisory positions, and aid when they move into self-employment. These opportunities represent the other part of the "wage" received by enclave workers. The informal mobility ladders thus created are, of course, absent in the secondary labor market where there is no primary bond between owners and workers or no common ethnic community to enforce the norm of reciprocity.

Paternalistic labor relations and strong community solidarity are also characteristic of middleman minorities. Although both modes of incorporation are similar and are thus frequently confused, there are three major structural differences between them. First, immigrant enclaves are not exclusively commercial. Unlike middleman minorities, whose economic role is to mediate commercial and financial transactions between elites and masses, enclave firms include in addition a sizable productive sector. The latter may comprise agriculture, light manufacturing, and construction enterprises; their production, marketed often by coethnic intermediaries, is directed toward the general economy and not exclusively to the immigrant community.

Second, relationships between enclave businesses and established native ones are problematic. Middleman groups tend to occupy positions complementary and subordinate to the local owning class; they fill economic niches either disdained or feared by the latter. Unlike them, enclave enterprises often enter in direct competition with existing domestic firms. There is no evidence, for example, that domestic elites deliberately established or supported the emergence of the Jewish, Japanese, Korean, or Cuban business communities as means to further their own economic interests. There is every indication, on the other hand, that this mode of incorporation was largely self-created by the immigrants, often in opposition to powerful domestic interests. Although it is true that enclave entrepreneurs have been frequently employed as subcontractors by outside firms in such activities as garmentmaking and construction, it is incorrect to characterize this role as the exclusive or dominant one among these enterprises.

Third, the enclave is concentrated and spatially identifiable. By the very nature of their activities, middleman minorities must often be dispersed among the mass of the population. Although the immigrants may live in certain limited areas, their businesses require proximity to their mass clientele and a measure of physical dispersion within it. It is true that middleman activities such as moneylending have been associated in several historical instances with certain streets and neighborhoods, but this is not a necessary or typical pattern, Street peddling and other forms of petty commerce require merchants to go into the areas where demand exists and avoid excessive concentration of the goods and services they offer. This is the typical pattern found today among middleman minorities in American cities.

Enclave businesses, on the other hand, are spatially concentrated, especially in their early stages. This is so for three reasons: first, the need for proximity to the ethnic market which they initially serve; second, proximity to each other, which facilitates exchange of information, access to credit, and other supportive activities; third, proximity to ethnic labor supplies on which they crucially depend. Of the four immigrant groups discussed above, only the Japanese partially depart from the pattern of high physical concentration. This can be attributed to the political persecution to which this group was subjected. Originally, Japanese concentration was a rural phenomenon based on small farms linked together by informal bonds and cooperative associations. Forced removal of this minority from the land compelled their entry into urban businesses and their partial dispersal into multiple activities.

Physical concentration of enclaves underlies their final characteristic. Once an enclave economy has fully developed, it is possible for a newcomer to live his life entirely within the confines of the community. Work, education, and access to health care, recreation, and a variety of other services can be found without leaving the bounds of the ethnic economy. This institutional completeness is what enables new immigrants to move ahead economically, despite very limited knowledge of the host culture and language. Supporting empirical evidence comes from studies showing low levels of English knowledge among enclave minorities and the absence of a net effect of knowledge of English on their average income levels.

Table 1 summarizes this discussion by presenting the different modes of incorporation and their principal characteristics. Two caveats are necessary. First, this typology is not exhaustive, since other forms of adaptation have existed and will undoubtedly emerge in the future, Second, political refugees are not included, since this entry label does not necessarily entail a unique adaptation path. Instead, refugees can select or be channeled in many different directions, including self-employment, access to primary labor markets, or confinement to secondary sector occupations.

Having discussed the characteristics of enclaves and middleman minorities, a final word must be said about the third alternative to employment in the lower tier of a dual labor market. As a mode of incorporation, primary sector immigration also has distinct advantages, although they are of a different order from those pursued by "entrepreneurial" minorities. Dispersal throughout the receiving country and career mobility based on standard promotion criteria makes it imperative for immigrants in this mode to become fluent in the new language and culture. Without a supporting ethnic community, the second generation also

Table 1 Typology of modes of incorporation

Variable	*Primary sector immigration*	*Secondary sector immigration*	*Immigrant enclaves*	*Middleman minorities*
Size of immigrant population	Small	Large	Large	Small
Spatial concentration, national	Dispersed	Dispersed	Concentrated	Concentrated
Spatial concentration, local	Dispersed	Concentrated	Concentrated	Dispersed
Original class composition	Homogeneous: skilled workers and professionals	Homogeneous: manual laborers	Heterogeneous: entrepreneurs, professionals, and workers	Homogeneous: merchants and some professionals
Percent occupational status distribution	High mean status/ low variance	Low mean status/low variance	Mean status/ high variance	Mean status/ low variance
Mobility opportunities	High: formal promotion ladders	Low	High: informal ethnic ladders	Average: informal ethnic ladders
Institutional diversification of ethnic community	None	Low: weak social institutions	High: institutional completeness	Medium: strong social and economic institutions
Participation in ethnic organizations	Little or none	Low	High	High
Resilience of ethnic culture	Low	Average	High	High
Knowledge of host country language	High	Low	Low	High
Knowledge of host country institutions	High	Low	Average	High
Modal reaction of host community	Acceptance	Discrimination	Hostility	Mixed: elite acceptance/ mass hostility

becomes thoroughly steeped in the ways of the host society. Primary sector immigration thus tends to lead to very rapid social and cultural integration. It represents the path that approximates most closely the predictions of assimilation theory with regard to (1) the necessity of acculturation for social and economic progress and (2) the subsequent rewards received by immigrants and their descendants for shedding their ethnic identities.

Clearly, however, this mode of incorporation is open only to a minority of immigrant groups. In addition, acculturation of professionals and other primary sector immigrants is qualitatively different from that undergone by others. Regardless of their differences, immigrants in other modes tend to learn the new language and culture with a heavy "local" content. Although acculturation may be slow, especially in the case of enclave groups, it carries with it elements unique to the surrounding community – its language inflections, particular traditions, and loyalties. On the contrary, acculturation of primary sector immigrants is of a more cosmopolitan sort. Because career requirements often entail physical mobility, the new language and culture are learned more rapidly and more generally, without strong attachments to a particular community. Thus, while minorities entering menial labor, enclave, or middleman enterprise in the United States have eventually become identified with a certain city or region, the same is not true for immigrant professionals, who tend to "disappear," in a cultural sense, soon after their arrival.

Awareness of patterned differences among immigrant groups in their forms of entry and labor market incorporation represents a significant advance, in our view, from earlier undifferentiated descriptions of the adaptation process. This typology is, however, a provisional effort. Just as detailed research on the condition of particular minorities modified or replaced earlier broad generalizations, the propositions advanced here will require revision. New groups arriving in the United States at present and a revived interest in immigration should provide the required incentive for empirical studies and theoretical advances in the future.

"Spatial Disparities in the Expansion of the Chinese Ethnoburb of Los Angeles"

from *Geojournal* (2005)

Jan Lin and Paul Robinson

Editors' Introduction

Lin and Robinson consider the emergence of the Chinese ethnic suburb or "ethnoburb" of the San Gabriel Valley of Los Angeles. The presence of ethnoburbs challenges traditional assumptions in urban sociology regarding the spatial and social assimilation of immigrants to American life, namely that immigrant enclaves would disperse with the residential outmovement to the middle-class suburbs. They associate the rise of ethnoburbs with convergent historical processes including the passing of civil rights legislation, the liberalization of U.S. immigration policy, and the globalization of the urban economy. The rise and expansion of the Chinese ethnoburb also took place against the backdrop of white flight to exurban residential areas in Los Angeles County and Orange County.

Through statistical analysis and mapping of U.S. Census data from 1990 to 2000, the authors found continuing expansion of the Chinese ethnoburb, indicating ethnic persistence rather than spatial and cultural assimilation. They further determined spatial disparity in the ethnoburb between a lower-class core and two middle- to upper-class fringe districts. In one fringe district around the city of Pasadena, they found evidence of cultural assimilation, with Chinese individuals exhibiting higher rates of speaking only English and lower rates of linguistic isolation (as measured by limited English speaking ability). In the ethnoburban core and the other fringe district in the eastern San Gabriel Valley, however, they found ethnic persistence evident in lower rates of speaking only English and higher rates of linguistic isolation. Linguistic isolation was not a barrier to educational attainment and middle-class socioeconomic status in the east district.

The expansion of the Chinese ethnoburb in the 1980s gave rise to a slow growth movement (led by the Residents Association for Monterey Park) that overlapped with an English-only movement that sought to regulate the proliferation of Chinese language signage on commercial corridors. This nativist upsurge gave way subsequently to a managed growth movement and the emergence of Chinese political leaders by the 1990s. The racial/ethnic succession from white to Chinese residents and businesses in the San Gabriel Valley of Los Angeles has similarly occurred in cities such as Cupertino in the Silicon Valley area of northern California's Santa Clara County. These cases shed new perspectives on the racial/ethnic dynamics of white flight beyond the traditional case of white–black succession.

The growth of the Chinese ethnoburb indicates that immigrants are able to gain educational attainment and socioeconomic mobility to buy suburban homes and attain the American dream of homeownership. Asian and Latino homebuyers counterbalanced economic fluctuation in the regional housing market and purchased homes filtered down after the flight of the white population to outlying areas. Ethnoburbs are also appearing in transitional inner-ring suburbs of other immigration gateway cities.

INTRODUCTION

Suburban Chinatowns are intriguing subjects for study and comparison. Enclaves of middle and upper class immigrants, they have emerged since the 1960s to coexist or compete with the older downtown Chinatowns found in American central cities. Wei Li (1998) has characterized these immigrant settlements as "ethnoburbs." Chinese ethnoburbs are the end result of convergent historical and urban processes. Historical shifts include the civil rights movements, United States immigration policy reform, and the globalization of the world-economy. Urban-level trends include Chinese upward mobility leading to migration out of traditional Chinatowns, general processes of U.S. suburbanization, and the phenomenon of white flight to the outer city. These Chinese ethnoburbs are emerging in locations such as New York's outer boroughs, Bellaire city southwest of Houston, and the San Gabriel Valley of Los Angeles County. Their emergence since the 1960s challenges original propositions regarding the spatial and cultural assimilation of immigrants into American life, namely that immigrant enclaves such as Chinatown will tend to disperse with residential outmovement into the suburbs and that immigrants will adopt the English language as they adjust to American society.

We tested these original propositions through examination of recent growth trends in the Chinese ethnoburb of the San Gabriel Valley region of Los Angeles County. We found that the ethnoburb emerged against the backdrop of white population flight out of the San Gabriel Valley from 1960 to 2000. Focusing our research on 1990 to 2000, we found that the Chinese population continued to grow in every San Gabriel city we sampled (except Montebello), with the percentage of Chinese exceeding 40% in the cities of Monterey Park and San Marino in the year 2000, suggesting continued growth rather than dispersal. The greatest growth in the last ten years has occurred at the fringes of the ethnoburban region in cities such as Arcadia, Diamond Bar, Rowland Heights, Temple City, and Walnut. Through analysis of population density and educational attainment, as well as tabular analysis, we determined a core–fringe structure, with a lower-educated core centered in Monterey Park/Alhambra, and two higher-educated fringe zones, a northwest district centered in Pasadena and an east district in the eastern San Gabriel valley. Further analysis of linguistic data determined that Chinese are linguistically assimilating in Pasadena while the Chinese language is persisting in the east district and the rest of the cities of the ethnoburban core. Thus we found evidence of cultural assimilation against a broader backdrop of ethnic persistence in the Chinese ethnoburb.

ETHNOBURBS AS A CHALLENGE TO ASSIMILATION THEORY

In the early twentieth century, the "human ecology" school of urban sociology conceptualized Chinatowns, like other immigrant enclaves, such as Little Italy, the Jewish "ghetto," and Kleindeutschland, as proliferating in the "zone-in-transition" which rings the central business district in American cities. These often congested immigrant colonies were seen to serve as "decompression chambers" for first generation arrivals, who would eventually move with upward social mobility and spatial mobility into outer concentric zones of the city. Robert E. Park, the founder of the "Chicago School," described a "race relations cycle" of contact, competition, conflict, accommodation, and assimilation, which through immigrant groups would be incorporated into American society (Park and Burgess 1921).

[. . .]

Human ecology theory predicted the spatial assimilation of immigrants into the suburbs at a time when cities like Chicago were primarily white, with large numbers of recent immigrants from southern and eastern Europe, when federal policy restricted new immigration, when financial transactions were tightly regulated, and when relative isolation reigned in U.S. geopolitics. Since the 1960s, however, we have seen the promoting of neo-liberal economic and trade policies and the expansion of American geopolitical influence. The liberalizing of immigration and financial controls brought unexpected flows of capital and labor from Asia, Latin America, and the Caribbean, through particular entrepots. We have thus seen the emergence of new immigrant enclaves such as Little Havana, Koreatown, and Chinatown, in immigration gateway cities such as Miami, Los Angeles, New York, and Houston. Many of the ethnic enclaves are appearing in suburban areas of these metropolitan regions.

The emergence of ethnoburbs presents some challenges to the assimilation paradigm. They are simultaneously real-world crucibles and conceptual test objects in the investigation of intergroup relations and globalization processes in American cities. They contradict some fundamental assumptions of spatial assimilation theory, namely that ethnic enclaves spatially entrap immigrants and subject them to disadvantages such as residential overcrowding, poor quality housing, and linguistic isolation (Massey 1985). Spatial assimilation is understood in this paper to mean dispersal of the ethnic population and the disappearance of ethnic enclaves. We examine language use as a good indicator of cultural assimilation, although cultural assimilation includes other dimensions of culture life, such as acquisition of American styles of dining, dress, and other forms of social behavior, such as intermarriage with Americans.

Rather than disappearance and dispersal, we find evidence that the Chinese ethnoburb continues to expand and grow. We now find growing commercial prosperity, suburban affluence, and transculturalism in ethnoburban regions such as the San Gabriel Valley of Greater Los Angeles. The expansion has occurred against a backdrop of white population flight.

BACKGROUND OF WHITE FLIGHT IN GREATER LOS ANGELES

We first examined broad demographic patterns in Los Angeles County over the last forty years. There has been a general process of white flight from 1960 to 2000, as white population in Los Angeles County dropped 38% from 4.88 to 3.05 million and the number of Latinos grew to 4.24 million, Asians to 1.21 million, and blacks to 967,000 in 2000 (Allen and Turner 2002). While leaving central city and older suburbs of Los Angeles County, whites have moved to newer housing developments on the urban fringe in the five-county area of Greater Los Angeles. These affluent exurban enclaves include mountain, canyon, or coastal areas in Ventura County, southern Orange County (such as Laguna Hills and Mission Viejo), and scenic areas of San Bernardino and Riverside Counties. Some have moved to newly developed areas offering cheaper housing, such as Palmdale and Lancaster in the Antelope Valley, and the Inland Empire valleys areas of San Bernardino and Riverside Counties. Many whites left the Los Angeles urbanized region altogether, as the white population of the consolidated five-county region grew by only 3% from 1960 to 2000 (from 6.35 to 6.54 million), total population in Greater Los Angeles increased 113% (from 7.75 to 16.48 million).

GROWTH OF THE ETHNOBURB IN THE SAN GABRIEL VALLEY

We then turn to the San Gabriel Valley, which is a large sub-region in the northeast foothills of Los Angeles County. It's comprised of 31 separate municipalities and some unincorporated regions, and holds nearly one-fifth of the county's population. The Chinese ethnoburb of the San Gabriel Valley emerged alongside other Asian and Latino settlement in a process of racial/ethnic succession during a period of white population flight to outer suburbs and housing estates.

For our data study area, we selected 22 of the 31 cities in the San Gabriel Valley in which Chinese made up at least 2% of the population in 2000 (see Tables 1 and 2). Two percent represents a certain critical mass of Chinese people. The Chinese population in these cities grew by 75,000 persons between 1990 and 2000, from 124,375 in 1990 to 199,376 in 2000. Chinese grew at a 60% rate, ten times higher than the general population growth rate of 5.7% in the study area. Latinos also grew by nearly 65,000 persons, at a rate of 14.5%. The influx of Asians and Latinos effectively counter-balanced the flight of the non-Latino white population, which dropped by 125,536, a 32% decline during the same period. The Asian population comprised nearly 30% of the population in our study area in 2000; Asians are growing faster than Latinos, but Latinos are still the majority at 43%. The San Gabriel Valley is more heavily Asian than Los Angeles County in general, however, where Asians comprise 13% of the population, and Latinos are 45%. The Chinese population increased from 11% of the San Gabriel Valley population in 1990 to nearly 17% in 2000.

The ethnoburb began in earnest in the 1970s with the dynamic activity of a real estate developer named Frederick Hsieh, who actively promoted

Table 1 Race/ethnic profile of San Gabriel Valley, 1990–2000

	Population 1990	*Population 2000*	*Population Change 1990–2000*	*Growth Rate 1990–2000 (%)*	*% of Total Population 1990*	*% of Total Population 2000*
White	391317	265781	–125536	–32.1%	34.6%	22.3%
Black	49916	41193	–8723	–17.5%	4.4%	3.4%
Am. Indian	3124	2742	–382	–12.2%	0.3%	0.2%
Asian/Pacific	235874	347306	111432	47.2%	20.9%	29.1%
Other	2218	1763	–455	–20.5%	0.2%	0.1%
Multiracial	0	23651	23651		0.0%	2.0%
Latino	447211	511967	64756	14.5%	39.6%	42.9%
Total	1129660	1194403	64743	5.7%	100.0%	100.0%
Chinese	124375	199376	75001	60.3%	11.0%	16.7%

Source: U.S. Census of Population and Housing, Summary File 1.
Note: These data for the San Gabriel Valley comprise those 22 cities in which Chinese constituted at least 2% of the total population.

Monterey Park as kind of Chinese "Beverly Hills" for the overseas Chinese of Taiwan and Hong Kong. Many of these Chinese were uncertain about their political futures and investment interests with the retrocession of Hong Kong to the People's Republic of China (PRC) and ongoing diplomatic disputes between Taiwan and the PRC. New immigration was already overspilling the boundaries of the old Chinatown north of downtown Los Angeles. In the 1980s, Chinese spread beyond Monterey Park into the adjacent cities of Alhambra, Arcadia, El Monte, Pasadena, Rosemead, San Gabriel, San Marino, and South Pasadena, to form a clear concentration in the western part of the San Gabriel Valley. Some Chinese began to "leapfrog" the industrial and poorer Latino residential districts that straddle the communities surrounding the I-605 freeway into the eastern San Gabriel Valley areas of Diamond Bar, Hacienda Heights, Rowland Heights, and Walnut, an area many Chinese dub the "Eastern District." A symbol of the migration to the Eastern district is the large Taiwanese Buddhist temple located in Hacienda Heights.

Table 2 reports on the Chinese population changes in the San Gabriel Valley ethnoburb from 1990 to 2000. The percentage of Chinese increased between 1990 and 2000 in all 22 cities of our study area, except for Montebello, indicating persistence of an ethnic enclave rather than dispersal. In the year 2000, the proportion of Chinese exceeded 40% in Monterey Park and San Marino, and reached 33% in Alhambra, 34% in Arcadia, and nearly 34% in the City of San Gabriel. The greatest population increases (exceeding 130%) between 1990 and 2000 occurred in fringe districts. The greatest increase was in Rowland Heights, where the Chinese population tripled. Growth rates were slowest during the same period in Monterey Park (12.7%) and Baldwin Park (12.9%). The population of Chinese dropped by 4121 persons (6.1%) from 1990 to 2000 in the City of Los Angeles, representing a departure from the traditional Chinatown in the central business district, while it grew by 60.3% in the San Gabriel Valley ethnoburb, and 34.4% across the broader County of Los Angeles.

EMERGENCE OF A CORE-FRINGE PATTERN

We turn our analysis on the fifteen San Gabriel Valley cities that contained above a threshold of 3500 Chinese persons. These cities are outlined in Figure 1 with a dot-density population map of Chinese in the San Gabriel Valley. Reducing the sample in this way allows us to focus on the cities in the San Gabriel Valley in which the ethnoburb is particularly concentrated. We then created a

Table 2 Chinese population in San Gabriel Valley ethnoburb, 1990–2000

City	*Chinese Population 1990*	*Chinese as % of City Population 1990*	*Chinese Population 2000*	*Chinese as % of City Population 2000*	*Change in Chinese 1990–2000*	*Chinese Growth Rate 1990–2000*
Alhambra	21303	25.9%	28437	33.1%	7134	33.5%
Arcadia	7180	14.9%	18041	34.0%	10861	151.3%
Baldwin Park	2751	4.0%	3106	4.1%	355	12.9%
Claremont	1032	3.2%	1500	4.4%	468	45.3%
Covina	1007	2.3%	1361	2.9%	354	35.2%
Diamond Bar	4355	8.1%	10091	17.9%	5736	131.7%
El Monte	6781	6.4%	11972	10.3%	5191	76.6%
Hacienda Heights	7839	15.0%	11921	22.4%	4082	52.1%
La Puente	567	1.5%	844	2.1%	277	48.9%
Montebello	3516	5.9%	2722	4.4%	–794	–22.6%
Monterey Park	21971	36.2%	24758	41.2%	2787	12.7%
Pasadena	3122	2.4%	4393	3.3%	1271	40.7%
Rosemead	10832	21.0%	15678	29.3%	4846	44.7%
Rowland Heights	4647	10.9%	14057	29.0%	9410	202.5%
San Gabriel	7649	20.6%	13376	33.6%	5727	74.9%
San Marino	3355	25.9%	5260	40.6%	1905	56.8%
South El Monte	549	2.6%	924	4.4%	375	68.3%
South Pasadena	3103	13.0%	3795	15.6%	692	22.3%
South San Gabriel	1142	14.8%	1616	21.3%	474	41.5%
Temple City	3631	11.7%	9322	27.9%	5691	156.7%
Walnut	3645	12.5%	8590	28.6%	4945	135.7%
West Covina	4398	4.6%	7612	7.2%	3214	73.1%
22-city ethnoburb total	124375	11.0%	199376	16.7%	75001	60.3%
City of Los Angeles	67196	1.90%	63075	1.70%	–4121	–6.1%
County of Los Angeles	245033	2.80%	329352	3.50%	84319	34.4%

Source: U.S. Census of Population and Housing, 2000, Summary Tape File 1.
Note: Our ethnoburb comprises those 22 (of a total of 31 cities) cities in the San Gabriel Valley in which the Chinese comprised at least 2% of the total population in the year 2000; Chinese Growth Rate is computed as change in Chinese 1999–2000 divided by the Chinese population in 1990.

map of educational attainment as percentage of the population that completed a college education (see Figure 2), and made the research discovery of a core-fringe pattern. There is an ethnoburban core exhibiting low levels of educational attainment centered in Monterey Park and Alhambra. There are two fringe zones of ethnoburban settlement where the Chinese population exhibits higher rates of educational attainment, a district around Pasadena as well as a district in the eastern San Gabriel Valley. The educational attainment level exhibited in the fringe districts are a sign of middle-to-upper class status as compared to the poorer ethnoburban core.

To further test the core/fringe structure, we derived other data such as household income, homeownership rate, home value, and average household size. We also measured indicators of language use in order to measure the level of linguistic assimilation.

Table 3 presents comparative data on a range of socioeconomic indicators across fifteen main

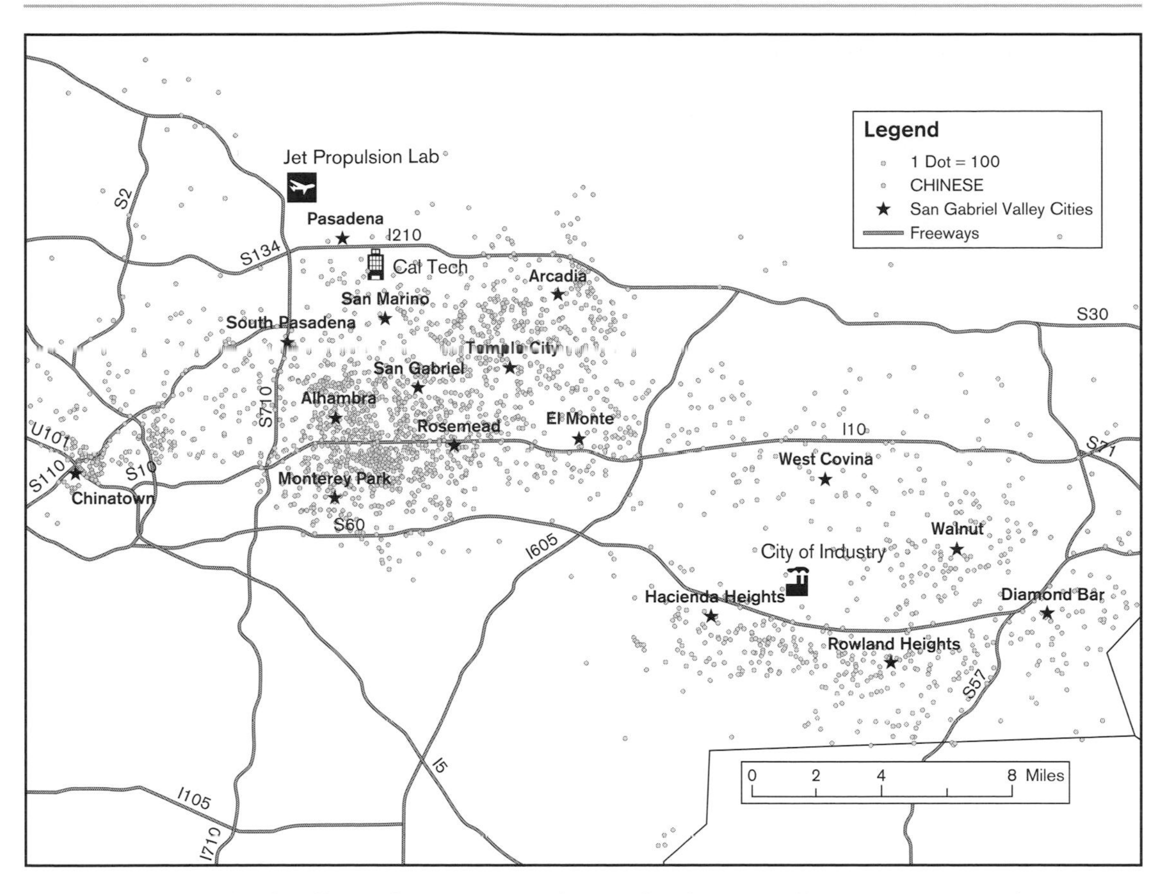

Figure 1 San Gabriel Valley Chinese Settlement Areas (cities with at least 3500 Chinese persons in 2000)
Source: U.S. Census of Population and Housing 2000, Summary File 1.

populations of the Chinese ethnoburb. They confirm the presence of a core–fringe pattern. The ethnoburb has grown beyond an initial colony in Monterey Park to a cover sprawling metropolitan area that includes a densely settled and lower-to-middle class ethnoburban core and two middle-to-upper income fringe zones. One fringe zone is in the northwest and the other in the eastern part of the San Gabriel Valley. The ethnoburban core comprising seven cities had 64.9% of the population of the ethnoburb. The Chinese population in these cities exhibits low-to-medium levels of household income, median home value, homeownership rate, and educational attainment. The ethnoburban core cities exhibit relatively low levels of linguistic assimilation.

The northwest district (comprising the cities of Pasadena, San Marino, South Pasadena) had 7.2%. The northwest district exhibits middle-to-high income levels. The Chinese population exhibits the highest home values and educational attainment in the ethnoburb. The Chinese in the northwest district are the most linguistically assimilated in our study area.

The east district cities of Diamond Bar, Hacienda Heights, Rowland Heights, Walnut, and West Covina, had 52,271 or 27.9% in 2000. These cities exhibit medium levels of household income, median home value, and educational attainment. Like in the ethnoburban core, these cities exhibit relatively low levels of linguistic assimilation.

ETHNOBURBAN CORE

As can be seen in Figure 1, the cities of what we call the ethnoburban core occupy a broad belt in the San Gabriel Valley. Since the 1970s, the

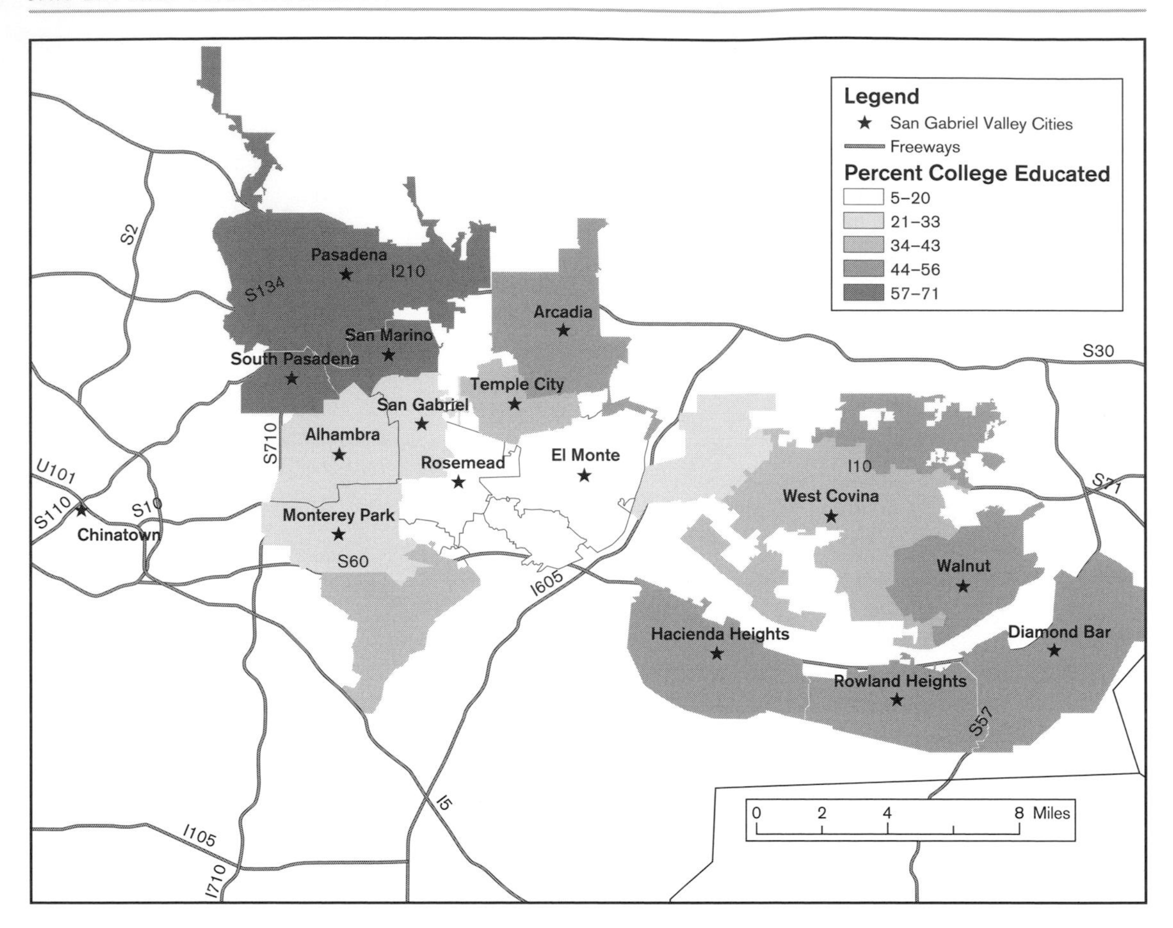

Figure 2 San Gabriel Valley Chinese Settlement Areas by Education Level
Source: U.S. Census of Population and Housing 2000, Summary File 1.

initial Chinese settlement area in Monterey Park has fanned out to include the adjoining cities. There are thousands of immigrant enterprises in retail trade, personal services, and business services generating jobs and revenues for the San Gabriel Valley Chinese population. The major commercial arterials and intersections are packed with mini-malls and shopping centers abounding with Chinese restaurants, grocery stores, book and video stores, salons, clothing and department stores, and other retail shops and services, as well as banks, certified public accountants, insurance and real estate companies, and international trading offices. There is also a considerable concentration of Chinese media, including 15 Chinese daily newspapers, several journals, six Chinese language television stations, many Chinese ethnic organizations, language schools, and social programs. This is the primary economic, cultural, and entertainment center for the Chinese population of the San Gabriel Valley. Because of the predominance of Taiwanese immigrants in the early days, Monterey Park was initially dubbed "Little Taipei," and then "Little Beijing," but the population is now more panethnic with immigrants from all parts of the greater Chinese diaspora including immigrants from Hong Kong, the PRC, Vietnam, and other parts of Southeast Asia.

[. . .]

The socioeconomic class status of Chinese households in the seven cities of the ethnoburban core is generally lower-middle class, exhibiting lower median household incomes, median home values, and homeownership rates than other districts in the ethnoburb. Temple City and Arcadia represent two areas of upward mobility. Incomes,

Table 3 Selected social indicators for Chinese population living in ethnoburban settlement clusters of the San Gabriel Valley

	Median Household Income 1999	*Median Home Value 1999*	*Home Owner (%)*	*Average Household Size*	*College Degree or Higher (%)*	*Speak Only English (%)*	*Linguistic Isolation* (%)*
Ethnoburban core							
Rosemead	$35,703	$180,300	55.2%	4.2	14.8%	4.1%	47.9%
Alhambra	$35,833	$197,500	42.3%	3.07	32.9%	5.5%	52.5%
Monterey Park	$36,155	$217,000	51.4%	3.12	32.3%	7.1%	53.6%
San Gabriel	$37,442	$222,100	48.3%	3.32	30.2%	5.0%	55.6%
El Monte	$38,647	$152,900	66.9%	3.77	17.9%	5.6%	54.9%
Temple City	$48,445	$222,700	76.5%	3.44	43.2%	6.2%	47.0%
Arcadia	$58,319	$381,300	73.0%	3.35	55.8%	5.5%	41.4%
Northwest district							
Pasadena	$46,532	$320,000	44.2%	2.01	66.3%	20.6%	35.3%
South Pasadena	$62,727	$399,600	67.1%	2.77	63.0%	22.4%	32.5%
San Marino	$104,046	$650,700	92.7%	3.67	70.8%	11.4%	27.1%
East district							
Rowland Heights	$46,164	$275,800	71.6%	3.32	48.4%	4.5%	47.9%
West Covina	$50,043	$192,400	75.6%	3.4	37.4%	7.5%	41.4%
Hacienda Heights	$52,198	$248,800	76.2%	3.39	50.2%	5.8%	45.9%
Diamond Bar	$62,253	$270,500	86.9%	3.41	56.1%	4.7%	41.9%
Walnut	$69,604	$292,300	92.1%	3.63	51.4%	6.5%	39.7%

Source: U.S. Census of Population and Housing 2000, Summary File 4.
Note: For this analysis of selected social indicators, we selected the 15 cities in the ethnoburb in which Chinese were at least 2% of the population and comprised at least 3500 persons in 2000. These data are specifically for Chinese persons.
* Linguistic isolation measures the percent of households in which all members 14 years of age and over speak a non-English language and also speak English less than "very well."

home values, and homeownership rates are higher in these two cities than the other cities of the ethnoburban core. Chinese educational attainment is also generally lower in these seven cities. The lowest rates of college completion are found in Rosemead and El Monte. Households in the core generally display low rates of speaking only English. They also display relatively high rates of linguistic isolation and many households have at least one member that does not speak English at all.

The experience of Chinese settlement in the San Gabriel Valley created some rocky political conflicts. An anti-foreigner homeowners movement (Residents Association of Monterey Park) emerged in the early 1980s in Monterey Park in response to residential and traffic congestion caused by the rapid immigrant influx, rising property taxes, and the proliferation of Chinese language signage on main commercial arterials such as Atlantic Avenue and Valley Boulevard. An English-only referendum was proposed and four nativist slow-growth candidates were elected to City Council in 1986 who favored immigrant restriction and a building moratorium. Slow-growth interests felt threatened by the growing density of development with the conversion of single-family housing sites to multi-unit condominium, apartment, and commercial projects. These conflicts were somewhat defused by the emergence of a multiracial coalition called Citizens for Harmony in Monterey Park (CHAMP), which promoted a pro-immigration stance and a "managed growth" policy surrounding the candidacy of Judy Chu, who was elected mayor in 1990 (Horton 1987; Fong 1994; Saito 1998).

Anti-Chinese sentiment linked with slow-growth politics has flared more recently in Arcadia, which had the largest increase in Chinese persons (10,861) and the third highest growth rate (over 151%) between 1999–2000 among the 22 cities of the ethnoburb. There were incidents of vandalism against Asian American homeowners in the 1990s, particularly where modest single-family homes were converted into mansion-style housing.

NORTHWEST DISTRICT

The northwest district comprising the cities of Pasadena, South Pasadena, and San Marino displays middle-class to affluent household incomes. Median home values are among the highest in the ethnoburban region. The lower median household incomes and homeownership rates in Pasadena reflect a greater number of single and childless households. San Marino has the most affluent city in the ethnoburb.

The Chinese of the northwest district are the most culturally assimilated in the ethnoburb. The three cities exhibit the highest levels of educational attainment and linguistic assimilation. Chinese households in these three cities spoke more English and were less linguistically isolated. The linguistic isolation variable measures the percentage of households in which all members 14 years of age and over speak a non-English language and speak English less than "very well."

Pasadena, the largest city in the San Gabriel Valley with a population of over 130,000 in 2000, is the economic engine of the northwest district. The city has over 5000 firms, including four of the six largest corporations in the San Gabriel Valley. Pasadena possesses a large retail and governmental employment sector, but also has a distinct educational, technology, and health complex. The California Institute of Technology is located in San Marino and the Jet Propulsion Laboratory is in northwest Pasadena. There is a surrounding technology and education complex. Hospitals and associated advanced medical services are also clustered in Pasadena. Employment in the Pasadena educational, technology, and health complex requires high levels of educational attainment and familiarity with the English language. The northwest district represents a zone of upward mobility and cultural assimilation into American society.

EAST DISTRICT

The east district cities are generally upper middle class. Homeownership rates are relatively high. Chinese persons in the east district generally exhibit higher levels of educational attainment than the ethnoburban core, but not as high as in the northwest district. They are not as culturally assimilated as the northwest district Chinese populations. The proportion of persons who speak only English and the proportion of households that are linguistically isolated are comparable with those rates found in the ethnoburban core.

There is a cluster of Chinese importing and technology distribution firms around the City of Industry near the east district fringe settlement, stretching out along the State 60 Highway. These importers are directed less at the ethnoburb, but instead at American consumers in general. These firms are involved in trade and marketing of cell phones and other consumer electronics, autoparts, and food imports. Taiwanese import–export firms are increasingly being joined by companies established by entrepreneurs from the PRC, importing goods such as shoes, toys, and heavy machinery. Officials from the PRC government now regularly send trade delegations to meet with the City of Industry Chamber of Commerce. During the 1990s, 12,700 jobs were created in the City of Industry, the most among all San Gabriel Valley Cities. The warehousing and factory districts in Industry and other cities that straddle the I-605 have easy access to the Port of Long Beach for transshipment to the Pacific Rim.

In contrast to the northwest district, the east district of Chinese settlement in the San Gabriel Valley represents a zone of upward mobility without cultural assimilation. East district Chinese people are more highly educated in general than in the ethnoburban core, but are less educated than Chinese in the education and technology district around Pasadena. The technology complex of the City of Industry and the east district is linked more to Pacific Rim trade. The east district is a zone of ethnic persistence and economic globalization.

THE ETHNOBURB IN NATIONAL AND GLOBAL CONTEXT

The entry of Chinese immigrants into the San Gabriel suburbs suggests the fulfillment of the typical American dream of working hard, saving money, and buying a home. Homeownership represents a pinnacle in the ladder of upward social mobility. A 1995 study by the Joint Center for Housing Studies of Harvard University found that Chinese immigrants moved very quickly into homeownership. While native-born white homeownership rates rose from 57.4% in 1980 to 72.1% in 1990, Chinese immigrants grew from 36.8% in 1980 to 82.8% in 1990. Immigrant Asians along with Latinos drove much of the housing market in southern California during the last two decades. Some analysts have suggested that immigrant homebuyers with growing families have counterbalanced some of the decline in housing demand caused by declining fertility among established American households. Chinese immigrants have purchased older homes that filtered down from previous owners in the older cities of the western San Gabriel Valley as well as in newer housing estates in the rapidly developing cities of the East District.

[. . .]

The contemporary wave of Chinese immigration to Los Angeles County for the most part bypassed the discriminatory experiences of Asians and other nonwhites who migrated into the area prior to the 1960s civil rights movements. The Chinese immigrants who settled in Los Angeles in the 1970s, 80s, and 90s encountered an urban America that was vastly different from that experienced by the existing residents of the core Chinatown. The decades-long struggle by the National Association for the Advancement of Colored People (NAACP) and other African American civil rights organizations for equal access to urban residential space culminated with the 1968 Fair Housing Act. Up until that legislation, upwardly mobile Chinese who wanted to move into more spacious housing areas were forced to buy homes in the same areas where middle-class African Americans were allowed to purchase homes because in Los Angeles, restrictive covenants on residential property applied to Asians, as well as blacks and Latinos. It is critical to recognize the timing of the immigration reforms that opened the way for the growth of the ethnoburb, in that the battles waged during the civil rights movement of the 1950s and 1960s, were a necessary precursor to the growth of the ethnoburb.

[. . .]

CONCLUSION AND COMPARISONS

[. . .]

A comparison of the three settlement districts leads us to our major research finding as well as suggesting directions for future research. There are two channels of social mobility in the Chinese ethnoburb of Los Angeles. One is through higher educational attainment and cultural assimilation into the northwest district cities around Pasadena surrounding the technology/education technopole that includes several large employers like the Jet Propulsion Laboratory, California Institute of Technology, and Huntington Memorial Hospital. This technopole is oriented towards the American economy.

The other trajectory of upward mobility is through the ethnoburban core and east district cities, where Chinese people exhibit lower levels of higher educational attainment and lower rates of cultural assimilation. Proficiency in the English language is not as important to success in the Chinese ethnic economy, where immigrants may speak Chinese with employers, co-workers, and customers. There is trade in a diversified range of Chinese goods and services. There is another technopole of enterprises connected with import and export of computer and other high-technology goods and services clustered around the City of Industry. This technopole is associated more with trade in technology products in the global-economy.

[. . .]

Chinese homeowners in the fringe districts represent upwardly mobile local Chinese who have generated and saved their income from local economic activities, or new middle-to-upper class Chinese immigrants to America. Chinese immigration will continue because of high economic growth rates in China, and the significance of Los Angeles in Pacific Rim trade, with the ethnoburb of the San Gabriel Valley being a principal gateway for the ongoing influx of labor, capital, and commodities.

The Chinese ethnoburb of Los Angeles' San Gabriel Valley has emerged during a time of white flight to the outer city or out of the metropolitan region completely. Immigrant Chinese have settled with other Asian and Latino immigrants in these middle-ring suburbs of the Los Angeles metropolitan area. Across America, immigrants have replaced whites in older inner or middle-ring suburbs built in the 1950s and 60s in a process of racial-ethnic succession. In "Sunbelt" metropolitan areas such as Santa Ana/Orange County, San Jose, San Diego, Houston, Dallas, Atlanta, Miami, and suburban northern Virginia, older suburbs that were previously considered somewhat parochial, white, and segregated, have become new settlement areas and trade zones for Latino and Asian immigrants.

[. . .]

REFERENCES

Allen, James P. and Eugene Turner. 2002. *Changing Faces, Changing Places: Mapping Southern Californians*. The Center for Geographic Studies, California State University, Northridge.

Fong, Timothy. 1994. *The First Suburban Chinatown: The Remaking of Monterey Park*. Philadelphia: Temple University Press.

Horton, John. 1987. *The Politics of Diversity: Immigration, Resistance, and Change in Monterey Park*. Philadelphia: Temple University Press.

Kotkin, Joel. 2000. *The New Geography*. New York: Random House.

Li, Wei. 1998. "Anatomy of a New Ethnic Settlement: The Chinese Ethnoburb in Los Angeles," *Urban Studies* 35 (3): 479–501.

Massey, Douglas. 1985. "Ethnic Residential Segregation: A Theoretical Synthesis and Empirical Review," *Sociology and Social Research* 69: 315–350.

Park, Robert E. and Ernest, W. Burgess. 1921. *Introduction to the Science of Sociology*. Chicago: University of Chicago Press.

Saito, Leland. 1998. *Race and Politics: Asians, Latinos and Whites in a Los Angeles Suburb*. Urbana: University of Illinois Press.

"Men Without Property: The Tramp's Classification and Use of Urban Space"

from *Antipode* (1983) [1978]

James S. Duncan

Editors' Introduction

James S. Duncan provides a fascinating account of the tactics and strategies employed by homeless people to find secure spaces for rest and residence while eluding the police, imprisonment, and the moral authority of the state. The public fear of the tramp, Duncan asserts, comes from their status as a willful negation of established social order. Vagrancy laws restrict their movement because they reflect the prevailing ideology concerning private places and freedom of movement. As long as the tramp adopts a low profile, he can occupy marginal spaces such as alleyways, spaces under bridges, dumps, railroad yards, and other no-man's lands of the city. Sometimes urban districts appear that are "jurisdictional voids" completely ceded to the homeless. As a way of containing the homeless, police accept these "skid rows" and charitable or state agencies may congregate their shelters and transitional facilities there.

Tramps are generally driven out of prime public and private places as eyesores, public nuisances, and threats to the social order. Full citizenship rights are not extended to those without property. Their presence may be tolerated in some public spaces, however, such as parks, libraries, and transportation stations. By employing "props" such as newspapers and books casually draped over their faces, they may catch a snooze or read under the tolerant surveillance of the policeman. These props help legitimize the tramp while submerging his stigmata. Prostitutes and gay hustlers may cruise in the park, as "the coppers will let you whistle low, but not loud." Tramps, hustlers, drug dealers, and other socially marginalized people may perform a kind of jurisdictional etiquette or interaction ritual with the forces of law and social order. This negotiation process helps reproduce middle class moral order in the city.

Duncan alludes to human ecologist Robert Park in his references to the moral order of the city. Duncan also alludes to critical legal theory when he notes that Vagrancy Laws emerged as early as the fourteenth century as an attempt on the part of landowners to expropriate labor at a time when it was in short supply and its value was high. Twentieth-century policies to regulate the homeless, on the other hand, control people with no labor value. Skid row deprives people of their "natural rights" rather than being an outgrowth of urban "natural selection." Tramps would not be regulated in a truly free society. Duncan has a view of symbolic interaction that recognizes inequalities of power and access to property and movement.

The recent gentrification and redevelopment process in many American cities has given the plight of the homeless new visibility. Marginal spaces of the inner city that were previously ceded to tramps are increasingly subject to police surveillance, and interest on the part of developers, planners, and public officials. There has been new outcry over the homeless as a problem of mental illness and a threat to public health and order. With the return of jobs and people to the central city, there are fears about a link between tramps and criminality. The police do "sweeps" through skid row under trumped-up charges of parole violation and other illegal activity.

There has been a series of classic urban ethnographic studies on tramps and the homeless. They include: Elliot Liebow, *Talley's Corner: A Study of Negro Streetcorner Men* (Boston: Little, Brown, 1967), Ulf Hannerz, *Soulside: Inquiries into Ghetto Culture and Community* (New York: Columbia University Press, 1969), David A. Snow and Leon Anderson, *Down on Their Luck: A Study of Homeless Street People* (Berkeley, CA: University of California Press, 1993), Mitchell Duneier, *Sidewalk* (New York: Farrar, Straus, and Giroux, 1999).

James Duncan is currently a University Lecturer in Geography and Fellow of Emmanuel College, University of Cambridge. He previously taught at the University of British Columbia and Syracuse University. He obtained his B.A. at Dartmouth College and his M.A. and Ph.D. at Syracuse University. His books include *Housing and Identity: Cross-Cultural Perspectives* (New York: Holmes and Meier Publishers, 1982), *Place/Culture/Representation*, edited with David Ley (London: Routledge, 1993), *Landscapes of Privilege: The Politics of the Aesthetic in an American Suburb*, written with Nancy G. Duncan (New York: Routledge, 2004), and *A Companion to Cultural Geography*, edited with Nuala C. Johnson and Richard H. Schein (Oxford: Blackwell, 2004).

[. . .]

Differential mobility, access to space and inequalities in power to influence others' use of space reflect the interrelationships between social groups in the city. In fact, land is divided up and access to space is limited in such ways that land can be said to constitute a relationship between men. Power and other social relationships between men [*sic*] are in various ways enacted in the use of land and in restrictions of others' use of land.

This paper is about the difficulty that tramps, being members of an extremely marginal group, encounter when they try to carve out a niche for themselves in the American city whose moral orders have little place for them. It concerns the strategies they employ to exist in the nooks and crannies of the urban world whose moral order denies the legitimacy of their nomadic existence. The tramp roams from city to city by freight train, on foot, or by broken down car. He attempts to adapt as best he can to the hostile urban environment. He wanders the skid rows, hides in alleyways, sleeps under bridges, on the sidewalks, in garbage dumps and in parks. He is perpetually on the move to avoid arrest, to look for a job or a handout, and to find a place to sleep.

In the first half of this essay I will discuss the relationship between the two groups under consideration, the tramp and the host population, the latter being those who have control over the various areas of the city through which the tramp wanders. Also mentioned will be the dominant ideology which provides a framework within which the host group operates. Next I consider the strategies the tramp employs to make a place for himself in the spatial order of the city. In particular, I will discuss the way the tramp takes the role of the host group in order to derive a classification of the city pertinent to his attempts to survive there.

THE MORAL ORDER OF THE LANDSCAPE

The city is composed of more or less well defined social areas each of which is controlled by one or more groups who sustain a moral order there. The moral order of an area is the public order. I use the term moral order here to capture the feeling of a group that the way it organizes its world is inherently correct. It is defined here as the set of customary relations in an area and the etiquette governing its landscape; it constitutes what is believed by the dominant group to be the proper arrangement and use of artifacts and the proper form that interaction in that landscape should take. It stipulates what people under what circumstances are allowed to engage in what activity in what places.

This moral order is not arrived at, however, without some negotiation on the part of the participants. The negotiation process is not only of a formal political nature but also arises out of routine social interaction. Superimposed upon local moral orders is the largely middle class moral order of the city as a whole. This is codified in laws and enforced by the police and other official agencies. Indirect attempts at control are also made by architects and planners who design buildings and outdoor spaces to keep order. The local moral orders also have their guardians in the form of residents (peer pressure), shopkeepers, gangs, and others who enforce the local etiquette. There is a certain tension between the local and city-wide moral orders that arises from the differences between them.

The police are expected to keep this tension under control. They accept the fact that local moral orders do exist and often take precedence over the city-wide order as long as they do not grossly contradict it or spill over into other areas.

[...]

The police tolerate certain deviant behavior as normal on skid row, which is one of the few areas the tramp may occupy with a minimum of risk. In this sense they respect the moral order of the row; they find, however, that other behavior there so flagrantly violates the larger moral order that they must intervene.

THE MARKET PLACE IDEOLOGY OF THE HOST GROUP

In a society such as ours, whose organization is based on individual property rights, a poor person will be viewed as a problem for the group controlling the area in which he lives. He possesses little property and hence has little stake in the existing order which functions primarily to protect property and ensure that orderly market relations take place. The tramp poses all of the same problems for the controlling group that the poor local person does but these problems are magnified as he, being an itinerant, has even less stake in the area. Not only does he lack property but he rarely has any ties with local residents. Thus the tramp feels no obligation to maintain the moral order and furthermore his mobility makes it difficult to force him to comply with it. The locals view the tramp as a threat and do their best to drive him away.

Social value in our society is based primarily upon property, real property or labor which one is willing and able to sell. While the tramp occasionally has a few possessions, a watch, a ring, a little money made on a temporary job or panhandled, perhaps a radio purchased in good times to be sold when the times get rough, these possessions come and go. The only property a tramp has on a fairly regular basis is his labor power which he can sell if he is willing, though often he is not because the wages offered are often less than can be made panhandling. Furthermore, his lack of skill or his frequent inability to hold a job because of alcoholism makes his labor worth very little to society.

[...]

According to the dominant ideology in society one must have sufficient property to be able to own or rent a place to sleep if he is not to be charged with vagrancy. Vagrancy laws, which restrict the movement of individuals from place to place, have been in effect in England since the fourteenth century and were adopted by the colonies and subsequently the United States. The social problem that they were originally intended to alleviate has long since disappeared while other totally unrelated problems have become their target. Vagrancy laws should be of special interest to geographers because they reflect the prevailing ideology concerning private places and freedom of movement. For example, it is illegal not to own the right of access to some private place. Every person who wanders about without access to a private place to sleep, who can show no means of support "... shall be deemed a tramp, and shall be subject to imprisonment..." It is such access that provides "... evidence that a man has a legitimate relationship to the social structure." The vagrancy laws are selectively applied to contain tramps within skid row or to banish them entirely from the city. Thus, those who are penniless or without normal occupation in the broad sense of the term may be severely restricted in their spatial mobility and even declared criminal.

STRATEGIES FOR GETTING BY

Let us now turn to a consideration of the strategies that are adopted by the tramp who finds that the moral order of the host group has no place for him. There are a variety of possible strategies, e.g., mobility, and eschewing cumbersome possessions and attachments to members of the local population. I will restrict myself here to a discussion of those strategies that make use of the landscape.

In order to avoid the guardians of the host landscape, the tramp must adopt a low profile. Often this is not easy to accomplish because the tramp wears old, worn clothes that are easily recognizable. It is not easy to change this because it involves an outlay of resources that is usually not available. He would also severely diminish his chances of being able to panhandle successfully because his clothes are a sign, often the only visible one, that he needs a handout.

> I believe that the dirtier the man is the better for the street make. If I am dirty, a man will give me a coin rather than have me walk down the street and have people think that I might be one of his friends.

Since passing as a member of the established moral order is not a viable method of adopting a low profile, the tramp must resort to another tactic, that of using the landscape as a cover. This method requires a large amount of environmental knowledge. He must know for example where he can find a warm place to spend a winter's night unobserved.

> You've got to have ingenuity. You've got to know New York, its people, how to get around. I sleep in the subways nowadays. It works out fine. . . . You can't sleep there when an officer's on duty. That's from eight p.m. to four a.m. So I go there about ten minutes till four; sleep to noon, usually . . . I like the Eighth Avenue line. Less stops. I sleep on one of the front or back cars; never the middle. Too many disturbances.

At times he must out of necessity sleep on the street and is therefore potentially open to harassment from the police and jackrollers (muggers). He finds ingenious ways to hide himself in this kind of exposed landscape. "There's garbage cans back there . . . I'd lay there with half my body in a garbage can and the upper half in a pasteboard box . . . that way no one knows." Once a tramp finds a place that provides good cover he may wish to protect it from others. Often some ingenious tactics are employed to attempt to claim a bit of public space.

> Snead, a dope addict, slightly past middle age, has been living for some time in the doorway of an unoccupied building. This doorway consists of a sort of vestibule and has plenty of room for a person to lie down. His only furniture is a broom and a box of broken glass. When he leaves he scatters the broken glass over the floor to keep others away. And when he returns he uses the broom to sweep up the glass.

Most of this environmental knowledge is very specialized. It is of interest to the tramp alone although certain members of the local population, notably the police, may also be tangentially interested. He knows certain ethnic streets where begging is especially good, specific street corners where truckers stop to pick up cheap labor, and that public libraries are good places to sleep during the day. He knows many different spots where he can sleep and keep warm on a cold winter night, such as in bus depot toilet stalls, near city steam pipes, in heated sandboxes in railroad yards, in warm stacks of bricks in brickyards, and in open churches. In fact Spradley discovered that tramps could name over one hundred categories of sleeping places.

Whereas most tramps are aware of these relatively stable elements of the urban environment, the more astute quickly capitalize on changes in the environment. For example, shortly after a chapter of Alcoholics Anonymous started holding meetings in a church near the Bowery in New York, a number of tramps congregated about the church during meetings taking advantage of the fact that members of A.A. are a "soft touch" for a handout.

[. . .]

Alongside the marketplace ideology (which guides relations between most strangers within the city) struggles the tramp's non-market ideology, which is based on reciprocity and is termed "brotherhood of the road" by its adherents. The maintenance of this alternative ideology may be considered a strategy for survival. A tramp shares with others whether he knows them or not, knowing that he may be taken care of when he is in need. He will share his bottle, flop (any of a wide variety of places to sleep), clothes, food, and information about local police practices and availability of spot jobs, etc. A tramp may give his jacket to another inmate when he leaves jail and then go out into the street to "cut in" on someone's bottle. According to informal rules one must never refuse to share his bottle.

A tramp's environmental knowledge must include all the intricacies of how the host group classifies space and specifically the social value that it attaches to different landscapes. Gaining such knowledge is the most important of all his strategies.

PRIME AND MARGINAL SPACE

Most citizens find public versus private the primary constraining classification in their use of space: one ought to be able to use any public space by virtue of his being a citizen provided he behaves in a fairly "normal" and legal fashion for the time and place. Private spaces, of course, are far more exclusive. The distinction between public and private is usually clear although it can be a little fuzzy around the edges. The propertyless are generally excluded from private places (they can sometimes stay in flophouses or go into bars or missions but not on a regular basis). What is more, they are also driven out of public places, for full citizenship rights are apparently not extended to the propertyless, a notion which has survived from the eighteenth century when citizenship was legally extended only to those who had property. Public property apparently, then, belongs to the citizenry as a whole which has the right to exclude the tramp.

[. . .]

In contrast to the propertied population and the relatively simple public versus private distinction which restricts its spatial movements, tramps are forced to use a different system of determining usable space within the city. This system is based on an unstated scale of social value which the host groups apply to different areas. The scale ranges from what I would term prime to marginal space. This implies no inherent value in the space itself; on the contrary the value is assigned to the space by a given group on the basis of how it uses the space. The social value of space, furthermore, is an ever-evolving phenomenon which is based upon varying degrees of consensus. Thus the tramp must learn the social value that the host group assigns to different types of urban space and to the regular temporal variations in these values; he must then resign himself to spending as much time as possible in marginal space.

Marginal space includes alleys, dumps, space under bridges, behind hedgerows, on the roofs of buildings, and in other no man's lands such as around railroad yards, which are not considered worth the cost of patrolling. As one tramp put it,

> If you are under the bridge down below Pike Street Market you are safe – the police don't walk down there, just too lazy.

Skid row and a few other very poor residential or commercial areas are also considered marginal. Here the tramp can achieve a low profile because the shabbiness of his dress does not stand out as it does elsewhere. Skid row is ceded to the tramps because the authorities realize that tramps must stay somewhere and there are definite advantages from the point of view of social control to keeping them together in one place.

[. . .]

The terms prime and marginal refer not to a dichotomy but a continuum. It must also be noted that they are relative terms. By looking at the city as a whole and grossly dividing it into space which is prime and marginal in the eyes of the host group I have in effect lumped all the citizens of the city except the tramps into one group, which is an obvious oversimplification. Whether a place is prime or marginal depends upon the perspective from which one views the situation. The tramp is aware of the diversity of perspectives within the host population. In day-to-day decisions he must see things from their perspective. He must take into consideration the fact that what one person considers prime space another considers marginal. He takes the perspective of some into account more than others because he realizes the relative inequalities of power among these groups to enforce their perspective. It must be remembered, however, that there is some general agreement as to prime and marginal in the gross sense for the lower class citizen is forced to a certain extent to accept the perspective of the middle class citizen and to realize that effectively his area is considered marginal even if it is of prime importance to him.

Likewise the tramp must "take the role of" the police officers in their subdivision of marginal space into more or less marginal. For example, the police categorize abandoned buildings as extremely marginal space.

> The patrolman knows these buildings are in use, for he constantly sees evidence . . . (but) he does not examine these places on his route. . . . They are no longer buildings in the formal sense. They are like the alleys the patrolman does not go into because he does not see them as being either public or private places . . .

These places, which are neither public nor private, have been classified by the patrolman as places with no property value and hence as jurisdictional voids. The tramp is quick to see the personal benefits to be derived from the patrolman's classification of marginal space. He also learns to divide marginal space on the basis of accessibility and inaccessibility to the police, who, for matters of safety and convenience like to patrol their beat in cars. In effect the police have given up jurisdiction over certain areas in which they cannot drive a car.

[. . .]

Similarly tramps take advantage of the policeman's classification of alleys as inaccessible space and use them for drinking and other illicit activities.

THE USE OF PRIME SPACE

The tramp is faced with a problem that really has no solution. In order to survive without being arrested he must occupy marginal areas, while in order to secure the wherewithal to survive he must often venture into high risk, prime areas. The vast majority of the city is prime space and therefore space which the tramp finds dangerous. Prime space, however, does not have a uniformly high social value. The tramp knowing the relative value assigned to different spaces by the host group minimizes his risk by using prime space which has moderate social value.

> . . . the central business district, and certain main thoroughfares, are the only places in the metropolis (with the exception of skid row) where the bum can successfully ply his trade. If he tries door-to-door begging . . . householders call the police.

It is extremely difficult to adopt a low profile in any prime area while one is trying to panhandle. On most city streets there are few if any places that provide cover. Furthermore, interaction with strangers violates the norms of conduct on such public streets, and panhandling is therefore easily noticed. "Normal" and hence acceptable conduct on such a street consists of walking purposefully and not interacting with others unless they are walking with you.

> In Chicago, an individual in the uniform of a hobo can loll on "the stem," but once off this preserve he is required to look as if he were intent on getting to some business destination.

The practice of picking up cigarette butts from the street, "snipe shooting," illustrates some of the problems inherent in using various types of prime space. To compete for these snipes one must have a detailed knowledge of the good snipe hunting areas.

> He told me the best place for snipes was outside a church – the Sunday folk always left long ones before they entered. On a good day you could pick up enough to make a pack of almost new ready-rolls.

However, although one can find good snipes on the Loop one must be careful of the police there.

> I wouldn't advise you to hunt snipes in the Loop, for while the butts are as long as this (measuring with his hands the length of a lead pencil), you are almost sure to get picked up by the police.

The advantages to be gained from using prime space, therefore, must be weighted against the risks.

There are certain other prime spaces that the tramp uses upon occasion in order to keep warm or to catch some sleep where he may not be noticed by the police. Two such places are public libraries, and train stations.

> Thus in some urban libraries, the staff and the local bums may reach a tacit understanding that dozing is permissible as long as the dozer first draws out a book and props it up in front of his head.

Likewise it appears that

> . . . newspaper readers never seem to attract the attention (of the station guards) and even the seediest vagrant can sit in Grand Central all night without being molested if he continues to read a newspaper.

In both cases the tramp adopts the strategy of using a prop, in this case reading material, in order to legitimize his presence in this prime space. These props appear to submerge his stigmatized identity sufficiently to stop it from spreading to others in the place. This can only be accomplished in impersonal, low interaction places where the stigmatized person can easily be disattended by others. These are places and props which isolate individuals, "remove them" as it were from others. It is as if they were required to build a wall around themselves in order that their presence might be tolerated in prime space.

For matters of convenience I have been speaking of the host group as if it were a single monolithic group which of course it is not. Within it there are many social worlds and factions with more or less conflicting interests. Some are reasonably friendly to tramps and occasionally will form temporary alliances and make deals with them against other segments of the host population. For example, tramps have a certain working relationship with homosexuals who exchange food, shelter, a chance to get cleaned up, clothes and occasionally money, for sexual relations. A third party to this relationship is often the police who tolerate such activity if it remains reasonably unobtrusive. Tramps whistle to get the attention of homosexuals who are "cruising" the park.

> They whistle low. They cannot whistle loud. The coppers will hear them. The coppers will let you whistle low, but not loud.

Such agreements are the product of implicit bargaining, or negotiation, between the police and tramps. The tramps continually try to make gains in freedom to use certain prime space to make contact with members of the host population while the police attempt to hold them in check without expending too much extra effort.

The fact that the tramp is forced to enter prime space on occasion should not obscure the fact that his principal strategy is to occupy marginal space. Excursions into prime space constitute a small but dangerous part of the tramp's existence.

ON THE SEPARATION OF NORMAL AND SPOILED IDENTITIES

There is a belief in our society that the social value of an individual must be roughly commensurate with the social value of the place he frequents. In a society which is socially mobile the objects and settings that one surrounds oneself with are an inseparable part of one's self. This argument logically leads to the notion that if one shares a setting with another then, to a degree, he shares his identity thus allowing stigma to spread by spatial association. A tramp does not have a "normal," that is propertied, identity but to put it in Goffman's terms, one that is "spoiled." Such people must be separated out for their spoiled identities can spoil the setting and by extension the "normal" people who are a part of it. Hence the hierarchical division of space according to social value is of critical importance to the host population.

Private spaces are the most closely associated with an individual's identity. The home in particular is an important part of one's self. Impersonal, highly public places do not usually constitute an integral aspect of an individual's identity. Sharing such a place does not entail sharing a part of one's social identity to the extent it does in a private place. Thus the separation of spoiled and normal identities is more easily accomplished. If the tramp walks briskly and purposefully through such an area he runs a fairly low risk of being arrested. By his rapid movement he makes little claim on space, thereby reducing the stigma he might otherwise spread. Sitting or standing, however, constitute much more of an involvement of one's identity with a place, i.e., it is an implicit claim that one somehow "belongs" there. The tramp therefore is rarely allowed to remain stationary in such an area for very long unless, as I have mentioned, he restricts himself to a few places such as the library and uses a prop to disassociate himself from others, so as to minimize the spread of stigma.

One could say that the prime areas with the highest social value are those residential sections with which middle to upper class individuals associate their identity. These areas convey positive information about such people. Other prime space includes the highly public places which are fairly neutral in terms of association with individuals' identities. In contrast, marginal areas are negative in the sense that they convey stigmatizing information about their inhabitants.

It is interesting to note that not all highly public areas are neutral in the above sense. There appears to be a generalized association of the identity of many propertied citizens with certain key places which are thought to represent the city. This association is generally termed civic pride. These key places are thought to be an important part of the city's presentation of self to outsiders. This is especially true when large numbers of tourists come to the city for special events or during the tourist season. At these times the city is on display and must be cleared of its unaesthetic elements.

> Drunks are also arrested in significant numbers before any large conventions, celebrations, or fairs. This is done primarily to "clean up" the area so it will not be unnecessarily offensive to visitors.

Thus the social value of prime space ebbs and flows at different times. The tramp cannot afford to be unaware of the temporal nature of the value of space.

Certain areas of the central business district of large cities, such as the Loop in Chicago, are thought to be continuously on display and therefore tramps are discouraged from ever going there.

> The main emphasis of the court is stay out of the Loop. That statement was made again and again, in fact in some instances a judge even stated, if you have to beg do it anywhere but in the Loop.

Skid row is thought of as an area which is morally bankrupt. The characterization of its inhabitants as morally defective leads to the conception of the row as an open asylum. People with "normal" propertied identities therefore should be kept out. The police have little sympathy for tourists who want to see the row or locals who want to take a "moral holiday" thereby engaging in illicit activity. These people cause trouble for the police and as such are discouraged from entering this marginal space.

CONCLUSION

Classification and use of space are intimately tied. One can describe the tramp's use of urban space; one can map it; however, one cannot begin to explain it without reference to the tramp's classification of various areas within the city. To understand his classification, one must examine the process by which he derives it. His classification of areas represents a plan of action, a plan which takes into account the action of others. Reconciling these lines of action requires negotiation which must be viewed in the context of power relationships and the ideology or concept of society held by the more powerful parties to the relationship. Occupying a very marginal place in the prevailing concept of society, tramps are in a very poor position from which to negotiate for rights to use space. Their classification of areas within the city is largely shaped by the prime/marginal distinction of the host group. Similarly their strategy of occupying marginal space is a direct result of the host's strategy of containment. The tramp, however, pays a price for using what is defined by the host group as marginal space. To quote sociologist Erving Goffman:

> A status, a position, a social place is not a material thing, to be possessed, and then displayed, it is a pattern of appropriate conduct, coherent, embellished, and well articulated . . . it is . . . something that must be enacted and portrayed.

By occupying marginal space, the tramp acts out and reconfirms his social marginality in the eyes of the host group. His strategy merely reaffirms the host's perspective and causes only minimal adjustments to be made in the host's moral order.

The division of the city into prime and marginal space as I have outlined it here is not in itself as important as the idea that the classification and use of urban areas by any group must be viewed in the context of that group's relation to other groups. Any group, including the most powerful, must negotiate with others, and this inevitably leads to compromise in their perspective. Thus the classification of prime and marginal, or relatively unsafe and safe, to look at the same distinction from the perspective of the tramp, should serve to indicate the relative or interactional nature of any regionalization of the city by any individual or social group.

"Race, Class, and Hurricane Katrina: Social Differences in Human Responses to Disaster"

from *Social Science Research* (2006)

James R. Elliott and Jeremy Pais

Editors' Introduction

In August 2005 Hurricane Katrina devastated the city of New Orleans and the Gulf Coast. Because of Katrina, 1 million people evacuated from the Gulf Coast, half of them being from New Orleans, and 1.36 million people filed for federal assistance. Compared to other natural disasters that have hit the United States, Katrina brought a new degree of devastation and out-migration, resulting in 707 temporary shelters in 24 states across the country. In this article Elliott and Pais look at how race and class affected people during and after the storm, and how the human response was affected by these variables.

Research from Hurricane Andrew in the late 1990s showed that class was the major factor in human response. The reason given for this is that race and ethnicity are linked to class, as they are determinants for resources such as income. This research showed that ethnic minorities were less likely to evacuate because of economic conditions. There is also evidence for race being the major factor in human disaster response. Research in the Deep South shows that even without bigotry, friendships, neighborhoods, churches, and other social groups are still ethnically divided.

Elliott and Pais compared race, gender, homeownership, and income from the Gallup polls with 2000 Census data in New Orleans. Their study empirically shows that the majority of residents evacuated before Katrina hit, but that there was a strong racial difference in evacuation timing. Black people were 1.5 times as likely to evacuate after the storm. Their evidence also showed that people whose income was $10,000–20,000 were twice as less likely to leave before the storm. Based on this, Elliott and Pais show that low-income blacks were more likely to remain in New Orleans (as opposed to people that were either black or low income but not both). Elliott and Pais also found that wealthier people were more likely to stay in hotels or rentals than with friends or family outside of the Gulf Region. Race was also found to have a large effect on disaster stress levels, with black people showing higher levels of stress than their white counterparts. Finally, Elliott and Pais addressed the difference of social support during the storm, where they found evidence that black people used religion for support most commonly, while white people used friends and family for support.

When Hurricane Katrina struck the Gulf Coast on August 29, 2005 with winds of 145 miles/hour, it damaged an estimated 90,000 square miles of housing throughout southern Louisiana, Mississippi, and Alabama. After the storm passed, news sources placed the total number of evacuees, or "internally displaced persons," at approximately one million, with nearly half coming from the city of New Orleans, which remained under full mandatory evacuation for weeks following the storm due to failed levee systems and subsequent flooding. The US Federal Emergency Management Agency (FEMA) has since reported that 1.36 million people filed for federal assistance as a direct result of the hurricane.

In the US history, there are no precedents for this degree of sudden devastation and out-migration from a major urban region. The Galveston Hurricane of 1900 killed an estimated 10,000 residents – the highest total of any US natural disaster – but the spatial impact remained highly localized. The Great Mississippi Flood of 1927 displaced nearly 500,000 persons, but it did not demolish a major city, instead ravaging wide, often disconnected swaths of rural hinterland. Other contenders – the 1906 San Francisco earthquake, which displaced 200,000 residents, and the 1871 Chicago fire, which displaced 100,000 residents – were catastrophic in their own rights, but neither resulted in near complete evacuation of their local populations, nor did they send so many people to so many places for relief. Two weeks after Hurricane Katrina struck the Gulf Coast, the American Red Cross reported operating 707 temporary shelters for Katrina evacuees in 24 states and the District of Columbia.

The devastating and seemingly arbitrary nature of disasters such as Hurricane Katrina can reinforce the popular notion that such events are random in their social dimensions. After all, if the physical infrastructure of our communities cannot withstand such catastrophe, how can the social infrastructure that also gives them shape?

The primary objective of this paper is to contribute to this line of research by examining the extent to which racial and class differences influenced human responses to Hurricane Katrina – the costliest natural disaster ever to hit the United States.

These historic developments have coalesced to produce a peripheral region characterized by deep and complex relations of racial and class division. Because comparably few "outsiders" of either native or foreign birth have moved into this area during recent decades, these relations have been left to unfold largely of their own inertia, undisturbed by mass in-migration from other parts of the country and the world. These events mean that the region devastated by Hurricane Katrina is very different from, say, San Francisco prior to the massive earthquake of 1989, or Miami prior to Hurricane Andrew of 1992, or Los Angeles prior to the brush fires of 1993, or Chicago prior to the heat wave of 1995.

Prior research leads us to expect that although residents of the Gulf South share a common region, their responses to Hurricane Katrina varied in non-random ways reflective of racial and class divisions that have taken root and grown in the area over time. In the present paper, we treat this expectation as a matter for empirical investigation. Focusing specifically on race and class, we ask which, if either, dimension of social life most differentiated human responses to Hurricane Katrina before and shortly after it hit the Gulf coast. Logically, the answer to this question can take one of three general forms: (1) class differences were more prominent than racial differences; (2) racial differences were more prominent than class differences; or (3) neither dimension was more prominent. We review grounds for each hypothesis below.

For decades, social scientists have debated which is more salient for explaining observed inequalities in US society: race or class? They have also wrestled with whether this "race versus class" framework is too simplistic for understanding the intricacies of social inequality, since race and class, while analytically distinct, constitute overlapping systems of social stratification that remain experientially entangled and causally circular. Our position is that both approaches – the analytically simple and the theoretically cautious – are useful: the first for identifying basic patterns of variation; the second for interpreting them.

A "class-trumps-race" perspective has appeared in recent research on hurricane response. For example, in their study of Hurricane Andrew, which hit the Miami area in 1992. Racial segregation creates communities of fate that can take on added salience in a disaster context. Race and ethnicity are linked to housing quality – not because of ethnically based cultural variations in housing preferences but because race and ethnicity are still important determinants of economic resources, such as income and credit, critical for obtaining housing. When faced with hurricane warnings, "Ethnic minorities are less likely to evacuate than Anglos probably as a result of economic conditions rather than race or ethnicity per se." In other words race "matters" but through more proximate, or direct, factors associated with class resources.

This perspective is echoed in urban sociologist Harvey Molotch's commentary on events immediately following Hurricane Katrina. Answering his own opening question, "Would so many white people struggling for life be ignored for so long?" Molotch writes that, "Racism explains some of what went on, but its route was indirect." Raising several possibilities, Molotch explains that, "One of the race-based explanations is that those left behind are consistently the most deprived. The legacy of slavery, exclusion, and segregation corrals those with least resources into a vulnerable space, natural, and economic."

Proponents of this perspective are not arguing that racial differences are inconsequential to human behavior and outcomes, even in times of natural disaster. Instead, they are asserting that what look like racial differences are more fundamentally class differences that are difficult to see without informed analysis. Supporting this perspective is the reality that thousands of working-class whites in St. Bernard Parish also suffered terribly from Hurricane Katrina. However, sparse settlement, difficult surface access, and imposition of a military no-fly zone helped to render their plight less visible to national media following the storm, thereby magnifying apparent racial divisions in human response to the storm and its immediate aftermath.

Critics of the "class" perspective contend that although economic resources certainly influence individual opportunities and outcomes, racial differences persist and shape how people organize, interpret, and respond to opportunities and outcomes around them. From this perspective, emphasis falls less on material differences among individuals and more on distinctive affiliations, institutions, and world views that inform modes of thinking and knowing and doing. We might call this influence loosely "culture."

In US society, especially the Deep South, few "axes of variation" are as salient as racial identities, especially those contrasting white from black. Research continues to show that while white bigotry and overt discrimination may be on the decline, close friendships, neighborhoods, churches, and social clubs remain highly segregated by race. These divisions are important for understanding human response to natural disasters because people respond to disasters not as isolated individuals but as members of these overlapping forms of social affiliation, which interpret, affirm and support particular definitions and responses to the situation. Moreover, research shows that these social units, particularly the family, are not restricted to local life but also influence extra-local networks called upon in times of crisis.

From this perspective, it is unsurprising that Kanye West, a wealthy pop star, who also happens to be black, seized the opportunity during a televised fund raiser for Hurricane Katrina victims to proclaim to the nation that, "[President] George Bush doesn't like black people."

For our purposes, the veracity of West's claim is unimportant. What matters is the fact that even if racial hatred ceases, persistent social patterns can endure over time, affecting whom we marry, where we live, what we believe, and so forth. These patterns, in turn, bind racial subgroups across class lines, helping to forge common responses to life events, including natural disasters, in ways that differ from racial "others" in the same region. From this perspective, "communities of fate" are bound as much by racial experiences and affiliations as by common material resources. This is not to say that class differences are unimportant, but rather that in times of crisis, class differences are likely to shrink and racial differences expand as individuals define, interpret and respond to the situation before them.

A final possibility is that racial and class differences mattered equally in short-term responses to Hurricane Katrina either because neither mattered or because both mattered to more or less the same degree.

During recent years this "race–class" debate has become particularly acute in the emergent literature on environmental justice. In this literature, researchers begin with the common and well-documented observation that low-income minority communities bear a disproportionate share of environmental hazards in our society, particularly when it comes to the siting of toxic facilities. They then proceed to disagree over the primary cause of these patterns. On the "race" side of the debate, scholars who charge "environmental racism" contend that industry and government officials locate environmental hazards in low-income, minority communities because these communities lack the social, political, and economic power to resist such treatment. On the "class" side of the debate, scholars counter that low-income minority communities often emerge and solidify around environmental hazards because property values are lower and opportunities for homeownership greater than in other parts of town. Under the first scenario, the key mechanism is racial discrimination in siting; under the second scenario, it is economic inequalities and market pressures that encourage "minority move-in."

In the Gulf South, researchers have documented ample evidence of both processes in the siting and development of toxic facilities (see Roberts and Toffolon-Weiss, 2001). However, as Hurricane Katrina attests, these are not the only environmental hazards in the region. In addition to polluting industries and toxic landfills, the region has long been spatially uneven with respect to elevation and flood protection, especially in and around the city of New Orleans. This unevenness, however, has never neatly conformed to racial or economic lines.

During the early 1900s, for example, the development and proliferation of new pumping stations allowed developers to drain and build new communities in New Orleans' traditional "low-land swamps" at the historic rear of the city, where significant African-American neighborhoods subsequently grew and solidified despite high vulnerability to flooding and levee failure. The same system of environmental modifications also enabled middle-class whites to expand toward the city's northern lakefront, where they used restrictive covenants to block African-Americans from the newly drained but still low-lying, and thus vulnerable, subdivisions abutting Lake Pontchartrain. Similar modifications later facilitated residential expansion into St. Bernard Parish, where working-class whites have since developed a strong and lasting attachment to land that remains largely below sea-level and compromised environmentally by the Mississippi River-Gulf Outlet (a.k.a. "Mr. Go"), a 76-mile navigational channel cut through local wetlands to permit large-hulled ships that now rarely traverse it.

These historic and geographic developments mean that exposure to the environmental hazards of hurricanes and flooding, while geographically uneven throughout the region, are not race- or class-exclusive. Large numbers of affluent and working-class whites and blacks all lost homes, jobs and community when Hurricane Katrina hit and the levees failed. To assess their responses to this disaster, and the extent to which race and class differentiated these responses, we now turn to the data.

To gauge this assumption empirically, we compared race, gender, age, homeownership, and income distributions for Gallup respondents from the City of New Orleans with data available for the City of New Orleans in the 2000 census. We chose this spatial comparison because it provided more reliable spatial boundaries and estimates than comparisons made across the entire affected region. Results (not shown) suggest that the Gallup survey oversampled women in the affected region by a rate of roughly 5% and oversampled blacks, non-senior citizens, and non-homeowners by roughly 15%. Results also indicate that the average household income among respondents in the Gallup survey is approximately 14% lower than that recorded by the 2000 census. These comparisons suggest that minorities and less affluent residents were indeed more likely to be sampled by the Gallup survey than whites and more affluent residents. However, they also indicate that this bias is not extreme and that the sheer number and diversity of respondents in the Gallup survey is sufficient to estimate accurately basic social differences in human responses to the storm. The fact that we statistically control for many of these differences in our regression analyses further minimizes this bias, leaving us with conservative estimates of race and class differences among residents of the affected region region (see Winship and Radbill, 1994).

To test general hypotheses regarding racial and class differences in human responses to Hurricane Katrina, we use logistic and ordinary least squares regression to estimate the effects statistically attributable to indicators of each. Because the City of New Orleans experienced not just the immediate wind and rain brought by Hurricane Katrina but also the after-effects of levee failure that subsequently flooded 80% of its land mass, events inside and outside the city differed markedly. For this reason, we estimate our models of evacuation timing and current housing and job statuses separately for respondents from the City of New Orleans (n = 331) and for respondents from outside the city (n = 963). Thereafter for models of stress, sources of emotional support, and likelihood of return, we pool all respondents and include a dummy indicator of city residence for theoretical reasons discussed below.

Our study confirms that the vast majority of residents in the affected region evacuated before the storm and that this response was actually more common in the City of New Orleans than outside it (70% compared with 66%). Consistent with subsequent levee failures, results also indicate that 96% of New Orleanians eventually left their homes, compared with only 80% of survivors residing outside the city. Our research reveals strong racial, but not class, differences in evacuation timing outside New Orleans. Blacks outside the city were 1.5 times more likely than similar whites to evacuate after, rather than before, the storm. Within the city, there were also strong racial differences, but these were restricted largely to a small subpopulation that reported never leaving the area – a subpopulation comprised almost entirely of African-Americans. Beyond this small group (less than 5% of the city's sample), household income played a strong and consistent role in predicting evacuation timing from the city. Appropriate calculations show that independent of other factors, New Orleanians with household incomes in the $40,000–50,000 range were nearly twice as likely as those in the $10,000–20,000 range to evacuate before, as opposed to after, the storm. This class difference climbs to nearly threefold when predicting odds of not evacuating the city at all.

Overall, these findings support both the race and class hypotheses and confirm that it was specifically low-income blacks, not blacks or low-income folks generally, who were most likely to remain in New Orleans through the disaster. Media accounts have offered numerous explanations for this type of response: inadequate personal transportation; limited spatial networks; lack of hotel reservations; and desire to remain in the city to collect government checks dispensed at the end of the month. Follow-up questions in the Gallup survey indicate that the most common reason that New Orleanians gave for not evacuating prior to the storm was that they did not believe the hurricane would be as bad as it eventually was (49%) – more than double the share reporting that they were too poor or lacked the necessary transportation to leave (21%). Further analysis indicates that this disbelief did not vary by race or class within the City of New Orleans, which suggests that income played a key role in predicting not only who left prior to the storm but also who was able to leave during and after the storm hit: wealthier stragglers eventually left the city under their own power, often from unflooded parts of town; poorer stragglers awaited help that was slow in coming to their deeply flooded districts.

Regardless of how and when residents evacuated, they need to recover, and housing and employment are critical to this process. Beginning with housing, our analysis shows that one month after the storm, only 5% of New Orleanians reported being back in their own homes, compared with two-thirds of non-city residents. Among remaining New Orleanians, most reported living in an apartment, hotel or temporary shelter (53%) or in someone else's home (42%).

Home ownership played a strong and consistent role in predicting respondents' housing situation a month after the storm. Outside the City of New Orleans, renters and boarders were 3.5 times more likely than homeowners to report living in someone else's home and 2.9 times more likely to report living in an apartment, hotel or other temporary shelter. But for New Orleanians, these differences are more stark, with renters and boarders 6.8 times more likely to be in someone else's home and 7.9 times more likely to be in some sort of post-storm rental. Results also indicate that the latter odds – most notably hotel and apartment rental – were higher among wealthier than poorer New Orleanians, that is, among those who could more readily afford market alternatives. This finding is consistent with prior research, which has found that the lower the social class of the family, the greater the tendency to evacuate to homes of relatives.

Seventy-five percent of all homeowners who reported their properties as "damaged but still livable" had already returned to them one month after the storm. Among renters, the same return rate was less than 25%. This lower rate may reflect not only lower attachment to place among renters and boarders but also less power over decisions about whether to re-enter and/or re-develop damaged properties. Landlords, for example, may be reluctant to readmit renters for a variety of reasons, including a fear of lawsuits involving unsafe conditions, a lack of capital and/or labor needed to repair damaged units, a desire to break leases and re-rent at higher rates in a high-demand market, and/or plans to demolish and rebuild alternative housing on the same space.

Of course, employment dynamics will also affect people's return to pre-storm communities. Among New Orleanians who were employed at the time of the storm, only a quarter report having the same job one month later, compared with over two-thirds of respondents from outside the city. Among workers from outside New Orleans, renters faced greater disadvantage than homeowners. Renters were 1.9 times more likely than homeowners to report having lost their jobs, although some have since secured alternative employment. Within the city, where home ownership rates are generally lower, this indicator of class resources mattered little for post-storm employment outcomes. Instead, the chief factors were race and income. All else equal, black workers from New Orleans were 3.8 times more likely to report having lost their pre-Katrina jobs than white workers. Moreover, if these same black workers had a household income of $10,000–20,000, they were nearly twice as likely to have lost their jobs as black workers from household incomes of $40,000–50,000. In other words, as prior labor market research has repeatedly documented, low-income blacks specifically – not blacks or low-income workers generally – have the most tenuous hold on their jobs, leaving them highly vulnerable to joblessness in times of disaster.

When such disasters do occur, individuals understandably become stressed, and prior research suggests that this stress tends to be higher in technological disasters than in natural disasters. This pattern is pertinent to Hurricane Katrina because many observers now view events within the City of New Orleans as primarily a technological disaster (levee failure) and events outside the city as primarily a natural disaster (wind, rain, and storm–surge destruction). To the extent that this general distinction is meaningful, it would suggest that New Orleanians may experience greater post-disaster stress than residents outside the city.

Race, not class, has a strong influence on post-disaster stress associated with Hurricane Katrina, with blacks generally reporting higher stress levels than whites, all else being equal.

In addition to these core concerns, our findings also indicate several patterns worth acknowledging: women report more stress than men following the storm, as do parents, residents with severely damaged housing, those not yet back home, and those without jobs. Results also indicate that, all else equal, residents who evacuated express more concern about the future than those who did not evacuate, perhaps raising important issues for future evacuation planning.

With this in mind, it is useful to examine where respondents turned for emotional support during the weeks immediately following the disaster.

Results once again reveal strong racial, but not class, differences. In general, whites are much more likely to report relying on family or friends to get through these difficult emotional times, whereas blacks are much more likely to report relying on religious faith. Further investigation indicates that this reporting of religious faith (and downplaying of family and friends) was high even among blacks who reported staying in someone else's home after the storm. Among whites, the opposite is true: those who reported religious faith for emotional support were unlikely also to report staying with or relying on family or friends.

Among blacks in the region, religious faith seems to be part of a communal glue that individuals and families use to cement social relationships and to ensure the emotional support that such relationships provide. So even when helping one another to cope, African-Americans from the area are likely to report that it is God that has made this type of network assistance possible, not family or friends themselves. Among whites, by contrast, self-reports about sources of emotional support seem to be more exclusive or dichotomous. If family or friends helped with evacuation, then whites are likely to report them as an important source of emotional support, and not religious faith. However, if family or friends did not help with evacuation, then whites are more likely to report religious faith, and not family and friends, as an important source of emotional support.

Like other disasters before it, Hurricane Katrina offers a unique laboratory in which to study the social infrastructure of its affected region. In this case, it is a region cross-cut by deep and complex divisions of race and class that have hardened over time without direct, excessive interference from outsiders. As media images streamed from the region to the nation and the world following the storm, a public debate emerged over the relative importance of class and race for shaping individual, as well as institutional, responses to the disaster.

In the present study, we have sought to fill some of this empirical gap by using data from the largest, most comprehensive survey of Hurricane Katrina survivors currently available. Overall, results indicate that both race and class played important roles in shaping these responses and that neither can be readily reduced to the other. Thus the real issue is not either/or, but where and to what degree.

With respect to race, there are two broad areas where racial differences seem to have mattered. The first involves timing of evacuation and sources of emotional support, that is, behavior more or less under the control of individuals themselves. Our findings indicate that blacks across the region were less inclined than whites to evacuate before the storm, mostly because they did not believe that the hurricane would be as devastating as it eventually was. Previous experience and public assurances suggest that this personal risk assessment may not have been as irrational as it now appears. Reports indicate that had the levees been built and inspected with the integrity typically expected of the Army Corps of Engineers, they would have likely survived the storm, sparing the city from the massive flooding that eventually covered 80% of its area.

As for emotional support, our findings indicate that blacks and whites differed, at least over the short term. Specifically, blacks were more likely to report "leaning on the lord" while whites were more likely to report relying on friends and family. We have suggested that this difference might be more a matter of interpretation and world view than actual differences in network support. Another possibility is that blacks' friends and family were more likely to be adversely affected by the storm and even more widely dispersed than whites', making them more a source of concern than support. Both scenarios could easily have worked together to produce the strong racial differences observed throughout the region.

The second and more troubling set of racial difference involves something largely outside survivors' control, namely job security. Our findings indicate that black workers from New Orleans were four times more likely than white counterparts to lose their jobs after the storm, all else equal. But of course, all else is not equal. When we factor income differences and their effects into the equation, results indicate that the "average" black worker in New Orleans is actually closer to seven times more likely to have lost his or her job than the "average" white worker. This disparity will certainly have a strong effect on who is able to return to the city as it rebuilds and who is not.

This issue of return is also where class standing, specifically home ownership, exhibits its strongest and most consistent effect. We suspect that this effect cuts two ways. On the one hand, home ownership provides survivors power over when and to what extent personal return and rebuilding will occur; on the other hand, it can also create a financial weight in the form of mortgage obligations that limit resettlement options elsewhere. This interpretation is supported by the consistent effects of home ownership across the region and by aggregate analyses which indicate that less affluent homeowners are more likely to say they will return than more affluent homeowners. This pattern is also consistent with findings from Hurricane Andrew in the Miami area during 1992.

Overall, these findings refute the apparent randomness of natural disasters as social events as well as the notion that racial differences are somehow reducible to more "fundamental" class divisions when considering human responses to such disasters. Both "axes of variation" – race and class – appear to have mattered in response to Hurricane Katrina, and while the entire region will continue to require the nation's ongoing support for years to come, results here indicate that it is low-income homeowners in particular who will need the most assistance in putting their lives and the region back together again.

In addition to these and related assistance programs, more general efforts to improve policy and planning initiatives for future disasters may benefit from considering the following possibilities. First, with respect to evacuation, our results affirm that poor inner-city residents are often the least likely to heed formal evacuation warnings, some because they lack transportation and others because they fail to take such warnings seriously. Our findings regarding the centrality of religious faith for racial minorities, women, and the elderly coupled with the negative association of this centrality for early evacuation, suggest that emergency planning initiatives can be improved by assisting local civic and faith-based organizations in developing a coordinated, grass-roots system of hazards education and warning dissemination. The basic idea would be to buttress top-down warnings with ongoing planning and preparedness orchestrated through trusted local associations, similar to how school teachers help to educate and evacuate their own groups of students when an ominous but distant fire bell sounds and the entire school must evacuate. At a regional level, such efforts would require a great deal of organizational creativity, money, and time, but if communities are serious about disaster mitigation and saving lives, such investments seem well worth the expense, effectively reinforcing official proclamations in times of emergency with bottom-up planning and social organization.

Surely these will not be the last or only policy lessons proffered in the wake of Hurricane Katrina, but one thing does appear certain: How the nation responds to this current and ongoing problem will help to define it not only as a society but as a civilization capable of communal expression of awakened conscience.

PART FOUR

Gender and Sexuality

Plate 6 Commuters in Hong Kong MTR Train, by Minjoo Oh.

Plate 7 Gay Pride Parade, New York City, USA, 1998 by Nikos Economopoulos. Reproduced by permission of Magnum Photos.

INTRODUCTION TO PART FOUR

Cities are not just landscapes of socioeconomic and racial/ethnic inequality; they are also terrains of inequality by gender and sexuality. While the urban underclass and homeless experience spatial isolation and containment in the marginal spaces of the inner city, women have since the postwar era experienced spatial entrapment in the suburbs. Suburbanization may be seen as a form of segregation. Only more recently have women found greater spatial mobility with their entry into the labor force and the widespread gentrification movements occurring in many central cities. The "return to the city" has been correlated with a set of demographic transitions surrounding a decline in traditional male-led households with children, and the growth of double-income no-children households, and double-income gay and lesbian households, as well as single households. The growing visibility of gays and lesbians in American society is partly correlated with the growing visibility of gay and lesbian communities in the central city.

Susan Saegart, in an essay titled "Masculine Cities and Feminine Suburbs" (*Signs* [Spring 1981]: S92–S111), establishes the structural and symbolic dichotomy of women in the suburbs and men in the cities. The symbolism of suburban women has been promoted in a range of cultural representations from advertising, literature, to film and television. This kind of symbolism reinforces inequalities of power in economic and urban life. While the suburbs are the domain of women, family, and private life, cities are the domain of men, work, and public life.

In her essay, published in the same 1981 issue of *Signs*, Ann Markusen elaborates further on the patriarchal structure of our cities. The separation of residential life from the central city in single-family households in the suburbs is the creation of a patriarchal capitalist society that differentiates and reproduces an understanding of men's work as waged work and household work as unpaid women's work. Economists define households as places of buying power and consumption, not production. Even the notion of the "journey to work" implies the household is not an area of production values.

Markusen blames patriarchy for the spatial segregation of women in the suburbs, not just the benign operation of the real estate market, the growth of automobile transportation, and the activity of the Federal Housing Administration in the postwar era. Suburban life leads to inefficiency and waste through increased commuting time and energy consumption, and alienation in household life. Suburbs erode extended family and community networks and other forms of social capital. They militate against the collectivization of household work and replace parks and collective play space with private yards. Patriarchy further serves capitalism by dividing the workforce and blunting class-consciousness. Patriarchy does reinforce the power of men as heads of households and primary decision makers, and eaters of better home-cooked meals. The suburbs reinforce the power of American individualism and privatism in the social and political culture.

Markusen points out that new urban trends are disrupting the old dichotomy of masculine cities and feminine suburbs, including: a) the onset of gentrification along with the breakdown of the traditional patriarchal household, b) the growth of retired households and retirement communities, and c) the growth of small and medium-sized towns as households flee large congested cities. She supports a range of urban social policies to promote the interests of women and families, including: a) fostering child care in extended family arrangements or neighborhood cooperatives, b) fostering the development of non-patriarchal collectivized affordable housing, and c) enhancing the efficiency of the household production sector, through combined live/work spaces, as seen in the "loft spaces" of urban artistic colonies. We may speculate also on the impact of the Internet on urban space and participation of women in our society. Electronic connectivity through the World Wide Web breaks down distances of time and space and permits economic participation from places far from urban central places through telecommuting and the growth of non-place-based Internet enterprises. The Internet affords some opportunity for women to "cocoon" in residential space while participating in economic and cultural life.

Melissa Gilbert offers a revision of the "spatial entrapment" thesis through her concept of "rootedness," which suggests that spatial limitations on women can confer resources as much as creating constraints. Where white feminists are typically more focused on improving the social and spatial mobility of women, Gilbert's perspective may be associated with the "womanism" of Alice Walker and the writings of bell hooks on home places. Gilbert finds that for low-income women of color living in the inner city, there is a certain power of place that they procure due to the density of social capital networks. Mobility is normally associated with empowerment for women, but relative "place-based" immobility may have certain advantages. In her fieldwork with low-income women of color in Worcester, Massachusetts, Gilbert finds that being close to a dense network of kin and friends improves their ability to procure child care. These social networks provide access to information about a variety of other social capital resources, including employment, housing, education, health care, and church. She finds that race intersects with gender to shape the spatiality of daily life and determine economic security.

Sy Adler and Johanna Brenner address some related issues in their article on lesbians and gay men in the city. Whereas gay men have recently raised their spatial and political visibility in communities such as Greenwich Village, New York, and the Castro district of San Francisco, lesbians have been relatively invisible, with less of a presence in terms of bars, businesses, festivals, and politics. This lesser visibility is not a symptom of absence, Adler and Brenner point out, but an expression of a different form of social networking and identity politics. They point out that women have historically played an important role in community politics and urban social movements. This activity may involve non-territorially based interpersonal networks and non-placed-based political issues.

Donald Donham explores the transformation of gay urban life in Soweto, South Africa with the end of the apartheid system of racial segregation. During the apartheid period, gay South African black laborers commuted from the townships in order to work in the white-dominated economy, and took up *skesanas*, sexually submissive males, as "wives." While racial segregation fostered a sexual culture in which these black laborers closeted their homosexuality, they have become liberated in the post-apartheid period to publically assume their gay identity, bringing what Donham calls a "modernization" of sexuality.

Finally, Sirpa Tani examines the complex range of relations between gender and sexuality stemming from the informal sex economy in Helsinki, Finland. Tani examines how media attention and the government's official representations of prostitution affected the everyday lives and ordinary social interactions of area residents, especially women. Indeed, female residents were represented in the media either as non-sexual victims of prostitution or "overtly sexual beings" (that is, as prostitutes). Partly in response to such claims and in an effort to reclaim control over the identity of their neighborhood, residents mounted campaigns to combat their community as being defined solely as a sexualized space. Spaces of prostitution are socially constructed, Tani informs us, but they are also individually experienced. In this instance, community activism on the issue of prostitution was less a product of the shared interests of residents and more the consequence of the experiences of women residents stigmatized in the course of their everyday lives.

"City Spatial Structure, Women's Household Work, and National Urban Policy"

from *Signs* (1981)

Ann R. Markusen

Editors' Introduction

The Women's Room is Marilyn French's 1977 classic tale of suburban isolation, in which the main character, Mira Ward, questions the accepted social norms regarding women's limited place in society. "The school had been planned for men, and there were places, she had been told, where women were simply not permitted to go. It was odd. Why? she wondered. Women were so unimportant anyway, why would anyone bother to keep them out?" After many twists and turns, Mira Ward's journey ends up embracing the relevance of women's work – unappreciated and unnoticed as it is – to the world at large.

In this selection, Ann Markusen directs our attention to work within the (mostly suburban) home – the "household reproduction of labor" – which has largely been ignored by classical urban scholars. She argues that the single-family home, geographically isolated in suburbs and from workplaces, is a product of the patriarchal (male-dominated) organization of household production. Hence, she leads us to the analysis of gender relations, not as *shaped by* urban form, but as *shaping how* cities and suburbs are built and ordered.

As Markusen points out, social reproduction (the production of the workforce) within the household should be considered an economic activity because it contributes to the overall reproduction of society. The cooking of meals and the provision of personal services, education, and child care are all labor activities that contribute to the maintenance of society. However, the compensation for labor time of the household workers – primarily women – is quite different from the terms under which other forms of labor are hired and organized. Household reproduction rests on the persistence of patriarchical gender relationships. More than half of all adult women under age sixty-five work for wages but they still bear the primary responsibility for household work. The economic nature of this activity, Markusen argues, is largely hidden by the informal economic contract involved (marriage). Patriarchy is the organizing principle of household labor and social production, hence urban spatial structure is a product of patriarchy as much as it is of capitalism.

Identity-based social movements, such as the women's movement, have challenged patriarchy and the power dynamic between men and women it entails. And in doing so, the isolated household domain akin to suburbanization has given way to alternative spatial developments. Women have mobilized to change land-use laws and homeownership financing and have been instrumental in the planning of urban mass transportation. Nonetheless and despite such efforts, Markusen concludes, patriarchy continues, as does the dominant cultural preference for the suburban cul-de-sac.

While it may be commonplace for us now to think of household labor as productive labor (if not hard work), we should recognize Ann Markusen's work as an early call to urbanists to include gender as an important variable in their research. Her work inspired a generation of urbanists, especially feminist geographers, to take up the task of critically interrogating the interplay between gender and spatial forms. Other early key texts on gender and urban space include Dolores Hayden's "What Would a Non-Sexist City Be Like? Speculations on Housing, Urban Design, and Human Work"; Gerda Wekerle's "Women in the Urban Environment"; and Susan Saegert's "Masculine Cities and Feminine Suburbs: Polarized Ideas, Contradictory Realities" – all of which appear in *Women and the American City*, edited by Catherine R. Stimpson, Elsa Dixler, Martha J. Nelson, and Kathryn B. Yatrakis (Chicago, IL: University of Chicago Press, 1981).

Ann Markusen is Professor of Planning and Public Policy at the University of Minnesota. She specializes in the areas of regional economics and planning and the high tech and defense industries as stimulants of regional development. Her books include *From Defense to Development? International Perspectives on Realizing the Peace* (London: Routledge, 2004) and *Second Tier Cities: Rapid Growth Outside the Metropole in Brazil, Korea, Japan and the United States* (Minneapolis, MN: University of Minnesota Press, 1999).

This chapter investigates the interrelationship between city spatial structure, women's household work, and urban policy. It first differentiates between two types of work in urban space: wage-labor production and household reproduction of labor power, generally and incorrectly ignored in analyses of urban spatial structure and dynamics by neoclassical location theorists and Marxist urbanologists alike. I contend that social reproduction, organized within the patriarchal household where an unequal internal division of labor favors men, profoundly affects and explains the use of urban space. The paper then presents a theoretical argument regarding the evolution of contemporary urban spatial structure. It argues that the dominance of the single-family detached dwelling, its separation from the workplace, and its decentralized urban location are as much the products of the patriarchal organization of household production as of the capitalist organization of wage work. While this arrangement is apparently inefficient and onerous from the point of view of women, it offers advantages to men and poses contradictions for capitalism.

Challenging such patriarchal structuring of urban space are residential choices that certain demographic groups are currently making. New spatial developments – such as retirement communities, gentrification (urban renewal for upper- and middle-income households), and the growth of small towns – suggest that the dominant urban decentralized form of housing and land use may pose major obstacles to efficient household production. A second, and less anarchic, challenge is from the women's movement. Since the 1960s women have organized to change land-use laws, to restructure the patterns of housing ownership and housing-unit structure, to form urban cooperatives for child care and other sharing of household work, and to restructure the urban transportation system. Nevertheless, current patterns of homeownership, real estate construction practices, and the permanency of urban physical structure are formidable barriers to the nonpatriarchal restructuring of urban space.

[. . .]

PRODUCTION AND SOCIAL REPRODUCTION IN URBAN SPACE

Human energy is largely spent in one of two activities: the production of commodities for market exchange and the reproduction of labor power. The former is organized, at least in capitalist society, within the institution of plants, shops, or offices, where employers hire workers (i.e., buy their labor power) for wages in order to produce commodities such as food, clothing, and shelter, machines, insurance, and so on. These are then sold to consumers or other employers for prices that more than cover wages and costs of production. Both neoclassical and Marxist economic analysts, studying the relationships among employer, worker, and output, seek to explain the conditions under which workers enter the labor market (labor force participation), under which their labor power may not be purchased by employers (unemployment), and by which certain levels of output are forthcoming from the use of certain production processes (productivity).

Similar studies look at the composition of output as a function of both demand (consumers' purchases) and supply (production conditions), and at the savings and spending behavior of workers, who receive wages, and of corporations and capitalists, who receive profits. Such studies contribute to the construction of aggregate Keynesian indicators, such as Gross National Product, and to interpretations of their causation that will guide a government intervention meant to perfect the operation of the production sphere. Marxist accounts emphasize different features of the same production process, particularly the exploitation of labor power and the inherent tendency toward crisis. Marxists consequently derive opposing views of the causes of capitalist economic problems, predicting that a socialist transformation is required to solve production sphere problems.

In the urban setting, these production concerns take on a spatial dimension. As urban economic problems, production problems become those of the transportation of employees to work (the journey to work), of the availability of desirable production space (including rights to pollute), of the fit of skills of a local population to regional industrial structure, and the multiplier process whereby locally generated income and locally stimulated production result in further increases in output and income in secondary and tertiary sectors. Neoclassical urban economic studies, at both the micro and macro level, aim at so characterizing the production process in urban space that government policy can substitute in cases of market failure (pollution control), strengthen or subsidize the market where it is weak (transportation), and stimulate the aggregate level of activity in a local economy when it is recessed (countercyclical revenue sharing). In Marxist versions, urban problems stem from the very structure of capitalism, because both economic crisis and class conflict produce contradictions in urban structure and governance.

While these are generally considered to be the "economic" realities of urban life, the sphere of social reproduction, or reproduction of the production sector workforce, is equally economic in the sense that it, too, requires labor time. Social reproduction involves the activities of both government and households that reproduce both current and future generations of labor power. These include the direct provision of the conditions of physical and mental health, cooked meals, personal services, education, maintenance of living conditions, and child care. However, the organization of social reproduction, meaning the context surrounding the use and compensation for labor time of the household workers involved, is quite different from the terms under which labor is hired and organized in the capitalist wage-labor sector.

Of the two basic institutions involved in social reproduction, government hires workers in much the same way that private businesses do. However, government does not compensate laborers strictly on the basis of productivity, does not aim to make a profit on their labor, and does not derive revenues from the sale of labor-produced output. It instead raises funds via a complex taxation system. Demands for public sector outputs are registered, not by a market process, but by a political system where voter-elected representatives determine levels of output and tax payments. Much more needs to be understood about the way in which the public sector operates, and the ways in which the structuring of urban space conditions the efficiency and equity of the local public sector. In minor ways, the recently proposed national urban policy addresses certain of these conditions by proposing incentives to metropolitan areas to adopt tax-sharing schemes and by requiring neighborhood citizen participation in certain public sector urban programs. Another subject worthy of research is the interrelationship between household production and local public sector structures.

Of greater interest to us now, however, is the structuring of social reproduction in the household realm of urban society. Most economists and other social scientists treat the household as a social and economic unit without disaggregating it to its individual members. Conventional economic analysis treats the household as a consuming and resource-owning unit, whose sole relationships to wage-labor and commodity markets consist of supplying labor to and consuming the output of the commodity production process. Similarly, Marxist analyses of the household under capitalism have emphasized its function as a private sphere absorbing the alienation of the capitalist workplace and even as a triumph of the working-class family preventing complete proletarianization. Both types of analysis obscure the role that household members, chiefly women, perform in the reproduction of labor power and the social relationships of the household in which it takes place.

In urban space, the location and organization of households are generally considered of economic interest only as the locational source of labor power and buying power. Neoclassical studies hypothesize that household-location decisions are primarily a function of the journey to work and secondarily reflect preferences for accessibility to open space, good public services, retail markets, and housing quality. Even the term "journey to work" suggests, incorrectly, that no work is done in the household. The locational outcomes are always presented as utility-maximizing decisions undertaken by the household as a whole. They have not been analyzed as decisions per se about production input required in the reproduction of labor power (e.g., minimizing travel time to schools, to health care, to markets, and to recreation). Nor has household-location analysis considered conflict among the members of the household, particularly between those who are primarily engaged in the capitalist labor market and those primarily responsible for social reproduction activities (especially when working as wage labor also), that is, conflict between men and women or husbands and wives.

The Marxist feminist literature is beginning to correct these omissions. It argues that the household is not a passive consumption unit, but one in which people reproduce their labor power, of both current and future generations, through a process that involves considerable male/female division of labor, extensive expenditure of labor time, and particular composition of output that has its own quality and distributional patterns. Even though more than half of all adult women under age sixty-five work for wages, they still bear the primary responsibility for household work. The products of their labor are meals, clean and mended clothes, home health care, preschool education of children, financial and transportation services, and so on. Yet, the economic nature of this activity is largely hidden by the informal economic contract involved. Marriage, this view argues, is an implicit rather than an explicit contract for the exchange and organizational control of labor power in the household.

HOUSEHOLD PRODUCTION, WOMEN'S ROLES, AND THE STRUCTURING OF U.S. URBAN SPACE

The most striking aspects of modern U.S. city spatial structure are the significant spatial segregation of residence from the capitalist workplace, the increasing low-density settlement, and the predominant single-family form of residential housing. Most contemporary analysts variously ascribe these developments to the rise of mass production techniques, to the automobile, to FHA mortgages, and to class and racial segregation. None of the analyses mentions women's household work, household social relations, or patriarchy as primary determinants of this structure. Feminists critical of urban structure claim that current forms are inefficient for social reproduction and reinforce women's roles as household workers and as members of the secondary labor force. Yet no one has systemically critiqued the myopia of both neoclassical location theory and Marxist urban spatial theory by documenting the centrality of the patriarchal form of household organization as a necessary and causal condition responsible for contemporary urban structure and its problems.

[. . .]

Our more specific concern here is with the spatial form produced by the conjuncture of patriarchy and capitalism in advanced industrial countries. By patriarchal structuring of social reproduction, I follow the definition of Hartmann, in which the labor power of women within a capitalism system is employed partially or wholly in the service of men in the household and where the returns to both women's and men's labor are contained in the family wage, which the man controls. Women's work involves the same basic activities that occur in capitalist production, but organized differently. A woman produces a meal for the household by purchasing raw material inputs at the grocery store or by growing a garden, combining them with her labor time, energy, and machine power provided by kitchen equipment, and serving them to household members. In contrast, her hired counterpart in the restaurant may do the same things, but her service is sold for a price that covers both her own wages and her employer's required return on investment. The productivity and efficiency of household production in the former case are just as important as in the latter, except that in the latter the market test is the willingness of the consumer to buy the restaurant meal at the price covering costs, while the quality and costliness of the household meal are directly "tested" by household members' satisfaction, and the implicit price is represented by the expenditure of household income and time.

The organization of household production has not been without its students and efficiency-promoting studies. Beginning in the middle of the nineteenth century, home economics specialists investigated ways of increasing the efficiency and quality of household production. They introduced mechanized techniques (sewing, washing, and dishwashing machines; vacuum cleaners; etc.), and they rationalized organization of women's time, using methods reminiscent of scientific management in the factory. Such efficiency concerns, however, have accepted the nuclear household as the unit of analysis. Most home economists have confined their prescriptions to changes within the household, rather than to changes in the size, composition, or spatial organization of households in urban places, including their juxtaposition to other institutions that supply inputs into the household production process. A few considered collective kitchens and apartment hotels, but for the most part dramatic changes in household organization were championed by utopians: Charlotte Perkins Gilman, for instance, in *Herland*, her feminist utopia originally published in 1915, and Melusina Fay Peirce in her 1868–71 campaign for cooperative housekeeping in Cambridge, Massachusetts.

Similarly, few treatments of optimal household location in urban space try to account for the maximization of efficiency in household production. Most location models assume that the household chooses between "*the* journey to work" (implying only one wage worker) and various consumption goods such as housing, open space, and public sector services. Gravity models (models that weight the multiple spatial orientations of the firm, e.g., toward sources of inputs and markets for outputs, by the transportation costs between each in order to predict the profit-maximizing location) have not been developed for intraurban household location. The exclusion of household production from the calculus for urban spatial location is damaging, because it may lead to erroneous conclusions about the ability of certain urban-policies to affect spatial form or to a blindness regarding the vulnerability of urban form to changes in household structure.

The fundamental separation between "work" spheres and home corresponds roughly to the division of primary responsibility between adult men and women for household production and wage labor, at least historically. Since patriarchy is the organizing principle of the former sphere, urban spatial structure must be as much a product of patriarchy as it is of capitalism. Patriarchy may thus contribute to and condition urban problems. The recent literature by feminists critiquing urban structure, both the single-family household and its spatial decentralization, argues that current patterns are inefficient for women because they result in wasted labor time and curtailed access to jobs and other facets of urban life. Popular accounts, such as Betty Friedan's critique of women's household experience in *The Feminine Mystique* and Marilyn French's detailing of suburban housework in the bestselling novel *The Women's Room*, also document the waste and alienation inherent in the current structure of urban housing, suburban neighborhoods, and intraurban transportation networks.

[. . .]

If these contentions regarding the inefficient structuring of urban space are true, why do women continue to choose household production roles in the existing urban structure? Primarily because their choices are so limited. In household decisions men and women are conditioned by the limited range of their options, a fact which conventional location analysis obscures. The major force encouraging women to remain in the household sector is the inaccessibility of jobs in the capitalist labor market, and the occupational segregation that keeps women's wages at strikingly low levels and women confined in low-skill, no-advancement jobs. These conditions, in turn, have been traced to patriarchal compromises struck between male workers and capitalist employers, where the primary incentive for male workers to oppose women's incorporation into wage labor was the level of service and power men commanded in the household. Even when women work for wages, and a majority do, their lower pay reinforces their role as supplemental breadwinners for the household, and they continue to bear the primary responsibility for housework.

A second, and tempering, observation must be made. Given the quality and paucity of the alternatives, women's control over their own working conditions may be greater in the household workplace than in a factory or service establishment. Of course, this may not be universally true; some husbands require extraordinary cleanliness and service on command, and some use physical violence to enforce their demands. Household production organized in single-family units permits some variation of tasks and development of a higher level of skills, compared to the specialization and deskilling generally enforced in the lower ranks of the labor market, a station to which most women would be assigned if working as wage labor. This consideration of quality of working conditions cautions us against concluding that household production per se is inefficient. A feminist society, in which oppressive household divisions of labor were eliminated, might still choose to perform many tasks in the reproduction of labor power within a modified form of household.

But, if contemporary urban spatial structure is dysfunctional for most women, then it must be functional or efficient from the point of view of capitalism or patriarchy or both. To distinguish between the two, we must look at the ways in which the conditions governing women's household production in urban settings have evolved with developments in the capitalist workplace and in the patriarchal household. Capitalist structuring of twentieth-century society has profoundly influenced the household, both directly through its requirements for the type of labor power to be reproduced and indirectly through its organization of commodity production. Increasingly, because of its hierarchical elaboration of the labor force, the capitalist production sphere required dramatically different conditions for the rearing of children. This has resulted in class-differentiated neighborhoods with tremendous responsibility for child care devolving upon women in individual households. Women may collaborate in the decision to live in suburbia, because, given the options, suburban locations offer safer, less crime-ridden, and less racially tense environments for both blacks and whites for bringing up children. Furthermore, since the types of household labor involved in caring for children are dissimilar from those involved in reproducing the labor power of adults, it may be that differentials in quality, productivity, and required labor time result in the relatively high incidence of households with small children in metropolitan suburban locations.

Second, capitalist organization of land and housing as commodities brokered by the real estate industry, built by the construction industry, and financed by the banking sector, has found current patterns of residential decentralization and single-family dwellings profitable. However, such profitability need not require this particular spatial form. For instance, in the absence of the patriarchal nuclear family, these same industries might have made just as much profit off the construction, sale, and financing of high-density, urban apartment-like living quarters accommodating various forms of household composition. It is true that the single-family form of housing is more amenable to the type of recycling of urban space that occurs with block-busting, speculation, and continual construction and destruction of housing values. Nevertheless, this argument cannot explain the single-family unit; more accurately, it reflects its popularity. The housing and land use recycling process has, however, indirectly promoted the isolation of the suburban household worker by eroding the extended family and community network that previously helped informally to collectivize household work in urban areas. The dynamics of urban land use increasingly result in the wholesale slummification or renewal of neighborhoods, a process that undermines entire ethnic communities and disperses their residents. This is also a function of the increasing labor mobility required by employers, which produces individual household migration patterns, both intraurban and interregional, that disrupt long-term community ties.

A final factor in the individualization of urban household work has been the restructuring of the income basis for retired people toward social security and pensions and away from inclusion in family support systems. This development has eliminated the need for generations to occupy the same household or to settle near each other. Incomes of the elderly are frequently insufficient to support residence in the same neighborhood as those of their sons and daughters. As a result of age segregation of urban and even regional space (seniors moving to Florida and Arizona), women have lost the propinquity of parents and other older neighbors as child carers.

[. . .]

It is unnecessary to attempt to unravel completely the relationship between patriarchy and capitalism in promoting contemporary labor spatial structure. While eliminating capitalism does not necessarily end patriarchy, as the socialist countries show, it is not clear that we could eliminate patriarchy without transforming capitalism. Some argue that patriarchy indirectly serves capitalism by dividing the workforce and blunting class consciousness. However, acceptance of this view does not require a corollary that *all* forms and products of patriarchal relationships are functional for capitalism. Even if the current urban structure is efficient for capitalism as a whole, one can make a case that it is functional for patriarchy, and that the widespread indifference to women's concerns about its inefficiency can be attributed mainly to patriarchal, rather than capitalist, ideology.

How do men benefit from contemporary urban spatial structure? I can offer several hypotheses. First, the "ideal" single-family, detached, urban or suburban dwelling embraces the contemporary patriarchal form of household organization. Within it, the man (when present) is generally considered the head of household (until the current census, he was officially so), the primary wage earner, and the major support of the household. This form discourages extended family or community sharing of housework; deploys the machinery of housework in individual units, which makes sharing difficult; and replaces collective play space (parks) with individual yards. The journey to work of the husband tends to dominate location decisions, and the distance to jobs, combined with inadequate public transportation, discourages women from working for wages. The publicly subsidized (FHA) propagation of homeownership entails significant commitments on the part of both women and men to maintenance work on the individual structure, toil that is confined and controlled within the household unit.

Second, the apparently wasteful expenditure of women's labor time under such circumstances underscores the nature and maintenance of power and privilege of men in the patriarchal household unit. Thorstein Veblen argued years ago that men receive satisfaction from the conspicuous consumption of women's labor time in the household: the more waste apparent, the more social status accruing to them. If this motive prevails, productivity of women's time in the household is not important, and a concern with it would directly conflict with accrual of status. Feminist historians have observed that working-class women as well as men aspired to replicate the upper-class family, where women did not have to work for a wage, but would instead perform the role of household manager and servant. The lack of interest in household efficiency in the larger spatial context also reinforces the illusion that only men "really" work and "provide" for their families. At any rate, we can conclude that in general, American men have not registered complaints about the waste of women's labor time, or inefficiency, of dispersed, residential, single-family dwelling patterns.

A third, and nonconflicting, hypothesis concerns the quality of output achievable under the patriarchal form of household organization. The advantages of this form of organization include the flexibility of scheduling of certain services (meals, recreation, etc.) and the potentially higher quality of home-cooked meals, homegrown vegetables, and personally tailored services. (Of course, such personally crafted services could be worse, as well!) In this sense, the current form of household production may not be inefficient, that is, wasteful of women's labor time. This argument would hold if women's preferences coincide with men's preferences; given the alternatives, women may indeed enjoy their own home-cooked meals and may prefer taking care of their own children to cheap restaurants or for-profit custodial child care. But it may also benefit men at women's expense. Flexible and quality meals for husbands may require the squandering of women's labor time.

We are, of course, dealing with shifting definitions of efficiency. In the argument regarding quality of household output, the "efficiency" of a home-cooked meal depends on which side of the dinner table you occupy. While it appears that men as a rule gain from the current organization of household production, these gains are qualified by their position as wage-earning workers. It is possible that such personal service could become very expensive with certain changes in capitalist workplace organization, commodity composition, and women's resistance. For instance, if women's wages were to rise enough to make their wage labor attractive; if women were to demand full partnership in household work; and if the quality and price of market-produced childcare, meals, and so on were to improve substantially, men might agree more readily to women's wage work, with a diminution of the quality and privilege associated with full-time household work, and to restructuring of the household in urban space. Nevertheless, the single most significant factor in the structure of the patriarchal household is the derogation of primary responsibility for household labor upon women. If men were to assume their share of responsibility for housework and childcare, then we might experience significant changes in household structure and spatial location in urban areas. Such an event might show us just how inefficient contemporary household production patterns really are. Similarly, the question of whether current patterns are functional for capitalism depends upon the net effects on profit produced through such arrangements and, further, on the way in which patriarchal concessions to male workers are necessary to the maintenance of a docile workforce.

Of course factors other than the patriarchal form of household production influence location decisions and housing choices, particularly race and class characteristics and the relationship between households and wage-labor work locations. Furthermore, American individualism has undoubtedly coincided with patriarchal household structure to promote the single-family suburban housing that is more common in U.S. cities than in European ones. Yet, the central theoretical point stands: patriarchy profoundly shapes American urban spatial structure and dynamics. It implies that the dismantling of patriarchal household arrangements might call for dramatic restructuring of cities.

THE RESHAPING OF URBAN SPACE

Patriarchy is not a static system. Currently under widespread attack by the women's movement, it also must accommodate itself to changes in capitalism, particularly the increasing availability of commodity substitutes for household labor (child care, restaurant meals, and so on). In unorganized as well as organized ways, changes are taking place in urban living arrangements that foreshadow the future evaluation of urban spatial structure. For instance, three significant demographic changes have occurred in the 1970s in the United States that involve new forms or locations of household production. Their emergence contributes substance to the argument that decentralized urban single-family households are not the most efficient workplaces for the reproduction of labor power and human life. Each instance involves a different segment of women in the population, but all three suggest significant efficiencies deriving from a more integrated use of urban space or a more collective form of household or neighborhood structure.

The first phenomenon is gentrification, that is, the reverse migration of middle- and upper-income residents to urban centers, bringing housing rehabilitation or rebuilding with them. While no urban analysts have pursued this explanation, it seems clear that gentrification is in large part a result of the breakdown of the patriarchal household. Households of gay people, singles, and professional couples with central business district jobs increasingly find central locations attractive. Particularly important has been the success of both gays and women in the professional and managerial classes in gaining access to decent-paying jobs. Gentrification must also be ascribed to the growth of high-income professional and managerial jobs per se in big cities, where the needs of contemporary large-scale corporations have concentrated jobs associated with the control and management of large-scale economic enterprises and with government-related activity.

Gentrification in large part corresponds to the two-income (or more) professional household that requires both a relatively central urban location to minimize journey-to-work costs of several wage earners and a location that enhances efficiency in household production (stores are nearer) and in the substitution of market-produced commodities (laundries, restaurants, child care) for household production. In some areas, such as the SoHo district in New York City, large networks of women-headed households share child care and other household production activities in a well-organized manner. The flexibility in designing living space out of converted lofts enhances this collectivization and permits variation in household composition.

Second is the movement of retired households to relatively nonurban settings, all the way from southern New Jersey and the Upper Peninsula in Michigan to retirement colonies like Leisure World in Florida and in the Southwest. While these households generally remain nuclear, the division of labor between men and women frequently breaks down. One striking feature of these communities is that they provide relatively easy access to the normal range of inputs in the household production process, eliminating the difficulties posed by urban and suburban traffic, parking, and high taxes. Condominiums reduce the household maintenance tasks. Such communities are experiencing the revival of noncommodity production and barter among residents, women and men alike, with a more explicit emphasis on the quality of household production. Retirement housing is also frequently more collective in shared space and facilities. Buildings or complexes may include group dining facilities, group recreational facilities, and small health-care and therapy centers.

Finally, there is striking evidence that the fastest growing American communities are small and medium-sized towns, not urban areas. Although suburbs continue to be developed, the total suburban population is not rising rapidly, and many urban areas as a whole are experiencing depopulation. While decentralization of employment and the relatively lower cost of living are most often cited as causes, my own informal observations suggest that women are frequently strong proponents of such moves, because they provide greater possibilities for shared child care and greater community involvement in social reproduction. Such places also involve easier access to jobs, even though they may not offer better pay than those available in urban areas. I have noticed, for example, that wives of construction workers who are regionally mobile often choose to live permanently in small towns rather than cities because they make household life easier while their husbands migrate to seasonal jobs.

These latter two examples are not cases of the breakdown of patriarchy per se but of household and urban spatial patterns that have evolved from a breakdown of the patriarchal structuring of household production. In fact, both of the latter cases may be explained as reactions to the extraordinary success of that form in dominating the housing stock and the spatial array of urban residences. In the former case, older people frequently find it inefficient to continue to live in, maintain, and pay taxes on a large house, which previously operated as the workplace for rearing children. The homogeneity of single-family neighborhoods makes it nearly impossible to find alternative housing in the same area. The choice to migrate to another region entirely may be more a hallmark of the destruction of any meaningful or accessible social unit larger than the family than a product of the cessation of wage work or the search for a warmer, healthier climate. Similarly, the migration to small towns of nuclear families not tied to urban labor markets signals the inefficiency of household production imposed by the patriarchal urban form, particularly its destruction of access for women to collective help with household production, to jobs, and to urban amenities.

However, just as patriarchy will not disappear without organized resistance, the trends we have just discussed cannot be counted on anarchically to undermine the patriarchal structuring of households and urban space. The sobering reality of this form of urban spatial structure is its permanence. It is literally constructed in brick and concrete. Therefore, its existence continues to constrain the possibilities open to women and men seeking to form new types of households and to reorder the household division of labor. The fact that housing, the primary workplace for social reproduction, is also the major asset for many people tends to reinforce the single-family, patriarchal shape of housing and neighborhoods. Builders, and people buying from them, worry that innovative housing forms will not have a resale value. People interested in suprafamilial communal living situations have generally had to either migrate to rural areas where they could construct their own housing or to older central city areas where they could convert large old housing units (originally built for extended families or boarders). Efforts by lesbians and other organized groups (e.g., a church group in Detroit) to take over entire neighborhoods are severely hampered by the individualization of land-ownership patterns and the legal sanctity of property. Developers, through urban renewal, may use public domain powers to clear entire sections of urban land for private business development, but community and neighborhood groups have no such access to eminent domain for efforts to collectivize living space for the tasks of social reproduction. The dominance of the single-family, detached dwelling in a decentralized urban spatial structure reinforces people's ideas of what forms of household structure are possible and penalizes them materially if they wish to pursue other visions.

Since the resurgence of the women's movement in the late 1960s, women who wish to change their living arrangements and household responsibilities, or to increase their options for doing so, have found it necessary to attack the structural determinants of patriarchal household urban form directly.

[. . .]

In many urban areas, women have set up cooperative child care and other forms of cooperatives that help alleviate the burdens of housework. Women have found regulations of the welfare state that presume the patriarchal family is the normal household unit and that are frequently oppressive to men as well as women.

The widespread activism of women in urban struggles around housing, child care, and neighborhood preservation has generally been neglected in the literature on urban social movements. Feminist critics of this literature point out that because women's struggles have frequently been over issues bearing on conditions of social reproduction, they are invisible to the students of urban social movements, who identify urban problems as capitalist, not patriarchal, phenomena. Furthermore, the literature frequently overemphasizes the role of men in such movements and misrepresents the goals and strategy of the organizations. While leadership may be male, the main organizing in many urban struggles has been accomplished by women, who tend to know their neighborhoods better, who build collectivization of household labor into many community group organizing efforts, and who seem to opt for a form of organization that permits them to continue their household work.

[. . .]

TOWARD A FEMINIST NATIONAL URBAN POLICY

Since urban spatial structure and housing form is such an important constraint on women's ability to change household work roles, national urban policy should be a major target for feminists. The Department of Housing and Urban Development should rank high with departments concerned with labor, health, education, and welfare as a major agency charged with policy responsibilities central to women. HUD's programs should be scrutinized for their impacts on women and new initiatives should be proposed and demanded.

What would a national urban policy that addressed women's issues look like? Perhaps if we could map this out, we could better assess current national urban policy. The answer grows out of our criticisms of the current structuring of urban space for household production and from a more general critique of social structures that impede women's progress and that could be addressed at the urban level. First of all, the most pressing issue for both women's involvement in the labor force and for economizing on household production involves child care. Without some form of socialized child care, women with young children will remain tied primarily to household production locations. Three solutions are possible: sharing informally through extended-family arrangements and neighborhood co-ops; publicly producing child care; and privately producing, for profit, child care. Each currently exists on a limited basis. Each has serious consequences for both the quality of child-care services and the labor-force participation of women. No thorough study of the implications of each as a prototype system has been done. Clearly, the problem is urban in nature: Should child care be provided cooperatively in small neighborhood complexes or single-family homes, by the public sector in public buildings (like elementary education) on a larger neighborhood scale, or in private enterprises in neighborhood or regional shopping centers? The alternative location – at the plant, office, or shop – has received little attention, even though it has interesting possibilities for parent/worker involvement in child care and for efficiency in journey-to-work trips.

A second issue, also of significance, is the type of housing available in urban areas. Women living without men but with children (a rapidly growing group), and groups of single adults, have difficult times both in adapting existing housing to their needs and in obtaining access to it. Restricted credit opportunities are only one of several discriminating barriers. We have no way of knowing what types of households people would choose to live in, given a choice, but we can say that these choices have not existed in the past, nor do they exist in the present. Federal policy might (as it does in the energy business) invest in research and development and experimental projects with various forms of collective and nonpatriarchal housing.

A third set of policies would encourage and subsidize the reintegration of uses in urban space to enhance the efficiency of the household production sector. These policies might pioneer and provide incentives to small-scale commercial development; to the decentralization of jobs in small establishments; to efficiencies in the use of urban space, such as more park space in place of endless private front yards; fine-grained transportation systems; etc. While I leave the design of these to my imaginative sisters, I might suggest one criterion that could be used to judge the desirability of such projects (net of other costs): the elimination of unnecessary labor time expended in individual travel (excluding public transportation time which can be used to read newspapers or books or to socialize), in individual yard improvement and grooming, in individual meal preparations, child care, and so on. Other criteria must address both the quality of output and the quality of women's working conditions.

[. . .]

Finally, a feminist urban policy would establish a new research agenda. This would explore the theoretical understanding of the relationship between patriarchal household form, urban housing, and spatial structure. It would pursue extensive empirical work to document the cost in women's labor time and the working conditions within the household resulting from various aspects of urban structure. It would develop and introduce a new type of cost/benefit calculus for judging the efficiency and social welfare of such public projects as transportation systems, housing, and urban infrastructure (parks, streets, utilities). This calculus would evaluate the effects of such projects on the household division of labor and its productivity. It would design, propose, and experiment with policies and projects that would directly address household production concerns, increase women's choices, and alleviate the onerous division of household labor. It would investigate the proper levels of government or form of collective organization that should be charged with reshaping elements of urban structure. Such a research agenda, which would undoubtedly uncover many creative possibilities, would end the invisibility of women's concerns within both the academy and the agencies that shape urban policy.

"'Race,' Space, and Power: The Survival Strategies of Working Poor Women"

from *Annals of the Association of American Geographers* (1998)

Melissa R. Gilbert

Editors' Introduction

The readings in Part Three addressed the spatial dimensions of post-World War II poverty in U.S. cities, pointing in particular to the serious decline in employment opportunities in the cores of older cities, where mostly poor minorities live. The rise of offshore manufacturing, suburbanization, misguided urban renewal policies, and racially biased practices of the real estate industry and lending institutions contributed to what Loïc Wacquant calls the "impacted ghetto" – a space disconnected from social, economic, and political opportunities.

In this reading, Melissa Gilbert further examines the consequences of the "disconnectedness" of the contemporary ghetto from the point of view of working women. Gilbert takes issue with the "spatial entrapment" thesis, which claims poor urban women are "cut off" – in terms of distance and mobility – from the better-paying jobs to be found in the suburbs. Trapped within the confines of the inner city, poor women have access to jobs that pay low wages and offer little, if any, advancement. Poor women who live in urban ghettos and who have no reliable means of transportation (to the suburbs) are rendered powerless in the face of structural economic change.

Gilbert concurs that structural economic factors have produced severe disadvantages for poor women and others who call ghettos home. She disagrees, however, with the theory's portrayal of residents as trapped – as passive subjects of economic forces with little or no recourse to any situation but despair. Gilbert draws on feminist perspectives of intersectionality, which see race, class, gender, and sexuality as mutually interacting, as opposed to discrete, separate spheres of social identity. The lives of poor women are more complicated than the spatial entrapment and other, similar labor theories allow. Poor women might be economically "trapped" in ghettos, isolated from jobs in the suburbs, and deprived the mobility to get to them. But not necessarily: They may be "rooted" in inner city neighborhoods as well and, as Gilbert claims, that rootedness can be a resource for women in their active efforts to secure economic, social, and personal well-being.

Women use personal networks of family members and friends to learn about jobs, places to live, and facilities for childcare. Because they have limited mobility and access outside the inner city, the networks for poor, mostly minority women tend to be more integrated and overlapping (work-based, church-based, etc.) than those for their suburban, mostly white counterparts. Drawing on a case study of African American and white women in Worcester, Massachusetts, Gilbert examines how poor women's rootedness in the urban core and their use of networks as a survival strategy are both enabling and constraining. Localized and intensive social networks are often crucial to finding jobs, homes, and childcare that fit the realities of limited mobility. Because these networks tend to be insular and recurring, however, the range of opportunities tends to be narrow and constricted.

Melissa Gilbert is Associate Professor in the Department of Geography and Urban Studies at Temple University in Philadelphia. She has written in the areas of labor market strategies, theoretical and policy debates concerning urban poverty, and labor and community organizing. Her publications include "Identity, Difference, and the Geographies of Working Poor Women's Survival Strategies," in K. Miranne and A. Young, eds, *Gendering the City: Women, Boundaries and Visions of Urban Life* (Lanham, MD: Rowman and Littlefield, 2000: 65–87); "Place, Politics, and the Production of Urban Spaces: A Feminist Critique of the Growth Machine Thesis," in A. Jonas and D. Wilson, eds, *Twenty-One Years After: Critical Perspectives on the Growth Machine* (Albany, NY: SUNY Press, 1999: 95–108); and "The Politics of Location: Doing Feminist Fieldwork at 'Home,'" *The Professional Geographer* 46, 1 (1994): 90–96.

The Personal Responsibility and Work Opportunity Reconciliation Act of 1996, more commonly referred to as "welfare reform," eliminated the federal guarantee of cash assistance to poor people and replaced it with a system that contains stringent work requirements and time-limited assistance. One of the underlying assumptions of welfare reform is that poor women with children will become economically self-sufficient through employment. Yet against this optimistic projection are the facts that most women are in sex-segregated occupations with the attendant low-wages and lack of opportunity for advancement, that there is still a significant gender wage gap, and that many women's wages are less than adequate to support a family. Indeed, the restructuring of the U.S. economy since the 1960s has led to a polarization of wages, an increase in low-wage service-sector jobs, and increases in part-time employment, all of which have further disadvantaged many women. The result of these trends is that many women who work for wages are little or no more financially secure than they would be if they had received payments from the now-defunct program, Aid to Families with Dependent Children (AFDC).

[. . .]

It is within this larger context that feminist geographers have made important contributions to the analysis of how labor-market inequalities are reproduced through space and in places by collecting overwhelming evidence that white women are more spatially constrained than white men. This is exemplified by women's shorter commutes to work, which negatively affect their employment opportunities. Kim England (1993) has termed this argument the "spatial-entrapment" thesis. Research examining the effects of "race" on women's levels of spatial containment, however, has demonstrated that many African-American women have longer commutes to work than white women and that this also negatively affects their employment opportunities. Clearly the fact that African-American women are more economically disadvantaged points to a limitation in the spatial-entrapment thesis. I suggest that this paradox requires us to reconceptualize the links between space and power underlying the thesis.

First, the longer commutes of many African-American women make clear that equating mobility with power and immobility with powerlessness is too simplistic to capture the spatiality of women's daily lives. Instead, I argue that no spatiality is inherently with or without power. Second, conceptualizing power as having a single source, as in, for example, gender relations, flattens out differences among women and minimizes the complexity of their lives. Rather, I argue, power should be conceptualized in terms of a multiplicity of interconnected, mutually transformative, and spatially constituted social relations. This reconceptualization of the links between space and power suggests that we move beyond the duality that equates mobility with power and immobility with powerlessness, exploring instead how mobility and immobility are related to multiple power relations. We can then begin to examine how rootedness, a potential outcome of spatial boundedness, may be a resource as well as a constraint, depending on the constellation of power relations. In doing so, we can provide a more nuanced analysis of the opportunities for and barriers to women's economic security than is now provided by the spatial-entrapment thesis.

To explore how the spatial boundedness of women's lives and its consequences vary, I examine the role of place-based personal networks in the survival strategies of working poor African-American and white women with children in Worcester, Massachusetts. The study illustrates how gender intersects with "race" to shape the spatiality of women's lives and the ways in which rootedness may be a resource as well as a constraint. Using the data from Worcester, I first evaluate the women's experiences in terms of the spatial-entrapment thesis, determining the spatial extent of African-American and white women's daily lives and its impact on their employment opportunities. Then I examine how women use the spatial boundedness of their lives to develop place-based personal networks – in the family, workplace, community, and neighborhood, and as one important aspect of their survival strategies – to ensure the health, safety, and security of themselves and their families. I show how women use personal networks to connect them to employment, housing, and childcare, and more broadly to ensure their economic and emotional well-being, while also demonstrating how these same networks can operate as constraints in terms of the kinds of jobs, childcare, and housing to which they are connected. I conclude by arguing that to understand the differences in the spatial boundedness of African-American and white women requires moving beyond the spatial-entrapment thesis to explore how the relationships among rootedness, networks, and survival strategies are shaped by racism.

[...]

RECONCEPTUALIZING THE LINKS BETWEEN SPACE AND POWER IN WOMEN'S DAILY LIVES

People are spatially bounded, and geographers and others conceptualize the degree of boundedness in terms of mobility and immobility. Associating mobility with power and immobility with powerlessness, however, is too simplistic to capture the spatiality of daily life.

Theorizing spatiality as a potential resource as well as a constraint highlights women's agency and adds complexity to our analyses of their lives. It also suggests that we need a conceptualization of power that will allow us to better explore differences among women. Feminists have paid increasing attention to the ways in which gender interacts with other power relations, such as sexuality, age, ethnicity, class, ableism, and "race," thereby conceptualizing power in terms of multiple axes of interconnected and mutually transformative relations.

[...]

Using this conceptualization, it becomes impossible to understand power as a dualism (all or nothing), and therefore to simplistically "read off" a given set of spatial outcomes for a particular set of power relations.

[...]

Building on this insight, I will examine how poor women – women who are often seen as either the victims of patriarchal structures or as passive and atomized individuals dependent on government subsidies – use rootedness in the construction of their survival strategies. I use the term "strategies" to refer to the everyday decisions and practices of poor women attempting to ensure the economic and emotional well-being of themselves and their families. By comparing the strategies of African-American and white women in Worcester, I will show the ways in which women's survival strategies are shaped by their location within a constellation of power relations.

[...]

GENDER, "RACE," AND THE GEOGRAPHY OF SOCIAL NETWORKS

There is an extensive literature documenting the significance of personal networks that people draw upon for emotional, social (including childcare and transportation), and financial support, as well as for housing and employment. Additionally, networks have been evaluated in terms of their existence and potential utility, or social capital, rather than their actual use. Women's networks tend to be focused on kin and neighbors, while men's networks tend to be focused on non-kin, particularly coworkers. These differences are due to the different structural locations of women and men in the family and labor market.

In addition, there is extensive literature examining the social networks of African-Americans, including comparisons of African-Americans' and whites' networks. Particular attention has been paid to the role of the extended family and church in the lives of African-Americans, because both have historically been important institutions in mitigating the effects of living in a racist society.

[. . .]

While some sociologists have conceptualized people's embeddedness in social networks, feminist geographers have explicitly analyzed the spatial dimensions of embeddedness, such as the impact of local context on spatially grounded networks and the role of space in shaping information flows.

[. . .]

An important aspect of spatial rootedness is the social networks that people develop in places. Research has shown that networks are a significant component of people's survival strategies; that the use of networks, as well as the type of networks used, can vary by gender, class and "race"; and that the spatial dimensions of networks have important consequences for women's economic opportunities and survival strategies.

This paper builds on these insights to more thoroughly examine the different contexts in which women create networks, the ways in which different networks are used for different purposes, and their place-based interrelatedness. In doing so, I explore the ways in which the spatiality of women's networks may be enabling as well as constraining, and the impact of racism on these processes.

THE STUDY AREA

Worcester, Massachusetts is a propitious context for this study because economic restructuring has had widespread impacts on its working-class people's economic opportunities. Worcester, like many cities in old industrial regions of the U.S., has experienced a decline in manufacturing jobs and an increase in service jobs. Research has also shown that Worcester has high levels of occupational sex segregation and wage disparities. These economic conditions have had severe impacts on African-American families, who are disproportionately likely to fall below the poverty line. Worcester has also disproportionately experienced a rise in female-headed households.

[. . .]

Worcester has always had a diversity of ethnic groups, with Irish, French Canadians, Scandinavians, Italians, and Poles being particularly prevalent. More recent arrivals to the city include Puerto Ricans, Asians, and African-Americans from the U.S. South. African-Americans were historically located in the neighborhood of Main South and along Belmont Street. Two historically African-American churches and a more recent congregation are located in or near these neighborhoods. When Interstate 290 was built through the city in the 1960s, it displaced many African-American families living in the Belmont Street area. Furthermore, a public-housing project, Plumly Village, was built in 1970 in the same area. Community leaders saw these actions as a deliberate attempt to break up the African-American community after the riots of the 1960s in many other U.S. cities.

DATA AND METHODS

Exploring how gender intersects with racism to shape the spatial boundedness of working poor women's lives and how rootedness may be a resource as well as a constraint in their survival strategies required collecting primary data. My goal was to interview African-American and white women living in Worcester who were working in low-waged jobs and who had or were raising children. I sought to interview enough women in each racialized group to allow a quantitative analysis of differences between racialized groups of women. . . .

[. . .]

I interviewed 26 African-American and 27 white women living primarily within the city of Worcester during the summer and fall of 1991. The average age of the women was 34, and their average years of education were 13.5. African-American women had significantly fewer years of education. With the exception of three who had lost their jobs between the time the interview was arranged and when it was conducted, all were employed. Most of the women worked full-time as clerical workers or in service occupations. In terms of income, there were no significant differences by racialized group, but African-American women had more children and larger households on average than white women. Most of the women had children requiring some kind of childcare. Seventy-eight percent of white women and 50 percent of African-American women were single parents. Only a few of these women received child support.

The nature of the research questions required that I obtain detailed qualitative information about the daily lives of the women in the study and their analyses of the meaning of, and reasoning behind, their activities and strategies. Therefore, I chose to conduct in-depth, interactive interviews. The semistructured interviews, averaging approximately two hours, contained both closed- and open-ended questions aimed at determining the spatial extent of daily life, the use of personal networks, and the importance of these networks in their survival strategies. Moreover, detailed employment, childcare, and residential histories were collected for each woman, thereby adding historical depth to a small sample. The interviews were tape recorded and transcribed.

[. . .]

THE SPATIAL EXTENT OF WOMEN'S DAILY LIVES

An analysis of the job histories of the women in this study revealed that the average travel time to work for all jobs held (N=186) was 14.9 minutes; women often traveled 10 minutes or less to their jobs and rarely traveled more than 30 minutes. Therefore, the experiences of the women in my study support the empirical claims of the spatial entrapment thesis; women with children generally have shorter commutes to work than do men.

Yet African-American women in this study appear to be more spatially entrapped than are white women. Interestingly, and contrary to previous research, African-American women traveled significantly less time to work than did white women. African-American women traveled an average of 11.9 minutes to their jobs (N=81), while white women traveled an average of 17.8 minutes to their jobs (N=105).

As further evidence of the spatial entrapment of women with children, we can examine women's childcare trips which, although neglected by the spatial-entrapment thesis, help to explain why many women with children have short commutes to work. Women's journey-to-work times increase substantially when the journey-to-childcare is included, suggesting that childcare trips limit women's ability to travel farther distances to work by increasing their overall commuting time.

By subtracting the direct travel-time-to-work from the travel time including childcare trips, we get a clear picture of the considerable time that the childcare trip adds to the journey-to-work. In an analysis of all jobs requiring a childcare trip, the average increase was 18.2 minutes. African-American women's childcare trips added less time to the journey-to-work than did white women's trips. White women spent significantly more time commuting (39.05 minutes), including childcare, than did African-American women (27.13 minutes), providing further evidence of the latter's greater spatial entrapment.

[. . .]

Susan Hanson and Geraldine Pratt (1995) have demonstrated that the length of the journey-to-work is strongly associated with a woman's occupation type. They found that women in female-dominated occupations are significantly more likely to work closer to home than women in other occupations. My findings support theirs and additionally show that women in female-dominated occupations added less time to their work trips because of childcare trips than did women in gender-integrated occupations. More than 58 percent of the women in my study said they would not spend more time traveling to work if they could find a job with higher wages. Women in female-dominated occupations were least likely to say that they would travel farther for higher wages, suggesting that they were more spatially entrapped. There were no significant differences, however, between African-American and white women.

[. . .]

THE INTERRELATEDNESS OF HOUSING, EMPLOYMENT, AND CHILDCARE DECISIONS

[. . .]

Many women, regardless of the presence or absence of a partner, must fulfill the multiple roles of employee, mother, and family provider. Consequently, many women's employment, childcare, and housing decisions are complex and often interrelated. Most of the women in my study made their housing decisions prior to their employment and childcare decisions, so these were determined by their residential location. Nearly all of the women I interviewed (97 percent) were making their employment decisions from a fixed residential location. Childcare factors, however, were sometimes part of women's housing decisions; nearly 13 percent of the responses as to why someone was sharing a residence were because they needed help with their children.

The employment decisions of women were strongly influenced by their responsibilities as mothers and family providers, including the hours worked, the type of work, and the specific job. The necessity of fulfilling multiple roles is evident in the hours women chose to work. While most of the women worked full-time, they overwhelmingly would have preferred working part-time due to their domestic and child-raising responsibilities. Nearly all of the women (94 percent) who worked full-time, but wanted to work part-time, could not do so because of financial reasons. Moreover, women's responsibilities as mothers and family providers affected the kinds of jobs that they do. Seventeen percent of the women chose their type of work because of their children. Some women selected certain fields because the hours allowed them to be home after school or because the job would not require overtime or travel. A number of women became family daycare providers in order to combine work and child raising or because they could not find jobs that would pay them enough to be able to afford childcare.

[. . .]

An additional 17 percent of the women chose the type of work because of the need to provide for their children. For example, some women would choose certain fields or employers they believed to be secure, or they would pick a growth field that did not require much training so that they could get off public assistance as quickly as possible. Most women wanted to do different jobs (83 percent), but could not because of their family responsibilities. When asked why they could not get their desired jobs, 58 percent of the responses were because of child responsibilities, while 30 percent said that they did not have the money to go back to school. Their stated reasons included that they could not afford the childcare necessary to attend school, were waiting for their children to be grown, or that the job did not pay enough or have the benefits that the women needed to support their families.

[. . .]

The intertwined nature of women's employment, housing, and childcare solutions sets spatial limits on their daily lives. Access to transportation is another factor in the spatial extent of women's daily-activity patterns and the nature of their employment, housing, and childcare decisions.

[. . .]

While 70 percent of the women employed at the time of the study drove themselves to work, only 12 percent had access to a car all of the time; 71 percent had access to a car sometimes, and 17 percent never had access to a car; here there were no significant differences between African-American and white women.

In sum, the women's experiences generally support the spatial-entrapment thesis. Interestingly, African-American women appeared to be more spatially entrapped than white women. This finding may be due to their more disadvantaged labor force position or to the fact that the African-American women in this study had on average more children than the white women.

[. . .]

ROOTEDNESS AS A RESOURCE AND/OR CONSTRAINT IN WOMEN'S SURVIVAL STRATEGIES

[. . .]

Women's networks

Women's networks are multifaceted; the community, workplace, and neighborhood are all places in which women create networks. I characterize women's networks as having four components: kin-based, work-based, community-based, and neighborhood-based. Women met their community-based contacts in a variety of contexts in the process of going about their daily lives. A common way for women to meet people in the community was through their children's friends or activities. Other places that women met people in their larger community included local bars, childbirth classes, and the welfare office. The kinds of community contacts mentioned most frequently were among people with whom the respondents had grown up or attended school, as well as church-based contacts. While approximately half of the women in both racialized groups attended church regularly, African-American women (26.9 percent) were more likely than white women (16.7 percent) to have said that they were actively involved in church activities. An African-American woman not originally from Worcester explained the importance of her church networks:

> I was always told, and anyplace I go, I have to find and establish a church family and that's what you do. Because people, I'm not going to say all people in the church are good, but you know once you establish yourself . . . I think that you get a lot of help.

[. . .]

The workplace is also an important context in which women build networks. More than two-thirds of the women had people at their workplace that they considered to be friends. Most of their friends were other women, although, not surprisingly, women in gender-integrated occupations had more male friends at work than did women in female-dominated occupations. Most of the friends were people that the women had met at the workplace. Interestingly, the workplace appears to be an important site of integration for African-American women. African-American women were most likely to meet women from other racialized groups at work, and most of their workplace friends were white. Few of the white women's workplace friends, however, were from other racialized groups. Some white women may be more open to interracial friendships than others, but occupational segregation by "race" can limit the opportunities for interaction.

[. . .]

While many women formed friendships that involved talking about personal problems with women from other racialized groups, they were more likely to form close friendships and socialize outside of work with women from the same racialized group.

As an extension of women's traditional gender roles, women traditionally have been seen as active participants within the neighborhood. Yet the neighborhood was less likely than the community or workplace to be a place in which the Worcester women developed networks, most likely because the women I interviewed were employed. As one African-American woman said, "I'm too busy. I'm not here often enough to make real friends." White women, however, were more likely than African-American women to develop neighborhood-based networks. African-American women (63 percent) were much more likely than white women (36.7 percent) to say that they did not feel part of a neighborhood. African-American women were more likely than white women to say that they knew no one in their current neighborhoods (40 percent versus 20 per cent), and were less likely to say that they had friends (44 percent versus 56 percent) or acquaintances (16 percent versus 24 percent) in the neighborhood. Nearly one-half of all of the women, however, had neighbors that they considered to be more than acquaintances. Most of their neighborhood friends were other women.

[. . .]

In sum, women's networks, which are multifaceted and often interrelated, are embedded in daily life. For both African-American and white women, the workplace was more important than the neighborhood as a site of social networks. African-American women, however, were more likely to develop church-based networks, while white women were more likely to develop neighborhood-based networks. It appears that the nature of residential segregation in Worcester contributes to this pattern. In addition, the nature of residential segregation and occupational segregation by "race," combined with the small racialized minority population in Worcester, contributes to the fact that African-American women's networks generally were more integrated than those of white women.

[. . .]

The impact of networks on women's survival strategies

While the spatial boundedness of women's lives helps to determine where and with whom women develop networks, it also affects women's survival strategies indirectly because their use of place-based networks can be enabling as well as constraining. Personal contacts, which connect women to jobs, childcare, and housing, and more generally are used for economic and emotional support, can be a resource as well as a constraint, often simultaneously.

The use and type of personal contacts played an important role in connecting women to employment. More than one-third of all jobs (N=213) were found through personal contacts. The use of personal contacts, however, was somewhat more likely to lead to female-dominated occupations. While family-based personal contacts were used most often, work-based contacts were also important. Such contacts affected women's employment outcomes. Work-based contacts led to jobs in male-dominated or gender-integrated occupations, while those obtained through kin contacts were more likely to lead to female-dominated jobs.

African-American women relied somewhat more on personal contacts than did white women, and they tended to rely more heavily on kin to connect them to jobs than did white women, who, in turn, relied on a variety of personal contacts, including community- and neighborhood-based contacts. African-American women would seem to be more disadvantaged through their use of networks. Thus, while personal networks are a resource women use to find jobs, the type of network used affects their employment opportunities.

[. . .]

There are important differences in the impersonal strategies African-American and white women used to attain employment. African-American women were most likely to change jobs with the same employer, while white women were most likely to respond to a newspaper advertisement.

[. . .]

Sixty percent of all childcare arrangements were found through personal contacts. While these were most likely to be kin-related, neighbors and community-based contacts also connected women to childcare, while work-based contacts rarely did. The use of contacts in general, and the type of contacts used, tended to connect women to different kinds of childcare. Impersonal contacts (e.g., newspapers, yellow pages, etc.) not surprisingly, led to formal childcare arrangements, while personal contacts overwhelmingly resulted in informal arrangements, particularly when kin-based. Such networks can operate as either a resource or a constraint. On the one hand, family members were most likely not to charge for their services, and they may have more flexible hours.

[. . .]

Informal childcare can also be a constraint, however, sometimes offering limited hours and no subsidies, as neither local nor federal governments subsidize unlicensed childcare.

[. . .]

Women used personal contacts to find housing to a greater extent than they did to find jobs and childcare. For all women, family members predominated and were likely to lead to the least expensive housing. African-American women were somewhat more likely than white to use personal contacts and to rely on kin, while white women, although most likely to rely on kin, were more likely than the other group to rely on neighbors.

[. . .]

In sum, an analysis of women's employment, childcare, and housing strategies highlights two points. First, personal networks created by women locally and intensively may be the key to a woman finding a job-housing-childcare combination that will allow her to be self supporting. At the same time, the local insularity of the network, relative to more widespread impersonal sources of information, may constrain the set of opportunities considered. Second, there are differences in the strategies African-American and white women developed to find jobs, childcare, and housing. African-American women relied more heavily on personal contacts, particularly kin, to connect them to jobs, childcare, and housing than did white women, who tended to use a variety of strategies, for example, newspapers or the yellow pages, in addition to personal contacts. These strategies can help African-American women deal with the effects of living in a racist and segregated society.

[. . .]

In addition to connecting women to jobs, childcare, and housing, the people in women's personal networks, primarily other women, are an important part of their economic and emotional survival more generally. Women were most likely to rely on family members for support in emergencies and for both financial and non-financial assistance. African-American women were less likely to rely on family members for financial support, suggesting, given their generally higher reliance on family members, that their families have less resources.

[...]

The African-American women who said that they participated heavily in church activities also said that they had or could rely on the church for different kinds of economic support (e.g., financial assistance or childcare in an emergency) as well as emotional or spiritual sustenance.

[...]

[M]any women use the spatial boundedness of their everyday lives to develop networks in place, which is an indication of rootedness. Women's use of place-based networks shows that rootedness can be both enabling and constraining, often simultaneously, suggesting that it is too simplistic to equate immobility with powerlessness. Furthermore, we can see that the manner in which rootedness can be a resource or constraint depends upon the constellations of power relations. African-American and white women's strategies differed in important ways because African-American women's strategies in Worcester reflect the constraints they experience from individual and institutionalized racism in the housing and labor market. African-American women's use of rootedness – their reliance on networks, particularly kin-based, and some African-American women's use of the church in their strategies – supports the previous research that has documented the importance of the extended family and church in many African-Americans' lives as strategies for living in a racist society. While other African-Americans play a crucial role in African-American women's networks, the fact that their networks are often segregated can constrain their opportunities.

EXPLAINING THE SPATIAL ENTRAPMENT OF AFRICAN-AMERICAN WOMEN

African-American women's daily activity patterns were more spatially bounded than were white women's due to a complex and interrelated set of factors, including residential segregation, the "racial" composition of women's networks, and the relationship between women's daily activity patterns and the nature and spatial extent of their networks.

[...]

The fact that African-American women were spatially segregated means that they were making their employment and childcare decisions from a more limited residential base than were white women.

[...]

The level of "racial" segregation of women's networks also contributes to differences in the spatial extent of women's daily activity patterns. Most of the personal contacts women used to find employment and childcare were of the same racialized group. Furthermore, 87 percent of both African-American and white women's childcare providers were of the same racialized group. Since most African-Americans in Worcester live in a few areas of the city, the spatial extent of African-American women's networks and their daily activity patterns, which are mutually constituted, are limited. This highlights the significance of place and context in shaping the relationship between space and multiple relations of power, in this case, racism and gender. Therefore it becomes important to ask how mobility and immobility are related to historically and geographically situated constellations of power relations. As my analysis of African-American and white women's spatial entrapment in Worcester has illustrated, we can better answer this question by moving beyond the spatial-entrapment thesis to examine how women use rootedness in their survival strategies.

CONCLUSIONS

I have presented an analysis of the role of personal networks in the survival strategies of working poor African-American and white women with children in Worcester, Massachusetts in order to demonstrate two limitations of the spatial-entrapment thesis. First, equating mobility with power and immobility with powerlessness is too simplistic to capture the spatiality of women's daily lives. Second, conceptualizing power as having a single source masks differences among women and impoverishes our analyses of the spatiality of women's lives. Instead, I have demonstrated that while patriarchal structures of inequality often result in the spatial entrapment of women, the spatial boundedness of women's lives can be both enabling and constraining, as women actively use rootedness in the construction of their survival strategies. Important aspects of rootedness are the networks that people develop in places. By examining women's use of personal networks in the creation of their survival strategies, it becomes apparent that the relationship between power and space is more complicated than suggested by the spatial-entrapment thesis.

I have also demonstrated that racism intersects with gender to shape the spatial boundedness of women's daily lives and the ways spatial rootedness is used in women's survival strategies. African-American and white women differed in terms of their use of personal networks, the kinds of personal networks used, and the spatial boundedness of their daily lives. Analyzing the differences in the spatiality of African-American and white women's survival strategies contributes to a better understanding of the ways that racism intersects with gender and class in shaping the spatiality of daily life in different places.

[. . .]

"Gender and Space: Lesbians and Gay Men in the City"

from *International Journal of Urban and Regional Research* (1992)

Sy Adler and Johanna Brenner

Editors' Introduction

In 1983, urban sociologist Manuel Castells published *The City and the Grassroots: A Cross-Cultural Theory of Urban Social Movements*. The book is a series of case studies of collective action undertaken by Castells and his students to demonstrate how social identity and cultural lifestyles may form the basis of urban social movements. Scholars have long noted that urban conflicts, such as those between renters and landlords, are rooted in larger class-based struggles. Castells set out to show that conflicts around ethnic and national movements and sexuality and gender relationships are important sources of urban change as well. When, for example, particular territories, such as neighborhoods, become associated with specific social group identities, movements may organize to claim a stake in conventional district electoral politics. Such was the case of the Castro district as discussed in the book's chapter, "Cultural Identity, Sexual Liberation, and Urban Structure: The Gay Community in San Francisco." Castells shows that the Castro's space-based social and cultural activities – the clustering of gay bars and ensuing nightlife, the surfeit of gay-owned and gay-friendly businesses and gay clientele, and numerous festivals and celebrations – operated as the basis for successful political organizing. Openly gay Castro businessman Harvey Milk was elected to the city's Board of Supervisors in 1977, largely due to organized efforts of the gay community in the Castro area (Milk and Mayor George Moscone were shot and killed by conservative supervisor Dan White in 1978).

Castells' study of San Francisco became the benchmark study for later work that addressed the importance of cultural identity to the study of urban politics and movements. It also served as the standard for case studies of sexuality, community, and urban social change that have since been examined in cities across the globe.

In this selection, Sy Adler and Johanna Brenner replicate Castells' methods in a study of a lesbian community in an intentionally unnamed U.S. city. Castells found lesbian communities less territorially visible and distinct than their gay male counterparts, due primarily to gender differences in the use and importance of urban space. Adler and Brenner depart from Castells' gender-difference explanation, noting that other variables must be taken into account in the study of lesbian communities. They see gender differences in the use of space as reflective of deeper, more significant causes: "the capacity to dominate space or claim territory depends on available wealth and restrictions placed by male violence on women's access to urban space."

Adler and Brenner suggest that the so-called "invisibility" of the lesbian community and, hence, its absence from territorial-based grassroots politics is a function of scholars looking in the wrong places (i.e., formal, organized, public expressions of community and representational/electoral politics). Women have long provided membership and leadership in social movements, the authors point out. The kinds of political issues that women address, however, tend to involve organization and mobilization through interpersonal networks (which may not be territorially clustered – think of the use of the Internet). Many of these issues – personal/household concerns about parenting and public policy, discrimination in the workplace, and gender-based violence – are not unique to the lesbian community; they are also the concerns of heterosexual feminists.

Sy Adler is Professor of Urban Studies and Planning and Johanna Brenner is Professor of Sociology and Coordinator of the Women's Studies Program at Portland State University. Adler conducts research on local, regional, and state institutions, and processes involving land use, transportation, health care, and natural resources, particularly in the Pacific Northwest. Johanna Brenner is author of *Women and the Politics of Class* (New York: Monthly Review Press, 2000). Her research interests are women and welfare and the impact of recent welfare reform on low-income women.

Recommendations on the issue of sexuality and the city include: David Bell and Gill Valentine, eds, *Mapping Desire: Geographies of Sexualities* (London: Routledge, 1995); Beatriz Colomina, ed., *Sexuality and Space* (Princeton: Princeton Architectural Press, 1992); Nancy Duncan, ed., *BodySpace: Destabilizing Geographies of Gender and Sexuality* (London: Routledge, 1996); Gordon Brent Ingram, Anne-Marie Bouthillette, and Yolanda Retter, eds, *Queers in Space: Communities, Public Places, Sites of Resistance* (Seattle: Bay Press, 1997); and Elizabeth Lapovsky Kennedy and Madeline D. Davis, *Boots of Leather, Slippers of Gold: The History of a Lesbian Community* (London: Routledge, 1993).

In the past two decades gay neighborhoods have become familiar parts of the urban landscape. Although these areas may include lesbians, gay men dominate their distinct subcultures, their businesses and their residences, their street life and their political activities. In his 1983 book, *The City and the Grassroots*, Manuel Castells argues that the predominance of gay men in the creation of distinctly homosexual urban neighborhoods reflects a profound gender difference. In relationship to space, gay men and lesbians, he says, behave first and foremost as men and women. Men seek to dominate space, while women attach more importance to networks and relationships, rarely having territorial aspirations: "Lesbians, unlike gay men, tend not to concentrate in a given territory, but establish social and interpersonal networks." Gay men require a physical space in order to conduct a liberation struggle, while lesbians are "placeless" and tend to create their own rich, inner world.

Lesbians are also politically different from gay men, according to Castells. They do not acquire a geographical basis for urban political objectives, because they create a political relationship with higher, societal levels. Lesbians "are far more radical in their struggle . . . [and] more concerned with the revolution of values than with the control of institutional power."

Castells' analysis makes several assumptions that we question. First, is it true that lesbians do not concentrate in a given territory? Second, does the absence of a publicly identifiable lesbian neighborhood reflect gender differences in interests, needs and values, or differences in resources available to gay men and lesbians? Third, do differences in the political orientation of politically active gay men and lesbians reflect gender differences in relationship to space or differences in political alliances, specifically the involvement of lesbians in feminist politics that include straight women?

The literature on differences between gay men and lesbians in relation to urban space is generally ambiguous about the existence of lesbian spatial concentrations. In support of his argument, Castells cites Deborah Wolf's 1980 study of the lesbian community in San Francisco, which, according to Wolf, "is not a traditional community in the sense that it has geographical boundaries." However, Wolf also noted that lesbians did tend to live in particular parts of town, and that "These areas bound each other and have in common a quality of neighborhood life." Wolf pointed out that since lesbians tend to be poor they live in older, ethnically mixed working-class areas, in low-rent apartment buildings or small, low-rent houses. "Since much of the socializing in the [lesbian] community consists of visiting friends, women without cars try to live near each other, so that gradually, within a small radius, many lesbian households may exist." Indeed, Castells himself noted that lower incomes mean less choice regarding the location of home and work, and that lesbian concentrations were emerging in the kinds of areas described by Wolf, particularly low-rent neighborhoods with counter-cultural communities.

[. . .]

Denyse Lockard (1985), who studied lesbians in a large south-western city, argues that lesbians create communities without a territorial base or geographical boundaries. However, she also argues that one of the four defining features of a lesbian community is

> its institutional base, the gay-defined places and organizations which characterize a community, and provide a number of functions for community members. It is this institutional base that provides the means by which a community can mobilize its members for action as a minority group in the larger society, and that is visible, at least to some degree, to the outside world.

Lockard does not explore whether these crucial institutional bases are themselves spatially concentrated, creating a semi-visible lesbian urban space, as is the case for the gay male community.

H.P.M. Winchester and Paul White (1988) did find discernible concentrations of lesbian facilities, with the most exclusively lesbian institutions located in a poor district of the inner city. Although these lesbian places are not public, they are known to those in contact with the lesbian community. Winchester and White assume, based only on anecdotal evidence, that there are no residential concentrations. Yet Wolf noted that in San Francisco most of the lesbian bars and community projects were located in the same neighborhoods as lesbian household concentrations. Thus, we would argue that there is at least some indication that there is a concentration of lesbian residence and activity space in urban areas.

In addition to questioning whether the differences in relationship to space between lesbians and gay men are as great as the literature assumes, we also question the explanations put forward for the differences that do exist. Mickey Lauria and Lawrence Knopp (1985) critique Castells' argument that the more visible concentration of gay men in cities expresses an innate male territorial imperative. However, they adopt what we consider to be an equally problematic approach. They argue that there has been a greater tolerance of relationships of depth, physicality and affection between women than between men, and even acceptance of lesbian sexuality in certain circumstances. Therefore, "gay males, whose sexual and emotional expressiveness has been repressed in a different fashion than lesbians, may perceive a greater need for territory."

While not denying this possibility, we would argue that before concluding that social-psychological needs and interests explain the absence of visible lesbian urban neighborhoods, we need to consider differences in capacity to dominate space, a variable reflecting available wealth as well as restrictions placed by male violence on women's access to urban space. The creation of visible, distinct neighborhoods requires more than residential concentration and the development of a network of voluntary and service organizations. To take over urban space requires also the control of residential and business property. Castells documents the role of gay businessmen, especially owners of taverns and retail stores and gay real estate entrepreneurs, in the development of the Castro district. It may be that gender differences in access to capital account, at least in part, for the fact that tendencies among lesbians in San Francisco towards concentration in residence and community institutions have not been followed by lesbian-owned businesses and real estate. In addition, as Gill Valentine (1989) argues, "women's fear of male violence . . . is tied up with the way public space is used, occupied and controlled by different groups at different times." These restrictions on the use of public space, which lesbians experience as women, would also account for the lack of visibility of their neighborhoods.

Finally, we question Castells' argument that lesbians relate to space primarily as women, that they are "placeless," uninterested in local politics, and therefore more radical than gay men. Localist politics and place-based organizing are hardly foreign to urban women, as Castells himself recognizes in his discussion of other urban communities. Martha Ackelsberg (1984) has noted that "Much recent research . . . has documented the prominent, if not predominant, role of women in urban struggles over what have been termed 'quality of life' issues." Women have historically provided the membership and leadership of neighborhood-based urban movements. Through community-based networks of friends and neighbors, many urban women become activists on issues at the interface of personal/household concerns and employer and public policy decisions, developing political consciousness in the process. Women are often brought into political activity as an extension of their family responsibilities. We would argue that at least a significant minority of lesbians share such concerns. They are certainly far more likely than gay men to carry such responsibilities, to have children, and thus be likely to confront the same kinds of private and collective consumption issues that bring heterosexual women into local politics. We disagree with Castells' argument that the more radical dimension sometimes present in lesbian political activism is an expression of fundamental differences between men and women in core values and preferences. Lesbian activists' interest in more than "representation" within the existing political system reflects the influence and organization of radical lesbian feminism within the community.

LOCATING THE LESBIAN NEIGHBORHOOD

Our main purpose in this paper is to clarify the ambiguities surrounding the existence and significance of spatial concentration among lesbians in urban communities. In order to do so, we attempted to replicate Castells' method of analysis as closely as possible in a study of a lesbian community in another United States city. In constructing our study, we sought help from lesbians knowledgeable about the lesbian community and from lesbian organizations who were asked for permission to use their membership lists. While wanting to cooperate, they were reluctant to give information, especially about residential locations. Their response is understandable, given the political environment. As in many other parts of the country, gay rights activism is countered by well-organized right-wing activists who are powerful enough in the state to overturn legislation protecting gays and lesbians from discrimination. In addition, attacks on gay men and lesbians by youth gangs are not infrequent in the city, and some of these gangs are connected to right-wing organizations. In this climate, while individuals were willing to identify themselves, they felt they could not risk identifying others who might or might not wish to be "out." Therefore, our informants, and the organizations who shared their mailing lists with us, did so only under the condition that we would do as much as we could to disguise the location of the community. For this reason, we have not identified the city, nor the street boundaries of the lesbian neighborhood.

When beginning our study, we recognized that the requirement of anonymity would create problems: our study would be more difficult to replicate and other scholars familiar with the area would not be able to suggest unique qualities of the city that we overlooked. We decided to proceed because it is generally so hard to obtain the kind of direct evidence of gay and lesbian residence which mailing lists provide. In our judgment, the value of the data outweighed the problems that disguising the location of the lesbian neighborhood would impose.

Castells sought to determine the areas in which gay men were concentrated, and to characterize these parts of San Francisco in socio-spatial terms. He then related this characterization to the social and urban characteristics of the rest of the city in order to understand the factors that influenced the settlement pattern of gay men. We sought first to establish the existence of concentrations of lesbians, and then to characterize the areas within which we found concentrations. We used the explanatory variables uncovered by Castells to see if these would also explain the lesbian settlement pattern that we found.

Castells used five different sources of information to locate areas of concentration of gay men: (1) key informants from the gay male community; (2) the presence of multiple male households on voter registration files; (3) gay bars and other social gathering places; (4) gay businesses, stores and professional offices; and (5) votes for gay candidate Harvey Milk in a local electoral contest. Census tract-based maps were prepared utilizing each of these sources, showing the distribution of concentrations of gay men across the city. Castells found that all five maps displayed a similar spatial pattern, permitting him to conclude that gay residential areas had been validly identified.

To identify areas of lesbian concentration, we used: (1) key informants from the lesbian community; (2) the location of lesbian bars and other social gathering places; (3) the location of lesbian businesses, professional services and social service agencies; and (4) two mailing lists of lesbian organizations.

Our 10 key informants unanimously agreed that spatially defined concentrations of lesbians did, in fact, exist. One neighborhood was identified by all informants as *the* lesbian community or "the ghetto," while some identified another neighborhood as a second lesbian community. This neighborhood was close to the first. Our informants identified street boundaries and also located these areas on a census tract map of the city.

Lesbian bars and social gathering places, businesses, professional and service agencies were identified by informants and from a variety of listings and advertisements in local women- and gay-oriented handbooks and resource guides, newspapers and magazines. The addresses from these sources were assigned to census tracts.

The addresses from the mailing lists of the two lesbian organizations, a total of 1150 names, were similarly assigned, making the assumption that the spatial distribution of people on these lists mirrors the distribution of the lesbian population. It might be argued that these lists are a biased sample of the lesbians in the city, since they include only those women who identify as lesbians. While this is true, such a bias does not undermine a test of our hypothesis, since we are arguing that lesbian-identified women do seek to establish spatially based communities. On the other hand, if Castells is correct in arguing that politically active lesbians tend to be more "placeless" than politically active gay men, then this bias would work against our hypothesis.

We chose not to use Castells' measure of household composition based on voter registration lists. We think that household composition is less valid an indicator of lesbian households than of gay male households, particularly when it is impossible to control for age. Mother–daughter households, for example, are not uncommon, particularly where the mother is elderly. A neighborhood with a high proportion of households in which two men live together can more reliably be assumed to be a gay male neighborhood than can a neighborhood with a high proportion of households in which two women live together.

We assigned the addresses on the mailing lists to census tracts, totaled addresses within census tracts, and then rank-ordered the tracts. The rank-order correlation for the two lists was (0.7), indicating a substantial agreement between them. In both cases 15 per cent or less of the tracts contained 50 per cent of the addresses. We selected for further analysis 12 tracts (constituting 10 per cent of the tracts in the city) that appeared among the top 13 on both lists, containing about 40 per cent of the addresses.

Our concerns about bias in the mailing lists were further allayed when we examined the spatial pattern that emerged from other sources, including the informant interviews and the locations of businesses, services and gathering-places. As in Castells' study of San Francisco, the various sources converged. Based on these diverse sources we were able to locate a major concentration of lesbian households in a set of 11 contiguous inner city census tracts east of the central business district, and another concentration in one inner city tract north-west of downtown. This tract is located in the city's "gay" neighborhood.

THE CHARACTERISTICS OF THE LESBIAN NEIGHBORHOOD

In order to further indicate the existence of lesbian concentration and to identify the features of this neighborhood that might account for this concentration, we compared the 12 tracts we found to contain concentrations of lesbian households with the rest of the city, using 1980 census data. Castells argued that gay settlement in San Francisco was opposed by "property, family, and high class: the old triumvirate of social conservatism." However, Castells did not address the extent to which life-cycle stage and related financial circumstances shaped the pattern of gay location. The gay men who were moving into the Castro area when Castells did his study were predominantly younger men, whose limited financial means would make it very difficult to purchase homes or pay high rents. We discuss below the ways in which financial limitations shape the pattern of lesbian concentration. Following Castells, we also hypothesized that areas with high proportions of home owning, of traditional family households and of high rents would be more difficult for lesbians to move into.

We found that the census tracts where lesbians concentrate have significantly lower levels of owner-occupied housing, lower rent levels and lower proportions of traditional families than the remainder of the city. Of the 12 tracts, only three have a higher proportion of home ownership than the city average, and only one is significantly higher: 75 per cent compared to 53 per cent for the city as a whole. This apparent anomaly is explained, in our view, by the fact that this tract is one-third black and has the highest median household income of all tracts with 400 or more blacks. An interracial, middle-income tract with relatively higher quality housing, this area would be more open to lesbians than one economically similar but socially more homogeneous. Lesbians with relatively higher incomes can purchase homes of higher quality than those available in the other tracts without having to locate outside the lesbian neighborhood.

While "property, family and wealth" are barriers to lesbian settlement, counter-cultural neighborhoods appear to be most open. Castells and others all found lesbians located in counter-cultural areas in the cities they studied. We found this to be true in our city also. In addition to the obvious cultural features of the area covered by our 12 census tracts – particularly the presence of counter-cultural institutions such as theatres, coffee shops, studios that present progressive/fringe entertainment, alternative businesses like food and bicycle repair coops, radical and feminist bookstores, etc. – census data indicate its counter-cultural character. Compared to the rest of the city, this area has a much higher proportion of people living in non-traditional households.

The other striking feature of this area is the high proportion of women living alone and of female-headed households. It is impossible to know, of course, whether or not these are lesbian households. And indeed it is impossible to know whether lesbians are more likely to live in this neighborhood because it is a "women's community," or whether it is a "women's community" because large numbers of lesbians live there, or both.

In our 12 census tracts, women from young to middle age (20–54) were significantly more likely to be living alone than women of the same age elsewhere in the city. "Female, non-family householders not living alone" may of course be living with a man, but we think it reasonable to assume that in the majority of cases, non-married heterosexual couples would identify the man as household head when forced to choose. Alternatively, these may be households of room-mates rather than lesbian households. Still, given the other evidence, the higher proportion of such households indicates that this is much more a women's community than the rest of the city.

Although there were fewer families with children living in the lesbian tracts than in the rest of the city, a significantly greater proportion of these families were female-headed. Given that these are also relatively low-rent areas, it might be argued that this figure only reflects the impact of single mothers' poverty.

[...]

However, the median income of female-headed households, no spouse present with children under 18 in our 12 tracts, is not significantly different from the income of similar female-headed households elsewhere in the city. This seems to us to indicate that single mothers are locating in these tracts for reasons other than (or at least in addition to) the relatively low rents.

Within the 12 tracts, three show a particularly high concentration of residences, businesses and services, accounting for 50 per cent of the total addresses located in the 12 tracts. These three tracts overlap the area which our key informants identified as the lesbian neighborhood. They appear to constitute a spatial "core" for the lesbian community. They rank very high on indirect measures of lesbian households.

To explain this configuration we hypothesized that these "core" tracts would be especially high in those features that encourage lesbian settlement: low proportion of owner-occupied housing, low rents, non-traditional households and families. Two of the tracts, as expected, are ranked low in proportion of home ownership and in their median rents, and are relatively low in their proportion of traditional family households. While they have more traditional family households than some tracts, they are also relatively high in the proportion of women-headed families. The third tract is somewhat anomalous, having low rents but a relatively high proportion of traditional families and owner-occupied housing compared to the other tracts. On the other hand, it is still more counter-cultural than the city as a whole and its owner-occupied houses are among the least expensive in the 12 tracts. Perhaps the low cost of homes has encouraged lesbians wishing to become home owners to risk moving into a more family-oriented area in order to be in the neighborhood core. Lesbians with children may be attracted to such an area as well.

THE HIDDEN NEIGHBORHOOD

It seems clear that there is a spatial concentration of lesbians in our city, a neighborhood that many people know about and move into to be with other lesbians. But the neighborhood has a quasi-underground character; it is enfolded in a broader counter-cultural milieu and does not have its own public subculture and territory.

To claim an urban territory takes more than residential concentration. It requires, in addition to this: (1) visibility – gay places, especially retail businesses and services run by and for gay people; (2) community activity – fairs, block parties, street celebrations etc., some kind of public, collective affirmation of the people who live in the neighborhood, even if it is only strolling out in the evening; (3) organization – of businesses and residents to defend the neighborhood's interests, relate to city government, financially support the community activities which create and maintain the urban subculture, giving the neighborhood its distinct character. Businesses and churches are often key providers of the financial and organizational infrastructure for community activity. Castells' history of the Castro district demonstrated the importance of gay businesses in developing the street life of San Francisco.

Castells also noted that the presence of gay men in the real estate and other professions in San Francisco was a major factor facilitating the development of a gay neighborhood. Lauria and Knopp (1985) similarly point to the importance of more affluent gay men in providing the capital necessary for gentrification. As gay men became landlords, the amount of rental housing available to gay men increased, drawing more residents to the area, increasing the clientele for gay retail businesses and the participants in neighborhood street life. Knopp (1990) also analyses the leading role played by gay real estate speculators and developers in the emergence and consolidation of New Orleans's gay community. If lesbians have not yet taken the next steps toward creating a distinct neighborhood, perhaps this reflects their lack of capital more than their lack of interest.

Castells and Lauria and Knopp argue that gay men use their spatial concentration as a base for political mobilization. Castells argues that lesbians, whose political values and interests are both more global and more radical than those of gay men, have less need for this kind of place-based organizing. It is true that lesbians are well represented in radical feminist political activities and organizations. But lesbians are also active in mainstream politics such as the Gay/Lesbian Task Force of the American Civil Liberties Union, the Right to Privacy Political Action Committee, the Women's Political Caucus and so forth. In the November 1988 election a major lesbian community organization contributed fully 50 out of 200 volunteers who worked to elect a pro-lesbian and gay (although not openly lesbian) candidate to the state assembly from their district. That lesbians have not yet fielded a lesbian candidate may reflect more their lack of visibility than their lack of interest in affecting local politics. Without visibility it is difficult to assess the extent of their voting base and thus their potential for successfully running an openly lesbian candidate. On the other hand, they are able to use their spatial concentration to mobilize around issues of concern to them. In the November election an anti-gay ballot measure barely lost in the city as a whole (52.7 per cent "no"), but in the lesbian neighborhood (our 12 tracts) it was defeated by a 46 per cent margin (73 per cent "no").

CONCLUSION

While we reject Castells' characterization of the gender differences between gay men and lesbians, we do think there are important dimensions along which gay men and lesbians may relate differently to urban space and urban politics. These differences do reflect the fact that lesbians are women. First, as we have already argued, lesbians, like other women, tend not to have access to capital. Second, lesbians are more likely than gay men to be primary caretakers of children. Their choices about where to live will have to take their children's needs into account. When they do locate in a lesbian neighborhood, they bring a set of interests and styles of sociability very different from those of people without children. This diversity, while a strength for the community as a whole, might also militate against the development of a distinctive urban subculture. Third, lesbians share with other women a vulnerability to male physical and sexual violence. Of course, it is true that gay men have had to defend themselves from attacks on their places and persons. And lesbians have ably resisted assaults, for example on lesbian bars. Still, it seems to us that women are more at risk in public places than are men, and that this in turn may limit lesbians' interest in or ability to create the kind of vibrant street culture that makes an urban community.

Finally, we agree with Castells that, compared to the politics of the gay male community, lesbian politics has tended to be more global and less restricted to representation in the existing political system, therefore less "place-based." But in explaining this difference, we would emphasize, in addition to restrictions on lesbians' access to space, the impact of feminism on lesbian politics and culture (and the impact of lesbians on the women's movement and its culture of resistance). Lesbians organize as lesbians, just as gay men organize as gays. But lesbians also organize as oppressed women. Their politics and their culture reflect this double vision. Lesbians contributed a significant core of activists to the women's movement from the beginning. Lesbian writers, poets, artists and musicians have played a major role in the development of a women's culture which draws audiences of both straight women and lesbians. Pressure from lesbians in feminist organizations has made the women's movement less homophobic, while issues of lesbian rights have become more integrated into feminist politics. Lesbians who created lesbian culture and urban communities in the early 1970s did so in connection with a movement against male domination whose critique went far beyond simple demands for incorporation into the existing society. While the relationship between straight women feminists and lesbian feminists has sometimes been more conflictual than collaborative, participation in the feminist movement has supported the more radical political tendencies within the lesbian community. (And, correlatively, lesbian feminist separatism constituted one of the more radical political tendencies within the women's movement.) In contrast, gay communities and culture preceded the political organization of gay men. The gay men's community is not homogeneous and does include political tendencies that favor more radical politics than the "interest group" strategies that are dominant. And lesbian communities are more heterogeneous, less counter-cultural and political than in the 1970s. Still, in so far as lesbian urban communities continue to express commitments and world views defined by feminist politics, which draw on but do not simply reflect gendered experience, differences in the urban culture and politics of lesbians and gay men will remain.

REFERENCES

Ackelsberg, M. 1984. "Women's Collaborative Activities and City Life: Politics and Policy." In J. Flammang, Ed., *Political Women: Current Roles in State and Local Government.* Beverly Hills, CA: SAGE.

Castells, M. 1983. *The City and the Grassroots: A Cross-Cultural Theory of Urban Social Movements.* Berkeley, CA: University of California Press.

Knopp, L. 1990. "Exploiting the Rent Gap: The Theoretical Significance of Using Illegal Appraisal Schemes to Encourage Gentrification in New Orleans." *Urban Geography*, 11: 48–64.

Lauria, M. and L. Knopp. 1985. "Toward an Analysis of the Role of Gay Communities in the Urban Renaissance." *Urban Geography*, 6: 152–69.

Lockard, D. 1985. "The Lesbian Community: An Anthropological Approach." *Journal of Homosexuality*, 11: 83–95.

Valentine, G. 1989. "The Geography of Women's Fear." *Area*, 21: 385–90.

Winchester, H. P. M. and P. E. White. 1988. "The Location of Marginalized Groups in the Inner City." *Environment and Planning D: Society and Space*, 37–54.

Wolf, D. 1980. *The Lesbian Community.* Berkeley, CA: University of California Press.

"Freeing South Africa: The 'Modernization' of Male–Male Sexuality in Soweto"

from *Cultural Anthropology* (1998)

Donald L. Donham

Editors' Introduction

As we have learned from the readings in Part Four, the interplay between gender and sexual identities and urban space is multifaceted. Expressions of identity – be they performative, political, cultural, or some combination of the three – structure the use of urban space, its representation and direction of change. Likewise, urban space – its location, built environment, zoning, and other legal statutes – influence the formation and expressions of social identity. This latter point is made clear in the following selection, in which anthropologist Donald Donham examines how apartheid-era segregation of settlement space in South Africa shaped both collective and individual understandings of homosexuality and what effect apartheid's demise had on such understandings.

Racial segregation defined the spatial arrangements of apartheid. Yet black workers often needed to be located within commuting distance in order to provide labor to the white-dominated economy. In places like Soweto rural black men, who otherwise had no right to live there, were permitted to reside without their wives in all-male hostels. Donham explains how apartheid-era spatial segregation and the separation of sexes coalesced over time to strongly influence the localized construction of homosexuality. A same-sex sexual culture developed around the urban hostels, in which black male workers took on *skesanas* – effeminate, sexually submissive males – as "wives." While the sexual or gendered identity of the sexually dominant male workers remained uncontested, the *skesanas* were collectively thought of as ostensibly girls (gender) or some biologically mixed "third" sex. Ceremonies were conducted to induct new *skesanas* as wives and an elaborate sexual-social hierarchy existed within the hostels. This same-sex culture continued for as long as hostels housed black workers laboring in white-dominated industries, businesses, and households.

The end of apartheid in the early 1990s swept away the institutional practices that sustained the sexual culture that developed around the hostels, thereby challenging localized notions of homosexuality. Wives joined their husbands. Hostels became sites of political infighting. New meanings and ways of looking at sexuality were required, including the notion that both the sexually dominant and submissive partners were to be considered gay. In addition, new opportunities, such as the annual gay pride march in Johannesburg, allowed for the public, right-based expression of sexual identity in post-apartheid South Africa. Gay South Africans made connections with counterparts in Europe and the United States, where political mobilization matured earlier. Donham concludes on a positive, yet cautious, note. The undoing of apartheid, especially the spatial segregation of "races," brought forth a cultural redefinition (a modernization, if you will) of sexuality. Yet the facts remain that blacks are overwhelmingly poor and global gay culture is represented as white and middle class. Women, especially lesbians, are culturally restricted, prohibiting a fully westernized notion of sexual emancipation.

Donald Donham is a cultural anthropologist who has written extensively on the effects relationships between culture and economy have on changing forms of power. He is Professor of Anthropology at the University of California, Davis. He was a Fellow at the Woodrow Wilson International Center for Scholars in Washington, DC in 1999–2000.

In February 1993, a black man in his mid-thirties named Linda (an ordinary male name in Zulu) died of AIDS in Soweto, South Africa. Something of an activist, Linda was a founding member of GLOW, the Gay and Lesbian Organization of the Witwatersrand. Composed of both blacks and whites, GLOW was and is the principal gay and lesbian organization in the Johannesburg area. Because Linda had many friends in the group, GLOW organized a memorial service at a member's home in Soweto a few days before the funeral.

Linda's father, who belonged to an independent Zionist church, attended and spoke. He recalled Linda's life and what a good person he had been, how hard he had worked in the household. But then he went on, in the way that elders sometimes do, to advise the young men present: "There was just one thing about my son's life that bothered me," he said. "So let me tell you, if you're a man, wear men's clothes. If you're colored, act colored. Above all, if you're black, don't wear Indian clothes. If you do this, how will our ancestors recognize [and protect] you?" Linda had been something of a drag queen, with a particular penchant for Indian saris.

To Linda's father and to his church, dress had ritual significance. One might even say that there was an indigenous theory of "drag" among many black Zionist South Africans, albeit one different from that in North America. To assume church dress not only indicated a certain state of personhood, it in some real sense effected that state. Writing on Tswana Zionists, who like the Zulu have been drawn into townships around Johannesburg, Jean Comaroff asserted, "The power of uniforms in Tshidi perception was both expressive and pragmatic, for the uniform instantiated the ritual practice it represented."

If dress had one set of associations within Zionist symbolism, it had others for a small group of young black South African activists who saw themselves as "gay." To the members of GLOW present, most of whom were black, Linda's father's comments were insulting. Most particularly, they were seen as homophobic. As the week wore on, GLOW began to organize to make their point and to take over the funeral.

As Saturday neared – nearly all Soweto funerals are held on the weekend – tensions rose. There were rumors that there might be an open confrontation between the family and GLOW. Along with Paul, a member of GLOW from Soweto, I attended, and the following is a description of what transpired, taken from a letter that I wrote home a few days later to my American lover:

> The funeral was held in a community center that looked something like a run-down school auditorium. There was a wide stage on which all of the men of the church, dressed in suits and ties, were seated behind the podium. In front of the podium was the coffin. And facing the stage, the women of the church were seated as an audience – dressed completely in white. (Independent churches have distinctive ways of dressing especially for women, but also sometimes for men.)
>
> To the right (from the point of view of the seated women) was a choir of young girls – again all in white: white dresses and white hats of various kinds (most were the kind of berets that you have seen South African women wear). I stood at the very back of the hall, behind the seated women, along with most of the members of GLOW and various other men and women, most dressed up. This last group was apparently made up of friends and relatives who were not members of the church.
>
> I had arrived late, about 9:30 in the morning (the service had begun at about 9:00). I was surprised to see, behind the coffin, in front of the podium, a GLOW banner being held by two members. There were flowers on the coffin and around it. Throughout the service, including the

sermon, the two GLOW members holding the banner changed periodically. From the back, two new people marched up through the ladies in white to take the place of the two at the banner. Then those who had been relieved came back through the congregation to the back of the church.

One GLOW member videotaped the funeral from the back. About six or seven of the members who had come were white. It was hard to tell exactly, but there were probably 10 or 15 black members. Quite a few, both white and black, wore GLOW T-shirts (the back of which said, "We can speak for ourselves"). Finally, two or three of the black members wore various stages of drag. One, Jabu, was especially notable in complete, full regalia – a West African-style woman's dress in a very colorful print with a matching and elaborately tied bonnet, one edge of which read, "Java print." Wearing a heavy gold necklace, she walked up and down the aisle to hold the banner at least twice – in the most haughty, queenly walk. It was almost as if she dared anybody to say anything. She made quite a contrast with the stolid, all-in-white ladies seated in the audience (one of whom was heard to comment to a neighbor, "She's very pretty, isn't she [referring to Jabu]? But look at those legs!").

When we arrived, Simon Nkoli, one of the first black gay activists in South Africa, was speaking. Simon was dressed in an immaculately white and flowing West African (male) outfit with gold embroidery. He spoke in English, and someone translated simultaneously (into Zulu). His speech was about gay activism in South Africa and the contributions that Linda, his dead friend, had made. At points in his speech, Simon sang out the beginning lines of hymns, at which point the congregation immediately joined in, in the [style] of black South African singing, without instruments and in part-harmony.

After Simon, there were other speeches by the ministers of the church. They emphasized that Linda was a child of the church, that his sins had been forgiven, and that he was in heaven. Diffusing any trace of tension between the church and GLOW, one of the ministers rose and apologized on behalf of Linda's father for offending the group earlier in the week.

[. . .]

Because there are so many funerals in Soweto on the weekend (probably 200 at Avalon cemetery alone) and because the cemetery had only one entrance (the better to control people), the roads were clogged and it took us an hour to go a few miles. The members of GLOW got out of the bus and *toi toied*, the distinctive, punctuated jogging-dancing that South African blacks have developed in antiapartheid demonstrations.

[. . .]

There was something about the routinized way that so many people had to bury their dead and leave (others were waiting) that brought home to me, in a way that I had not anticipated, what apartheid still means in many black people's lives. . . . The South African police stood in the background. Continually, another and another group arrived, and as each rushed to its gravesite, red dust began to cover us all. The sun got hotter.

[. . .]

After the graveside service, GLOW members gathered at one of the member's houses in Soweto and proceeded to get drunk. I had had enough. Driving back to Johannesburg (it's a little over 30 minutes), I almost had an accident. Tired and with my reflexes not working for left-hand-of-the-road driving, I turned into oncoming traffic. By the end of the day, I felt overwhelmed. Another gay man dead – yet another. And his burial had brought together, for me, a mind-numbing juxtaposition of peoples and projects, desires and fears – Zionist Christians and gay activists, the first, moreover, accommodating themselves to and even apologizing to the second. Could anything comparable have happened in the United States? A gay hijacking of a funeral in a church in, say, Atlanta?

APARTHEID AND MALE SEXUALITY

Although engaged in another research project, in my free time with friends like Paul, I thus stumbled onto a series of questions that began to perplex me: Who was Linda? In the letter quoted above, I had unproblematically identified Linda as "gay." But in *his* context, was he? And if so, how did he come to see himself as so? And I quickly confronted questions of gender as well. Did Linda consider himself as male? And if so, had he always done so?

As issues like these began to pose themselves, I soon realized that for black men in townships around Johannesburg, identifying as gay was both recent and tied up, in unexpectedly complex ways, with a much larger historical transformation: the end of apartheid and the creation of a modern nation; in a phrase, the "freeing" of South Africa.

This story, more than any other, constitutes for most South Africans (certainly black South Africans) what Stuart Hall referred to as a "narrative of history." It structures identity, legitimates the present, and organizes the past. There are indeed few places on earth in which modernist narratives of progress and freedom currently appear so compelling. This undoubtedly results, at least in part, because apartheid itself was an antimodernist project that explicitly set itself against most of the rest of the "developed" world.

[. . .]

So how did Linda become gay? I never met or interviewed Linda, but fortunately, for the purposes of this article, before he died Linda wrote an extraordinarily self-revealing article with Hugh McLean entitled "*Abangibhamayo Bathi Ngimnandi* (Those Who Fuck Me Say I'm Tasty): Gay Sexuality in Reef Townships." The collaboration between Hugh and Linda – both members of GLOW – was itself a part of the transformations I seek to understand: the creation of a black gay identity, Linda's "coming out," and the "freeing" of South Africa.

To begin with, Linda did not always consider himself – to adopt the gender category appropriate at the end of his life – to be "gay." If anything, it was female gender, not sexuality as such, that fit most easily with local disciplinary regimes and that made the most sense to Linda during his teenage years. Indeed, in apartheid-era urban black culture, gender apparently overrode biological sex to such a degree that it is difficult, and perhaps inappropriate, to maintain the distinction between these two analytical concepts below.

Let me quote the comments of Neil Miller, a visiting North American gay journalist who interviewed Linda:

> Township gay male culture, as Linda described it, revolved around cross-dressing and sexual role-playing and the general idea that if gay men weren't exactly women, they were some variation thereof, a third sex. No one, including gay men, seemed to be quite sure what gay meant – were gay men really women? men? or something in between? . . . When Linda was in high school word went out among his schoolmates that he had both male and female sex organs. Everyone wanted to have sex with him, he claimed, if only to see if the rumors were true. When he didn't turn out to be the anatomical freak they had been promised, his sexual partners were disappointed. Then, there was the male lover who wanted to marry Linda when they were teenagers. "Can you have children?" the boy's mother asked Linda. The mother went to several doctors to ask if a gay man could bear a child. The doctors said no, but the mother didn't believe them. She urged the two boys to have sex as frequently as possible so Linda could become pregnant. Linda went along with the idea. On the mother's orders, the boys would stay in bed most of the weekend. "We'd get up on a Saturday morning, she'd give us a glass of milk, and she'd send us back to bed," Linda told me. After three months of this experiment, the mother grew impatient. She went to yet another doctor who managed to convince her that it was quite impossible for a man, even a gay man, to bear a child. Linda's relationship with his friend continued for a time until finally the young man acceded to his mother's wishes and married a woman, who eventually bore the child Linda could never give him.

The description above uses the word gay anachronistically. In black township slang, the actual designation for the effeminate partner in a male same-sex coupling was *stabane* – literally, a hermaphrodite. Instead of sexuality in the Western sense, it was local notions of sexed bodies and gendered identities – what I shall call sex/gender in the black South African sense – that divided and categorized. But these two analytical dimensions, gender and sex, interrelated in complex ways. While she was growing up, Linda thought of herself as a girl, as did Jabu, the drag queen at Linda's funeral about whom I shall have more to say below. Even though they had male genitalia, both were raised by their parents as girls and both understood themselves in this way.

If it was gender that made sense to Linda and Jabu themselves (as well as to some others close to them, such as parents and "mothers-in-law"), strangers in the township typically used sex as a classificatory grid. That is, both Linda and Jabu were taken by others as a biologically-mixed third sex. Significantly, as far as I can tell, neither ever saw themselves in such terms.

[...]

If an urban black South African boy during the 1960s and 1970s showed signs of effeminacy, then there was only one possibility: she was "really" a woman, or at least some mixed form of woman. Conversely, in any sexual relationship with such a person, the other partner remained, according to most participants, simply a man (and certainly not a "homosexual").

This gendered system of categories was imposed on Linda as she grew up:

> I used to wear girls' clothes at home. My mother dressed me up. In fact, I grew up wearing girls' clothes. And when I first went to school they didn't know how to register me.

Miller recorded the following impressions:

> Linda didn't strike one as particularly effeminate. He was lanky and graceful, with the body of a dancer. The day we met he was wearing white pants and a white cotton sweater with big, clear-framed glasses and a string of red African beads around his neck. But even as an adult, he was treated like a girl at home by his parents. They expected him to do women's jobs – to be in the kitchen, do the washing and ironing and baking. "You can get me at home almost any morning," he told me. "I'll be cleaning the house." There were girls' shopping days when he, his mother, and his sister would go off to buy underclothes and nighties. Each day, he would plan his mother's and father's wardrobes. As a teenager, Linda began undergoing female-hormone treatments, on the recommendation of a doctor. When he finally decided to halt treatments, his father, a minister of the Twelfth Apostle Church, was disappointed. It seemed he would rather have a son who grew breasts and outwardly appeared to be a girl than a son who was gay. Even today, Linda sings in the choir at his father's church – in a girl's uniform.
>
> "What part do you sing?" I asked him.
>
> "Soprano, of course," he replied. "What did you think?"

The fact that Linda wore a girl's uniform in church into the early 1990s offers some insight into his father's remarks that caused such a stir in GLOW. His father was not, it seems, particularly concerned with "cross-dressing." Phrasing the matter this way implies, after all, a naturally given bodily sex that one dresses "across." To Linda's family, he was apparently really a female. What the father was most upset about was dressing "across" race, and the implications that had for ancestral blessings.

In sum, black townships during the apartheid era found it easier to understand gender-deviant boys as girls or as a biologically mixed third sex. By the early 1970s, a network of boys who dressed as girls existed in Soweto, many of whom came to refer to themselves in their own slang as skesana.

[...]

Skesanas dressed as women and adopted only the receptive role in sexual intercourse. Here is Linda speaking:

> In the township they used to think I was a hermaphrodite. They think I was cursed in life to have two organs. Sometimes you can get a nice *pantsula* [tough, macho guy] and you will find him looking for two organs. You don't give him the freedom to touch you. He might discover that your dick is bigger than his. Then he might be embarrassed, or even worse, he might be attracted to your dick. This is not what a *skesana* needs or wants. So we keep up the mystery. We won't let them touch and we won't disillusion them ... I think it makes you more acceptable if you are a hermaphrodite, and they think your dick is very small. The problem is, the *skesanas* always have the biggest dicks. And I should know ...

It would be a mistake to view this system of sex/gender categories as *only* being imposed upon skesanas. In adopting their highly visible role, skesanas sometimes used the traditional subordinate role of the woman to play with and ultimately to mock male power. According to Linda,

> On a weekend I went to a shebeen [informal drinking establishment] with a lady friend of mine. I was in drag. I often used to do this on the weekends – many *skesanas* do it. We were inside. It seemed as if four boys wanted to rape us, they were *pantsulas* and they were very rough. One of them proposed to my friend and she accepted. The others approached me one by one. The first two I didn't like so I said no! I was attracted by the third one, so I said yes to him. As we left the shebeen, my one said to me, "If you don't have it, I'm going to cut your throat." I could see that he was serious and I knew I must have it or I'm dead. So I asked my friend to say that she was hungry and we stopped at some shops. I went inside and bought a can of pilchards [inexpensive fish]. I knew that the only thing the *pantsula* was interested in was the hole and the smell. *Pantsulas* don't explore much, they just lift up your dress and go for it. We all went to bed in the one room. There were two beds. The one *pantsula* and my friend were in one and I was in the other bed with this pantsula. . . . Sardines is one of the tricks the *skesanas* use. We know that some *pantsulas* like dirty pussy, so for them you must use pilchards, but not Glenrick [a brand] because they smell too bad. Other *pantsulas* like clean pussy, so for them you can use sardines. For my *pantsula* I bought pilchards because I could see what kind he was. So before I went to bed I just smeared some pilchards around my anus and my thighs. When he smelled the smell and found the hole he was quite happy. We became lovers for some months after that. He never knew that I was a man, and he never needed the smell again because he was satisfied the first time.

Although the connection would have been anathema to the Puritan planners of apartheid, skesana identity was finally tied up with the structure of apartheid power – particularly with the all-male hostels that dotted Soweto. In these hostels, rural men without the right to reside permanently in Soweto and without their wives lived, supposedly temporarily, in order to provide labor to the white-dominated economy. From the 19th century onward, there is evidence that at least some black men in these all-male environments saw little wrong with taking other, younger workers as "wives." In these relationships, it was age and wealth, not sex, that organized and defined male-male sexual relationships; as boys matured and gained their own resources, they in turn would take "wives." This pattern has been described among gangs of thieves on the Rand in the early 20th century and among gold mine workers into the 1980s.

Certainly, in Soweto in the 1960s, hostels populated by rural men had become notorious sites for same-sex sexual relations. Township parents warned young sons not to go anywhere nearby, that they would be swept inside and smeared with Vaseline and raped. To urban raised skesanas like Linda, however, these stories apparently only aroused fantasy and desire. Linda described a "marriage ceremony" in which she took part in one of the hostels, as follows:

> At these marriage ceremonies, called *mkehlo*, all the young *skesanas* . . . sit on one side and the older ones on the other. Then your mother would be chosen. My mother was MaButhelezi. These things would happen in the hostels those days. They were famous. The older gays [sic] would choose you a mother from one of them. Then your mother's affair [partner] would be your father. Then your father is the one who would teach you how to screw. All of them, they would teach you all the positions and how to ride him up and down and sideways . . .

MODERNITY AND SEXUALITY IN THE "NEW" SOUTH AFRICA

By the early 1990s, a great deal had changed in South Africa and in Linda's life. Nelson Mandela had been released from prison. It was clear to everyone in South Africa that a new society was in process of being born. This clarity had come, however, only after more than a decade and a half of protracted, agonizing, and often violent struggle – a contest for power that upended routines all the way from the structures of the state down to the dynamics of black families in Soweto. As a result, the cultural definitions and social institutions that supported the sex/gender system in which Linda had been raised had been shaken to their roots.

By the 1990s, Linda and his friends no longer felt safe going to the hostels; many rural men's compounds in the Johannesburg region had become sites of violent opposition to the surrounding black townships, the conflict often being phrased in terms of the split between the Inkatha Freedom Party and the ANC. Also, as the end of apartheid neared, rural women began to join their men in the hostels, and the old days of male–male marriages were left behind. Looking back from the 1990s, Linda commented,

> This [male–male marriage] doesn't happen now. You don't have to be taught these things. Now is the free South Africa and the roles are not so strong, they are breaking down.

I will make explicit what Linda suggested: with the birth of a "free" South Africa, the notion of sexuality was created for some black men, or more precisely, an identity based on sexuality was created. The classificatory grid in the making was different from the old one. Now, both partners in a same-sex relationship were potentially classified as the same (male) gender – and as "gay."

Obviously, this new way of looking at the sexual world was not taken up consistently, evenly, or completely. The simultaneous presence of different models of same-sex sexuality in present-day South Africa will be evident by the end of this article. Whatever the overlapping ambiguities, it is interesting to note who took the lead in "modernizing" male–male sexuality in black South Africa: it was precisely formerly female-identified men like Linda and Jabu. But if female-identified men seem to have initiated the shift, a turning point will be reached when their male partners also uniformly identify as gay. It is perhaps altogether too easy to overstate the degree to which such a transformation has occurred in the United States itself, particularly outside urban areas and outside the white middle and upper classes.

If one sexual paradigm did not fully replace another in black townships, there were nonetheless significant changes by the early 1990s. Three events, perhaps more than others, serve to summarize these changes. First was the founding of a genuinely multiracial gay rights organization in the Johannesburg area in the late 1980s – namely, GLOW. Linda was a founding member. Second, around the same time, the ANC, still in exile, added sexuality to its policy of nondiscrimination. As I shall explain below, the ANC's peculiar international context – its dependence on foreign support in the fight against apartheid – was probably one of the factors that inclined it to support gay rights. According to Gevisser,

> ANC members in exile were being exposed to what the PAC's [Pan-Africanist Congress] Alexander calls "the European Leftist position on the matter." Liberal European notions of gender rights and the political legitimacy of gay rights had immense impact on senior ANC lawyers like Albie Sachs and Kader Asmal, who have hence become gay issues' strongest lobbyists within the ANC.

Finally, a third event that heralded change was the first gay pride march in Johannesburg in 1990, modeled on those held in places like New York and San Francisco that celebrated the Stonewall riots of 1969. Linda and his friends participated, along with approximately one thousand others. This annual ritual began to do much, through a set of such internationally recognized gay symbols as rainbow flags and pink triangles, to create a sense of transnational connections for gay South Africans.

How was Linda's life affected by these changes? Exactly how did sexuality replace local definitions of sex/gender in her forms of self-identification? According to Linda himself, the black youth uprising against apartheid was the beginning:

> Gays are a lot more confident now in the townships. I think this happened from about 1976. Before that everything was very quiet. 1976 gave people a lot of confidence . . . I remember when the time came to go and march and they wanted all the boys and girls to join in. The gays said: We're not accepted by you, so why should we march? But then they said they didn't mind and we should go to march in drag. Even the straight boys would wear drag. You could wear what you like.

As black youth took up the cause of national liberation and townships became virtual war zones, traditional black generational hierarchies were shaken to the core. Black youth came to occupy a new political space, one relatively more independent of the power of parents. But as such resistance movements have developed in other times and places, gender hierarchies have sometimes been strengthened. In resisting one form of domination, another is reinforced. In the black power movement in the United States during the 1960s, for example, masculinist and heterosexist ideals were sometimes celebrated.

Why did this reaction, with respect to gender, not take place in South Africa? One respect in which the South African case differs, certainly compared to the United States in the 1960s, is the extent to which the transnational was involved in the national struggle. Until Mandela was released, the ANC was legally banned in South Africa. Leaders not in prison were based outside the country, and there can be no doubt that the ANC could not have accomplished the political transition that it did without international support. In this context, the international left-liberal consensus on human rights – one to which gay people also appealed – probably dampened any tendency to contest local racial domination by strengthening local gender and sexuality hierarchies. Any such move would certainly have alienated antiapartheid groups from Britain to Holland to Canada to the United States.

But the significance of the transnational in the South African struggle was not only material. The imaginations of black South Africans were finally affected – particularly, in the ways in which people located themselves in the world. And it was precisely in the context of transnational antiapartheid connections that some skesanas like Linda, particularly after they were in closer contact with white gay people in Johannesburg, became aware for the first time of a global gay community – an imagined community, to adapt Benedict Anderson's phrase, imaginatively united by "deep horizontal bonds of comradeship."

How did this occur? Perhaps the incident, more than any other, that catalyzed such associations, that served as a node for exchange, was the arrest of Simon Nkoli. Nkoli, by the 1980s a gay-identified black man, was arrested for treason along with others and tried in one of the most publicized trials of the apartheid era – the so-called Delmas treason trials. After Nkoli's situation became known internationally, he became a symbol for gay people in the antiapartheid movement across the globe. For example, in December of 1986, while he was in prison, Nkoli was startled to receive more than 150 Christmas cards from gay people and organizations around the world.

According to Gevisser,

> In Nkoli, gay anti-apartheid activists found a ready-made hero. In Canada, the Simon Nkoli Anti-Apartheid Committee became a critical player in both the gay and anti-apartheid movements. Through Nkoli's imprisonment, too, progressive members of the international anti-apartheid movement were able to begin introducing the issue of gay rights to the African National Congress. The highly respectable Anti-Apartheid Movements of both Britain and Holland, for example, took up Nkoli's cause, and this was to exert a major impact on the ANC's later decision to include gay rights on its agenda.

These cultural connections and others eventually helped to produce changes in the most intimate details of skesanas' lives. To return to Linda, gay identity meant literally a new gender and a new way of relating to his body. In Linda's words,

> Before, all skesanas wanted to have a small cock. Now we can relax, it does not matter too much and people don't discuss cocks as much. . . . Before, I thought I was a woman. Now I think I'm a man, but it doesn't worry me anyway. Although it used to cause problems earlier.

In addition to how he viewed his body, Linda began to dress differently:

> I wear girl's clothes now sometimes, but not so much. But I sleep in a nightie, and I wear slippers and a gown – no skirts. I like the way a nightie feels in bed.

Consider the underneath-of-the-iceberg for the intimacies that Linda described: it is difficult to reconstruct the hundreds of micro-encounters, the thousands of messages that must have come from as far away as Amsterdam and New York. Gevisser outlines some of the social underpinnings of this reordering:

> The current township gay scene has its roots in a generalized youth rebellion that found expression first in 1976 and then in the mid-1980s. And, once a white gay organization took root in the 1980s and a collapse of rigid racial boundaries allowed greater interaction between township and city gay people, ideas of gay community filtered into the already-existent township gay networks. A few gay men and lesbians, like Nkoli, moved into Hillbrow. As the neighborhood started deracializing, they began patronizing the gay bars and thus hooking into the urban gay subculture – despite this subculture's patent racism. GLOW'S kwaThema chapter was founded, for example, when a group of residents returned from the Skyline Bar with a copy of Exit [the local gay publication]: "When we saw the publicity about this new non-racial group," explains Manku Madux, a woman who, with Sgxabai, founded the chapter, "we decided to get in touch with them to join."

The ways in which an imagined gay community became real to black South Africans were, of course, various. In Jabu's case, he had already come to see himself differently after he began work in a downtown hotel in Johannesburg in the late 1970s:

> Well, I joined the hotel industry. I started at the Carlton Hotel . . . There was no position actually that they could start me in. I won't say that being a porter was not a good job; it was. But they had to start me there. But I had some problems with guests. Most of them actually picked up that I'm gay. How, I don't know. Actually . . . how am I going to word this? People from foreign countries, they would demand my service in a different way . . . than being a porter . . . We had Pan Am, British Airways, American Airways coming. Probably the whole world assumes that any male who works for an airline is gay. I used to make friends with them. But the management wasn't happy about that, and they transferred me to the switchboard.

The assumption of a gay identity for Jabu affected not only his view of the present but also of the past. Like virtually all forms of identification that essentialize and project themselves backward in history, uniting the past with the present (and future) in an unchanging unity, South African black gay identity does the same. According to Jabu, being gay is "natural"; gay people have always been present in South African black cultures. But in his great-grandmother's time, African traditional cultures dealt with such things differently:

> I asked my grandmother and great-grandmother (she died at the age of 102). Within the family, the moment they realized that you were gay, in order to keep outside people from knowing, they organized someone who was gay to go out with you, and they arranged with another family to whom they explained the whole situation: "Okay, fine, you've got a daughter, we have three sons, this one is gay, and then there are the other two. Your daughter is not married. What if, in public, your daughter marries our gay son, but they are not going to have sex. She will have sex with the younger brother or the elder brother, and by so doing, the family will expand, you know." And at the end of the day, even if the next person realizes that I am gay, they wouldn't say anything because I am married. That is the secret that used to be kept in the black community.

[. . .]

It goes without saying that the category of gay people in South Africa is hardly homogeneous, nor is it the same as in Western countries. Being black and gay and poor in South Africa is hardly the same as being black and gay and middle-class, which is again hardly the same as being white and gay and middle-class, whether in South Africa or in North America. Despite these differences, there is still in the background a wider imagined community of gay people with which all of these persons are familiar and, at least in certain contexts, with which they identify. How this imagined community becomes "available" for persons variously situated across the globe is a major analytical question.

In Paul Gilroy's analysis of the black diaspora, he writes suggestively of the role of sailors, of ships, and of recordings of black music in making a transnational black community imaginatively real. As black identity has been formed and reformed in the context of transnational connections, black families have typically played some role – complex to be sure – in reproducing black identity. Gay identity is different to the degree that it does not rely upon the family for its anchoring; indeed, if anything, it has continually to liberate itself from the effects of family socialization.

This means, ipso facto, that identifying as gay is peculiarly dependent upon and bound up with modern media, with ways of communicatively linking people across space and time. In North America, how many "coming out" stories tell about trips to the public library, furtive searches through dictionaries, or secret readings of novels that explore lesbian and gay topics? A certain communicative density is probably a prerequisite for people to identify as gay at all, and it is not improbable that as media density increases, so will the number of gay people.

In less-developed societies of the world today, then, transnational flows become particularly relevant in understanding the formation of sexual communities. Sustained analysis of these connections has hardly begun, but I would suggest that we start not with ships but with airplanes, not with sailors (although they undoubtedly played their role here as well, particularly in port cities) but, in the South African case, with tourists, exiled antiapartheid activists, and visiting anthropologists; and finally not with music, but with images, typically erotic images – first drawings, then photographs, and now videos, most especially of the male body.

Given the composition of the global gay community, most of these images are of the *white* male body. For black men, then, identifying as gay must carry with it a certain complexity absent for most white South Africans. Also, the fact that international gay images are overwhelmingly male probably also affects the way that lesbian identity is imagined and appropriated by South African women, black and white. In any case, it could be argued that these kinds of contradictory identifications are not exceptional under late capitalism; they are the stuff of most people's lives. And lately the flow of images has been greatly accelerated; South African gays with access to a modem and a computer – admittedly, a tiny minority so far – can now download material from San Francisco, New York, or Amsterdam.

[...]

The overlapping of the national and gay questions means that gay identity in South Africa reverberates – in a way that it cannot in the United States – with a proud, new national identity. Let me quote the reaction of one of the white gays present at Linda's funeral:

> As I stood in Phiri Hall behind the black gay mourners behind the hymn-singing congregants, I felt a proud commonality with Linda's black friends around, despite our differences; *we were all gay, all South African.*

[...]

"Whose Place is this Space? Life in the Street Prostitution Area of Helsinki, Finland"

from *International Journal of Urban and Regional Research* (2002)

Sirpa Tani

Editors' Introduction

Most cities, both large and small, house some form of a red-light district where street prostitution functions within a local, marginalized economy. In recent decades, urbanists have taken an interest beyond simply mapping the prostitution urban economy and focused more on the complex interrelationships of gender, sexuality, and urban public space. In this reading, Sirpa Tani examines how representations of life in the prostitution area of Helsinki are constructed and contested. As an ethnographer living in her area of study, Tani became interested in the range of interactions between official and media-based notions of the sex industry and the subjective experiences for those residents (especially women) who lived in the neighborhood. The major opposition to prostitution was a grassroots organization called "Prostitution Off The Streets" that formed to combat Helsinki neighborhoods from becoming sexualized spaces. The organization created a petition to bring an end to the prostitution and mounted tactics such as publishing license plate numbers of automobiles used in solicitation. Her study first examines the media representations of women in Helsinki's prostitution zone, followed by a thick description of battles between "kerb-crawlers" (men seeking prostitutes) and women on the streets. Finally, Tani explains how policy and grassroots efforts to control prostitution reflect local or national norms over the tolerance of difference in urban public space.

Sirpa Tani is Professor of Geography and Environmental Education at the University of Helsinki, Finland. Her research interests are cultural geography, urbanism, and the meanings of environmentalism for young people.

The collapse of the Soviet Union and the economic problems there had a strong impact on the sex industry in Finland. It became easier for Russians and Estonians to come to Finland with tourist visas and, at the same time, sex tourism across the border from Finland to Russia increased. The first Estonian and Russian prostitutes came to Finland as tourists and worked individually, but soon the sex industry became more organized. Some motels and hostels in remote places had brothel-like activities, which attracted male clients from near and far. In Helsinki, foreign prostitutes started to work not only in erotic restaurants but also on the streets. Street prostitution and kerb-crawling became concentrated in a densely built-up residential area near the city centre, causing confusion, anger and fear among some local residents, who started to protest against prostitution on the streets and in the media. I happened to move to the neighbourhood before it turned into a prostitution area and therefore felt I had a rare opportunity to follow the changes from the beginning.

Street prostitution has often been seen either as a question of sexuality and space or as a struggle between different interest groups. The majority of studies have been based on secondary data or, when the viewpoint has been that of the prostitutes, in-depth interviews. These two types of data have seldom been used together in one study. The anti-prostitution campaigns have been interpreted using media coverage as the source material, but the opinions of the residents affected by the issue have not been heard.

The empirical analysis is based on qualitative data gathered over a period of six years (1993–8), but the debate has been followed up to the end of 2000. The first newspaper articles concerning contemporary street prostitution in Helsinki were published in 1993. Public debate was occasional until 1995 and I started a more detailed analysis at that time. Of the 89 articles I analysed, 14 were published in 1995, 47 in 1996, 11 in 1997 and 17 in 1998. There were also 33 letters to the editor, which I analysed.

In order to interpret subjective experiences in the street prostitution debate, I interviewed twenty-one people from 1996 to 1998. Twelve of them were interviewed alone and nine in small groups (2–4 persons per interview). Fourteen were women and seven men, aged between 24 and 64. Two fought actively against prostitution and kerb-crawling in the area, and two worked in the neighbourhood at night. Eighteen interviewees lived in the research area or in the close vicinity. The remaining three were social workers, two working among the prostitutes and one with the youth of the area. It can be said that all these people had a personal relationship to the research area. The interviews were tape-recorded and later transcribed.

I lived in the research area from 1993 to 1999, and therefore observed the environment in my daily life. Because the aim of the study was to interpret subjective and media images of the street prostitution debate, I found it very fruitful to live in the neighborhood during the research period. Sometimes this "closeness", however, was disturbing: I could not close my eyes and draw a line between my fieldwork and my leisure time in the area.

In my research, I also encountered some problems. It was difficult to find a way of communicating with prostitutes on the streets because I felt like an intruder in their space, disturbing their work. Because I lived in the area and walked there regularly, I was aware that the prostitutes could easily regard me as one of the women "activists" who occasionally came into conflict with prostitutes on the streets. Therefore I abandoned the idea of conversations with them on the streets. I tried to make contact with some of them through the help of social workers, but without success.

Kerb-crawlers form another group in my research that has no voice of its own. It is problematic to make contact with male drivers in order to carry out an academic interview when they are kerb-crawling in the prostitution area. Having to exclude the prostitutes and kerb-crawlers from the interviews clearly places limitations on this study. Bearing this in mind, I am not trying to explain the whole issue of street prostitution in Helsinki, but focus instead on two aspects of it: the subjective experience of local residents and media images.

STREETS AS SPACES OF RESISTANCE

It has been estimated that when street prostitution in Helsinki reached its peak in the 1990s, there were around 50 street prostitutes, while the total number of prostitutes in the city was between 1,000 and 2,000, 200 of whom were foreigners (Koskela *et al.*, 2000). Although most of the prostitutes in Finland are Finnish, the majority of the *street prostitutes* were foreigners, and therefore the whole issue of prostitution could easily and conveniently be interpreted as a problem caused by foreigners (Turunen, 1996). Street prostitution has operated mainly as a small-scale business in Finland, but because of its visibility it has attracted more attention in the media than other branches of the sex industry.

Besides the strong emphasis on the foreigner issue in the public debate, prostitution in Finland has also been very strongly connected to heterosexuality, to straight women as prostitutes and straight men as their clients. Male prostitutes, either straight or gay, have been left out of the official statistics, and therefore their number can only be guessed at. What is obvious, though, is the fact that there are practically no male street prostitutes in Helsinki; during my stay in the research area from 1993 to 1999, I saw only *one* transvestite working in the area.

Prostitution legislation in western societies varies from strict criminalization to legalization, and the Finnish case falls between those two extremes. For example, in Eire prostitution is illegal in every respect (Goodall, 1995), and in Sweden prostitutes' clients became criminalized in 1999. In the USA prostitution is criminalized everywhere except in Nevada, where 10 of the 15 counties have legalized brothels (Stetson, 1997; Hausbeck and Brents, 2000; Weitzer, 2000). In the majority of countries where brothels are illegal and prostitution is unlawful, the practice is still tolerated and has been allowed to concentrate in certain areas where it can be controlled. These "toleration zones" can be found, for instance, in Germany (Goodall, 1995; Hubbard, 1997). The pivotal role of the well-known red-light district in Amsterdam changed in 2000, when brothels were legalized throughout the Netherlands. In Britain, prostitution itself is not illegal, but soliciting, living on "immoral earnings" and procuring, as well as kerb-crawling, are (O'Neill and Barberet, 2000). In practice, prostitution is "tolerated" in specific areas in the big cities.

The Vagrancy Act, which had controlled prostitution in Finland, was repealed in 1986. Since then, prostitution, unlike procuration, has not been illegal and there have been no official means of regulating it. Until the end of the 1980s, prostitution remained "invisible" (Høigârd and Finstad, 1992; Turunen, 1996). There was a "silent compromise" prevailing: prostitution was not interfered with so long as it did not "disturb the peace in society" (Järvinen, 1990). In Finland, the prostitution debate of the 1990s has raised the question of whether to criminalize or legalize prostitution, but no decisions have yet been made.

In Helsinki, the initiators of local activism were some female residents who were disturbed by prostitution and kerb-crawling in their neighbourhood. During the summer of 1994, they contacted their neighbourhood association in order to get some help to put an end to kerb-crawling. A local free newspaper published the car registration numbers of the suspected kerb-crawlers. Some female residents started to patrol the area at night to make observations and to try to prevent kerb-crawling. They founded a movement called "Prostitution Off The Streets". Its aim was to fight against street prostitution in general, and to defend the right of the local residents to a peaceful environment. The members of the movement organized the signing of a petition against street prostitution, discovered the identity of the kerb-crawlers and telephoned either their wives or employers to reveal that their cars had been seen in the street prostitution area. This kind of activity captured the interest of the national press.

The activists made an appeal for their right to "come home in peace" as they called it. As a result, all the through traffic on one street was prohibited from 11pm to 6am as an experiment. Because of the positive experience – a decrease in disturbing traffic – the prohibition was later made permanent and extended to the neighbouring streets. The prostitutes and kerb-crawlers moved one street corner to the north.

In the public debate, the viewpoint of the local activists was portrayed as a question of their moral right to protect their neighbourhood from "outsiders". It was a question of social exclusion (Sibley, 1995) and the perceived "otherness" of prostitutes and kerb-crawlers. The activists felt that the neighbourhood was *their* space and that they had every right to fight for it. In their speeches they created the notion of shared values in the area by promoting the idea that when they were acting against prostitution, they were acting *for the whole community*.

Moral questions have been attached to prostitution throughout its history. Those people who are against prostitution have often viewed prostitutes either as poor victims exploited by pimps and clients, or as women who "exercised choice in entering prostitution". In the Finnish debate, a kind of "moral panic" was reported in the press when different parties tried to define the rights and the wrongs for the whole neighbourhood, and often from completely different angles. Prostitution was often associated with crime, drugs and asocial (or immoral) behaviour and it was represented as one of the first signs of neighbourhood decline.

The same kind of local reaction has been reported in other countries. For example, in the USA, having become frustrated by the ineffectiveness of the authorities' response to local prostitution problems, local residents have begun to take direct action. Their tactics include citizen patrols, recording car number plates, and "public shaming", such as publishing the names of clients in newspapers or on television, or putting up posters where the clients are named. In San Francisco, for instance, the anti-prostitution campaigns have become highly organized. Since 1995, there have been classes for male clients to discourage or to "shame, educate and deter" them from future contacts with prostitutes.

WOMAN AS THE "SIGN" OF A PROSTITUTE

The Finnish press portrayed the situation on the streets by focusing on the confrontations between different interest groups, such as the local residents and the prostitutes. The tabloid *Iltalehti* (4 September 1995) interviewed one of the residents who described her feelings in the following way:

> If you are going to have an evening out, you have to take a taxi right from your doorstep. You cannot walk in the street in peace – You do not even have to have dressed conspicuously, and still the kerb-crawlers jeer at you.

The tabloid portrayed her as "a pleasant, normal young woman" who did not "look like a prostitute at all". The quotation is a typical example: the press constructs simplified and stereotyped images of women on the streets. They are represented either as non-sexual victims (local residents) or "overtly sexual beings" (prostitutes) (Bondi, 1998: 191). The residents are described as ordinary-looking housewives carrying shopping bags, or some other means is used to express the idea that they cannot be interpreted as prostitutes by their appearance.

Most of the journalists who wrote the articles and constructed the polarized images were men, but the activists on the streets were female residents who demanded peace in their neighbourhood. In my interviews, many women, both the activists and other residents, made a clear distinction between themselves and the prostitutes. They argued that although they did not look like prostitutes, the kerb-crawlers treated them as if they were. The street had turned into *a sexualized space*, where woman – the non-sexual victim – seemed to have become *a sign of a prostitute*.

> Me, who's like a ten ton weight, not a little thing, but everything seems to go . . . when everything sells, even someone who looks like me . . . I am not any little miss in a mini-skirt (Katri, lives in the area).

Most of the female interviewees described themselves as casually dressed, and some of them even defined themselves as unattractive in appearance compared to the prostitutes. Some of them said that they felt it was impossible to go out dressed up:

> What annoyed me the most at some point was maybe that I went along with it myself, with the idea that when you're walking down the street you think before you go what you are going to wear (Suvi, social worker).

Women's awareness of being objects of the male gaze on the streets is reflected in these interviews. They emphasize that being dressed up is not permissible in the area – that is, if they are not willing to run the risk of being taken for a prostitute by kerb-crawlers. Here, "women can be seen as (unwittingly) reproducing gendered power relations by policing their own dress". If they were harassed on the streets, some of them turned the blame on themselves:

> You start to examine if you look a lot like a hooker. You feel that way about yourself (Tuula, lives in the area).

Women are used to dressing down in order to give the right impression – or they are advised to do so to prevent crime or sexual harassment, and therefore they can easily worry about looking too feminine. Not all the interviewed women, however, reacted in this way. Some of them consciously wanted to maintain their feminine look, even in the street prostitution quarters:

> I wear mini-skirts just the same as before and I do not let it affect me. That would be kind of like giving in (Anna, social worker).

In the media, the prostitutes were represented as "opposite" to the female residents. In an article in which a female resident was portrayed as "pleasant" and "normal", prostitutes were described as follows: "They differ from the local residents. Most of them are Estonians or they come from somewhere else in the East. They have high-heeled shoes, dyed hair and heavy make-up. The men watch them closely". This kind of stereotyped vision of the prostitute is common not only in the press, but also in other cultural artifacts [sic]: for example in literature, film and television, as well as in the history of prostitution itself. It has been said that street prostitutes have to adopt "specific codes" to mark themselves "for sale" – to make themselves distinct from other women on the street. The press emphasizes this by describing both the prostitutes and female residents solely in terms of their appearance.

Prostitutes were represented as the "other" in the area, as being totally different from local residents. Besides being said to look different, they were also reported to be foreigners. The majority of the street workers came from either Estonia or Russia, which made it even easier to turn the anger against "outsiders" and to make highly generalized distinctions like us/them, female residents/female prostitutes and Finns/foreigners. Some of the interviewees used this kind of "othering" in their descriptions, too:

> You can see it on the street all right. They walk about with wigs on, short skirts and high heels and Russian-style clothes. Of course, it catches the eye of someone like me (Tuula, lives in the area).

The Russian or Estonian accent was often seen as a way of identifying the prostitutes and differentiating them from other women on the streets. It was supposed that if street prostitution were to become a permanent part of urban culture in Helsinki, more Finnish women would recruit themselves to it. In other words, the foreign prostitutes were represented as a threat to the Finnish women. It was easy to blame them because they were foreigners and because they did not live in the neighbourhood.

In the interviews, most of those who described the prostitutes were female residents, but also one male (gay) resident told me about his first contact with a female prostitute and described her looks:

> From the corner of Aleksis Kiven katu and Fleminginkatu walked a woman who had a short skirt, hardly concealing anything, and shiny black leather boots. She started to stare at me and after that she walked straight towards me. I was almost shitting myself already, how can I escape? I could not think of anything but to start reading my monthly [bus] pass (Oskari, lives in the area).

This example shows an interesting viewpoint on the matter. Usually, the media and the literature on prostitution emphasize the subject/object relationship between men and women, representing men as active sex-hunters and women as passive victims. In this case, life on the streets seems to turn into the opposite. The gay man felt that the female prostitute sexually harassed him in his neighbourhood. The street had become not only a *sexualized* space, but also a very strongly *heterosexual* space.

In the press, the confrontation between local residents and prostitutes was heavily emphasized, but in my interviews, the views were much more diverse. Some people explained the situation on the streets from the prostitutes' point of view and used different levels of interpretation to explain the complexity of the debate. Some of those who saw the prostitutes as subjects, not objects, were social workers involved in giving support and advice to the prostitutes. They had undertaken fieldwork in the area and therefore reflected on their own position on the streets in a different way:

> Of course, you go quite humbly, in the sense that you are on someone else's territory in a way (Eira, social worker).
>
> Well, my work puts me in the same kind of marginal position my customers are in (Anna, social worker).

In my interviews not all the residents blamed the prostitutes for the disturbance. They felt that the situation on the streets was far more complicated than its image in the media, or they wanted to support the prostitutes' right to carry out their work in peace. Some were quick to condemn the actions of the local activists:

It is really paranoid when they [the local activists] think that these people, these girls, are not human beings, that they're like cattle. I can't understand that kind of attitude (Pekka, lives in the area).

SHRINKING SPACES?

In the newspapers, the third group besides the prostitutes and the female residents was the kerb-crawlers. Newspapers portrayed the situation much as a drama where the kerb-crawlers had one of the leading roles. The male drivers were represented as active hunters in the night, or "adventurers", who could easily take possession of the space despite the traffic restrictions.

In my interviews, the activists admitted that the kerb-crawling was the main problem for them, not the prostitutes. They directed their protest against the kerb-crawlers in order to restrict their driving in the area at night. Patrolling at night the same streets that prostitutes worked, the female activists also became objects of the male gaze. Because of the activists' strong opposition to kerb-crawling, they often had confrontations with the male drivers. Usually the confrontations were verbal but there were also some situations in which physical violence broke out. Some activists told me that drivers had physically threatened them:

They have threatened me back there, no bother. What small pieces they are going to tear me into, or how they are going to rape me or shoot me (Leena, local activist).

Although some women emphasized their right to dress up and take possession of the space, they were also aware of the possibility of being misinterpreted by the kerb-crawlers. They used several strategies to respond to the harassment. Some of them tried to ignore the kerb-crawlers on the streets and highlighted the importance of looking bold.

You put on this kind of a tough role, like, fucking hell, come near and I'll show you. You switch on a different persona or a different role when you want to go somewhere (Suvi, social worker).

I walk wherever I want. Like if I'm taking a walk in the evening and I want to go along Aleksis Kiven katu, I go and that's it. I don't let it affect me (Riikka, lives in the area).

In the case of kerb-crawling, the majority of the female interviewees described how in different situations they took different roles on the streets. Sometimes they felt bold, sometimes fearful. Boldness and fearfulness are often matters of choice, or at least matters of mood and situation. For many women the neighbourhood seemed to "become smaller" – to "shrink" – at night. Space was "elastic" for them, different in different circumstances. They felt that certain streets were not safe enough or were not "meant" for them. Many of them admitted that their awareness of the prostitution made them avoid certain streets at night. This shows how people attach different meanings to their environment depending on the time and the situation. Space is in essence a subjective place, the content of which is never at a standstill.

There were also occasions where women reacted more vocally against sexual harassment on the streets:

I tell them every time – See you in court – when they say anything rude. They go then (Leena, local activist).

I was shouting at the front door but this big-eared bloke arrived, I guess he was some pimp, some sad Estonian he was, who said you cannot stop us here. I blew up completely. I don't come to Estonia to sell myself at your door. Bugger off back to where you came from! (Minna, lives in the area).

Although the drivers were seen as an important party in the situation, they were seldom interviewed in the press. One reason for this is the fact that their cars effectively shielded them from the eyes of other people. The drivers seemed to stay anonymous, unlike their cars. In my interviews, the kerb-crawlers were discussed, but few of the interviewees had as detailed descriptions of them as they had of the prostitutes.

It is mainly, when you think of it, men in their forties there (Eira, social worker).

Both the prostitutes and the female activists walking on the streets were easy targets of the male gaze, and their sex appeal was judged by the kerb-crawlers. They were often said to be unable to distinguish the prostitutes from other female walkers on the street. A social worker, however, questioned this point:

> I do not believe a Finnish man is so stupid that he cannot tell a woman carrying shopping bags is not a sex worker. It's a question of something else, the will to repress, to dominate, if you have to call at a woman like that from the car window and make offers. The codes are clear enough as to who wants to get in that car and who does not (Anna, social worker).

Many female residents felt that the male drivers placed them in the position of sexual objects on the streets. It cannot be a question of the inability of the kerb-crawlers to "read signs of prostitution" correctly, but instead must have something to do with the images attached to specific areas of the city.

By noting the registration numbers and publishing them in local newspapers, the activists made the kerb-crawlers more visible. Most of them were considered to be outsiders, coming from further afield. In the press and in the interviews it was often stated that the majority of them came from the countryside and were thought not to be able to "behave in an appropriate way" in the urban environment. They – like the prostitutes – were defined as "others" and therefore it was easier to struggle against them.

Ironically, the number of cars cruising in the area was closely linked to the amount of public interest in the issue. The more the newspapers focused on street prostitution, the more curious outsiders became. The publicity increased the number of kerb-crawlers and "tourists" in the area and made the activists even angrier. In one interview an activist described how they tried to keep a low profile in their work in order to avoid "advertising" street prostitution. Sometimes, however, they used the media quite openly to inform the audience of their latest actions. For example, there were many journalists in the area when the activists sold apples – "forbidden fruits" – to the kerb-crawlers in order to highlight the point that even street vending in Finland is subject to licence, whereas prostitution is not (e.g. *Helsingin Sanomat*, 6 June 1999). Newspapers were quick to play their part in these actions by reporting them sensationally.

DISPLACEMENT AND OTHER STRATEGIES

Community protest caused successive *displacements* of the prostitutes and their clients. With the help of the traffic restrictions, the unwanted phenomena were relocated several times. The traffic signs, which indicated the limitations of through traffic, turned into *signifiers of the prostitution area* and made prostitution more visible and easier to attack.

Although the traffic restrictions had a calming effect on the area, not all the residents supported them. Some felt that the prohibitions made their life in the neighbourhood inconvenient; sometimes they were stopped by the police and even fined when they were simply driving around the block to find a parking place.

During the spring of 1996, the public debate focused on the idea of finding suitable locations for prostitution. The Chair of the City Government raised the possibility of creating a red-light district. She argued that because the Finnish legislation did not offer any effective means of restricting or prohibiting prostitution, new ways to deal with the problem were needed. She emphasized the importance of acquiring more knowledge about the subject and proposed that the City Government should visit red-light districts in the cities of Central Europe. These radical ideas raised conflicting views among city councillors. Some were accepting of the idea of a defined area for prostitution; others were vehemently opposed to it – some were ready to criminalize both the prostitutes and their clients.

In the media, the idea of establishing a red-light district polarized opinion. Nobody would want a red-light district in his or her neighbourhood, it was argued. Therefore, one possibility that was considered was that it could be located in a nearby uninhabited industrial site or warehouse area. That caused a strong negative reaction among the prostitutes who saw the possible new location as a threat to their safety. The debate over location locked horns over two opposing arguments: that of the prostitutes about *safe space* and that of the activists concerning *moral space*.

In my interviews, people expressed different opinions on the idea of creating a red-light district in Helsinki. Only one activist was strictly against it. She did not approve of prostitution because she felt it was invariably linked to disease, drugs and alcohol – to an "immoral" way of life. However, in the Finnish case the majority of the interviewees did not reject the idea of an established area for prostitution. They highlighted the point that it was difficult to decide what would be the most suitable area for prostitution. Some were eager to name areas where it could be located. The proposals varied from highly esteemed residential areas to uninhabited industrial and dock areas.

Red-light districts have given rise to contradictory opinions in other countries too. Their existence has often been justified by the argument that they make surveillance easier and more effective. In countries where prostitution is legal, the regulated industry includes registration for prostitution, compulsory medical checks and taxation. In countries where prostitution is technically illegal but where it is allowed within the bounds of "toleration zones" – for example in Germany and the United Kingdom – the situation is interesting. Although prostitution is not officially allowed, the police in those defined areas do not trouble prostitutes. Outside such areas, they can be prosecuted. However, red-light districts have often been linked to notions of danger, immorality, drugs and crime. The main reason for restricting prostitution to special areas has been the desire to protect public morality, to "shelter respectable citizens" from an immoral way of life. The climate of fear surrounding *visible prostitution* has been present in the Finnish case, promoted both by the media and by the local activists.

During the summer of 1996, the debate on prostitution in the Finnish media had dissipated somewhat and there was no burning issue to which journalists paid attention. Then it was reported that street prostitution had spread, not only in Kallio, but also to the city centre. This caused a new stir in the media and the level of the debate changed. The tabloid *Ilta-Sanomat* interviewed some female politicians who felt that prostitution in the main streets could be harmful to the *image* of the city. Not one of them expressed any concern about the possible stigmatization of the Kallio area. It appears there was a kind of "official" Helsinki for the local politicians, whose image they considered it important to protect. Kallio, on the other hand, has suffered a long history of poor reputation. Its working-class image is imprinted in people's minds so strongly that the defamation of the area does not seem to bother anybody except the local residents.

A major shift occurred when the City Government decided to prohibit prostitution in public places by municipal ordinance from the beginning of December 1999. This was interpreted to mean that selling – as opposed to buying – sex was not allowed on the streets. This interpretation was explained as being equivalent to street vending which – as the subject of licence – is restricted but customers are not sanctioned. Prostitutes can thus be fined for soliciting in Helsinki although prostitution is not illegal in Finland.

The situation changed dramatically after the prohibition. The police fined some prostitutes, but during the winter of 1999–2000 prostitution was said to have disappeared from the streets of Helsinki. For a while, this really seemed to be the case. According to my observations, though, before long the prostitutes and their clients returned to the area. Because the media had stopped reporting the phenomenon, the incidental "tourists" seemed to have lost their curiosity about the issue.

CONCLUSION

In geographical research, prostitution has often been seen as a question of gender and sexuality in urban public space. Prostitutes have been represented as the "other" in urban neighbourhoods and struggles over public space have been widely reported.

The aim of "purifying" the neighbourhood of unwanted phenomena has also been the case in Helsinki. The prostitutes have been represented as undesirable "others" in the media. Because the majority of street prostitutes were foreigners, it has been easy to emphasize the differences between the locals and the prostitutes and to create antagonism between "us and them" in the press. The local activists have used the stigmatization of the prostitutes to their advantage as well.

They have justified their actions by maintaining their right to define a suitable way of life in the neighbourhood. They have seen not only the prostitutes but also the kerb-crawlers as outsiders who have no right to use the space as their own. It has been a question of *ownership of the public space*. This leads to the idea of local interests, or the notion of community.

Residential activism is often seen as a manifestation of shared interests, but this case shows that there is no such thing as a collectively shared local experience. The activists' opinions are not shared by all the residents, but because of the publicity they have gained, it is easy to generalize them as collective attitudes. The residents experience and interpret their environment in the context of their daily life, and depending on the context they can reinterpret the same environment in various ways. Although spaces of prostitution are socially constructed, they are also individually experienced. The street prostitution area can simultaneously be seen as sexualized space, moral space, safe/unsafe space, power-space and as a space of struggle and resistance, but also as a unique and meaningful place.

In this article, street prostitution has been seen from two different points of view: that of the media and that deriving from the local level. The debate has been followed from the beginning to the present day, when prostitution, following several displacements, has almost disappeared from the streets of Helsinki and turned mainly to private places. The planned law will overrule the municipal ordinances and will cause difficulties in interpretation. It will raise some interesting questions: What kind of prostitution is disturbing? Who is disturbed? Who has the right to define the approved way of life in an urban environment? The prohibition of prostitution that causes a "disturbance" in public places will not end prostitution, but it could make it invisible to people outside the sex industry – and in doing so will make the marginalized prostitutes even more marginal in Finnish society.

PART FIVE

Globalization and Transnationality

Plate 8 Chinatown, New York City, USA, 1998 by Chien-Chi Chang. A newly arrived immigrant eats noodles on a fire escape. Reproduced by permission of Magnum Photos.

INTRODUCTION TO PART FIVE

This is the global age, as is often said. But what does globalization mean for cities? In this section, we include readings that address both structural political economic processes tied to globalization and the consequences of those processes for everyday urban life. The structural approach has been the mainstay of research on globalization. Recently, scholars have opened up the field of globalization studies to include factors other than economic and structural ones. The emphasis on cultural aspects of globalization addresses a different set of questions, such as, what does globalization mean to everyday life in the city?

The most prominent feature of structural analyses of globalization – with an impact on the ranking of cities and the economic, social, and political transformations within cities – is the movement or flow of international finance, foreign direct investments, and other forms of capital. The accelerated flow of capital is the defining attribute of globalization. These flows do not occur on their own, technological automation notwithstanding. There are key actors and institutions with vested interests in the mobility of capital that profit from making the flow of money and goods proceed as smoothly as possible. Transnational corporations are perhaps the most well known of these actors because their profits are directly tied to investment flows and favorable production and trade arrangements across the globe. Financial markets, such as the New York Stock Exchange, function to move financial resources to their potentially most profitable position (and, hence, location). Supranational organizations, such as the World Bank and the International Monetary Fund (IMF), serve to ensure the implementation of national economic policies that favor investment (and, conversely, disinvestment such as the ease of moving factories offshore). Finally, transnational trading blocs, such as the North American Free Trade Agreement (NAFTA) and the European Community (EC), work to lessen political obstacles that impede flows of capital across borders.

Within this framework of capital flows, the cities that fare best are those that "capture" capital, if only temporarily, and provide the crucial services that facilitate profit accumulation and continued investment. Global cities are "infrastructural nodes." John Friedmann and Saskia Sassen concur that a city's ranking in the global hierarchy is dependent upon its level of organizations and firms who command and control key financial decisions. Also important are telecommunications infrastructure and the availability of financial markets. Given the hypermobility of capital, there is a real incentive for cities to lower the economic and political barriers to doing business (e.g., real estate, payroll, revenue, and other taxes). Sassen's essay further explains why cities are important nodes in the global circulation of money, information, and people. Her work attests to the reality that globalization affects nearly every segment of urban politics, culture, and economy. The connections between global and local – clearly seen in local labor markets – also mean that cities are bound to each other in new and unanticipated ways. The economic imbalance between the core regions of the world and the periphery is heightened by capital flows in search of profits and cheap labor markets. The class structure within global cities is increasingly polarized as well. At the top of the class hierarchy sit the corporate elites and wealthy professional class whose fortunes are connected to investment capital, real estate transactions, and high-end service provisions. An army of service workers is found at the bottom of the hierarchy. These individuals, predominantly minorities, increasingly foreign-born, and more often than not female, hold low-skilled jobs with little chance of advancement. Sassen sees the possibility for a new urban politics based on these layered networks between cities.

Brenda Yeoh and T. C. Chang examine one city's effort to position itself favorably within the (hierarchical) network of cities and the transnational flow of people, goods, services, and money. Singapore is a small, urbanized island nation whose recent and vast economic fortunes are completely tied to its participation in the global economy. While we may attribute Singapore's success to its investment in education and creating an inviting business environment, Yeoh and Chang suggest that state economic and social policies have produced unintended consequences for Singaporean society. In their view, Singapore's urban landscape functions as a site of transnationalism, in which flows of people, money, and information overlap in ways both interesting and unanticipated.

The final two readings focus on the flow of people both within nations and regions and between them. In lesser-developed countries, globalization has fueled the growth of industrial parks and Enterprise Zones (manufacturing spaces where limited regulation encourage relocation of factories from other countries). The resulting employment opportunities (among other factors, such as agribusiness) have encouraged an influx of migrants from the countryside, overburdening many large cities. Global cities, too, have experienced a growth in migrants from the southern hemisphere (e.g., Latin America, the Caribbean, North Africa) as opportunities, both real and imagined, draw individuals and families from impoverished areas.

How is globalization actually experienced by those who migrate? The structural approaches to the study of globalization tend to paint a grim picture of hidden forces that compel changes in the way people live and work and meet little or no local resistance. As individuals living and socializing in different places, we rarely think of forces or processes occurring at different scales. Globalization, as the remaining authors inform us, is experienced locally in real circumstance and, therefore, exists only in its articulation locally. Stoller and McConatha provide insight on how West African immigrant traders "make do" in New York City. These migrants capitalize on transnational networks that connect New York and Africa to improve their collective and individual lot while selling their wares on the streets of Manhattan. Here we are privy to global processes at work in ways both innovative and unexpected. In the final reading, Peggy Levitt examines the importance of remittances for transnational migration. In the past, immigrants have routinely sent portions of their earned income (remittances) back to their relatives in the home country. But with globalization the ties between home and host countries remain strong and sustainable – thanks largely to the ease of communication and travel. Levitt develops the idea of social remittances to include flows of ideas and social capital between host and sending countries. Social remittances provide us with evidence of the everyday relevance of transnational flows to migrants in the global age.

"The World City Hypothesis"

from *Development and Change* (1986)

John Friedmann

Editors' Introduction

In what he calls a "world city hypothesis," John Friedmann lays out the architecture of the global system of cities, presenting a set of theses that explain how city form is connected to global economic forces. Cities are ranked – from core, global to peripheral, regional – according to their position within a "new international division of labor." The most important cities within this hierarchy are those that carry out advanced economic functions, such as serving as a global financial center and headquarters for multinational corporations. A city's standing within the global economic system, in turn, strongly influences a host of economic, social, and political changes that takes place within it, from types of employment to funding for parks or museums.

Friedmann notes that the organization of the world city system is far from haphazard; key patterns may be detected. Global capital tends to use certain cities throughout the world as "basing points" in which investments and the development of markets predominate. There is a linear character to the distribution of global cities along an East–West axis. The Asian sub-system of global cities stretches from Tokyo to Singapore. The core cities of the American subsystem are New York, Chicago, and Los Angeles, with linkages north to Toronto and south to Mexico City and Caracas. Finally, the European sub-system includes London, Paris, and cities in the Rhine Valley, with linkages to a few cities in the southern hemisphere.

The economic sector fueling the global economy and, hence, the organization of the global city hierarchy, is investment flows and the support services behind them, such as advertising, accounting, insurance, and legal services. Within global cities, a profound division of labor tends to exist: a workforce of highly educated professionals who specialize in making complex business decisions (what are often referred to as "command and control functions" in the global cities literature) co-exists with a larger labor force of low-skilled workers employed in the manufacturing, personal services, hospitality, and entertainment industries.

An additional feature of the global system of cities is the flow of people – migration. In secondary or semi-peripheral global cities, migrants tend to flock to inner city neighborhoods or suburban fringe settlements from the countryside out of economic necessity. As a result, these cities have experienced exponential growth rates that have deepened urban poverty, pollution, and environmental degradation. Because global cities are "growth poles" for a variety of industries at multiple levels, they also function as magnets for immigrants from developing countries in search of employment. In response to massive flows of immigrants or to prevent them, global cities have enacted a variety of legal restrictions to limit the numbers of potential arrivals.

The geography of the contemporary world system of cities is marked by socioeconomic polarization that may be mapped at three scales. The first scale is global in which a small number of core, very wealthy countries (and especially notable cities within them) exist in contradistinction to a larger number of peripheral, extremely poor countries. The second scale is regional, in which polarization is evident, where second-level global cities with high levels of wealth and urban amenities are surrounded by impoverished rural regions often lacking in basic infrastructure. The third scale of polarization exists within cities themselves, where enormous income gaps between corporate professionals and low-skilled workers predominate.

John Friedmann is a leading figure in research on urban planning. He founded the Program on Urban Planning in the Graduate School of Architecture and Planning at UCLA and served as its head for a total of 14 years between 1969 and 1996. He is a recipient of a Guggenheim Award (1976) and an American Collegiate Schools of Planning Distinguished Planning Educator Award (1988). He is currently Professor Emeritus of Urban Planning at UCLA. Friedmann is author of several books and articles, including *Urbanization, Planning, and National Development* (Beverly Hills, CA: Sage Publications 1973); *Life Space and Economic Space: Essays in Third World Planning* (Oxford: Transaction Press 1988); *Empowerment: The Politics of Alternative Development* (Cambridge, MA: Blackwell 1992); "World City Formation: An Agenda for Research and Action" (with Goetz Wolff), *International Journal of Urban and Regional Research* (1982); and, "Where We Stand: A Decade of World City Research," in Paul L. Knox and Peter J. Taylor, eds, *World Cities in a World-System* (Cambridge: Cambridge University Press 1995).

Some fifteen years ago, Manuel Castells and David Harvey revolutionized the study of urbanization and initiated a period of exciting and fruitful scholarship. Their special achievement was to link city forming processes to the larger historical movement of industrial capitalism. Henceforth, the city was no longer to be interpreted as a social ecology, subject to natural forces inherent in the dynamics of population and space; it came to be viewed instead as a product of specifically social forces set in motion by capitalist relations of production. Class conflict became central to the new view of how cities evolved.

Only during the last four or five years, however, has the study of cities been directly connected to the world economy. This new approach sharpened insights into processes of urban change; it also offered a needed spatial perspective on an economy which seems increasingly oblivious to national boundaries. My purpose in this introduction is to state, as succinctly as I can, the main theses that link urbanization processes to global economic forces. The world city hypothesis, as I shall call these loosely joined statements, is primarily intended as a framework for research. It is neither a theory nor a universal generalization about cities, but a starting-point for political enquiry. We would, in fact, expect to find significant differences among those cities that have become the "basing points" for global capital. We would expect cities to differ among themselves according to not only the mode of their integration with the global economy, but also their own historical past, national policies, and cultural influences. The economic variable, however, is likely to be decisive for all attempts at explanation.

The world city hypothesis is about the spatial organization of the new international division of labour. As such, it concerns the contradictory relations between production in the era of global management and the political determination of territorial interests. It helps us to understand what happens in the major global cities of the world economy and what much political conflict in these cities is about. Although it cannot predict the outcomes of these struggles, it does suggest their common origins in the global system of market relations. There are seven interrelated theses in all. As they are stated, I shall follow with a comment in which they are explained, or examples are given, or further questions are posed.

1. *The form and extent of a city's integration with the world economy, and the functions assigned to the city in the new spatial division of labour, will be decisive for any structural changes occurring within it.*

Let us examine each of the key terms in this thesis.

(a) City. Reference is to an economic definition. A city in these terms is a spatially integrated economic and social system at a given location, or metropolitan region. For administrative or political purposes, the region may be divided into smaller units, which underlie, as a political or administrative space, the economic space of the region.

(b) Integration with the world capitalist system. Reference is to the specific forms, intensity, and duration of the relations that link the urban economy into the global system of markets for capital, labour and commodities.

(c) Functions assigned to it in the new spatial division of labour. The standard definition of the world capitalist system is that it corresponds to a single (spatial) division of labour. Within this division, different localities – national, regional, and urban subsystems – perform specialized roles. Focusing only on metropolitan economies, some carry out headquarter functions, others serve primarily as a financial centre, and still others have as their main function the articulation of regional and/or national economies with the global system. The most important cities, however, such as New York, may carry out *all* of these functions simultaneously.

(d) Structural changes occurring within it. Contemporary urban change is for the most part a process of adaptation to changes that are externally induced. More specifically, changes in metropolitan function, the structure of metropolitan labour markets, and the physical form of cities can be explained with reference to a worldwide process that affects the direction and volume of transnational capital flows, the spatial division of the functions of finance, management and production or, more generally, between production and control, and the employment structure of economic base activities.

These economic influences are, in turn, modified by certain endogenous conditions. Among these the most important are: first, the *spatial patterns of historical accumulation*; second, *national policies*, whose aim is to protect the national economic subsystem from outside competition through partial closure to immigration, commodity imports and the operation of international capital; and third, certain *social conditions*, such as *apartheid* in South Africa, which exert a major influence on urban process and structure.

2. *Key cities throughout the world are used by global capital as "basing points" in the spatial organization and articulation of production and markets. The resulting linkages make it possible to arrange world cities into a complex spatial hierarchy.*

Several taxonomies of world cities have been attempted. In Table 1, a different approach to world city distribution is attempted. Because the data to verify it are still lacking, the present effort is meant chiefly as a means to visualize a possible rank ordering of major cities, based on the presumed nature of their integration with the world economy.

When we look at Table 1, certain features of the classification spring immediately into view.

(a) All but two primary world cities are located in core countries. The two exceptions are São Paulo (which articulates the Brazilian economy) and the city-state of Singapore which performs the same role for a multi-country region in South-East Asia.
(b) European world cities are difficult to categorize because of their relatively small size and often specialized functions. London and Paris are world cities of the first rank but beyond that, classification gets more difficult. By way of illustration, I have included as world cities of the first rank the series of closely linked urban areas in the Netherlands focused on the Europort of Rotterdam, the West German economy centred on Frankfurt, and Zurich as a leading world money market.
(c) The list of secondary cities in both core and semi-periphery is meant to be only suggestive. Within core countries, secondary cities tend on the whole to be somewhat smaller than cities of the first rank, and some are more specialized as well (Vienna, Brussels and Milan). In semi-peripheral countries, the majority of secondary world cities are capital cities. Their relative importance for international capital depends very much on the strength and vitality of the national economy which these cities articulate.

The complete spatial distribution suggests a distinctively linear character of the world city system which connects, along an East–West axis, three distinct sub-systems: an Asian sub-system centred on the Tokyo–Singapore axis, with Singapore playing a subsidiary role as regional metropolis in Southeast Asia; an American sub-system based on the three primary core cities of New York, Chicago and Los Angeles, linked to Toronto in the North and to Mexico City and Caracas in the South, thus bringing Canada, Central America and the small Caribbean nations into the American orbit; and a West European sub-system focused on London, Paris and the Rhine valley axis from Randstad and Holland to Zurich. The southern hemisphere is linked into this sub-system via Johannesburg and São Paulo (see Figure 1).

Table 1 World city hierarchy[a]

Core Countries[b]		*Semi-Peripheral Countries*[b]	
Primary	*Secondary*	*Primary*	*Secondary*
London* I	Brussels* III		
Paris* I	Milan III		
Rotterdam III	Vienna* III		
Frankfurt III	Madrid* III		
Zurich III			Johannesburg III
New York I	Toronto III	São Paulo I	Buenos Aires* I
Chicago II	Miami III		Rio de Janeiro I
Los Angeles I	Houston III		Caracas* III
	San Francisco III		Mexico City* I
Tokyo* I	Sydney III	Singapore* III	Hong Kong II
			Taipei* III
			Manila* II
			Bangkok* II
			Seoul* II

* National capital. Population size categories (referring to metro-region): I 10–20 million; II 5–10 million; III 1–5 million.

[a] Selection criteria include: major financial centre; headquarters for TNCs (including regional headquarters); international institutions; rapid growth of business services sector; important manufacturing centre; major transportation node; population size. Not all criteria were used in every case, but several criteria had to be satisfied before a city would be identified as a world city of a particular rank. In principle, it would have been possible to add third- and even fourth-order cities to our global hierarchy. This was not done, however, since our primary interest is in the identification of only the most important centres of capitalist accumulation.

[b] Core countries are here identified according to World Bank criteria. They include nineteen so-called industrial market economies. Semi-peripheral countries include for the most part upper middle income countries having a significant measure of industrialization and an economic system based on market exchange.

3. *The global control functions of world cities are directly reflected in the structure and dynamics of their production sectors and employment.*

The driving force of world city growth is found in a small number of rapidly expanding sectors. Major importance attaches to corporate headquarters, international finance, global transport and communications; and high level business services, such as advertising, accounting, insurance and legal. An important ancillary function of world cities is ideological penetration and control. New York and Los Angeles, London and Paris, and to a lesser degree Tokyo, are centres for the production and dissemination of information, news, entertainment and other cultural artifacts.

In terms of occupations, world cities are characterized by a dichotomized labour force: on the one hand, a high percentage of professionals specialized in control functions and, on the other, a vast army of low-skilled workers engaged in manufacturing, personal services and the hotel, tourist and entertainment industries that cater to the privileged classes for whose sake the world city primarily exists.

In the semi-periphery, with its rapidly multiplying rural population, large numbers of unskilled workers migrate to world city locations in their respective countries in search of livelihood. Because the "modern" sector is incapable of absorbing more than a small fraction of this human mass, a large "informal" sector of microscopic survival activities has evolved.

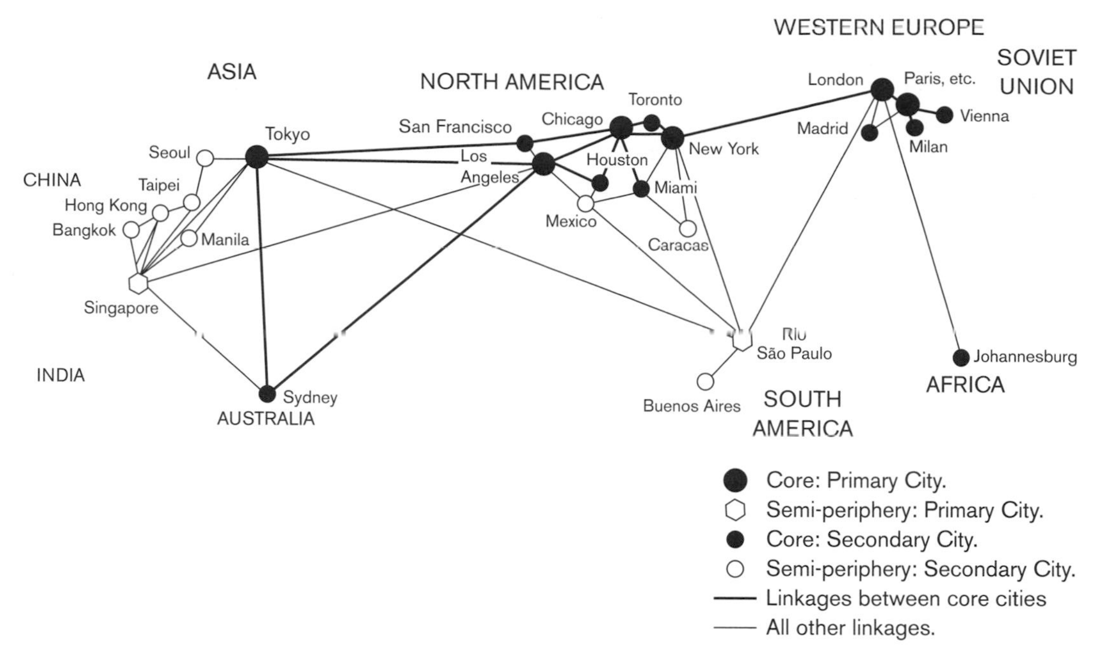

Figure 1 The hierarchy of world cities

4. World cities are major sites for the concentration and accumulation of international capital.

Although this statement would seem to be axiomatic, there are significant exceptions. In core countries, the major atypical case is Tokyo. Although a major control centre for Japanese multi-national capital, Japanese business practices and government policy have so far been successful in preventing foreign capital from making major investments in the city.

In the semi-periphery, the economic crisis since 1973 has led to massive international indebtedness, originally incurred in the hope of staving off economic disaster in the teeth of a world-wide recession of unprecedented depth and duration. A combination of declining per capita incomes, slow growth in the core of the world economy, IMF-imposed policies, the high cost of capital, capital repatriation, capital flight and obligatory loan repayments in some cases amounting to more than 35 per cent of export earnings have contributed in a number of Latin American countries to *a net export* of capital. If this trend, extraordinary for the postwar period, should persist, the semi-periphery is bound to backslide into peasant-peripheral status. Although strenuous attempts are being made to reverse this tidal drift into economic insolvency, declining living standards for the middle classes, immiseration for the poor and the collapse of the world economic system as it presently exists, the outcome is not at all certain.

5. World cities are points of destination for large numbers of both domestic and/or international migrants.

Two kinds of migrants can be distinguished: international and interregional. Both contribute to the growth of primary core cities, but in the semi-periphery world cities grow chiefly from interregional migration.

In one form or another, all countries of the capitalist core have attempted to curb immigration from abroad. Japan and Singapore have the most restrictive legislation and, for all practical purposes, prohibit permanent immigration. Western European countries have experimented with tightly controlled "guest worker" programmes. They, too, are jealous of their boundaries. And traditional immigrant countries, such as Canada and Australia, are attempting to limit the influx of migrants to workers possessing professional and other skills in high demand. Few if any countries have been as open to immigration from abroad as the United States, where both legal and illegal immigrants abound.

In semi-peripheral countries, periodic attempts to slow down the flow of rural migrants to large cities have been notably unsuccessful. Typically, therefore, urban growth has been from 1.5 to 2.5 times greater than the overall rate of population increase, and principal (world) cities have grown to very large sizes. Of the thirty cities in Table 1, eight have a population of 15 ± 5 million and another six a population of 7.5 ± 2.5 million. Absolute size, however, is not a criterion of world city status, and there are many large cities even in the peasant periphery whose size clearly does not entitle them to world city status.

6. *World city formation brings into focus the major contradictions of industrial capitalism – among them spatial and class polarization.*

Spatial polarization occurs at three scales. The first is global and is expressed by the widening gulf in wealth, income and power between peripheral economies as a whole and a handful of rich countries at the heart of the capitalist world. The second scale is regional and is especially pertinent in the semi-periphery. In core countries, regional income gradients are relatively smooth, and the difference between high and low income regions is rarely greater than 1:3. The corresponding ratio in the semi-periphery, however, is more likely to be 1:10. Meanwhile, the income gradient between peripheral world cities and the rest of the national economies which they articulate remains very steep. The third scale of spatial polarization is metropolitan. It is the familiar story of spatially segregating poor inner-city ghettos, suburban squatter housing and ethnic working-class enclaves. Spatial polarization arises from class polarization. And in world cities, class polarization has three principal facets: huge income gaps between transnational elites and low-skilled workers, large-scale immigration from rural areas or from abroad and structural trends in the evolution of jobs.

In the income distribution of semi-peripheral countries, the bottom 40 per cent of households typically receive less than 15 per cent of all income and control virtually none of the wealth. These data refer to countries that, overall, have low incomes when measured on the scale of a Western Europe or the US. In many of the primary cities of the core, however, the situation is not significantly better. In Los Angeles and New York, for example, huge immigrant populations are seriously disadvantaged.

In the semi-periphery, the massive poverty of world cities is underscored by the relative absence of middle-income sectors. The failure of semi-peripheral world cities to develop a substantial "middle class" has often been noted. Although there are important salaried sectors in cities such as Buenos Aires, their economic situation is subject to erosion by an inflationary process that is almost always double-digit and in some years rises to more than 200 per cent! Middle sectors have also become increasingly vulnerable to unemployment.

The basic structural reason for social polarization in world cities must be looked for in the evolution of jobs, which is itself a result of the increasing capital intensity of production. In the semi-periphery, most rural immigrants find accommodation in low-level service jobs, small industry and the "informal" sector. In core countries, the process is more complex. Given the downward pressure on wages resulting from large scale immigration of foreign (including undocumented) workers, the number of low-paid, chiefly non-unionized jobs rises rapidly in three sectors: personal and consumer services (domestics, boutiques, restaurants and entertainment), low-wage manufacturing (electronics, garments and prepared foods) and the dynamic sectors of finance and business services which comprise from one-quarter to one-third of all world city jobs and also give employment in many low-wage categories.

The whole comprises an *ecology* of jobs. As shown in Figure 2, the restructuring process in cities such as New York and Los Angeles involves the *destruction* of jobs in the high-wage, unionized sectors (EXODUS) and job *creation* in what Saskia Sassen-Koob calls the production of global control capability. Linked to these dynamic sectors are certain personal and consumer services (employing primarily female and/or foreign workers), while the slack in manufacturing is taken up by sweat shops and small industries employing non-union labour at near the minimum wage. It is this structural shift which accounts for the rapid decline of the middle-income sectors during the 1970s.

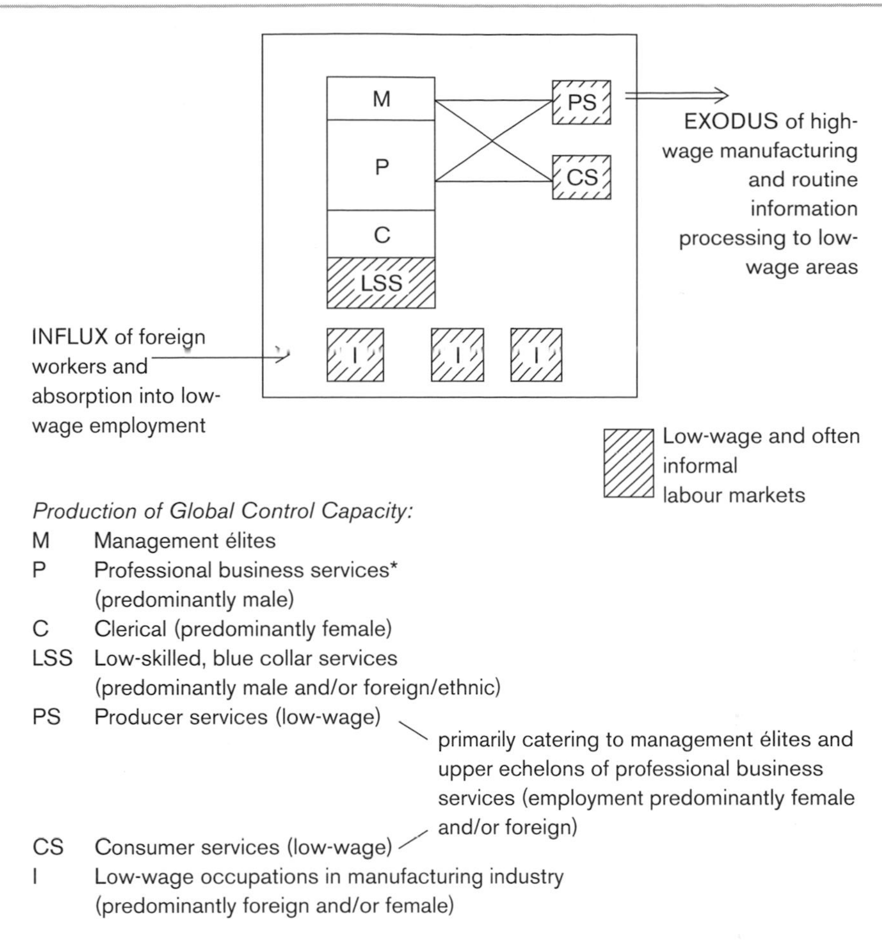

Figure 2 World city restructuring in core countries

* Many professional business services engage increasingly in international trade serving their clients, the transnational corporations, both at home and abroad. They include accounting, advertising, banking, communications, computer services, health services, insurance, leasing, legal services, shipping and air transport, and tourism. In 1981, US service exports equalled 50 per cent of merchandise exports and were still rising.

7. World city growth generates social costs at rates that tend to exceed the fiscal capacity of the state.

The rapid influx of poor workers into world cities – be it from abroad or from within the country – generates massive needs for social reproduction, among them housing, education, health, transportation and welfare. These needs are increasingly arrayed against other needs that arise from transnational capital for economic infrastructure and from the dominant élites for their own social reproduction.

In this competitive struggle, the poor and especially the new immigrant populations tend to lose out. State budgets reflect the general balance of political power. Not only are corporations exempt from taxes; they are generously subsidized in a variety of other ways as well. At the same time, the social classes that feed at the trough of the transnational economy insist, and usually insist successfully, on the priority of their own substantial claims for urban amenities and services. The overall result is a steady state of fiscal and social crisis in which the burden of capitalist accumulation is systematically shifted to the politically weakest, most disorganized sectors of the population. Their capacity for pressing their rightful claims against the corporations and the state is further contained by the ubiquitous forces of police repression.

"Whose City Is It? Globalization and the Formation of New Claims"

from *Public Culture* (1996)

Saskia Sassen

Editors' Introduction

"Globalization" and "global economy" are catch-all buzzwords that are often used to explain away a whole set of complex processes that link people and places around the world. In this reading, Saskia Sassen directs our attention beyond a mere and often deceptive description of "global cities" and asks why cities in particular are places of relevance to globalization and, consequently, what is the relevance of this strategic role to the everyday lives and politics of the people who live and work in cities. A focus on cities exposes "globalization at play" – that is, it permits us to see how the global information economy is embedded in the microeconomies and the diverse cultures of the city. In the globalized economy emphasis is given to circulation of capital, the invention and circulation of financial products, and the free flow of information. But Sassen examines other, less apparent aspects of globalization. For example, the multiethnic population of large cities is both a product of and a contributing factor to economic globalization; one localized dimension of economic globalization is the exploitation of a disadvantaged labor force comprised primarily of immigrants and people of color. Likewise, Sassen shows us how cities are bound to each other by the dynamics of globalization and how these connections may allow for a new form of local politics that transcends national borders. The transnational movement of migrants has disanchored the traditional sources of identity (ethnicity, country of origin, etc.), thus allowing for new forms of community politics and claims toward communal and individual social rights. The global city, then, is a contested terrain of multiple scales, where the labor participation of the disadvantaged has become intrinsically tied to the activities of high-salaried professionals.

Saskia Sassen is the Robert S. Lynd Professor of Sociology and co-Chair of the Committee on Global Thought at Columbia University and the Centennial Visiting Professor at the London School of Economics. Her research and many publications focus on the areas of globalization, immigration, and global cities.

The organizing theme is that place is central to the multiple circuits through which economic globalization is constituted. One strategic type of place for these developments, and the one focused on here, is the city. Including cities in the analysis of economic globalization is not without conceptual consequences. Economic globalization has mostly been conceptualized in terms of the duality national-global where the latter gains at the expense of the former. And it has largely been conceptualized in terms of the internationalization of capital and then only the upper circuits of capital. Introducing cities in an analysis of economic globalization allows us to re-conceptualize processes of economic globalization as concrete economic complexes situated in specific places. A focus on cities decomposes the nation-state into a variety of sub-national components, some profoundly articulated with the global economy and others not. It also signals the declining significance of the national economy as a unitary category in the global economy. And even if to a large extent this was a unitary category constructed in political discourse and policy, it has become even less of a fact in the last fifteen years.

Why does it matter to recover place in analyses of the global economy, particularly place as constituted in major cities? Because it allows us to see the multiplicity of economies and work cultures in which the global information economy is embedded. It also allows us to recover the concrete, localized processes through which globalization exists and to argue that much of the multiculturalism in large cities is as much a part of globalization as is international finance. Finally, focusing on cities allows us to specify a geography of strategic places at the global scale, places bound to each other by the dynamics of economic globalization. I refer to this as a new geography of centrality, and one of the questions it engenders is whether this new translational geography also is the space for a new translational politics. Insofar as my economic analysis of the global city recovers the broad array of jobs and work cultures that are part of the global economy though typically not marked as such, it allows me to examine the possibility of a new politics of traditionally disadvantaged actors operating in this new translational economic geography. This is a politics that lies at the intersection of economic participation in the global economy and the politics of the disadvantaged, and in that sense would add an economic dimension, specifically through those who hold the other jobs in the global economy – from factory workers in export processing zones to cleaners on Wall Street.

The centrality of place in a context of global processes engenders a transnational economic and political opening in the formation of new claims and hence in the constitution of entitlements, notably rights to place, and, at the limit, in the constitution of "citizenship". The city has indeed emerged as a site for new claims: by global capital which uses the city as an "organizational commodity," but also by disadvantaged sectors of the urban population, frequently as internationalized a presence in large cities as capital. The de-nationalizing of urban space and the formation of new claims centered in transnational actors and involving contestation, raise the question, Whose city is it?

I see this as a type of political opening that contains unifying capacities across national boundaries and sharpening conflicts within such boundaries. Global capital and the new immigrant workforce are two major instances of transnationalized categories that have unifying properties internally and find themselves in contestation with each other inside global cities. Global cities are the sites for the overvalorization of corporate capital and the devalorization of disadvantaged workers. The leading sectors of corporate capital are now global, in their organization and operations. And many of the disadvantaged workers in global cities are women, immigrants, people of color. Both find in the global city a strategic site for their economic and political operations.

The analysis presented here grounds its interpretation of the new politics made possible by globalization in a detailed understanding of the economics of globalization, and specifically in the centrality of place in a context where place is seen as neutralized by the available capacity for global communications and control. My assumption is that it is important to dissect the economics of globalization in order to understand whether a new transnational politics can be centered in the new transnational economic geography. Secondly, I think that dissecting the economics of place in the global economy allows us to recover non-corporate components of economic globalization and to inquire about the possibility of a new type of transnational politics. Is there a transnational politics embedded in the centrality of place and in the new geography of strategic places, such as is for instance the new worldwide grid of global cities? This is a geography that cuts across national borders and the old North–South divide.

Immigration, for instance, is one major process through which a new transnational political economy is being constituted, one which is largely embedded in major cities insofar as most immigrants, whether in the US, Japan or Western Europe are concentrated in major cities. It is, in my reading, one of the constitutive processes of globalization today, even though not recognized or represented as such in mainstream accounts of the global economy.

PLACE AND PRODUCTION IN THE GLOBAL ECONOMY

I think of the mainstream account of economic globalization as a narrative of eviction. Key concepts in that account – globalization, information economy, and telematics – all suggest that place no longer matters and that the only type of worker that matters is the highly educated professional. It is an account that privileges the capability for global transmission over the material infrastructure that makes such transmission possible; information outputs over the workers producing those outputs, from specialists to secretaries; and the new transnational corporate culture over the multiplicity of work cultures, including immigrant cultures, within which many of the "other" jobs of the global information economy take place. In brief, the dominant narrative concerns itself with the upper circuits of capital; and particularly with the hypermobility of capital rather than with that which is place-bound.

Massive trends towards the spatial dispersal of economic activities at the metropolitan, national and global level are indeed all taking place, but they represent only half of what is happening. Alongside the well-documented spatial dispersal of economic activities, new forms of territorial centralization of top-level management and control operations have appeared. National and global markets as well as globally integrated operations require central places where the work of globalization gets done. Further, information industries require a vast physical infrastructure containing strategic nodes with hyper-concentrations of facilities. Finally, even the most advanced information industries have a work process – that is, a complex of workers, machines and buildings that are more place-bound than the imagery of information outputs suggests.

Centralized control and management over a geographically dispersed array of economic operations does not come about inevitably as part of a "world system". It requires the production of a vast range of highly specialized services, telecommunications infrastructure, and industrial services. These are crucial for the valorization of what are today leading components of capital. A focus on place and production displaces the focus from the power of large corporations over governments and economies to the range of activities and organizational arrangements necessary for the implementation and maintenance of a global network of factories, service operations and markets; these are all processes only partly encompassed by the activities of transnational corporations and banks.

Global cities are centers for the servicing and financing of international trade, investment, and headquarter operations. That is to say, the multiplicity of specialized activities present in global cities are crucial in the valorization, indeed overvalorization of leading sectors of capital today. And in this sense they are strategic production sites for today's leading economic sectors. This function is reflected in the ascendance of these activities in their economies. I have posited that what is specific about the shift to services is not merely the growth in service jobs but, most importantly, the growing service intensity in the organization of advanced economies: firms in all industries, from mining to wholesale buy more accounting, legal, advertising, financial, economic forecasting services today than they did twenty years ago. Whether at the global or regional level, cities are adequate and often the best production sites for such specialized services. The rapid growth and disproportionate concentration of such services in cities signals that the latter have re-emerged as significant production sites after losing this role in the period when mass manufacturing was the dominant sector of the economy.

The extremely high densities evident in the downtown districts of these cities are the spatial expression of this logic. The widely accepted notion that agglomeration has become obsolete when global telecommunication advances should allow for maximum dispersal, is only partly correct. It is, I argue, precisely because of the territorial dispersal facilitated by telecommunication advances that agglomeration of centralizing activities has expanded immensely. This is not a mere continuation of old patterns of agglomeration but, one could posit, a new logic for agglomeration. Information technologies are yet another factor contributing to the new logic for agglomeration. These technologies make possible the geographic dispersal and simultaneous integration of many activities. But the distinct conditions under which such facilities are available have promoted centralization of the most advanced users in the most advanced telecommunications centers.

[...]

What we are seeing is a dynamic of valorization which has sharply increased the distance between the valorized, indeed overvalorized, sectors of the economy and devalorized sectors even when the latter are part of leading global industries. This devalorization of growing sectors of the economy has been embedded in a massive demographic transition towards a growing presence of women, African-Americans and Third World immigrants in the urban workforce.

We see here an interesting correspondence between great concentrations of corporate power and large concentrations of "others". Large cities in the highly developed world are the terrain where a multiplicity of globalization processes assume concrete, localized forms. A focus on cities allows us to capture, further, not only the upper but also the lower circuits of globalization. These localized forms are, in good part, what globalization is about. We can then think of cities also as one of the sites for the contradictions of the internationalization of capital. If we consider, further, that large cities also concentrate a growing share of disadvantaged populations – immigrants in Europe and the United States, African-Americans and Latinos in the United States – then we can see that cities have become a strategic terrain for a whole series of conflicts and contradictions.

A NEW GEOGRAPHY OF CENTRALITY AND MARGINALITY

The global economy materializes in a worldwide grid of strategic places, uppermost among which are major international business and financial centers. We can think of this global grid as constituting a new economic geography of centrality, one that cuts across national boundaries and across the old North–South divide. It signals the emergence of a parallel political geography, a transnational space for the formation of new claims by global capital.

[...]

The most powerful of these new geographies of centrality at the inter-urban level binds the major international financial and business centers: New York, London, Tokyo, Paris, Frankfurt, Zurich, Amsterdam, Los Angeles, Sydney, Hong Kong, among others. But this geography now also includes cities such as Sao Paulo, Buenos Aires, Bangkok, Taipei and Mexico City. The intensity of transactions among these cities, particularly through the financial markets, transactions in services, and investment has increased sharply, and so have the orders of magnitude involved. At the same time, there has been a sharpening inequality in the concentration of strategic resources and activities between each of these cities and others in the same country.

[...]

The growth of global markets for finance and specialized services, the need for transnational servicing networks due to sharp increases in international investment, the reduced role of the government in the regulation of international economic activity and the corresponding ascendance of other institutional arenas, notably global markets and corporate headquarters – all these point to the existence of transnational economic processes with multiple locations in more than one country. We can see here the formation, at least incipient, of a transnational urban system.

The pronounced orientation to the world markets evident in such cities raises questions about the articulation with their nation-states, their regions, and the larger economic and social structure in such cities. Cities have typically been deeply embedded in the economies of their region, indeed often reflecting the characteristics of the latter; and mostly they still do. But cities that are strategic sites in the global economy tend, in part, to disconnect from their region. This conflicts with a key proposition in traditional scholarship about urban systems, namely, that these systems promote the territorial integration of regional and national economies.

Alongside these new global and regional hierarchies of cities, is a vast territory that has become increasingly peripheral, increasingly excluded from the major economic processes that fuel economic growth in the new global economy. A multiplicity of formerly important manufacturing centers and port cities have lost functions and are in decline, not only in the less developed countries but also in the most advanced economies. This is yet another meaning of economic globalization.

But also inside global cities we see a new geography of centrality and marginality. The downtowns of cities and metropolitan business centers receive massive investments in real estate and telecommunications while low-income city areas are starved for resources. Highly educated workers see their incomes rise to unusually high levels while low- or medium-skilled workers see theirs sink. Financial services produce superprofits while industrial services barely survive. These trends are evident, with different levels of intensity, in a growing number of major cities in the developed world and increasingly in some of the developing countries that have been integrated into the global financial markets.

THE FORMATION OF GLOBAL RIGHTS FOR CAPITAL IN THE NEW URBAN GRID

[. . .]

A focus on a strategic subnational unit such as is the global city illuminates these two conditions that are at opposite ends of the governance challenge posed by globalization and are not captured in the more conventional duality national-global.

On the one hand, a focus on leading information industries in global cities introduces into the discussion of governance the possibility of capacities for regulation derived from the concentration of significant resources, including fixed capital, in strategic places, resources that are essential for participation in the global economy. The considerable place-boundedness of many of these resources contrasts with the hypermobility of information outputs. The regulatory capacity of the state stands in a different relation to hypermobile outputs than to the infrastructure of facilities, from fiber optic cable served office buildings to specialized workforces, present in global cities.

At the other extreme, the fact that many of these industries operate partly in electronic spaces raises questions of control that derive from key properties of the new information technologies, notably the orders of magnitude in trading volumes made possible by speed. Here it is no longer just a question of the capacity of the state to govern these processes, but also of the capacity of the private sector, that is, of the major actors involved in setting up these markets in electronic space. Elementary and well known illustrations of this issue of control are stock market crashes attributed to program trading, and globally implemented decisions to invest or disinvest in a currency or an emerging market which resemble a sort of worldwide stampede facilitated by the fact of global integration and instantaneous execution worldwide.

The specific issues raised by these two variables, i.e. place-boundedness and speed, are quite distinct from those typically raised in the context of the national–global duality. A focus on this duality leads to rather straightforward propositions about the declining significance of the state vis-à-vis global economic actors. The overarching tendency in economic analyses of globalization and of information industries has been to emphasize certain aspects: industry outputs rather than the production process involved, the capacity for instantaneous transmission around the world rather than the infrastructure necessary for this capacity, the impossibility of the state to regulate those outputs and that capacity insofar as they extend beyond the nation-state. And this is by itself quite correct; but it is a partial account of the implications of globalization for governance.

The transformation in the composition of the world economy, especially the rise of finance and advanced services as leading industries, is contributing to a new international economic order, one dominated by financial centers, global markets, and translational firms. Correspondingly we may see a growing significance of other political categories both sub- and supranational. Cities that function as international business and financial centers are sites for direct transactions with world markets that take place without government inspection, as for instance the euro-markets or New York City's international financial zone (International Banking Facilities). These cities and the globally oriented markets and firms they contain mediate in the relation of the world economy to nation-states and in the relations among nation-states.

A key component in the transformation over the last fifteen years has been the formation of new claims by global capital: the claim on national states to guarantee the domestic and global rights of capital. Transnational economic processes inevitably interact with systems for the governance of national economies insofar as those processes materialize in concrete places. National legal regimes are becoming more internationalized in some of the major developed economies and we are seeing the formation of transnational legal regimes. Transnational legal regimes have become more important and have begun to penetrate national fields hitherto closed. The hegemony of neo-liberal concepts of economic relations with its strong emphasis on markets, deregulation, free international trade has influenced policy in the 1980s in the USA and UK and now increasingly also in continental Europe. This has contributed to the formation of transnational legal regimes that are centered in Western economic concepts of contract and property rights. Through the IMF and IBRD (International Bank for Reconstruction and Development) as well as GATT (the WTO after 1995) this regime has spread to the developing world.

[...]

UNMOORING IDENTITIES AND A NEW TRANSNATIONAL POLITICS

Typically the analysis about the globalization of the economy privileges the reconstitution of capital as an internationalized presence; it emphasizes the vanguard character of this reconstitution. At the same time it remains absolutely silent about another crucial element of this transnationalization, one that some, like myself, see as the counterpart of that of capital: this is the transnationalization of labor. We are still using the language of immigration to describe this process. Secondly, that analysis overlooks the transnationalization in the formation of identities and loyalties among various population segments that explicitly reject the imagined community of the nation. With this come new solidarities and notions of membership. Major cities have emerged as a strategic site for both the transnationalization of labor and the formation of transnational identities. In this regard they are a site for new types of political operations.

[...]

The large Western city of today concentrates diversity. Its spaces are inscribed with the dominant corporate culture but also with a multiplicity of other cultures and identities. The slippage is evident: the dominant culture can encompass only part of the city. And while corporate power inscribes these cultures and identities with "Otherness" thereby devaluing them, they are present everywhere. For instance, through immigration a proliferation of originally highly localized cultures now have become presences in many large cities, cities whose elites think of themselves as cosmopolitan, that is transcending any locality. An immense array of cultures from around the world, each rooted in a particular country or village, now are reterritorialized in a few single places, places such as New York, Los Angeles, Paris, London, and most recently Tokyo.

Immigration and ethnicity are too often constituted as "otherness". Understanding them as a set of processes whereby global elements are localized, international labor markets are constituted, and cultures from all over the world are deterritorialized, puts them right there at the center of the stage along with the internationalization of capital as a fundamental aspect of globalization today. There has been growing recognition of the formation of an international professional class of workers and of highly internationalized environments due to the presence of foreign firms and personnel, the formation of global markets in the arts, and the international circulation of high culture. What has not been recognized is the possibility that we are seeing an internationalized labor market for low-wage manual and service workers. This process continues to be couched in terms of the "immigration story", a narrative rooted in an earlier historical period.

I think that there are representations of globality which have not been recognized as such or are contested representations. Among these is the question of immigration, as well as the multiplicity of cultural environments it contributes in large cities, often subsumed under the notion of ethnicity. What we still narrate in the language of immigration and ethnicity I would argue is actually a series of processes having to do with the globalization of economic activity, of cultural activity, of identity formation. Immigration and ethnicity are constituted as otherness. Understanding them as a set of processes whereby global elements are localized, international labor markets are constituted, and cultures from all over the world are de- and reterritorialized, puts them right there at the center along with the internationalization of capital as a fundamental aspect of globalization. This way of narrating the migration events of the post-war era captures the ongoing weight of colonialism and postcolonial forms of empire on major processes of globalization today, and specifically those binding emigration and immigration countries.

While the specific genesis and contents of their responsibility will vary from case to case and period to period, none of the major immigration countries are innocent bystanders.

MAKING CLAIMS ON THE CITY

These processes signal that there has been a change in the linkages that bind people and places and in the corresponding formation of claims on the city. It is true that throughout history people have moved and through these movements constituted places. But today the articulation of territory and people is being constituted in a radically different way at least in one regard, and that is the speed with which that articulation can change. One consequence of this speed is the expansion of the space within which actual and possible linkages can happen. The shrinking of distance and of time that characterizes the current era finds one of its most extreme forms in electronically based communities of individuals or organizations from all around the globe interacting in real time and simultaneously, as is possible through the Internet and kindred electronic networks.

I would argue that another radical form assumed today by the linkage of people to territory is the loosening of identities from what have been traditional sources of identity, such as the nation or the village. This unmooring in the process of identity formation engenders new notions of community of membership and of entitlement.

The space constituted by the global grid of global cities, a space with new economic and political potentialities, is perhaps one of the most strategic spaces for the formation of transnational identities and communities. This is a space that is both place-centered in that it is embedded in particular and strategic sites; and it is transterritorial because it connects sites that are not geographically proximate yet intensely connected to each other... An important question is whether it is also a space for a new politics, one going beyond the politics of culture and identity.

Yet another way of thinking about the political implications of this strategic transnational space is the notion of the formation of new claims on that space. Has economic globalization at least partly shaped the formation of claims? There are indeed major new actors making claims on these cities, notably foreign firms who have been increasingly entitled to do business through progressive deregulation of national economies, and the large increase over the last decade in international business people. These are among the new city users. They have profoundly marked the urban landscape. Their claim to the city is not contested, even though the costs and benefits to cities have barely been examined.

[...]

Perhaps at the other extreme of conventional representations are those who use urban political violence to make their claims on the city, claims that lack the de facto legitimacy enjoyed by the new "city users". These are claims made by actors struggling for recognition, entitlement, claiming their rights to the city.

There are two aspects in this formation of new claims that have implications for the new transnational politics. One is the sharp and perhaps sharpening differences in the representation of these claims by different sectors, notably international business and the vast population of low income "others" – African-Americans, immigrants, women. The second aspect is the increasingly transnational element in both types of claims and claimants. It signals a politics of contestation embedded in specific places – global cities – but transnational in character. At its most extreme, this divergence assumes the form of a) an overvalorized corporate center occupying a smaller terrain and one whose edges are sharper than, for example, in the post-war era characterized by a large middle class; and b) a sharp devalorization of what is outside the center, which comes to be read as marginal.

Globalization is a contradictory space; it is characterized by contestation, internal differentiation, continuous border crossings. The global city is emblematic of this condition. Global cities concentrate a disproportionate share of global corporate power and are one of the key sites for its overvalorization. But they also concentrate a disproportionate share of the disadvantaged and are one of the key sites for their devalorization. This joint presence happens in a context where (1) the globalization of the economy has grown sharply and cities have become increasingly strategic for global capital; and (2) marginalized people have found their voice and are making claims on the city as well. This joint presence is further brought into focus by the sharpening of the distance between the two. The center now concentrates immense power, a power that rests on the capability for global control and the capability to produce superprofits. And marginality, notwithstanding little economic and political power, has become an increasingly strong presence through the new politics of culture and identity, and an emergent transnational politics embedded in the new geography of economic globalization. Both actors, increasingly transnational and in contestation find in the city the strategic terrain for their operations.

[...]

In the three decades after WWII, the period of the Pax Americana, economic internationalization had the effect of strengthening the inter-state system. Leading economic sectors, especially manufacturing and raw materials extraction, were subject to international trade regimes that contributed to build the inter-state system. Individual states adjusted national economic policies to further this version of the world economy. Already then certain sectors did not fit comfortably under this largely trade-dominated inter-state regime: out of their escape emerged the euro-markets and off-shore tax havens of the 1960s.

The breakdown of the Bretton Woods system produced an international governance void rapidly filled by multinationals and global financial markets. Inside the state we see a further shift away from those agencies most closely tied to domestic social forces, as was the case during the Pax Americana, and towards those closest to the transnational process of consensus formation.

I have worked with the concept "regulatory fracture" rather than, say, violation, in order to name a specific dynamic: to wit, that the materialization of global processes in a place often produces a regulatory void. One result is that both "regulation" and "violation" become problematic categories and, at the limit, do not apply. We might think of it analytically as a border-land, rather than a borderline – a terrain for action/activity that remains underspecified at least from the perspective of regulation.

An issue that is emerging as significant in view of the spread of Western legal concepts is the critical examination of the philosophical premises about authorship and property that define the legal arena in the West.

This language is increasingly constructing immigration as a devalued process insofar as it describes the entry of people from generally poorer, disadvantaged countries, in search of the better lives that the receiving country can offer; it contains an implicit valorization of the receiving country and a devalorization of the sending country.

In the colonial era, it was the cities in the colonies which were probably the most internationalized.

There are many different forms such contestation and "slippage" can assume. Global mass culture homogenizes and is capable of absorbing an immense variety of local cultural elements. But this process is never complete. The opposite is the case in my analysis of data on electronic manufacturing, which shows that employment in lead sectors no longer inevitably constitutes membership in a labor aristocracy. Thus Third World women working in export processing zones are not empowered: capitalism can work through difference. Yet another case is that of "illegal" immigrants; here we see that national boundaries have the effect of creating and criminalizing difference. These kinds of differentiations are central to the formation of a world economic system.

[...]

The city remains a terrain for contest, characterized by the emergence of new actors, often younger and younger. It is a terrain where the constraints placed upon, and the institutional limitations of governments to address the demands for equity engenders social disorders. Urban political violence should not be interpreted as a coherent ideology but rather as an element of temporary political tactics, which permits vulnerable actors to enter in interaction with the holders of power on terms that will be somewhat more favorable to the weak.

"Globalising Singapore: Debating Transnational Flows in the City"

from *Urban Studies* (2001)

Brenda S. A. Yeoh and T. C. Chang

Editors' Introduction

In the age of globalization, cities are confronted by the need to "capture" their share from a constant transnational flow of people, goods, services, and money. In city-states, such as Singapore, social policies clearly reflect their full dependence upon participation in the global economy. In this reading, Brenda Yeoh and T. C. Chang examine the state's efforts at concerted economic development and, more importantly, address the often unintended consequences of transnationalism. Yeoh and Chang see transnational flows not as distinct or separate but as overlapping and interdependent categories. The urban landscape then functions as a space where multiple, global communities intersect in ways reflective of the local and historical context. Four categories of transnationals – transnational business, "third world" populations, expressive specialists, and tourists – are examined. For each category, the authors appraise the social policies designed to attract and manage "outsiders." In addition, the authors show how each group's interests and economic positions converge to shape Singapore's identity as a global city.

Brenda S. A. Yeoh has taught social and historical geography at the Department of Geography, National University of Singapore. She has written extensively on the politics of space in colonial and postcolonial cities, gender, migration, and transnational communities. T. C. Chang is an Associate Professor of Geography at the National University of Singapore. His research interests are in the areas of urban, social-cultural, and tourism geographies.

An important vein in the burgeoning literature examines the connectivities between what is, on the one hand, a rather abstract set of globalisation processes and, on the other, the processes of "transnationalism" which sustain "multi-stranded social relations" linking together "home" and "host" societies. It is argued that transnational practices and networks of capital, labour, business and commodity markets, political movements and cultural flows are both the products of, and catalyst for, contemporary globalisation processes. For example, transnational communities – communities that sit astride political borders and that, in a very real sense, are "neither here nor there", but in both places simultaneously – are both created in response to, and at the same time sustain and fuel, the process of globalisation.

An important site where the analytical lens can be trained to examine the way in which the material processes and discourses of globalisation and transnationalism intersect lies in dominant cities of the world urban hierarchy.

[. . .]

In this paper, we have chosen four groups of transnationals as our organising scheme, primarily because its range accords well with the framing of categories in local discourse in Singapore and at the same time allows us to put together side-by-side what are often treated as separate threads in global city debates.

(1) "transnational business", involving high-waged, highly skilled professional, managerial and entrepreneurial elites usually associated with finance, banking and business services;
(2) "third world populations", comprising low-waged immigrants who occupy insecure niches in the unskilled or semi-skilled sectors of the urban service economy;
(3) "expressive specialists", who participate in the cultural scene in areas such as art, fashion, design, photography, film-making, writing, music and cuisine; and
(4) "tourists", who are present in considerable numbers, attracted by the cosmopolitan intensity of the global city.

For each category, we discuss the government's notion of best practice in drawing in and negotiating with these transnational "outsiders" whose economic or social itineraries have converged, at least momentarily, in Singapore as a place. Each arena of debate has provoked new ways of thinking about the local, whether this be a questioning of the meaning of citizenship or a revalorisation of "local" or "traditional" heritage, or simply a concern for community benefits. Beyond examining the four different arenas of debate, we also give attention to the way discourses overlap, interconnect and reshape the parameters and contours of discussion. The paper points to the challenges to be addressed in Singapore's bid to develop a "cosmopolitan and creative" global city epitomising the essence of transnationalism while at the same time remaining a "home" distinguished by a strong sense of local identity and community.

[. . .]

Within the world city hierarchy, Singapore is ranked in the second tier, among those aspiring for "superleague" status. As Chua (1998, p. 995) points out, Singapore "eminently qualifies for a place in the collection of cities which are discursively grouped under [the term 'world cities']" having been, since its "birth" during the modern age of mercantile capitalism, intensely part of global service capitalism.

In Singapore, the construction of nationhood has been a major state-driven project since independence. The pressures of creating a nation of "one people" belonging to "one place", and associated manoeuvres to secure political legitimacy, build ideological consensus, discipline its industrial workforce and mould the consciousness of its new citizens have been important imperatives threaded into all major state policies – housing, education, language, community development, national service, economic development – governing various aspects of social and political life. The current vision to secure and enhance Singapore's position as a global city has further added a new dimension to these endeavours.

[. . .]

THE BUSINESS AND PROFESSIONAL CLASS: CREATING AN "OASIS OF TALENT"

As "the most powerful manifestations of the internationalization of capital in the world space economy", global cities distinguished by a high concentration of corporate headquarters, advanced telecommunication and R&D infrastructure and international financial services also contain disproportionately large clusters of high-waged professional and managerial expatriate workers (Beaverstock, 1996, p. 424). In aspiring towards global city status, the Singapore state has not only emphasised building up Singapore's total business and human resource capabilities to attract global transnational corporations to the city-state, but also has promoted Singapore as a major engine of foreign direct investments into the rest of Asia. As a space of flows, the vision for Singapore involves criss-crossing circulatory streams of people moving in multiple directions: not only is the city-state to draw in elite professionals and specialists from all corners of the globe, it is also to become a springboard for venturing into the region. The latter applies not only to expatriate workers, but also to Singaporeans who are continually exhorted to develop an entrepreneurial spirit and a global and regional outlook while remaining committed to "home". Another related initiative in the rhetoric of promoting Singapore as a global player is the promotion of Singapore as a "wired" world city, one deeply embedded in global circuits of information flows and equipped with advanced telecommunications facilities and the integration of wide-scale information technology strategies. Transforming the city-state into a knowledge-based economy and information society is a key strategy in Singapore's regionalization drive not only to become a "brains service node" for the region, but also in order to create a "virtual state" in which citizens abroad can remain "hooked up" to the nation.

The corollary of these major strategies to position Singapore as a significant node in the global space of flows is the development of a highly skilled human resource base on the premise that the "key success factor" in confronting a global future is not only the "hardware" such as technological infrastructure, but also the "software – the ideas and knowledge of its people" (the words of the Prime Minister, Goh, 1999a). Besides investing heavily in information technology and human capital to meet global competition, the state has emphasised the strategy of "gathering global talent" and "making Singapore a cosmopolitan city".

[...]

In short, the vision is to create "an oasis of talent" in the city-state which will serve as a hub for business, information and knowledge skills anchoring global networks linking the world's three economic growth engines of Asia, Europe and the Americas.

UNSKILLED IMMIGRANT WORKERS: THE "UNDERBELLY" OF GLOBAL CITIES

To date, "foreign talent" which deserves a place in Singapore society is defined primarily by their ability to fill particular economic niches. The government has been careful to add that this call to absorb "foreign talent" to create a more creative, cosmopolitan Singapore is not tantamount to opening the floodgates, but will be a highly selective process. Of the categories of foreigners to be brought in to fill the gaps identified in government pronouncements, selectivity is also exercised in terms of the degree of permanence accorded to each group. At the bottom of the pecking order, unskilled workers such as domestic workers and construction workers are relegated to the most transient of categories – work permit holders – subject to the "use and discard" philosophy.

However, it must be noted that the global city is not only a crucial node in the development of new geographies of skilled professional and managerial workers, but also is sustained by low-skilled, low-status migrants, who service the needs of the privileged in both residential and commercial settings. Indeed, it has argued that low-waged immigrant-sector labour is not a residual category in the economy of the global city, but a basic precondition, enmeshed in processes which represent "the underbelly of globalisation". As earlier noted, the number of foreign workers (the term usually applies to the unskilled and semi-skilled as opposed to "foreign talent" which is reserved for the highly skilled) had escalated at the same rate as skilled expatriate workers.

State policy with regard to foreign workers is conceived to ensure that they are no more than a transient workforce, "a buffer to even out the swings of the business cycle" and subject to repatriation during periods of economic downturn. To dampen the rising demand for foreign workers, stringent legislation has been put in place not only to restrict their number and ensure their short-term migrant status, but also to govern their employment.

[...]

Clearly, much of the state rhetoric about the creation of a cosmopolitan society with a "Big Singapore mentality" which welcomes foreigners is not intended to be all-encompassing. There are definite limits to cosmopolitanism in Singapore's vision of a global city. To date, public discourse on foreign workers has focused on issues such as the social problems of foreign worker enclaves, the impact of maids on the Singapore family and the need for quick solutions to repatriate foreign workers found abandoned in the streets to avoid the "issue of vagrancy" from tainting "clean and green" Singapore rather than their incorporation into the social fabric of the global city.

TOURISTS: VISUALISING A TOURISM CAPITAL

A third category of transnational people are tourists. Singapore's quest to be a global tourist centre is encapsulated by the STB's (Singapore Tourism Board) vision to develop Singapore as a "tourism capital". The STB explains that all great cities acquire the status of a "capital" by exerting their hegemony in fields like finance, business, communications, fashion and religion. In the new millennium, Singapore aspires to a pivotal position in tourism by serving three roles: a destination for visitors, a location for tourism investments and a tourism gateway to Southeast Asia. To be a tourism capital, three interrelated "best practices" are undertaken requiring Singapore to redefine its urban and tourism planning philosophies and realign its regional policies.

The first best practice is to redefine the very parameters of "tourism". In the new millennium, tourism will refer not only to the business of attracting visitors, but also attracting capital and entrepreneurs to invest in the country. Although it has successfully served as a destination for over three decades, Singapore's tourism future is less assured because of regional competition and local resource constraints. An ambitious plan is required where economic wealth is generated not just through the capricious flows of visitors, but also through stable investment streams. Transnational flows of visitors are therefore to be augmented by the transnational flows of capital, expertise and corporate elites. Singapore is thus to be developed as a destination, tourism business centre and tourist hub.

The redefinition of tourism to include "investments" as well as "people" has implications for local urban planning. A second best practice, therefore, is to reconfigure Singapore's tourism base to be dually attractive to visitors, investors, capitalists and tourism-related enterprises. To make Singapore both "visitable" and "investable", the urban environment will have to develop symbolic landscape cues of urban vivacity and cultural dynamism. In the light of this, the STB has outlined a wide-scale urban enhancement scheme in which 11 "experiential themes" were selected to accentuate Singapore's cultural and aesthetic resources, while also focusing urban redevelopment in strategic sites. For example, the "Night Zone" theme highlights the need to develop Singapore's nocturnal food and entertainment spots, while "Ethnic Singapore" emphasises the importance of multicultural diversity through the revitalisation of Chinatown and Little India.

[...]

Recent debates on the STB's proposed plans for a revamped Chinatown exemplify deep-seated antagonisms towards tourist-linked projects. Unveiled in 1998, the STB's plans to create themed streets (such as Food Street or Tradition Street); elemental gardens (featuring the five elements of Chinese mythology); and a village theatre featuring staged operas and performances have been vociferously attacked in the press and public forums for creating an Orientalist caricature of Chinatown, sanitised for foreign consumption. Against an upswell of local criticism, the STB was forced back to the drawing board to reconceive a blueprint that speaks directly to Singapore's immigrant Chinese past. Under its new plans, the elemental gardens have been erased and an interpretative centre celebrating the stories of the local inhabitants – including disenfranchised coolies and immigrants – will be constructed. The STB explains that the new Chinatown will be a "magnet for locals" and ongoing consultations with the public will be conducted to ensure that the place "delivers an experience that is authentic and heartfelt for Singaporeans and residents here".

[...]

The challenges sketched above compel us to acknowledge the local repercussions of Singapore's global tourism ambitions. The desire to attract tourists to the country and to serve as a gateway to the region certainly engenders widespread transformations of local landscapes and a reconfiguration of Singapore's position in the regional hierarchy of cities. As a tourism capital negotiating the transnational flows of people and capital, the Singapore state must be prepared for the challenges of regional competition from other would-be tourism capitals, as well as the voices of dissent emanating from the grassroots. Reconciling local and regional tensions with Singapore's international ambitions becomes an imperative in a globalising city-state.

CREATIVE SPECIALISTS: CREATING A SYMBOLIC ECONOMY THROUGH THE ARTS

The STB's goal of attracting tourism investments constitutes only one aspect of the EDB's (Economic Development Board) larger goal of spearheading a lifestyle services cluster in Singapore. According to EDB, five sectors constitute this cluster:

(1) tourism and leisure services;
(2) the arts;
(3) media services;
(4) medical services; and
(5) education services. (EDB, 1995, p. 34)

The aim is to develop Singapore as a "global city with total business capabilities" in tertiary sectors while simultaneously promoting the "softer aspects" of life. These "software" businesses serve as a form of "cultural capital" which imbues places with an image of social vibrance and a high quality of life. Developing a thriving cultural industry is also key to fulfilling the symbolic function of projecting the country multifariously as a global city, an exciting tourist destination, a viable investment venue and a good home in which to live. Three interrelated best practices are implicit in Singapore's quest to cultivate a creative lifestyle industry.

The first best practice is to market Singapore as a "Global City for the Arts", targeting sophisticated cultural tourists. Towards this end, many arts attractions and events have been staged in Singapore such as Tresors – the International Arts and Antiques Fair; Broadway productions of *Cats*, *Phantom of the Opera* and *Les Misérables*; as well as popular entertainment events ranging from DisneyFest to pop, jazz and new-age music concerts. Marketing these events regionally has attracted Malaysians, Indonesians and other Asian tourists. For example, tourist presence at *Phantom of the Opera* contributed S$1 million in revenues alone while Tresors 1996 attracted 50 tourist "arts packages" constituting 200 hotel room nights (STB, 1996c, p. 38). Capitalising on Western art forms, Singapore hopes to attract a burgeoning Asian market by providing an alternative to Western cultural hubs such as New York and London.

Apart from targeting tourists, a second best practice is to develop Singapore as an arts business hub. Replicating its successful hub concept in manufacturing, finance, transport and communications, a similar ambition is envisioned for arts and culture. Hence, new incentive schemes, subsidies and infrastructure have been developed to fuel the arts as a source of jobs, revenues and investment. Education in the arts was revised in 1998 to produce practitioners to meet the labour needs of the nascent industry. Ambitious plans have been drawn up to develop new venues for the performing arts, museums and galleries.

The third but perhaps least obvious best practice is the EDB's deployment of the arts as a vehicle to negotiate transnational communities. This negotiation process takes two forms: attracting global talent and retaining local community. On the one hand, the arts function as a cultural magnet to attract all forms of foreign expertise, expatriates and talents to work and live in Singapore. This includes creative specialists as well as foreign talents in other sectors lured by Singapore's vibrant cosmopolitan ambience. On the other hand, a lively cultural scene also serves as a social glue, retaining local residents by offering an improved standard of living rivalling the best of other global cities. The "global city for the arts" vision therefore fulfils economic goals and sociopolitical agendas.

The material bases of Singapore's cultural production have provoked two main strands of local debate. The "symbolic economy" of the arts is criticised because the arts and culture are being nurtured in Singapore for their economic worth rather than for their own intrinsic value. For example, the concept of arts hubbing emphasises massive infrastructural development, the import of foreign cultural specialists and recognises the needs of tourists. The benefits to local practitioners and experimental arts groups, and the attention to smaller-scale development projects, are less visible.

[. . .]

The rush to develop Singapore as a global city for the arts has therefore prioritised mega events, hallmark infrastructure and the "borrowing" of international artistes. More must certainly be done at the grassroots level to stimulate local talent and to provide modestly scaled infrastructure.

A second area of debate pertains to the "local peculiarities" of a "global city". As an entertainment hub, Singapore is expected to conform to international norms in artistic and cultural practices, a requirement which many creative specialists feel the government has been slow to adhere to. Recently aired debates on the NAC's strict rulings on public busking, the Police Authority's unconventional regulations on pop concerts and the government's censure of local writer Catherine Lim for her political views provoke questions on Singapore's readiness to be a vibrant cultural hub. The policing of pop concerts offers an illuminating case in point. Prior to the relaxing of regulations in October 1998, concert organisers had to follow strict rules such as not allowing performers to interact with the audience, preventing dancing by members of the public and installing barricades in front of stages. These regulations are hostile to international performers and tourists, and conflict with Singapore's ambition to be an entertainment hub. They have also given rise to criticisms that pop concerts in Singapore are "sub-standard" because performers are cowed by police regulations to practise self-censorship.

[...]

The goal of developing Singapore as a cultural and entertainment capital peopled with creative talents, tourists and skilled migrant workers has certainly raised important questions of local identity in a globalising city-state. Indeed, the quest to embrace Western art forms and entertainment events as part of Singapore's cultural resource base has prompted many to question the place (or lack thereof) of Asian arts and creative expressions in the country. The desire to promote Western art forms may run the risk of marginalising local talents, rendering Singapore as a global city for the "borrowed arts". In its quest to be a global cultural hub, local best practices must be questioned anew and realigned along international standards.

In this paper, we explore the construction of Singapore as a global city and meeting-place of transnational communities comprising four groups: high-skilled managerial elites; low-waged immigrant communities; global and regional tourists; and creative specialists in the field of arts. As Singapore aims to be a "global city" as well as the "best home" for its citizens, the tensions between its global aspirations and local assertions have to be constantly renegotiated.

In this context, the Singapore state has already begun engaging with a number of "best practices" to further Singapore's globalising vision. To position the city as an international hub for business, information and knowledge, as well as a springboard for venturing into the region, Singapore's economic architects have developed a concerted effort to augment and develop the pool of transnational talent. This includes both foreigners and Singaporeans who are equipped and willing to respond to globalising and regionalising forces with the transnational finesse of navigating successfully between being "home" and "away" at the same time. Global cities, however, are not constituted solely by transnational elites.

[...]

The distinctive character and problems of global cities emerge from their disparate communities comprising transients, sojourners, immigrants and citizens, and their interactions with each other. In many ways, Singapore has always been a community of locals and cosmopolitans, of insiders and outsiders. As an immigrant society, the earliest challenge of independence was to create a cohesive community among disparate ethnic groups.

Perhaps the most obvious tension lies in the local–foreign divide. While national identity is arguable in a nascent state in Singapore, it is clear that the onslaught of globalisation debates in recent years has honed a sense of who the "locals" are, in counterpoint to the "foreigners". This is evident in the polarisation of "local" and "foreign" in the various debates on the privileges of the citizen vis-a-vis the non-citizen and the constitution of an "ideal ratio" between locals and foreigners and is also clearly distilled in the portrayal of the "cosmopolitan" versus the "heartlander", terms which have gained currency as a means to define the "local". Even as Singaporeans tussle with how to define themselves, they have also to come to terms with the foreigners in their midst.

In envisioning a global city of flows, another issue which has to be grappled with is the whole notion of "home/citizenship". Marked by significant citizen-absences and non-citizen-presences, citizenship and the rights to participate in the public sphere within the global city-state can no longer be defined solely on the basis of presence in a residential community or place (Staeheli and Thompson, 1997). The category "citizen" (and its converse "non-citizen") has to be dismantled and reconstructed to remain relevant to the reality of the kaleidoscopic ethnoscape of the global city, perhaps by associating rights and responsibilities to different degrees of permanence rather than binary categories of citizen–non-citizen. This will be a difficult political manoeuvre, given the anxieties about foreigners who invade the terrain of citizenry (for example, those in the "in-between" Permanent Resident category who enjoy certain "citizen privileges", or foreign workers who are given access to "community" resources) embedded in the Singaporean psyche. Finally, an issue which lies at the heart of the global city vision which has been given far less attention in public discourse is the interconnections and interdependence of the different transnational streams whose itineraries meet in the global city. We have already argued that the social reproduction of the elite workforce including "foreign talent" in the global city is predicated on the services of the low-wage, low-skilled sector which often comprised other foreigners. The failure to give emphasis to the interdependence of the two groups of transnationals is partly the failure to take into account the significance of the reproductive sphere in the production and sustenance of the global city. The large inflow of foreign domestic workers into the global city is a response to the developing crisis in the reproductive sphere as more local women are encouraged to enter the paid workforce in a globalising economy as well as to the demands for (low) paid domestic service by the global (highly paid) expatriate labour-force which sustains the multinational business space in the city.

There is a need to acknowledge the different transnational flows outlined above as interdependent categories, and to understand global cities as meeting places for multiple communities whose intersection in place reflects their symbiotic though not always conflict-free relationship with each other. It is in this light that we conclude there is no standard "best practice" for "model cities" because the configuration, intensity and problems of transnational communities are different from place to place. Nevertheless, as the Singapore case testifies, site-specific "best practices" have been and will always be forged in cities with an eye towards being cosmopolitan, creative and humane all at the same time. How successful they are will provide the ground for further debates and analyses.

"City Life: West African Communities in New York"

from *Journal of Contemporary Ethnography* (2001)

Paul Stoller and Jasmin Tahmaseb McConatha

Editors' Introduction

A key facet of globalization is international migration – the movement of large numbers of people across borders. As in the past, contemporary immigrants to the United States build communities in their new cities and towns. Yet these communities differ from the ethnic enclaves of the late nineteenth and earlier twentieth century migrations. Due to relatively inexpensive airfares, modern communications, and the availability of goods and services on a global basis, many of today's immigrants readily construct "transnational communities" – spaces, both imagined and real, in which the worlds of homeland and host country intermingle. According to Peggy Levitt, who has written extensively on the topic, "transnational social expectations, cultural values, and patterns of human interaction . . . are shaped by more than one social, economic, and political system" (2001: 197).

In this selection, Paul Stoller and Jasmin Tahmaseb McConatha present a community study of West African street vendors in New York City. The traders' community is truly transnational. West African traders have constructed what Stoller and McConatha label, an "invented community," linked by shared economic pursuits and Africanity. Traders form networks based on common interests of religion and ethnicity. They are also linked to their families and to economic networks back home in West Africa. While living in New York City, the traders participate in an imagined community that draws on their collective memory of home and is reiterated through Islamic religious practices.

Stoller and McConatha organize their chapter to clearly show how West African traders "live" transnationalism. They address the social, cultural, and psychological dimensions of West African city life, showing how Muslim traders have built communities from their personal and economic networks that span nations.

West African traders live in different neighborhoods throughout New York City but for most their initial days and weeks in the United States are spent in "vertical villages" – a handful of low-rent single room occupancy (SRO) residences in Manhattan populated by immigrants from West Africa. There, new arrivals are immediately linked to an intact fellowship of individuals with knowledge of the City's economic, social, and cultural terrain. Yet, for all its benefits, fellowship cannot replace family. There is a palpable sense of loss of family connection among traders and their primary activity is to maintain contact with wives, parents, children, and siblings back home. Such contact includes letters, telephone calls, as well as remittances (money earned by traders and sent to family members in West Africa).

In addition to family ties, the practice of Islam ranks high among priorities for traders. Indeed, Islam bridges the social and cultural distance between West Africa and New York City – religious practices and cooperation and trust among fellow Muslims have empowered traders to cope with social isolation in America. Stoller and McConatha point out that Islam is the foundation of the traders' transnational communities: "personal networks in which kinship, ethnicity, and nationality affect the density of contact and degree of trust and co-operation are baseline community forms." Formal clubs and associations based on national origin also help sustain communities but, as the authors point out, these organizations tend to reflect class and status divisions from back home.

The existence of a transnational community clearly helps traders adjust to life in New York, to maintain ties with home, and to manage their businesses. It provides a necessary social structure for individual traders to tap into. How successfully or unsuccessfully individuals make use of the community is the final topic Stoller and McConatha address. They use the concept "cultural competence" to discuss the individual trader's ability "to adapt to an environment by making effective choices and plans and by controlling – as much as possible – the events and outcomes of daily life." Some traders have succeeded better than others due to unequal levels of cultural competence. Two forms of competence are key to managing life in New York City. Linguistic competence in English enables traders to construct transnational business, expand their trading operations, and "charm" customers. Cultural competence – knowledge of the diverse economic and social customs and value systems that operate in New York City – is as important, as some traders have learned the hard way.

Paul Stoller is Professor of Anthropology and Sociology at West Chester University. He has conducted research among Songhay-speaking peoples in the Republic of Niger and in New York City for more than two decades. He is the author of *Money Has No Smell: The Africanization of New York City* (Chicago: University of Chicago Press 2002).

Jasmin Tahmaseb McConatha is Professor of Psychology at West Chester University. Her research interests include life span development, cross-cultural psychology, and cultural influences on communication patterns.

See also Peggy Levitt, "Transnational Migration: Taking Stock and Future Directions," *Global Networks*, 1, 3 (2001), 195–216 and *The Transnational Villagers* (Berkeley, CA: University of California Press, 2001).

The notion of community has long been central to the ethnographic enterprise. In sociology, the landmark study of community was Robert and Helen Lynd's *Middletown* (1929). In their highly descriptive book, however, the Lynds did not attempt to construct a theory of community. Robert Redfield took up that task in the 1940s and 1950s when he published *The Folk Culture of Yucatan* (1941) and *The Little Community* (1955). Redfield depicted the community of Teplotzlan as bounded, harmonious, and homogenous. His work inspired the criticism of Oscar Lewis who had studied the same community but found a very different kind of society. Where Redfield uncovered harmony, homogeneity, and social adjustment, Lewis found violence, sociopolitical schisms, and social maladjustment (see Lewis 1951). As Sherry Ortner (1997) has recently pointed out, Lewis's criticism still rings true to contemporary scholars who have, for the past fifteen years, underscored the fragmentation and hybridity of sociocultural processes and organizations. Although the concept of community has been plagued with epistemological and conceptual problems, Ortner thinks it is a notion well worth preserving in the era of globalized, transnational settings. For her,

> the importance of community studies is this: such studies have the virtue of treating people as contextualized social beings. They portray the thickness of people's lives, the fact that people live in a world of relationships as well as a world of abstract forces and disembodied images.

In this article, we attempt to extend Ortner's (1997) insights to dispersed communities of West African street vendors in New York City. Several of the emergent "postcommunities" that Ortner identified conform to community forms found among West African traders in New York City. Congregating at the African market on 125th Street or now at the Malcolm Shabazz Harlem market, West African traders have constructed an "invented community" of "African brothers" linked by shared economic pursuits and Africanity. The traders also participate in what Ortner called *translocal communities.* Despite being dispersed over great distances, they are linked through common interests like religion or ethnicity. As mobile merchants who follow the circuit of African American and Third World cultural festivals, West African traders from New York City form periodic links of support and friendship with fellow Africans as they travel across the United States. They are also linked to their families and to economic networks back home in West Africa. Another form of community in which the traders participate is a *community of mind* – of memory – a community engendered primarily through Islam. No matter where they are, West African traders, most of whom are Muslims, try to pray five times a day and obey the various dietary and behavioral dictates of Islam. One of the central themes of Islam, in fact, is the "community of faithful," the *Ummah.* The themes of this community of the mind are reinforced through daily prayer, daily behavior toward others, and *Jumma* services during which localized communities of believers are asked to assemble at the Friday mosque. During these services, the imam delivers readings from the *Qu'ran* and sometimes speaks to the faithful about the values of their religion.

Membership in these various and fluid forms of community can enable West African traders to cope with the deep cultural alienation they experience in New York City. Consider briefly the case of Moussa Boureima, a West African trader from Niger, who has sold caps, gloves, and scarves at the Malcolm Shabazz Harlem market since 1994. He suffers from rheumatism and back pain but is hesitant to go to a hospital or see a physician. As an undocumented immigrant, he distrusts public agencies, which will ask him for identification and perhaps compromise his precarious immigration status. He puts off seeing a physician out of shame and embarrassment; his English is poor, and the doctor may not be able to understand him. In New York City, his life is socially, culturally, and linguistically alien to him. How does Moussa Boureima cope? In this article, we suggest that the degree of his adaptive success – his well-being – in New York City results from his cultural competence as well as from his ability to make use of support-system resources provided by various forms of West African communities.

Cultural competence, for us, refers to a person's ability to adapt to an environment by making effective choices and plans and by controlling – as much as possible – the events and outcomes of daily life. Although community support systems have been shown to diminish feelings of loneliness, help resolve social problems, and buffer cultural dislocation, our data indicate that they afford only the potentiality for these outcomes. It is up to the individual to tap community resources – however they might be structured – to foster his or her well-being. Boureima, for example, can find help for his medical bills from informal associations constructed by his compatriots but only if he chooses to seek treatment.

[...]

HOUSED FRUSTRATIONS

Since coming to New York City in September 1990, Boube Mounkaila, whose robust health obviates the need for modern or traditional medical treatment, has lived in three apartments. For six weeks, he lived in a single-room occupancy (SRO) hotel in Chelsea, which he disliked. "Too many roaches and bandits," he said, "and the place smelled bad, too." Since then, he has lived in Harlem, first in a building on 126th Street and then in an apartment on Lenox between 126th and 127th Streets. He lived alone in this rundown apartment building, the hallways of which stank of stale onions and cabbage. His neighbors included two other Nigerian immigrants who lived on his floor, a Senegalese immigrant married to a Puerto Rican woman who was a U.S. citizen, two Senegalese merchants, and a large Puerto Rican family. He lived in a one-room, second story walk-up that had no toilet or bath or shower facilities. Each floor featured one toilet and one bath/shower situated at the end of a dark hallway.

In 1994, Boube said he paid his landlord four hundred dollars per month for the room, a sum that did not include utilities. At that time, he sent a monthly money order of five hundred dollars to support his wife, daughter, and his mother. This sum also helped to support his two sisters, their husbands, and their children. His father had died some years earlier. These expenses created a considerable financial strain. Like that of most peddlers, Boube's concern about money has seasonal ups and downs. At the 125th Street market, he might take in seven hundred to eight hundred dollars on the weekends in the spring, summer, and fall. During the week, however, he might gross two hundred on a good day. In the winter, these sums decline precipitously. He also expressed concern over his living conditions.

> "They take advantage of us," he said. "Look at this place. A small room without a toilet. And I pay $400 per month plus utilities. They know that many of us don't have papers, and they expect us to pay in cash and not cause problems. What choice do we have?"

Boube wanted to move to a two- or three-room apartment with a bath and toilet, but his monthly expenses prevented him from doing so. "I want to leave here," he said, "but when I finish paying for food, gas, electricity, telephone, inventory, parking and car insurance, and family expenses, there's not much left."

Boube was finally able to move to his current two-room apartment, which is on Lenox Avenue between 126th and 127th Streets in January 1995. According to Boube, a local entrepreneur who respects Africans owns the building. In March 1995, Boube pointed out a man dressed in a black leather jacket, dress slacks, and a black Stetson who was getting into his Jaguar. "That's my landlord," Boube said. "He likes Africans." He explained that when he moved in two months earlier, he had had some cash-flow problems. He needed to get his phone turned back on and had been having difficulty paying his rent. The new landlord, however, was willing to accept four hundred dollars per month for a space considerably larger than his previous apartment. Sensing that Boube was hard working and would turn out to be a reliable tenant, he allowed him stay rent-free for the first month. Sometimes, he let Boube use his phone. "And he is always respectful," Boube stated emphatically. "He's a fine man."

Boube's new apartment was on the third floor of a tenement in an apartment that looks out on Lenox Avenue. The stairwell was dark and creaky and smelled fetid even in winter. Mailboxes were located on the first floor, but because mail was so frequently stolen, Boube had a post office box at the local post office. Above the mailboxes was a sign: "Anyone caught throwing garbage out the window will be punished." In March 1995, Boube's neighbors included his compatriots Sala Fari and Issifi Mayaki, who shared an apartment; several African American men; and a young white man who studied at Columbia University. His new apartment consisted of two rooms, perhaps seven feet by twenty feet. One room remained empty. Boube planned to make this space his salon but had not yet bought furniture for it. He installed a curtain to cordon off the other room. A new double bed and a chest of drawers had been placed in the front room near the window. Boube had adorned the bedroom walls with print portraits of African American women. His salon consisted of his old plastic deck chairs and several low coffee tables. A cheap cotton carpet with Oriental designs stretched between the chairs and tables. On one table was a large boom box. The second table supported a television and VCR. He explained that he had bought a multisystem television so he could play PAL as well as VHS videocassettes. Many of the French and African videocassettes, he said, could be viewed only on the PAL system. "I used to have a much larger TV. I paid $1,400 for it," he mentioned, "but I sent it to my older brother in Abidjan." In the small foyer that separates the two front rooms from the kitchen and bathroom, Boube had hung three images on the wall: a poster of the Dome of the Rock in Jerusalem, a print of a beautiful African American woman, and a picture of Jesus.

> "Why the picture of Jesus?" we asked during an interview in his apartment one afternoon. He replied, "A woman from Canada gave it to me. We spent one whole day talking at the market and she gave me the photo. She thought that because I am Muslim I might refuse the picture. But I like all religions."

Although Boube had tired quickly of his experience in a SRO hotel in Chelsea, many West Africans are forced to remain in SROs despite their deplorable living conditions. Perhaps the best known African "village" in New York City is the Park View Hotel at 55 West 110th Street. Francophone West Africans who live there call it the Cent Dix (the 110th). The building is in a state of advanced disrepair. In 1994, city hall cited it for a variety of code violations that included the presence of leaks, urine, feces, roaches, trash, and garbage in public access. The cracked and peeling plaster walls that lined the hallways have attracted numerous drug dealers and other hustlers.

[. . .]

In 1992, the building's owner, Joe Cooper, said that perhaps three-quarters of the residents were West Africans. In 1995, however, less than half the occupants remained West African. Cooper said the deteriorating conditions compelled West Africans who had the funds to either leave the building or return to West Africa. The owner complained that the more recent occupants in the hotel were destructive. As soon as he fixed something, he asserted, someone destroyed the repairs or created new problems. Despite Cooper's claims, however, the local police say that crime is not widespread at the Park View.

The juxtaposition of African and African Americans has led to a number of social tensions. Several Africans said they had been disappointed to encounter hostility from blacks in the neighborhood. They reported with bitterness that they were sometimes accused of selling the Americans' ancestors into slavery. Some of "the Africans, on the other hand, believed that African Americans were unwilling to work hard." Many African Americans at the Park View, in contrast, praised their African neighbors, saying that they were friendly, respectful, and hard working.

If the West African residents at the Cent Dix hate the building in which they live, why have so many remained there? They stay, in part, because they do not know where else to go and because, despite everything, the Cent Dix offers something very essential: fellowship. Charles Kone, from the Ivory Coast, summed up this painful mix: "It's misery . . . There's no security, no maintenance. But I knew that when I got to the airport, there was a place I could go. It's like a corner of Africa." In fact, at the Park View Hotel, West Africans have even set up convenience stores and established communal kitchens. From the vantage of many West African residents, it has become a "vertical village."

These vertical villages, moreover, are well known in far-away West Africa. Experienced West African businessmen, who routinely travel to New York City, instruct first-time travelers to look for African taxi drivers on their arrival at John F. Kennedy Airport. These drivers, so it is said, know to take the new arrivals to one of several SRO hotels in Manhattan. When Boube Mounkaila first came to America in September 1990, he found a West African taxi driver, a Malian, who took him to a SRO hotel in Chelsea.

The idea of West African social cooperation, a theme deeply embedded in the cultures of most West African societies, is not limited to shabby, run-down SRO hotels. Islamic practice is centered on cooperative economics and the establishment and reinforcement of fellowship in a community of believers. West African traders also have constructed elaborate personal and professional networks that create a sense of fellowship. Many West African traders at the Malcolm Shabazz Harlem market share information on the best product suppliers, exchange goods, and recommend clients to their colleagues.

CULTURAL ISOLATION

Although the vast majority of West African street vendors in New York City have expressed profound appreciation for the economic opportunities they enjoy and exploit in the United States, they have invariably complained of loneliness, sociocultural isolation, and alienation from mainstream American social customs. These conditions, moreover, seem to have an impact on the subjective well-being of men like Moussa Boureima, Boube Mounkaila, and Issifi Mayaki.

[. . .]

Immigration usually reinforces social isolation. Intensified by cultural difference, feelings of isolation from the larger sociocultural environment can have a significant impact on physical and psychological well-being. Isolation limits the range of activities and interactions in which people can participate; it also reduces feelings of control and competence.

[. . .]

This lack of control compels Moussa Boureima, who is sick, to avoid hospitals; it convinces Boube Mounkaila that he can do little to resolve regulatory dilemmas provoked by the city of New York or the INS.

Sustaining such social and emotional support systems as family may diminish some of the negative effects of immigration. One of the greatest detriments to feelings of well-being among many West African street traders in New York City is, indeed, the absence of family. Constructed as lineages, their families are usually their primary source of emotional and social support. Caught in regulatory limbo, Issifi Mayaki, a principal figure in our study, is unable to return to West Africa to see his family, whom he misses and longs for. He says this situation frustrates him and sometimes makes him mean spirited.

[. . .]

For most West Africans, the ideal, if not the reality, of a cohesive family that lives and works together is paramount. This ideal, however remote, has survived regional, national, and international family dispersion. It compels men like Moussa Boureima, Issifi Mayaki, and Boube Mounkaila to phone regularly their kin in West Africa; it compels them to send as much money as possible to help support their wives, their children, and their aging parents, aunts, uncles, and cousins.

The absence of an extended family has several psychological ramifications for West African traders. Besides support, families provide a sense of trust and feelings of competence. As Issifi Mayaki has said, one can usually trust her or his blood kin. Generally speaking, the closer the blood ties, the greater the degree of trust. Absence of family therefore creates an absence of trust, which leads to a considerable amount of stress and anxiety. For young men, the absence of wives also means that they are in a kind of sexual and social limbo. They share profound cultural and social mores with their wives in whom they place great trust. In Niger, for example, marriage, which sometimes involves cousins, ties family relations in webs of mutual rights and obligations. Men expect their wives, even during their long absences, to remain faithful to them. To avoid opportunities for infidelity during long absences, long-distance traders often insist that their wives live in the family compound, surrounded by observant relatives who not only enforce codes of sexual fidelity but also help to raise the family's children. Many of the men, on the other hand, believe it is their inalienable right to have sexual relationships with other women – especially if they are traveling. As Muslims, moreover, they have the right, if they so choose, to marry up to four women, although this practice is increasingly rare. These are some of the cultural assumptions that many lonely and isolated West African traders bring to social/sexual relationships with the women they encounter in New York City. To say the least, these assumptions clash violently with contemporary social/sexual sensibilities in America.

El Hadj Moru Sifi, like many Nigerians in New York City, talked of being socially and culturally isolated during his time in America. A rotund man well into his fifties who hailed from Dosso in western Niger, El Hadj Moru did not like the food, detested what he considered American duplicity, and distrusted non-Africans. Between 1992 and 1994, he sold sunglasses on 125th Street in Harlem. Work and sleep constituted much of his life. El Hadj supported two devoted wives in Niger. "Our women," he said in August 1994, just prior to his departure,

> "show respect for their men. They also know how to cook real food. None of these Burger King and Big Macs. They make rice, gumbo sauce with hot pepper, and fresh and clean meat. That's what I miss. I want to sit outside with my friends and kin and eat from a common bowl. Then I want to talk and talk into the night. I want to be in a place that has real Muslim fellowship."

During his two years in New York City, El Hadj said that he had remained celibate – by choice. He did not trust the women he met. The women, he said, often took drugs, slept with men, and sometimes even gave birth to drug-addicted babies. "Some of these women even have AIDS. Soon, I will be in Niger in my own house surrounded by my wives and children. I will eat and talk well again." El Hadj Moru's attitudes are not uncommon among West African traders in New York City. Many of his "brothers" have also chosen to remain celibate.

Boube Mounkaila has been anything but celibate during his time in New York City. Like his brother traders, he misses his family, including a wife whom he has not seen in eight years and a daughter born several months after his departure for America whom he has never seen. Sometimes, when he thinks of his family, says Boube, "my heart is spoiled. That's when I listen to the *kountigi* [one stringed-lute] music."

From the time he arrived in New York City in 1990 as a twenty-eight-year-old undocumented immigrant, Boube has attracted the attention of many women. He is a tall, good-looking man who can be charming. He also has become fluent in what he calls "street English." Because he sells handbags, most of his clients are women, old and young. On any given day, a young woman might be sitting in Boube's stall waiting patiently for him. In speaking of Boube, some of his compatriots shake their heads and say, "Ah Boube, he likes the women too much."

Boube's domestic circumstances are exceptional among West African traders in New York City. For most of the traders, life is much less dramatic; it follows the course of a man like El Hadj Moru Sifi – one works, eats, and sleeps, with occasional interludes or with longstanding relationships with one woman. Issifi Mayaki's situation has been more typical. Issifi is a forty-year-old handsome and well-dressed man who speaks good English. Between 1994 and 1997, Issifi had a girlfriend, a social worker who was a single African American woman with a ten-year-old daughter. Issifi met his girlfriend when he sold African print cloth on 125th Street in 1994. She expressed interest in him. He told her of his wife and children in Africa. She said that was okay; she appreciated his forthrightness. They began to see one another but maintained separate residences.

When Issifi began to travel to festivals far from New York City, his relationship with the woman began to unravel. She did not like the idea of him traveling to festivals. She became jealous of his wife in Africa. When he told her about plans to travel home to see his family, she did not want him to go. She did not want to share him with anyone. Issifi began to believe that American women wished to totally consume their men, which, he said, was not the African way. This cultural clash became the source of contention, and eventually they drifted apart.

Other traders have made other domestic arrangements. Abdou Harouna, who like El Hadj Sifi comes from Dosso, Republic of Niger, is not a trader but a "gypsy" cab driver. Abdou came to New York City in 1992. In 1994, he married an African American woman, not simply because he wanted a way to obtain immigration papers but because he had fallen in love with Alice, who is a primary school teacher. They now have a daughter and live in Harlem. "Alice," Abdou said, "has a pure heart. She is a good person, and I'm a lucky man."

One of the Nigerian traders, Sidi Sansanne, has two families: one in the South Bronx and another in Niamey, Niger. In his thirties, Sidi has become a prosperous merchant who runs a profitable import–export business, which requires him to travel between Niamey and New York City seven to ten times a year. Sidi is perhaps the ideal model of West African trader success. He came to the United States in 1989 and sold goods on the streets of midtown Manhattan. He invested wisely and realized that the American market for Africana was immense. He saved his money and went to Niger to make contact with craft artisans. He began to import to the United States homespun West African cloth, traditional wool blankets, leather sacks, bags, and attaché cases, as well as silver jewelry.

In time, he established a family in New York City, obtained his Employment Authorization Permit, and in 1994 became a permanent resident – a green-card holder. As a permanent resident, Sidi has been able to travel between the United States and West Africa without restriction. Because he has traveled to and from Africa so frequently, Sidi became a private courier. For a small fee, he has taken to Africa important letters or money earmarked for the families of various traders. From Africa, he has carried letters and small gifts to his compatriots in New York City. The freedom to travel has also enabled Sidi to find new craft ateliers in Niger. During his six-week sojourns in Niger, he, of course, has tended to his other family.

This pattern is a transnational version of West African polygamous marriage practices. In western Niger, for example, prosperous itinerant traders establish residences in the major market towns of their trading circuit. In this way, they attempt to pay equal attention to their wives and children and minimize the inevitable disputes that are triggered when cowives live in one compound.

FELLOWSHIP AND COMMUNITY

[. . .]

For the West African traders at the Malcolm Shabazz Harlem market and elsewhere in New York City, a central component of community is Islam. In Islam, any adherent is a member of the community of believers. Islam unquestionably structures the everyday lives of the traders and keeps alive their sense of identity in what, for most of them, remains an alien and strange place. During six years of conversations with West African traders, the subject of Islam was invariably raised, especially when the conversation broached the subject of the quality of life in the United States. They have said that in the face of social deterioration in New York City, Islam has made them strong; its discipline and values, they have said, have empowered them to cope with social isolation in America. It has enabled them to resist divisive forces that, according to them, ruin American families. But the greatest buffer to cultural dislocation is the perception, held by almost all the traders, that Islam makes them emotionally and morally superior to most Americans.

El Hadj Harouna Souley is a Nigerian in his forties. He made the expensive pilgrimage to Mecca when he was thirty-four, which is an indication of his considerable success in commerce. El Hadj Harouna embodies the aforementioned sense of Islamic moral superiority. Between 1994 and 1997, he sold T-shirts, baseball caps, and sweatshirts from shelves stuffed between two storefronts on Canal Street in Lower Manhattan. Like most West African traders, he is a member of a large family. Although his parents are dead, he has one wife, fourteen children, four brothers, five sisters, and scores of nieces and nephews.

On a rainy afternoon in December 1995, El Hadj Harouna sat under an awning on the steps of Taj Mahal, a radio and electronics store on Canal Street near West Broadway. He pointed out two street hawkers, both African Americans, employed by the owners of Taj Mahal. "You see those men there," El Hadj Harouna said, referring to the hawkers. "They only know their mother. Sometimes they don't even know who their father is. That's the way it often is in America. Families are not unified. Look at him," he said, referring to the older of the two hawkers.

> "He's from Georgia. His family sends him money every month, and as long as I have known him he has not returned there to visit them. Why do some people here not honor their parents? Why don't families stick together – at least in spirit? I want to get back to my family compound where we can all live together,"

El Hadj Harouna stated emphatically. He continued,

> "Can parents here depend on their children to take care of them when they are old? I don't know. I've seen children who sit at home and eat their parents' money, but they think that they owe their parents no obligation. My children phone me every week and ask me to come home. When I am old, even if I have no money, my children will look after me. I will do no work. I will eat, sleep and talk with my friends."

El Hadj Harouna continued his conversation but now concentrated on religion:

> "My Muslim discipline gives me great strength to withstand America. I have been to Mecca. I give to the poor. I rise before dawn so that I can pray five times a day, every day. I fast during Ramadan. I avoid pork and alcohol. I honor the memory of my father and mother. I respect my wife. And even if I lose all my money, if I am able, Inshallah, to live with my family, I will be truly blessed."

West African community structures in New York City take on several forms that are more concrete than the "community of believers." Personal networks in which kinship, ethnicity, and nationality affect the density of contact and degree of trust and cooperation are baseline community forms. In addition to these personal networks, there are translocal communities based on national origin. These are formal associations like L'Association des Nigeriens de New York, L'Association des Maliens aux USA, L'Association des Senegalais aux USA, and the Club des Femmes d'Affaires Africaines de New York (a New York African businesswomen's association). These associations are usually connected to, if not organized by, the diplomatic missions of the various Francophone African countries. Meetings are held once a month in the evenings, usually at a particular nation's United Nations Mission, at which issues of mutual concern are discussed. The associations hold receptions for major Muslim and national holidays. They collect funds to help defray a compatriot's unexpected medical expenses. In the case of a compatriot's death, they also contribute funds to ship the body back to West Africa for burial. L'Association des Nigeriens, for example, has raised money to buy food for hungry people in Niger.

It would be easy and perhaps facile to suggest that these West African communities – formal and informal, economic and personal, translocal or imagined – supply community adherents with financial and emotional support. Such support, it could be argued, also provides social harmony and a sense of belonging that protects members from the disintegrative stresses of cultural alienation. On one level, this statement is most certainly true. Belonging to the community of the faithful provides a religiously sanctioned set of explanations for the West African's situation in America. As participating members in the Association des Nigeriens aux USA, Nigerians engage in a mutually reinforcing set of rights and obligations based on mutual citizenship. This organization represents the interests of Nigerians in New York City. Participation in personal networks yields both economic benefits and, in some cases, the concrete fellowship desired by most West African traders in New York City.

Closer inspection of these community forms, however, reveals a more complex scenario. Although West African traders speak highly of their various "national" associations, their participation in the regular activities of the organizations – the monthly meetings – is infrequent. There are a few traders, of course, who are active members, but the majority of the traders have neither the time nor the inclination to attend association meetings or events. The meetings are held in the evenings at the Nigerian Mission on East 44th Street between 1st and 2nd Avenues. Many traders do not want to travel there from Harlem or the South Bronx after a long day at the market.

More important, the presence of the association in New York City brings into relief a primary tension in Nigerian society – one that exists between members of the Niger's educated elite and its peasants. In western Niger, peasants often express a distrust of the literate civil servants, whom they sometimes refer to as *anasarra*, which can mean, depending on the context, "European," "non-Muslim," or "white man." Less educated Nigerians, including village traders, sometimes say that civil servants who command state power, having learned the European's language and ways, have become foreigners in their own country. In Niger, this strong statement may well be a means of articulating class differences. A similar distrust has been expressed in New York City. In February 1994, a former Nigerian civil servant and no friend of the government of Niger, who sold goods on 125th Street on weekends only, claimed that the Association des Nigeriens had deceived the merchants. "There is a clear division," he said, "between educated and uneducated Nigerians in New York City. The Association recently collected money from the street merchants and stated that the money went to help people in Niger. In fact," he said, "the money helped to pay the electric bill at the Nigerian Mission to the UN and the traders didn't know."

By the same token, participation in economic networks can produce negative as well as positive results. In Issifi Mayaki's case, his participation in a transnational network of African art traders, one based on the trust of cooperative economics, led to betrayal and the theft of his inventory. Boube Mounkaila lost the entire contents of his Econoline Van, which had been parked in a secure, fenced-off space in Harlem. The complicity of one of his economic associates enabled thieves to enter the guarded space and steal his goods. Neither Issifi nor Boube reported these thefts to the police, whom they distrust.

Membership in the community of the faithful, the community of Muslims, creates a spiritual bond and provides a source of support as well as a buffer against the stress of city life in New York. As we have mentioned, Islam, like any religion, provides explanations to the traders about the absurdities of life. It supplies an always-ready set of explanations for problems encountered by Muslims in societies in which Islam is not a major sociopolitical force. For many West African traders in New York City, Islam, as a way of life, is morally superior to other faiths practiced in the United States. And yet, being a member of the community of the faithful does not dissipate a West African trader's financial difficulties, nor does it eliminate the stress of potential illness or existential doubts brought on by cultural alienation.

[. . .]

COMMUNITY AND COMPETENCE

[. . .]

In this article, we have briefly considered the social, material, historical, and psychological dimensions of West African city life in New York. As we have seen, West Africans have skillfully used their traditions as Muslim traders to build personal and economic networks that result in a variety of communities. These communities, in turn, provide them the potential for economic security, social cohesion, and cultural stability in an alien environment.

Despite this rich set of resources, however, some traders have succeeded better than others have, which brings us to the issue of competence. The issue of competence influences the perception of control among West Africans in New York City. There is, for example, a wide diversity of linguistic competence among the traders. Men like Boube Mounkaila and Issifi Mayaki speak English well. Boube's linguistic competence makes him socially confident. His facility in English enables him to construct transnational exchange networks with Asians, African Americans, and Middle Easterners. Since transnational transactions in New York City are usually conducted in English, his linguistic competence has enabled him to expand his operations. Using his skills in English, Boube arranged to purchase a vehicle, buy automobile insurance, and obtain a driver's license. Boube also employs his considerable linguistic skills to charm his mostly African American customers. Mastery of English, in short, has increased Boube's profits and expanded considerably his social horizons. That expansion has made Boube a keen observer of shifting social and economic environments, which, in turn, increases further his business profits. The same can be said for Issifi Mayaki and scores of other West African street vendors.

Lack of competence in English, however, results in missed opportunities. Even though men like Moussa Boureima and his roommate, Idrissa Dan Inna, have been in New York City for more than three years, they speak little if any English. In 1994, they enrolled in a night school course sponsored by a church in Harlem but dropped out after one week. "I don't know," said Idrissa, who sold West African hats and bags on 125th Street at the time, "I just can't learn English. I don't have the head for it. I know it hurts my business. I can't really talk to the shoppers about the goods." When confronted with various financial, social, or personal problems, men like Moussa and Idrissa have to rely on more linguistically competent traders, which, in accordance with the findings of social psychologists, affects their self-image negatively and makes them even more socially isolated.

Although West African street vendors in New York City display various linguistic abilities, they must all confront the problem of cultural competence. Many seem to have mastered the culture of capitalism, but their lack of a more general cultural competence has cost them dearly. In this important domain, one of the key issues is that of trust. According to Islamic law, traders are expected to be completely honest and trustworthy in the dealings with suppliers, exchange partners, and customers. Among West African traders, who are mostly Muslims, trust is paramount. Most of the traders, the majority of whom come from families and ethnic groups long associated with long-distance trading in West Africa, adhere to the Muslim principles of economic transactions. Not surprisingly, their trust has often been betrayed. An exchange partner stole Issifi Mayaki's textile inventory. Betrayals cost Boube Mounkaila his inventory. And yet, these men had the competence to use community resources to rebound from these defeats and move along their paths in New York City. Other traders who have suffered similar setbacks have drifted into isolation, changed occupations, or returned to West Africa.

[. . .]

The communities that West Africans in New York have constructed do not, then, define their city lives; rather, they provide resources – economic, social, and cultural – that dilute the stress of living in an alien environment. More specifically, these communities enable many, but not all, West African traders to enhance their subjective feelings of well-being and control. The sense and reality of community, as we have suggested, is no panacea for the ills associated with state regulation, poverty, and sociocultural alienation. For West African traders, then, city life is molded by the congruence of historical, material, social, and psychological factors. These factors not only define a sense of community but also shape the cultural competencies that affect their adjustment to an alien environment.

The results of our study also underscore Ortner's (1997) contention that the notion of community – however problematic – is one worth retaining in the social sciences. Given a refined framework of "community," the social scientist is able to demonstrate how macro-forces (globalization, immigration, informal economies, and state regulation) affect the lives of individuals living in the fragmented transnational spaces that increasingly make up contemporary social worlds.

“Social Remittances: Migration Driven Local-Level Forms of Cultural Diffusion”

from *International Migration Review* (1998)

Peggy Levitt

Editors’ Introduction

When immigrants settle in a new and different society, the most common response to assimilation pressures is to combine some of the social and cultural elements of their home country with those encountered in the host country. Advances in global communication have allowed immigrants to retain many elements of their “former” everyday life and remain in close contact with family members, friends, and associates who remain in their home country. The ease of opportunity for immigrants to inhabit both worlds – their old home and that of the new – has led to what scholars refer to as binational societies, where ties between home and host countries remain strong and sustainable. These ties are maintained in large part due to the ease of global travel and communication, the importance of migration to the economies of sending countries, the relocation services provided to migrants, and the marginal statuses immigrants tend to occupy in their new host countries. As the connections between sending and receiving countries strengthen, a transnational public sphere develops. In this reading, sociologist Peggy Levitt examines the concept of social remittance – the flow of cultural ideas, behaviors, social identities, and social capital from communities in receiving or host countries to those in sending or home countries. Social remittance, she argues, plays an important role in promoting immigrant entrepreneurship in cities across North America, in adjusting to new demands on community and family formation, and in the political integration of immigrants in their new home societies. By focusing on flows of social remittance, we can see firsthand how transnational the process of immigration has become in the global age

In this reading, Levitt examines social remittances between the United States and the Dominican Republic. Social remittance transfers are common because of the close proximity and ease of access between the two countries. But other important factors come into play, including the history of U.S. dominance over the economy of the Dominican Republic and a strong sense of national identity among Dominicans that helps to strengthen social ties and remittance transfers. As Levitt points out, not all immigrants engage in remittance and, for those who do, variations in their degree of participation abound. Like many other immigrant groups, Dominicans may abandon the social and cultural tools of their homeland, go about their lives without changing their Dominican-based social and cultural resources, or create new hybrid forms of cultural capital that blend the old with the new. Levitt shows how remittances of various kinds shape Dominicans’ views of the United States and help prospective migrants prepare for relocation. While social remittances may have transformed our earlier notions of assimilation, their importance to the transfer of innovative ideas and practices in areas of education, business, and health care have proved highly beneficial to both sending and receiving societies.

Peggy Levitt is Professor of Sociology at Wellesley College and a Research Fellow at the Weatherhead Center for International Affairs and the Hauser Center for Nonprofit Organizations at Harvard University, where she co-directs the Transnational Studies Initiative. She has written extensively on issues of international migration, transnational studies, and global cultural production and consumption. Her books include *God Needs No Passport: Immigrants and the Changing American Religious Landscape* (New Press, 2007) and *The Changing Face of Home: The Transnational Lives of the Second Generation* (Russell Sage, 2002).

Carmen Cardenas is a 27-year-old woman who lives with her mother and sister in Miraflores, a small village in the Dominican Republic. She is not married. She did not complete primary school. She does not have a steady job. She goes to Santo Domingo, the Dominican capital, about once a year but she has never traveled outside of the Dominican Republic. Like many Mirafloreños, she has two brothers and a sister living in Jamaica Plain, Boston, who support her. Though she has never been there, Carmen can vividly describe "La Center" and "La Mozart," or Center Street and Mozart Street Park, which are focal points for the Mirafloreño migrant community. She launches easily into a discussion of how life in Miraflores compares to life in the United States.

Much attention has been paid to the world-level diffusion of institutions, culture, and styles that arise from economic and political globalization. But this only partially explains Carmen's familiarity with a world she does not actually know. She can also envision a world beyond her direct experience because of social remittances – a local-level, migration-driven form of cultural diffusion.

Social remittances are the ideas, behaviors, identities, and social capital that flow from receiving- to sending-country communities. They are the north-to-south equivalent of the social and cultural resources that migrants bring with them which ease their transitions from immigrants to ethnics. The role that these resources play in promoting immigrant entrepreneurship, community and family formation, and political integration is widely acknowledged. Social remittances merit attention for several reasons. First, they play an important, understudied role in transnational collectivity formation. Second, they bring the social impacts of migration to the fore. And third, they are a potential community development aid. Because they travel through identifiable pathways to specific audiences, policymakers and planners can channel certain kinds of information to particular groups with positive results.

[. . .]

THEORETICAL ROOTS

Scholars of migration traditionally believed that most immigrants severed ties to their countries-of-origin as they assimilated into the country that received them. Recent work, however, suggests that at least some individuals remain oriented toward the communities they come from and the communities they enter. This sustained and constant contact between communities-of-origin and destination prompted scholars to speak of what they have alternatively termed transnational migration circuits, transnational social fields, transnational communities or binational societies.

These kinds of transnational relationships are not entirely new. Prior research indicates that earlier groups, such as the Irish and Italians in the United States, also remained involved in the affairs of their sending countries . . . Several factors, however, heighten the intensity and durability of transnational ties among contemporary migrants including: 1) ease of travel and communication, 2) the increasingly important role migrants play in sending-country economies, 3) attempts by sending states to legitimize themselves by providing services to migrants and their children, 4) the increased importance of the receiving-country states in the economic and political futures of sending societies, and 5) the social and political marginalization of migrants in their host countries. These factors mean that, in some cases, migrants may be active in their countries-of-origin and destination for extended periods.

As connections between sending and receiving countries strengthen and become more widespread, a transnational public sphere emerges. A public sphere is a space where citizens come together to debate their common affairs, contest meanings, and negotiate claims. What happens within the public spheres created by migration is by no means a foregone conclusion. They represent potential that may remain unused or take on a variety of forms. Public spheres may disintegrate if migrants incorporate easily into their host societies and sever their homeland ties or become arenas where nonmigrants and migrants come together periodically to articulate collective claims. Such public spheres may develop, as has Miraflores, into a transnational collectivity, where both migrants and nonmigrants enact some aspects of their lives simultaneously, though not equally, in multiple settings.

Miraflorenos formed one type of transnational collectivity. I call it a transnational community because of the strong, geographically-focused ties that almost two-thirds of the households in Miraflores share with family members in Boston. Both migrants and nonmigrants expressed a sense of consciously belonging to a group that spanned two settings. They formed an organization attesting to this. They described numerous scenarios in which they enacted roles or participated in organizations using ideas and behaviors gleaned from both places.

[...]

THE UNIQUENESS OF THE DOMINICAN CASE

The Dominican case is characterized by several unique qualities that make it particularly conducive to social remittance transfers. First, the nature of the Miraflorenos transnational community, and the social remittances that flow through it, are strongly influenced by the Dominican Republic's close proximity to the United States. It is easy and relatively inexpensive to travel back and forth. Migrant and nonmigrant family members speak regularly by phone. Social and cultural influences from the United States, already so prominent on the island, smooth the way for social remittance transfers.

Migration flows involving countries that are culturally and geographically farther apart would certainly produce different kinds of transfers.

Second, the social remittance exchanges described here take place within the context of an extensive history of U.S. dominance over Dominican economic and political affairs. They are part-and-parcel of the United States' long-term, deep involvement in Dominican national affairs. Migration flows involving more independent states or situations in which there is less of a power difference between countries of departure and reception would give rise to different transnational exchanges.

Finally, Radhames L. Trujillo, the dictator who ruled the Dominican Republic between 1930 and 1961, instilled a clear sense of national identity in Dominicans. Most migrants arrive strongly attached to their country of origin, which heightens the intensity of social remittance flows. In countries lacking such a highly developed sense of nationhood or in cases where migrants gladly abandon their homelands for fear of persecution or lack of economic opportunity, social remittance transfers will be weaker.

SOCIAL REMITTANCE CREATION

Migrants bring a set of social and cultural tools that aid their adjustment to their new lives. Studies of evolutionary institutional change suggest useful approaches for understanding how these resources are transformed into social remittances . . . I identify three broad patterns of interaction with the host society. Clearly, Miraflorenos did not fit precisely within these molds. I suggest them as tools to clarify how different levels of contact affect social remittance emergence.

At one end of the spectrum were recipient observers. Most of these individuals did not work outside their homes or worked in places where the majority of their coworkers were other Latinos. They tended to shop and socialize with other Latinos and reported few social contacts with Anglos. They did not actively explore their new world because the structure of their lives did not bring them close enough to it. Instead, they took in new ideas and practices by observing the world around them, listening to how others described it, or learning about it by reading the newspaper or watching television.

Other Miraflorenos participated more fully in U.S. life. Their interactions at work, on public transportation, or with medical/educational professionals forced them to shift their reference frames. They needed new skills to be able to get along. These instrumental adapters altered and added to their routines for pragmatic reasons. They readjusted their reference frames to equip them better to meet the challenges and constraints of migrant life.

Finally, some Miraflorenos were purposeful innovators. In contrast to recipient observers, they were sponges who aggressively searched for, selected and absorbed new things. Unlike instrumental adapters, they wanted to get ahead rather than just survive. They creatively added and combined what they took in with their existing ideas and practices, thereby expanding and extending their cultural repertoire.

Several patterns of social remittance evolution resulted. Each of these was most common among, though not restricted to, these three patterns of social interaction. In some cases, migrants abandoned elements of the social and cultural tools with which they arrived. They were irrelevant in the United States or structural and social constraints mitigated against their use. For instance, Dona Gabriela stopped holding Hora Santas, a popular, home-based religious ceremony to honor someone's death, because they were too difficult to organize in Boston. She could never be sure that enough people would attend since everyone was so busy and lived so much farther away from one another. Beliefs weakened and behaviors became unfamiliar when they were not used frequently. *Patrones* (benefactors) could make fewer claims on their clients because they could not bestow favors on a regular basis and because they had more limited access to distributory goods.

Migration left a second set of the social and cultural resources brought by migrants unchanged. This second pattern occurred most frequently in the recipient observer group. Many of their norms and practices went unchallenged because of their limited interaction with the host society. Their same repertoire worked because they recreated lives very similar to those they led in Miraflores.

A third pattern of social remittance emergence occurred when migrants added new items to their cultural repertoire that did not alter existing elements. They expanded the range of practices they engaged in without modifying old habits or ideas. This occurred most often among instrumental adapters. One example of this was the new skills Miraflorena women acquired at work. Since most migrant women did not work outside their homes before they came to the United States, they learned a new set of skills in the process of job seeking alone.

> When I got here, my sister tried to get me a job but there was no work at the company she was working for. I had to go down and speak to the supervisors at the places people told me about. I wasn't used to talking to people I didn't know. I had to use the telephone. In Miraflores, they had just gotten a phone at Carmen's house [her mother-in-law next door] and I wasn't used to talking to people that way. I had to find my way downtown on the subway. And I had hardly ever been to Santo Domingo by myself.
>
> (Gabriela, migrant Miraflorena, Boston)

The new skills Gabriela learned during her job search did not call into question her old ones. She added to her repertoire of skills and understandings but did not transform it.

In a fourth scenario, which was most common among purposeful innovators, migrants' ideas and practices combined with host-country norms. In these instances, cross-pollination occurred producing hybrid social forms. Dress is a good example of this, though its impact is most apparent in Miraflores, where the remitted practices have taken effect. Miraflorenas generally wear tight-fitting, brightly-colored clothing. They continued to dress this way in Boston with some modification, exchanging shorts for pants and sleeveless blouses for long-sleeved shirts. They also started wearing boots in the cold weather.

Nonmigrants observed these styles when migrants came back to visit. They also received clothing as gifts. Because young women, in particular, wanted to emulate these patterns, they combined elements of their own wardrobes with items from the United States and created a new hybrid style. Women wore boots with shorts. They wore long-sleeved clothing in 80 degree weather. Patterns of dress no longer reflected climate; rather, current fashion combined U.S. and Dominican elements.

[. . .]

WHAT IS EXCHANGED?

There are at least three types of social remittances – normative structures, systems of practice, and social capital.

Normative structures

Normative structures are ideas, values, and beliefs. They include norms for interpersonal behavior, notions of intrafamily responsibility, standards of age and gender appropriateness, principles of neighborliness and community participation, and aspirations for social mobility. They also include expectations about organizational performance such as how the church, state, or the courts should function. Norms about the role of clergy, judges, and politicians are also exchanged.

Several prior studies have described normative structure-type social remittances without defining them as such. The changing values and social ties that Polish immigrants wrote about to their non-migrant family members allegedly fostered greater individuality at home (Thomas and Znackieki, 1927). Return migrants to the West Indies repatriated change-inducing ideologies they learned from the Black Power movement in the United States (Patterson, 1988). Miraflorenos also communicated the values and norms they observed to those at home.

> When I go home, or speak to my family on the phone, I tell them everything about my life in the United States. What the rules and law are like. What is prohibited here. I personally would like people in the Dominican Republic to behave the way people behave here. The first time I went back to the Dominican Republic after nine years away, I arrived at the airport. I saw the floor was filthy and that the smokers threw their cigarette butts everywhere. And I said wait a minute. I even said it to the police who were there. How can this be, the gateway to our country is the airport. It should be clean and neat and people should be polite. When people put out their cigarettes they should use an ashtray. Tourists will get a bad impression when they see this mess. So when I smoked, I used an ashtray. It's not just saying things but doing them to provide a good example. When I'm in Miraflores, when I see people throwing garbage on the ground, I don't go and pick it up because that would be too much, but I get up and throw my own garbage away and everyone sees me do it. And those that have a little consciousness, without me saying anything, the next time they have to throw something out, they'll probably remember that they saw this, and I right thing to do, and they'll do it. These things and many more, the good habits I've acquired here, I want to show people at home.
>
> (Pepe, migrant, Boston)

Host societies offer both positive and negative role models and migrants are equally adept at emulating both.

> Life in the United States teaches them many good things but they also learn some bad things as well. People come back more individualistic, more materialistic. They are more committed to themselves than they are to the community. They just don't want to be active in trying to make the community better any more. Some learned to make it the easy way and they are destroying our traditional values of hard work and respect for the family.
>
> (Qavier Sanchez, nonmigrant, Miraflores)

Migrants' notions of identity also shifted with the changes in their social status brought about by migration. These revised concepts exposed non-migrants to a more ample range of possible self-concepts from which to choose. In Miraflores, notions of gender identity were particularly in flux. Migrant women modified their ideas about women's roles in response to their more active engagement in the workplace and with the public sector in Boston. They transmitted these new ideas about identity back to Miraflores. Nonmigrant women used these social remittances to construct new versions of womanhood. While their ideas were often somewhat romanticized, they still represented a marked change in thinking about male–female relations.

> I don't say that I won't get married to someone from here because he doesn't have a gold neck-lace. I want to marry someone from there

> because they have another mentality. They have changed the way they think. When they live there, they see that there isn't machismo. In the United States machismo doesn't exist.
> (Luisa, nonmigrant, Miraflores)

Systems of practice

Systems of practice are the actions shaped by normative structures. For individuals, these include household labor, religious practices, and patterns of civil and political participation. Within organizations, they include modes of membership recruitment and socialization, strategies, leadership styles, and forms of intraorganizational contact.

In Miraflores, these types of social remittances had far-reaching effects that ranged from altering the organization of social interaction to changing patterns of political participation. Miraflorenos spent much of their time outdoors due to the heat. They built homes with many windows and front *galerias* (porches). As a result, they socialized constantly with their neighbors while at work and at leisure and conducted much of their private lives in public view.

Miraflorenos in Boston were more isolated from those around them. Several respondents mentioned that they lived in the same building for several years but never met their neighbors across the hall. While some felt this deepened their loneliness, others became accustomed to living more independently, without "everyone knowing their affairs." When they returned to Miraflores, their desire to protect their privacy was one factor motivating the type of houses they built. Some eliminated the front *galeria* and oriented their homes toward a more private patio in the back. Others built homes surrounded by high walls. As more and more people imitate these behaviors, ties between neighbors are likely to weaken.

Patterns of organizational behavior were also socially remitted. Partido Revolucionario Dominicano (PRD) leaders claimed to transmit some of the vote-winning strategies they observed in the United States.

> Here are a series of concrete examples. With respect to publicity, we have begun to imitate North American political advertising with the slogans, posters, "bumper stick." We print up leaflets. We didn't use these things before – photographs of the president but not posters. This comes from the United States.
> (PRD vice-president, Santo Domingo)

Social capital

Both the values and norms on which social capital is based, and social capital itself, were socially remitted. Basch (1992) found this among Vincentian and Grenadian immigrant leaders and activists who were able to use the prestige and status they acquired in the United States to their advantage at home. This also occurred in Miraflores. When the nonmigrant sister of the Miraflores Development Committee (MDC) president in Boston became ill, her family went to the health clinic to ask the doctors to visit her at home, which they refused to do. Her relatives became angry and reminded the physicians that she was the MDC president's sister and that it was the MDC that had recently funded clinic renovations. When the doctors heard this, they suddenly became available. The president's family in Miraflores harnessed the social capital he accrued in Boston to help a family member at home.

Access to social capital also declined. When migrants did not contribute to community projects but were felt to be in a position to do so, nonmigrant family members in Miraflores suffered the consequences. For example, after Manuel became a supervisor in Boston, he started charging other Miraflorenos to place them in jobs. When word got back to Miraflores, committee members openly criticized his behavior and did not include his family in the next project.

MECHANISMS OF TRANSMISSION

Social remittance exchanges occur when migrants return to live in or visit their communities of origin; when nonmigrants visit their migrant family members; or through interchanges of letters, videos, cassettes and telephone calls. Social remittances are transmitted between individuals, within organizations by individuals enacting their organizational roles, or through the looser, informally-organized groups and social networks that are connected to the formal organizations. For instance, temporary political party working groups organized by PRD members for a specific campaign or informal transnational groups formed to organize popular religious ceremonies are also forums for social remittance transmission. The mechanisms of social remittance transmission differ from other types of global culture dissemination in several ways. First, the workings of social remittance transmission are specifiable while it is often difficult to distinguish how world-level institutions and global culture emerge and diffuse. Social remittances travel through identifiable pathways; their source and destination are clear. Migrants and nonmigrants can state how they learned of a particular idea or practice.

A second feature that distinguishes how social remittances are transmitted from other kinds of global transfers is that the latter are sometimes unsystematic and unintentional. Nonmigrants may also begin to hold their politicians accountable when they happen to hear on the radio that the President of the United States is being investigated for his real estate dealings. But these messages are not aimed at specific individuals. Many Miraflorenos are not listening or learn about them by accident. A comparable transfer of information that is a social remittance occurs when migrants speak directly to their family members about a different kind of politics and encourage them to pursue change. In this case, ideas are communicated intentionally to a specific recipient or group of recipients. Villagers can specify when and why they changed their minds about something or began to act in different ways.

A third distinguishing aspect of social remittance transmission is that it usually occurs between individuals who know one another personally or who are connected to one another by mutual social ties. Social remittances are delivered by a familiar messenger who comes "with references." This personalized character of the communication stands in contrast to the faceless, mass-produced nature of global cultural diffusion.

[. . .]

A fourth characteristic differentiating social remittance transmissions from other cultural flows is the timing with which they are communicated relative to other types of transfers. In many cases, a staged process occurs whereby macrolevel global flows precede and ease the way for social remittance transmission. Since many nonmigrants wanted to emulate the consumption patterns they were already familiar with from U.S. media, they were more receptive to the new political and religious styles that migrants subsequently introduced. Calls for more political and economic participation by women met with greater acceptance because they came on the heels of globally accepted discourses about women's equality. Social remittance flows do not arise out of the blue. They are part-and-parcel of an ongoing process of cultural diffusion. Gradual transmission sets the stage for future remittance transfers which then seem to make more sense.

Social remittances are just one of a number of change catalysts at work in sending-country communities. A variety of factors – including the remittance itself, the transnational system, the messenger, the target audience, differences between sending and receiving countries, and the transmission process – determine the nature and magnitude of social remittance impact.

The nature of the remittance itself

Remittance impact partially depends on how easy a particular remittance is to transmit. Some remittances are difficult to package. They do not lend themselves to becoming neat data packets but instead are slippery, unstable, and unwieldy to send. To be transmittable, they may have to be broken down into component parts, thereby heightening the potential for misinformation and confusion. In contrast, other types of remittances are fairly straightforward. They travel cleanly through transmission channels, after which they are either adopted or ignored.

For example, in the case of Miraflores, social remittances like member recruitment techniques or vote winning strategies were clear. They were either appropriated as is, modified and adopted, or disregarded. In contrast, values and norms fluctuated more easily. Immigrants constantly redefined and renegotiated them. Their unstable nature made them more difficult to simplify and express and therefore diminished their force.

The nature of the transnational system

[...]

Remittances flow more efficiently through tightly connected, dense systems because they tend to consist of similar parts and to involve similar technologies. Transfers within more open, informal systems are sloppier, less efficient, and more prone to interference by other cultural exchanges. They are like the child's game of "telephone" because each time a message is communicated it becomes more distorted in the translation.

The Boston-Dominican transnational religious system, for instance, arose primarily from personal relations between priests, parishioners, and seminarians. Because these transnational ties grew out of interpersonal connections, communication tended to be more circuitous, unsystematic, and leaky. In more structured settings, or in cases involving more tightly-woven social networks, the most efficient transmission channel was clear. The connections between organizational units or network members were closer and more well established, thereby heightening remittance impact.

[...]

COSTS AND BENEFITS OF SOCIAL REMITTANCES

The impacts of social remittances are both positive and negative. The ideas and practices migrants assimilated and refashioned in the United States prompted constructive change in some arenas and heightened problems in others. This comes across most clearly through a comparison of the realms of politics and the law.

About half of those interviewed saw migrants as catalysts for positive change. They argued that even though migrants were not citizens, they witnessed a fairer, more organized, and more equitable political system, about which they communicated to those in Miraflores.

> I have never been to Boston but my brothers say that the elections there are honest. Bill Clinton can't just tamper with votes because he wants to stay in power like Balaguer does here. In Santo Domingo, politics is a risk. Everything is personal. If I am from one party and you are from another, we can't share with one another. We can't discuss things. There you can say what you think. During the last elections, my brother told me how Bush and Clinton in a certain T.V. program said things to each other, and at the end they shook hands and one felt that the things that they said remained behind because it was a political thing. Here the same thing happens, but after the T.V. program is over they go outside and fight. Here, if there is a rally, the police are there and they hassle the parties on the Left. There the police have to supervise the entire rally and they can't favor one party over the other.
>
> (Freddy, nonmigrant, Miraflores)

Proponents of this view argued that the social remittances migrants sent engendered demands for a different kind of politics.

> There are greater demands for more democracy within the parties, that the justice system should be separate from the executive branch which is so corrupt . . . Emigration is a factor in the modernization of the political system in favor of a new type of establishment within the society. It is playing a role, since the people who come back come with these ideas. Even though they haven't participated in the political heart of the United States, they have lived there. They have a notion of the relationship between public and private and that these distinctions are clearer than those in the Dominican Republic. And these things, any person picks up on them because they see it when their kids go to school or when they pay taxes. If people live this in their daily lives, it produces a change in mentality. It is not that they have formed a movement in favor of the rights of citizens, but they have friends and neighbors, and they say to their cousins if you have a problem, go to a lawyer. Don't try to work it out through a friend.
>
> (PRD vice president, Santo Domingo)

As commented upon by one political leader, business as usual was no longer acceptable.

> Return emigrants say to us that if I go to get a form in the United States, I don't have to pay a cent because it is the state's obligation to give it to me. If I pay my taxes, I have the right to

> this, this, and this. If I don't consume 2,000 pesos worth of electricity, I don't pay a 2,000 peso electricity bill. Although they have menial jobs, they acquire a certain discipline that they didn't acquire here because here there weren't any jobs for these people . . . These are the things that are transmitted in the process of resocialization. This is a benefit that these people bring back to our society.
>
> (PRD leader, Santo Domingo)

Social remittances, then, prompted at least some Miraflorenos to seek change.

[. . .]

CONCLUSION

Social remittances are a conceptual tool for analyzing local-level cultural diffusion. Migrants do not absorb all aspects of their new lives unselectively and communicate these intact to those at home, who accept them as is. Instead, there is a screening process at work. Senders adopt certain new ideas and practices while filtering out others and receivers adopt particular elements while ignoring others. Furthermore, the nature of the social remittance itself, the way in which it is transmitted, the characteristics of the transnational organizational systems and networks through which its flows, and differences between individual and nation-state senders and receivers also influence social remittance impact. These factors ultimately shape who receives what kinds of remittances, how likely they are to adopt them, and their effects on social and political life.

[. . .]

The impact of social remittances is both positive and negative. There is nothing to guarantee that what is learned in the host society is constructive or that it will have a positive effect on communities of origin. Factors increasing social remittance effect are "ethics blind."

Positive social remittances will not, in and of themselves, bring about social reform. In communities like Miraflores, however, they are a fact of life. Social remittances represent a potential tool with which practitioners and planners can promote better outcomes. Because social remittances emanate from clear sources and travel through identifiable pathways to clear destinations, certain kinds of remittance flows can be purposefully stimulated. Deliberate exchanges about health and educational practices could contribute positively to social change. The transfer of new business skills or community organizing techniques could do so as well. The intentional transmission of more accurate information about working conditions and economic prospects in the United States might also create more realistic expectations of the migration experience. These transfers should, by no means, go one way. Information from educators, business owners, priests, and healthcare professionals serving the community in Miraflores could also help their counterparts in Boston be more effective in their tasks.

PART SIX

Culture and the City

Plate 9 Jazz Musician, New York City, 1958 by Dennis Stock. Reproduced by permission of Magnum Photos.

Plate 10 Times Square, Manhattan, New York City, 1999 by Raymond Depardon. Reproduced by permission of Magnum Photos.

INTRODUCTION TO PART SIX

Culture has always been a significant part of what cities are and do. The visual culture expressed in architectural styles, monuments, and the designs of parks, as well as the less formal culture offered by street musicians and artists in neighborhood festivals and fairs, contribute to how cities feel and are experienced. Until recently, however, the study of culture and its importance to urban form and change was relatively circumscribed. Over the years, a number of sociocultural and symbolic interactionist sociologists campaigned to push the study of culture to the forefront of urban studies. This includes the work of Anselm Strauss (*Images of the American City*. New York: Free Press of Glencoe, 1961) and Gerald Suttles ("The Cumulative Texture of Local Urban Culture," *American Journal of Sociology* 90, 2 [1984]: 283–304). But the discipline's preoccupation with demographic and political-economic changes left very little room for "culture" as a primary focus of urban theory or research.

That changed with the precipitous decline in the manufacturing-based economies of cities that took place in the 1970s and 1980s. With the relentless pace of deindustrialization, older cities, both large and small, refashioned their economies from the *production of things* (from consumer durables, such as automobiles and appliances, to the light manufacturing of toys, clothes, and books) to the *production of spectacles* (events, leisure, and cultural activities). We live in an age where the production and consumption of symbolic and cultural objects can be as profitable as the production and consumption of durable commodities.

Interestingly, many urbanists found a renewed interest in culture through the prism of political economic urban theories. David Harvey's influential book *The Condition of Postmodernity* (Oxford: Blackwell, 1989) outlined the connections between cultural changes (postmodernism) and political economic changes underway in western cities. Other efforts to integrate analyses of cultural production and consumption into urban theories include Sharon Zukin's study of New York's SoHo district and her work in the 1990s (parts of which are included in this section). Significant changes in how urban capitalism works – namely, the shift toward the production of services and spectacles – have placed the analysis of symbols and imagery on the cutting edge of urban studies. The material development of the urban built environment is not ignored but the analytical focus has shifted to the production and interpretation of cultural symbols and themes that inundate the modern city.

But the analysis of culture within urban studies has moved beyond simply seeing culture as linked to political economic or class-based concerns. As Sharon Zukin reminds us in her reading, individuals and groups make sense of and experience the places where they shop, socialize, and live, thereby shaping the urban environment. These experiences may be as diverse as the groups who inhabit the city. Thus, while certain powerful urban actors, such as planners and real estate developers, use culture to realize profits from urban development, their ability to shape how we experience the city is limited.

Richard Florida's work on the creative class further exemplifies this shift in viewing culture as a determining (as opposed to dependent) factor in shaping the form of cities. Florida asks why certain cities in the United States fare better in attracting talented professionals and creative people more than others. In light of the shift toward a postindustrial economy since the 1970s (a full four decades ago), most cities now boast, in varying degrees, the infrastructure of lifestyle amenities (high-end restaurant and shopping districts, entertainment zones, and renovated parks, esplanades, and other leisure spaces). Despite increasing similarities, not all cities are reputed as exciting, progressive places to live. Florida's research shows that the ranking of cities as attractive to talented and creative people is based on both objective factors (demographic diversity, urban infrastructure) and subjective factors (open-mindedness and tolerance of social difference and forward-thinking toward new ideas). That is, a city's "cultural feel" matters. According to Florida, those cities with "backwards" or provincial place identities or reputations will fare poorly in the highly competitive endeavor to attract and keep talent. His work has interesting implications for urban development policy. How do cities foster the intangibles of tolerance and acceptance?

The final two readings address another important topic within urban cultural studies – place identity. As indicated in the readings in Part Five, cities across the globe are consistently (re)defined by the production and circulation of information, services, and capital. Is it possible for a city to sustain a place identity? How are place identities shaped in the age of globalization? LiPuma and Koelble tackle these questions in their study of Miami, a city of quintessential ethnic and cultural diversity. LiPuma and Koelble show us that – quite expectedly – Miami's diversity often leads to tensions between ethnic groups but also produces surprising, hybrid forms of connections between cultures. Despite the complex and arguably chaotic social and cultural composition of the city, Miami manages to sustain a stable place identity by idealizing certain aspects of its diverse city life while downplaying others. In the final reading, by Karen Aguilar-San Juan, we also see how important agency is to the production of place identity. Focusing on a particular ethnic group – the Vietnamese – Aguilar-San Juan demonstrates how differently place identities can be formed, depending on the unique characteristics (the built environment, local history) of location. The unique characteristics of Boston and Orange County, California shape the very different Vietnamese enclaves that have formed over time in each. Whereas Fields Corner is more predisposed to the tight-knit and overlapping urban neighborhood culture found in the northeast United States, the place identity of Orange County's Little Saigon is clearly emblematic of Southern California's suburban car culture.

"Whose Culture? Whose City?"

from *The Cultures of Cities* (1995)

Sharon Zukin

Editors' Introduction

Sharon Zukin is a leading urban sociologist in the study of cities and culture. Her 1982 *Loft Living*, which examined New York City's SoHo neighborhood, is a landmark study of the intersection of culture and urban development. In it, she carefully presents the complementary and contradictory roles artists, tenants, manufacturers, real estate developers, and city officials play in the transforming of SoHo from a light manufacturing loft district in the 1960s to a trendy, increasingly upscale residential and commercial district. In the reading that follows, Zukin again addresses the interplay of various urban actors around issues of culture, which, she argues, has taken on greater significance in how cities are built and how we experience them.

Indeed, culture is the "motor of economic growth" for cities and forms the basis of what Zukin labels the "symbolic economy." The symbolic economy is comprised of two parallel production systems: the production of space, in which aesthetic ideals, cultural meanings, and themes are incorporated into the look and feel of buildings, streets, and parks, and the production of symbols, in which more abstract cultural representations influence how particular spaces within cities should preferably be "consumed" or used and by whom. The latter generates a good deal of controversy: as more and more ostensibly "public" spaces become identified (and officially sanctioned) with particular, often commercially generated, themes, we are left to ask "whose culture? whose city?"

We can easily see the symbolic economy at work in urban places such as Boston's Faneuil Hall, New York's South Street Seaport, or Baltimore's Harborplace. Here, cultural themes – mainly gestures toward a romanticized, imaginary past of American industrial growth – are enlisted to define place and, more specifically, what we should do there (shop, eat) and who we should encounter (other shoppers, tourists). Such places, although carefully orchestrated in design and feel, are popular because they offer a respite from the homogeneity and bland uniformity of suburban spaces. Local government officials and business alliances have turned toward manufacturing new consumption spaces of urban diversity (albeit narrowly defined) or showcasing existing ones – ethnic neighborhoods, revitalized historic districts, artist enclaves – as a competitive economic advantage over suburbs and other cities.

Culture, then, is purposefully used by developers and city officials to frame urban space to attract new residential tenants, to entice high-end shoppers, or court tourists and visitors from around the globe. But the fusing of culture and space is not limited to governments, corporations, and the real estate industry. The arguably less powerful inhabitants of the city – the ordinary residents, community associations, and block clubs – use cultural representations, too, to stamp their identity on place and to exert their cultural presence in public spaces. Ethnic festivals and parades mark the city and provide a cultural roadmap to what its spaces mean for certain groups and users. Every summer the city of Toronto hosts Caribana, the largest Caribbean festival in North America. The festival celebrates the vibrant ties of this Canadian city's large immigrant population to the Caribbean. Brilliantly costumed masqueraders and dozens of trucks carrying live soca, calypso, steel pan, reggae, and salsa artists enliven the city's streets. But the festival is by no means "local"; it attracts over a million participants annually, including hundreds of thousands of tourists, for whom Toronto "means" Caribana.

Finally, Zukin draws our attention to the increasing slippage between public and private spaces within the contemporary city. Historically, urban parks and streets have functioned as spaces where persons from different classes, ethnicities, and walks of life have intermingled and "rubbed elbows." Although always regulated and controlled, the identity of public spaces was seen as open and never exclusively defined by a single use or specific or preferred set of users. While parents and toddlers may "own" a corner of a city park on warm, sunny days, others – teenagers, lovers, or beer drinkers – lay claim to that same spot at different times of the day. As corporations increasingly sponsor public events and locally financed security forces police streets, the use of public spaces and their intended users are narrowed. Culture again becomes enlisted. Just as symbols, images, and other forms of representations may "define" shopping districts, restaurants, and theme parks, so they work in the (re)definition of public venues. Many urban plazas, waterfronts, and shopping streets have become "managed" by business associations, hence imposing their own identity on supposedly "public" space. The privatization (and militarization) of public space serves to reinforce the fear and conflict "between 'us' and 'them,' between security guards and criminals, [and] between elites and ethnic groups."

In addition to *Loft Living: Culture and Capital in Urban Change* (Baltimore, MD: The Johns Hopkins University Press, 1982), Sharon Zukin is the author of *Landscapes of Power: From Detroit to Disney World* (Berkeley, CA: University of California Press, 1991), *The Cultures of Cities* (from which this selection is excerpted) (Cambridge, MA: Blackwell, 1995), with Michael Sorkin, ed., *After the World Trade Center* (New York: Routledge, 2002), and *Point of Purchase: How Shopping Changed American Culture* (New York: Routledge, 2003). She is Broeklundian Professor of Sociology at Brooklyn College, CUNY.

Cities are often criticized because they represent the basest instincts of human society. They are built versions of Leviathan and Mammon, mapping the power of the bureaucratic machine or the social pressures of money. We who live in cities like to think of "culture" as the antidote to this crass vision. The Acropolis of the urban art museum or concert hall, the trendy art gallery and café, restaurants that fuse ethnic traditions into culinary logos – cultural activities are supposed to lift us out of the mire of our everyday lives and into the sacred spaces of ritualized pleasures.

Yet culture is also a powerful means of controlling cities. As a source of images and memories, it symbolizes "who belongs" in specific places. As a set of architectural themes, it plays a leading role in urban redevelopment strategies based on historic preservation or local "heritage." With the disappearance of local manufacturing industries and periodic crises in government and finance, culture is more and more the business of cities – the basis of their tourist attractions and their unique, competitive edge. The growth of cultural consumption (of art, food, fashion, music, tourism) and the industries that cater to it fuels the city's symbolic economy, its visible ability to produce both symbols and space.

In recent years, culture has also become a more explicit site of conflicts over social differences and urban fears. Large numbers of new immigrants and ethnic minorities have put pressure on public institutions, from schools to political parties, to deal with their individual demands. Such high culture institutions as art museums and symphony orchestras have been driven to expand and diversify their offerings to appeal to a broader public. These pressures, broadly speaking, are both ethnic and aesthetic. By creating policies and ideologies of "multiculturalism," they have forced public institutions to change.

On a different level, city boosters increasingly compete for tourist dollars and financial investments by bolstering the city's image as a center of cultural innovation, including restaurants, avant garde performances, and architectural design. These cultural strategies of redevelopment have fewer critics than multiculturalism. But they often pit the self-interest of real estate developers, politicians, and expansion minded cultural institutions against grassroots pressures from local communities.

At the same time, strangers mingling in public space and fears of violent crime have inspired the growth of private police forces, gated and barred communities, and a movement to design public spaces for maximum surveillance. These, too, are a source of contemporary urban culture. If one way of dealing with the material inequalities of city life has been to aestheticize diversity, another way has been to aestheticize fear.

Controlling the various cultures of cities suggests the possibility of controlling all sorts of urban ills, from violence and hate crime to economic decline. That this is an illusion has been amply shown by battles over multiculturalism and its warring factions – ethnic politics and urban riots. Yet the cultural power to create an image, to frame a vision, of the city has become more important as publics have become more mobile and diverse, and traditional institutions – both social classes and political parties – have become less relevant mechanisms of expressing identity. Those who create images stamp a collective identity. Whether they are media corporations like the Disney Company, art museums, or politicians, they are developing new spaces for public cultures. Significant public spaces of the late 19th and early 20th century – such as New York City's Central Park, the Broadway theater district, and the top of the Empire State Building – have been joined by Disney World, Bryant Park, and the entertaiment-based retail shops of Sony Plaza. By accepting these spaces without questioning their representations of urban life, we risk succumbing to a visually seductive, privatized public culture.

THE SYMBOLIC ECONOMY

[. . .]

Building a city depends on how people combine the traditional economic factors of land, labor, and capital. But it also depends on how they manipulate symbolic languages of exclusion and entitlement. The look and feel of cities reflect decisions about what – and who – should be visible and what should not, on concepts of order and disorder, and on uses of aesthetic power. In this primal sense, the city has always had a symbolic economy. Modern cities also owe their existence to a second, more abstract symbolic economy devised by "place entrepreneurs," officials and investors whose ability to deal with the symbols of growth yields "real" results in real estate development, new businesses, and jobs.

Related to this entrepreneurial activity is a third, traditional symbolic economy of city advocates and business elites who, through a combination of philanthropy, civic pride, and desire to establish their identity as a patrician class, build the majestic art museums, parks, and architectural complexes that represent a world-class city. What is new about the symbolic economy since the 1970s is its symbiosis of image and product, the scope and scale of selling images on a national and even a global level, and the role of the symbolic economy in speaking for, or representing, the city.

[. . .]

The entrepreneurial edge of the economy [has] shifted toward deal making and selling investments and toward those creative products that could not easily be reproduced elsewhere. Product design – creating the look of a thing – was said to show economic genius. Hollywood film studios and media empires were bought and sold and bought again. In the 1990s, with the harnessing of new computer-based technologies to marketing campaigns, the "information superhighway" promised to join companies to consumers in a Manichean embrace of technology and entertainment.

[. . .]

The growth of the symbolic economy in finance, media, and entertainment may not change the way entrepreneurs do business. But it has already forced the growth of towns and cities, created a vast new work force, and changed the way consumers and employees think. The facilities where these employees work – hotels, restaurants, expanses of new construction and undeveloped land – are more than just workplaces. They reshape geography and ecology; they are places of creation and transformation.

The Disney Company, for example, makes films and distributes them from Hollywood. It runs a television channel and sells commercial spinoffs, such as toys, books, and videos, from a national network of stores. Disney is also a real estate developer in Anaheim, Orlando, France, and Japan and the proposed developer of a theme park in Virginia and a hotel and theme park in Times Square. Moreover, as an employer, Disney has redefined work roles. Proposing a model for change in the emerging service economy, Disney has shifted from the white-collar worker to a new chameleon of "flexible" tasks. The planners at its corporate headquarters are "imaginers"; the costumed crowd-handlers at its theme parks are "cast members." Disney suggests that the symbolic economy is more than just the sum of the services it provides. The symbolic economy unifies material practices of finance, labor, art, performance, and design.

[. . .]

The symbolic economy recycles real estate as it does designer clothes. Visual display matters in American and European cities today, because the identities of places are established by sites of delectation. The sensual display of fruit at an urban farmers' market or gourmet food store puts a neighborhood "on the map" of visual delights and reclaims it for gentrification. A sidewalk cafe takes back the street from casual workers and homeless people.

[. . .]

Mass suburbanization since the 1950s has made it unreasonable to expect that most middle-class men and women will want to live in cities. But developing small places within the city as sites of visual delectation creates urban oases where everyone *appears* to be middle class. In the fronts of the restaurants or stores, at least, consumers are strolling, looking, eating, drinking, sometimes speaking English and sometimes not. In the back regions, an ethnic division of labor guarantees that immigrant workers are preparing food and cleaning up. This is not just a game of representations: developing the city's symbolic economy involves recycling workers, sorting people in housing markets, luring investment, and negotiating political claims for public goods and ethnic promotion. Cities from New York to Los Angeles and Miami seem to thrive by developing small districts around specific themes. Whether it is Times Square or el Calle Ocho, a commercial or an "ethnic" district, the narrative web spun by the symbolic economy around a specific place relies on a vision of cultural consumption and a social and an ethnic division of labor.

[. . .]

I see public culture as socially constructed on the micro-level. It is produced by the many social encounters that make up daily life in the streets, shops, and parks – the spaces in which we experience public life in cities. The right to be in these spaces, to use them in certain ways, to invest them with a sense of our selves and our communities – to claim them as ours and to be claimed in turn by them – make up a constantly changing public culture. People with economic and political power have the greatest opportunity to shape public culture by controlling the building of the city's public spaces in stone and concrete. Yet public space is inherently democratic. The question of who can occupy public space, and so define an image of the city, is open-ended.

Talking about the cultures of cities in purely visual terms does not do justice to the material practices of politics and economics that create a symbolic economy. But neither does a strictly political-economic approach suggest the subtle powers of visual and spatial strategies of social differentiation. The rise of the cities' symbolic economy is rooted in two long-term changes – the economic decline of cities compared to suburban and non-urban spaces and the expansion of abstract financial speculation – and in such short-term factors, dating from the 1970s and 1980s, as new mass immigration, the growth of cultural consumption, and the marketing of identity politics. We cannot speak about cities today without understanding how cities use culture as an economic base, how capitalizing on culture spills over into the privatization and militarization of public space, and how the power of culture is related to the aesthetics of fear.

CULTURE AS AN ECONOMIC BASE

[. . .]

Culture is intertwined with capital and identity in the city's production systems. From one point of view, cultural institutions establish a competitive advantage over other cities for attracting new businesses and corporate elites. Culture suggests the coherence and consistency of a brand name product. Like any commodity, "cultural" landscape has the possibility of generating other commodities.

[. . .]

In American and European cities during the 1970s, culture became more of an instrument in the entrepreneurial strategies of local governments and business alliances. In the shift to a post-postwar economy, who could build the biggest modern art museum suggested the vitality of the financial sector. Who could turn the waterfront from docklands rubble to parks and marinas suggested the possibilities for expansion of the managerial and professional corps. This was probably as rational a response as any to the unbeatable isolationist challenge of suburban industrial parks and office campuses. The city, such planners and developers as James Rouse believed, would counter the visual homogeneity of the suburbs by playing the card of aesthetic diversity.

[. . .]

Art museums, boutiques, restaurants, and other specialized sites of consumption create a social space for the exchange of ideas on which businesses thrive. While these can never be as private as a corporate dining room, urban consumption spaces allow for more social interaction among business elites. They are more democratic, accessible spaces than old-time businessmen's clubs. They open a window to the city – at least, to a rarified view of the city – and, to the extent they are written up in "lifestyle" magazines and consumer columns of the daily newspapers, they make ordinary people more aware of the elites' cultural consumption. Through the media, the elites' cultural preferences change what many ordinary people know about the city.

The high visibility of spokespersons, stars, and stylists for culture industries underlines the "sexy" quality of culture as a motor of economic growth. Not just in New York, Los Angeles, or Chicago, business leaders in a variety of low-profile, midsize cities are actively involved on the boards of trustees of cultural institutions because they believe that investing in the arts leads to more growth in other areas of the urban economy. They think a tourist economy develops the subjective image of place that "sells" a city to other corporate executives.

[. . .]

CULTURE AS A MEANS OF FRAMING SPACE

For several hundred years, visual representations of cities have "sold" urban growth. Images, from early maps to picture postcards, have not simply reflected real city spaces; instead, they have been imaginative reconstructions – from specific points of view – of a city's monumentality. The development of visual media in the 20th century made photography and movies the most important cultural means of framing urban space, at least until the 1970s. Since then, as the surrealism of *King Kong* shifted to that of *Blade Runner* and redevelopment came to focus on consumption activities, the material landscape itself – the buildings, parks, and streets – has become the city's most important visual representation. Historic preservation has been very important in this representation. Preserving old buildings and small sections of the city re-presents the scarce "monopoly" of the city's visible past. Such a monopoly has economic value in terms of tourist revenues and property values. Just an image of historic preservation, when taken out of context, has economic value. In Syracuse, New York, a crankshaft taken from a long-gone salt works was mounted as public sculpture to enhance a redevelopment project.

[. . .]

More common forms of visual re-presentation in all cities connect cultural activities and populist images in festivals, sports stadiums, and shopping centers. While these may simply be minimized as "loss leaders" supporting new office construction, they should also be understood as producing space for a symbolic economy.

[. . .]

Linking public culture to commercial cultures has important implications for social identity and social control. Preserving an ecology of images often takes a connoisseur's view of the past, re-reading the legible practices of social class discrimination and financial speculation by reshaping the city's collective memory. Boston's Faneuil Hall, South Street Seaport in New York, Harborplace in Baltimore, and London's Tobacco Wharf make the waterfront of older cities into a consumers' playground, far safer for tourists and cultural consumers than the closed worlds of wholesale fish and vegetable dealers and longshoremen. In such newer cities as Los Angeles or San Antonio, reclaiming the historic core, or the fictitious historic core, of the city for the middle classes puts the pueblo or the Alamo into an entirely different landscape from that of the surrounding inner city. On one level, there is a loss of authenticity, that is compensated for by a re-created historical narrative and a commodification of images; on another, men and women are simply displaced from public spaces they once considered theirs.

[...]

But incorporating new images into visual representations of the city can be democratic. It can integrate rather than segregate social and ethnic groups, and it can also help negotiate new group identities. In New York City, there is a big annual event organized by Caribbean immigrants, the West Indian-American Day Carnival parade, which is held every Labor Day on Eastern Parkway in Brooklyn. The parade has been instrumental in creating a pan-Caribbean identity among immigrants from the many small countries of that region. The parade also legitimizes the "gorgeous mosaic" of the ethnic population described by Mayor David N. Dinkins in 1989. The use of Eastern Parkway for a Caribbean festival reflects a geographical redistribution of ethnic groups – the Africanization of Brooklyn, the Caribbeanization of Crown Heights. More problematically, however, this cultural appropriation of public space supports the growing political identity of the Caribbean community and challenges the Lubavitcher Hassidim's appropriation of the same neighborhood. In Pasadena, California, African-American organizations have demanded representation on the nine-person commission that manages the annual Rose Parade, that city's big New Year's Day event. These cultural models of inclusion differ from the paradigm of legally imposed racial integration that eliminated segregated festivals and other symbolic activities in the 1950s and 1960s. By giving distinctive cultural groups access to the same public space, they incorporate separate visual images and cultural practices into the same public cultures.

Culture can also be used to frame, and humanize, the space of real estate development. Cultural producers who supply art (and sell "interpretation") are sought because they legitimize the appropriation of space. Office buildings are not just monumentalized by height and facades, or they are given a human face by video artists' screen installations and public concerts. Every well-designed downtown has a mixed-use shopping center and a nearby artists' quarter. Sometimes it seems that every derelict factory district or waterfront has been converted into one of those sites of visual delectation – a themed shopping space for seasonal produce, cooking equipment, restaurants, art galleries, and an aquarium. Urban redevelopment plans, from Lowell, Massachusetts, to downtown Philadelphia, San Francisco, and Los Angeles, focus on museums. Unsuccessful attempts to use cultural districts or aquariums to stop economic decline in Flint, Michigan and Camden, New Jersey – cities where there is no major employer – only emphasize the appeal of framing a space with a cultural institution when all other strategies of economic development fail.

Artists themselves have become a cultural means of framing space. They confirm the city's claim of continued cultural hegemony, in contrast to the suburbs and exurbs. Their presence – in studios, lofts, and galleries – puts a neighborhood on the road to gentrification. Ironically, this has happened since artists have become more self-conscious defenders of their own interests as artists and more involved in political organizations. Often they have been co-opted into property redevelopment projects as beneficiaries, both developers of an aesthetic mode of producing space (in public art, for example) and investors in a symbolic economy. There are, moreover, special connections between artists and corporate patrons. In such cities as New York and Los Angeles, the presence of artists documents a claim to these cities' status in the global hierarchy. The display of art, for public improvement or private gain, represents an abstraction of economic and social power. Among business elites, those from finance, insurance, and real estate are generally great patrons of both art museums and public art, as if to emphasize their prominence in the city's symbolic economy.

[...]

So the symbolic economy features two parallel production systems that are crucial to a city's material life: the *production of space*, with its synergy of capital investment and cultural meanings, and the *production of symbols*, which constructs both a currency of commercial exchange and a language of social identity. Every effort to rearrange space in the city is also an attempt at visual re-presentation. Raising property values, which remains a goal of most urban elites, requires imposing a new point of view. But negotiating whose point of view and the costs of imposing it create problems for public culture.

[. . .]

PUBLIC SPACE

[. . .]

It is important to understand the histories of symbolically central public spaces. The history of Central Park, for example, shows how, as definitions of who should have access to public space have changed, public cultures have steadily become more inclusive and democratic. From 1860 to 1880, the first uses of the park – for horseback riders and carriages – rapidly yielded to sports activities and promenades for the mainly immigrant working class. Over the next 100 years, continued democratization of access to the park developed together with a language of political equality. In the whole country, it became more difficult to enforce outright segregation by race, sex, or age.

During the 1970s, public space, especially in cities, began to show the effects of movements to "deinstitutionalize" patients of mental hospitals without creating sufficient community facilities to support and house them. Streets became crowded with "others," some of whom clearly suffered from sickness and disorientation. By the early 1980s, the destruction of cheap housing in the centers of cities, particularly single-room-occupancy hotels, and the drastic decline in producing public housing, dramatically expanded the problem of homelessness. Public space, such as Central Park, became unintended public shelter. As had been true historically, the democratization of public space was entangled with the question of fear for physical security.

[. . .]

Business Improvement Districts follow a fairly new model in New York State and in smaller cities around the United States, that allows business and property owners in commercial districts to tax themselves voluntarily for maintenance and improvement of public areas and take these areas under their control. The concept originated in the 1970s as special assessment districts; in the 1980s, the name was changed to a more upbeat acronym, business improvement districts (BIDs). A BID can be incorporated in any commercial area. Because the city government has steadily reduced street cleaning and trash pickups in commercial streets since the fiscal crisis of 1975, there is a real incentive for business and property owners to take up the slack. A new law was required for such initiatives: unlike shopping malls, commercial streets are publicly owned, and local governments are responsible for their upkeep.

[. . .]

What kind of public culture is created under these conditions? Do urban BIDs create a Disney World in the streets, take the law into their own hands, and reward their entrepreneurial managers as richly as property values will allow? If elected public officials continue to urge the destruction of corrupt and bankrupt public institutions, I imagine a scenario of drastic privatization, with BIDs replacing the city government.

[. . .]

In their own way, under the guise of improving public spaces, BIDs nurture a visible social stratification. They channel investment into a central space, a space with both real and symbolic meaning for elites as well as other groups. The resources of the rich Manhattan BIDs far outstrip those even potentially available in other areas of the city, even if those areas set up BIDs. The rich BIDs' opportunity to exceed the constraints of the city's financial system confirms the fear that the prosperity of a few central spaces will stand in contrast to the impoverishment of the entire city.

BIDs can be equated with a return to civility, "an attempt to reclaim public space from the sense of menace that drives shoppers, and eventually store owners and citizens, to the suburbs." But rich BIDs can be criticized on the grounds of control, accountability, and vision. Public space that is no longer controlled by public agencies must inspire a liminal public culture open to all but governed by the private sector. Private management of public space does create some savings: saving money by hiring nonunion workers, saving time by removing design questions from the public arena. Because they choose an abstract aesthetic with no pretense of populism, private organizations avoid conflicts over representations of ethnic groups that public agencies encounter when they subsidize public art, including murals and statues.

Each area of the city gets a different form of visual consumption catering to a different constituency: culture functions as a mechanism of stratification. The public culture of midtown public space diffuses down through the poorer BIDs. It focuses on clean design, visible security, historic architectural features, and the sociability among strangers achieved by suburban shopping malls. Motifs of local identity are chosen by merchants and commercial property owners. Since most commercial property owners and merchants do not live in the area of their business or even in New York City, the sources of their vision of public culture may be eclectic: the nostalgically remembered city, European piazzas, suburban shopping malls, Disney World. In general, however, their vision of public space derives from commercial culture.

[. . .]

SECURITY, ETHNICITY, AND CULTURE

One of the most tangible threats to public culture comes from the politics of everyday fear. Physical assaults, random violence, hate crimes that target specific groups: the dangers of being in public spaces utterly destroy the principle of open access. Elderly men and women who live in cities commonly experience fear as a steady erosion of spaces and times available to them. An elderly Jewish politician who in the 1950s lived in Brownsville, a working-class Jewish neighborhood in Brooklyn where blacks began to move in greater numbers as whites moved out, told me, "My wife used to be able to come out to meet me at night, after a political meeting, and leave the kids in our apartment with the door unlocked." A Jewish woman remembers about that same era, "I used to go to concerts in Manhattan wearing a fur coat and come home on the subway at 1 a.m." There may be some exaggeration in these memories, but the point is clear. And it is not altogether different from the message behind crimes against black men who venture into mainly white areas of the city at night or attacks on authority figures such as police officers and firefighters who try to exercise that authority against street gangs, drug dealers, and gun-toting kids. Cities are not safe enough for people to participate in a public culture.

"Getting tough" on crime by building more prisons and imposing the death penalty are all too common answers to the politics of fear. Another answer is to privatize and militarize public space – making streets, parks, and even shops more secure but less free, or creating spaces, such as shopping malls and Disney World, that only *appear* to be public spaces because so many people use them for common purposes. It is not so easy, given a language of social equality, a tradition of civil rights, and a market economy, to enforce social distinctions in public space. The flight from "reality" that led to the privatization of public space in Disney World is an attempt to create a different, ultimately more menacing kind of public culture.

[. . .]

Gentrification, historic preservation, and other cultural strategies to enhance the visual appeal of urban spaces developed as major trends during the late 1960s and early 1970s. Yet these years were also a watershed in the institutionalization of urban fear. Voters and elites – a broadly conceived middle class in the United States – could have faced the choice of approving government policies to eliminate poverty, manage ethnic competition, and integrate everyone into common public institutions. Instead, they chose to buy protection, fueling the growth of the private security industry. This reaction was closely related to a perceived decline in public morality, an "elimination of almost all stabilizing authority" in urban public space.

[. . .]

In the past, those people who lived so close together they had to work out some etiquette for sharing, or dividing, public space were usually the poor. An exception that affected everyone was the system of racial segregation that worked by law in the south and by convention in many northern states until the 1960s, when – not surprisingly – perceptions of danger among whites increased. Like segregation, a traditional etiquette of public order of the urban poor involves dividing up territory by ethnic groups. This includes the system of "ordered segmentation" that the Chicago urban sociologist Gerald Suttles described a generation ago, at the very moment it was being outmoded by increased racial and ethnic mixing, ideologies of community empowerment, and the legitimization of ethnicity as a formal norm of political representation. Among city dwellers today, innumerable informal etiquettes for survival in public spaces flourish. The "streetwise" scrutiny of passersby described by the sociologist Elijah Anderson is one means for unarmed individuals to secure the streets. I think ethnicity – a cultural strategy for producing difference – is another, and it survives on the politics of fear by requiring people to keep their distance from certain aesthetic markers. These markers vary over time. Pants may be baggy or pegged, heads may be shaggy or shaved. Like fear itself, ethnicity becomes an aesthetic category.

[...]

For a brief moment in the late 1940s and early 1950s, working-class urban neighborhoods held the possibility of integrating white Americans and African-Americans in roughly the same social classes. This dream was laid to rest by movement to the suburbs, continued ethnic bias in employment, the decline of public services in expanding racial ghettos, criticism of integration movements for being associated with the Communist party, and fear of crime. Over the next 15 years, enough for a generation to grow up separate, the inner city developed its stereotyped image of "Otherness." The reality of minority groups' working-class life was demonized by a cultural view of the inner city "made up of four ideological domains: a physical environment of dilapidated houses, disused factories, and general dereliction; a romanticized notion of white working-class life with particular emphasis on the centrality of family life; a pathological image of black culture; and a stereotypical view of street culture."

By the 1980s, the development of a large black middle class with incomes more or less equal to white households' and the increase in immigrant groups raised a new possibility of developing ethnically and racially integrated cities. This time, however, there is a more explicit struggle over who will occupy the image of the city. Despite the real impoverishment of most urban populations, the larger issue is whether cities can again create an inclusive public culture. The forces of order have retreated into "small urban spaces," like privately managed public parks that can be refashioned to project an image of civility. Guardians of public institutions – teachers, cops – lack the time or inclination to *understand* the generalized ethnic Other.

[...]

When Disneyland recruited teenagers in South Central Los Angeles for summer jobs following the riots of 1992, it thrust into prominence a new confluence between the sources of contemporary public culture: a confluence between commercial culture and ethnic identity. Defining public culture in these terms recasts the way we view and describe the cultures of cities. Real cities are both material constructions, with human strengths and weaknesses, and symbolic projects developed by social representations, including affluence and technology, ethnicity and civility, local shopping streets and television news. Real cities are also macro-level struggles between major sources of change – global and local cultures, public stewardship and privatization, social diversity and homogeneity – and micro-level negotiations of power. Real cultures, for their part, are not torn by conflict between commercialism and ethnicity; they are made up of one-part corporate image selling and two-parts claims of group identity, and get their power from joining autobiography to hegemony – a powerful aesthetic fit with a collective lifestyle.

[...]

How do we connect what we experience in public space with ideologies and rhetorics of public culture? On the streets, the vernacular culture of the powerless provides a currency of economic exchange and a language of social revival. In other public spaces – grand plazas, waterfronts, and shopping streets reorganized by business improvement districts – another landscape incorporates vernacular culture or opposes it with its own image of identity and desire. Fear of reducing the distance between "us" and "them," between security guards and criminals, between elites and ethnic groups, makes culture a crucial weapon in reasserting order.

"Cities and the Creative Class"

from *City and Community* (2003)

Richard Florida

Editors' Introduction

We often tend to think of cities and culture in terms of museums, art galleries, and concert halls, as well as high-end shopping districts and bohemian artist enclaves. In this selection, Richard Florida asks us to view these cultural amenities and others as powerful attributes that attract a highly desirable workforce he calls the "creative class." Florida is the author of the "creative capital thesis," which holds that a community of creative and talented people is fundamental to contemporary urban development and that there are underlying (and detectable) factors that favor certain cities ("creative centers") over others.

Cities have long been identified with diversity, creativity, and innovation. But, as Florida argues, urban scholars and policymakers have typically and exclusively identified these assets with corporations and firms. This is apparent when one examines the tax abatements and other financial incentives municipal and state governments offer companies to relocate to or remain in their cities. When we turn our explanatory gaze to municipal efforts to attract talented and creative individuals (and to keep them as residents), we see certain theories might apply but each has shortcomings.

Agglomeration and cluster theories offer one explanation of why certain industries find cities desirable places to do business. Most cities have legal districts, for example, where law offices and courthouses are in close proximity. Legal transactions tend to require a good deal of face-to-face contact. Other industries, such as investment firms, tend to benefit in varying degrees from the efficiency and ease of geographic proximity.

Social capital theory also values the immediacy of close ties and connections among groups and individuals. Its major proponent, Robert Putnam, argued that a community's prosperity rests on strong ties and associations among its members. Yet, as Florida points out, the social and technological changes of the past twenty years make Putnam's call for a return to small-town community a romantic notion at best.

Florida finds greater utility in human capital theory, which posits that urban and regional growth is increasingly dependent on the presence of higher education and similar institutions to attract talented individuals. Many states and cities have funded university-corporate partnerships as economic incubators to enhance high-tech, well-paying employment.

Finally, Florida discusses his own "creative capital" theory of urban and regional development. He agrees partly with the human capital approach – infrastructure (good schools, good housing, etc.) is indeed important to attract talented and creative people. But positive lifestyle factors are also necessary to attract and maintain the creative class. This class is comprised of professionals who work in knowledge-based occupations in high-tech sectors, financial services, the legal and health-care professions, and business management.

In this age of telecommunications, professionals are able to live and work in a variety of cities or small towns. Why, then, do certain cities tend to attract a large share of the creative class? Florida argues that creative centers provide an environment where all forms of creativity – artistic and cultural, technological and economic – take place. Cities are misguided, Florida argues, in their emphasis upon infrastructural attractions, such as sports stadiums and tourism-and-entertainment districts. In his research, he found that members of the creative class desire "high-quality experiences, an openness to diversity of all kinds, and, above all else, the opportunity to validate their identities as creative people." The cities that are currently faring best have an abundance of technology, talent, and tolerance. Each is a necessary, Florida tells us, but by itself an insufficient condition. To attract creative and talented people and stimulate urban development, a place must have all three. The benefits of a concentrated creative class are many: growth in the well-paying employment sector, a higher tax base, reinvestment in the built environment, and population growth.

Florida is leading an effort to compel scholars to think more broadly about factors that contribute to successful cities – to move away from our late-nineteenth- and early-twentieth-century view of the city in terms of production of goods to seeing cities as centers of experience, lifestyle, amenities, and entertainment. This shift is not only apparent in cities such as San Francisco, Seattle, and Boston, but in older industrial cities such as Chicago and Philadelphia, which are moving to become centers of entertainment and cultural experiences.

Richard Florida is the Heinz Professor of Economic Development at Carnegie Mellon University. He is the author of *The Rise of the Creative Class: And How It's Transforming Work, Leisure, Community and Everyday Life* (New York: Basic Books, 2002) in which he identifies the underlying changes and the structural economic transformations behind dramatic changes in work and lifestyle that are fueling urban change. The book was awarded the Political Book Award for 2002 by the *Washington Monthly* and was listed as one of the ten most influential books of 2002 by the *Globe and Mail*. He is founder of Creativity Group, a consulting firm that works with corporations and governments around the world.

From the seminal work of Alfred Marshall to the 1920 studies by Robert Park to the pioneering writings of Jane Jacobs, cities have captured the imagination of sociologists, economists, and urbanists. For Park and especially for Jacobs, cities were cauldrons of diversity and difference, creativity and innovation. Yet over the last several decades, scholars have somehow forgotten this basic, underlying theme of urbanism. For the past two decades, I have conducted research on the social and economic functions of cities and regions. Generally speaking, the conventional wisdom in my field of regional development has been that companies, firms, and industries drive regional innovation and growth, and thus an almost exclusive focus in the literature on the location and, more recently, the clustering of firms and industries. From a policy perspective, this basic conceptual approach has undergirded policies that seek to spur growth by offering firms financial incentives and the like. More recently, scholars such as Robert Putnam have focused on the social functions of neighborhoods, communities, and cities, while others, such as the urban sociologist Terry Clark and the economist Edward Glascar, have turned their attention toward human capital, consumption, and cities as lifestyle and entertainment districts.

[...]

AGGLOMERATION AND CLUSTER THEORIES

Many researchers, sociologists, and academics have theorized on the continued importance of place in economic and social life. An increasingly influential view suggests that place remains important as a locus of economic activity because of the tendency of firms to cluster together. This view builds on the influential theories of the economist Alfred Marshall, who argued that firms cluster in "agglomerations" to gain productive efficiencies. The contemporary variant of this view, advanced by Harvard Business School professor Michael Porter, has many proponents in academia and in the practice of economic development. It is clear that similar firms tend to cluster. Examples of this sort of agglomeration include not only Detroit and Silicon Valley, but the *maquiladora* electronics and auto-parts districts in Mexico, the clustering of disk-drive makers in Singapore and of flat-panel-display producers in Japan, and the garment district and Broadway theater district in New York City.

The question is not whether firms cluster but why. Several answers have been offered. Some experts believe that clustering captures efficiencies generated from tight linkages between firms. Others say it has to do with the positive benefits of co-location, or what they call "spillovers." Still others claim it is because certain kinds of activity require face-to-face contact. But these are only partial answers. More importantly, companies cluster in order to draw from concentrations of talented people who power innovation and economic growth. The ability to rapidly mobilize talent from such a concentration of people is a tremendous source of competitive advantage for companies in our time-driven economy of the creative age.

THE SOCIAL CAPITAL PERSPECTIVE

An alternative view is based on Robert Putnam's social capital theory. From his perspective, regional economic growth is associated with tight-knit communities where people and firms form and share strong ties. In his widely read book *Bowling Alone*, he makes a compelling argument that many aspects of community life declined precipitously over the last half of the 20th century. Putnam gets his title from his finding that from 1980–1993, league bowling declined by 40 percent, whereas the number of individual bowlers rose by 10 percent. This, he argues, is just one indicator of a broader and more disturbing trend. Across the nation, people are less inclined to be part of civic groups: voter turnout is down, so is church attendance and union membership, and people are less and less inclined to volunteer. All of this stems from what Putnam sees as a long-term decline in social capital.

By this, he means that people have become increasingly disconnected from one another and from their communities. Putman finds this disengagement in the declining participation in churches, political parties, and recreational leagues, not to mention the loosening of familial bonds. Through painstakingly detailed empirical research, he documents the decline in social capital in civic and social life. For Putman, declining social capital means that society becomes less trustful and less civic-minded. Putnam believes a healthy, civic-minded community is essential to prosperity.

Although initially Putnam's theory resonated with me, my own research indicates a different trend. The people in my focus groups and interviews rarely wished for the kinds of community connectedness Putnam talks about. If anything, it appeared they were trying to get away from those kinds of environments. To a certain extent, participants acknowledged the importance of community, but they did not want it to be invasive or to prevent them from pursuing their own lives. Rather, they desired what I have come to term "quasi-anonymity." In the terms of modern sociology, these people prefer weak ties to strong.

This leads me to an even more basic observation. The kinds of communities that we both desire and that generate economic prosperity are very different than those of the past. Social structures that were important in earlier years now work against prosperity. Traditional notions of what it means to be a close, cohesive community and society tend to inhibit economic growth and innovation. Where strong ties among people were once important, weak ties are now more effective. Those social structures that historically embraced closeness may now appear restricting and invasive. These older communities are being exchanged for more inclusive and socially diverse arrangements. These trends are also what the statistics seem to bear out.

[. . .]

Historically, strong-tied communities were thought to be beneficial. However, there are some theorists that argue the disadvantages of such tight bonds. Indeed, social capital can and often does cut both ways: it can reinforce belonging and community, but it can just as easily shut out newcomers, raise barriers to entry, and retard innovation.

[. . .]

Places with dense ties and high levels of traditional social capital provide advantages to insiders and thus promote stability, while places with looser networks and weaker ties are more open to newcomers and thus promote novel combinations of resources and ideas.

HUMAN CAPITAL AND URBAN-REGIONAL GROWTH

Over the past decade or so, a potentially more powerful theory for city and regional growth has emerged. This theory postulates that people are the motor force behind regional growth. Its proponents thus refer to it as the "human capital" theory of regional development.

Economists and geographers have always accepted that economic growth is regional – that is driven by, and spreads from, regions, cities, or even neighborhoods. The traditional view, however, is that places grow either because they are located on transportation routes or because they have natural resources that encourage firms to locate there. According to this conventional view, the economic importance of a place is tied to the efficiency with which one can make things and do business. Governments employ this theory when they use tax breaks and highway construction to attract business. But these cost-related factors are no longer as crucial to success.

The proponents of the human capital theory argue that the key to regional growth lies not in reducing the costs of doing business, but in endowments of highly-educated and productive people. The human capital theory – like many theories of cities and urban areas – owes a debt to Jane Jacobs. Decades ago, Jacobs noted the ability of cities to attract creative people and thus spur economic growth. The Nobel-prize-winning economist Robert Lucas sees the productivity effects that come from the clustering of human capital as the critical factor in regional economic growth, referring to this as a "Jane Jacobs externality."

[. . .]

THE CREATIVE CAPITAL PERSPECTIVE

The human capital theory establishes that creative people are the driving force in regional economic growth. From that perspective, economic growth will occur in places that have highly educated people. But in treating human capital as a stock or endowment, this theory begs the question: Why do creative people cluster in certain places? In a world where people are highly mobile, why do they choose some cities over others and for what reasons?

Although economists and social scientists have paid a lot of attention to how companies decide where to locate, they have virtually ignored how people do so. This is the fundamental question I have tried to answer. In my interviews and focus groups, the same answer kept coming back: people said that economic *and* lifestyle considerations both matter, and so does the mix of both factors. In reality, people were not making the career decisions or geographic moves that the standard theories said they should: They were not slavishly following jobs to places. Instead, it appeared that highly educated individuals were drawn to places that were inclusive and diverse. Not only did my qualitative research indicate this trend, but the statistical analysis proved the same.

Gradually, I came to see my perspective, the creative capital theory, as distinct from the human capital theory. From my perspective, creative people power regional economic growth and these people prefer places that are innovative, diverse, and tolerant. My theory thus differs from the human capital theory in two respects: (1) it identifies a type of human capital, creative people, as being key to economic growth; and (2) it identifies the underlying factors that shape the location decisions of these people, instead of merely saying that regions are blessed with certain endowments of them.

To begin with, creative capital begins most fundamentally with the people I call the "creative class." The distinguishing characteristic of the creative class is that its members engage in work whose function is to "create meaningful new forms." The super-creative core of this new class includes scientists and engineers, university professors, poets and novelists, artists, entertainers, actors, designers, and architects, as well as the "thought leadership" of modern society: nonfiction writers, editors, cultural figures, think-tank researchers, analysts, and other opinion-makers. Members of this super-creative core produce new forms or designs that are readily transferable and broadly useful, such as designing a product that can be widely made, sold, and used; coming up with a theorem or strategy that can be applied in many cases; or composing music that can be performed again and again.

Beyond this core group, the creative class also includes "creative professionals" who work in a wide range of knowledge-based occupations in high-tech sectors, financial services, the legal and health-care professions, and business management. These people engage in creative problem-solving, drawing on complex bodies of knowledge to solve specific problems. Doing so typically requires a high degree of formal education and thus a high level of human capital. People who do this kind of work may sometimes come up with methods or products that turn out to be widely useful, but that is not part of the basic job description. What they are required to do regularly is think on their own. They apply or combine standard approaches in unique ways to fit the situation, exercise a great deal of judgment, and at times must independently try new ideas and innovations.

According to my estimates, the creative class now includes some 38.3 million Americans, roughly 30 percent of the entire U.S. workforce – up from just 10 percent at the turn of the 20th century and less than 20 percent as recently as 1980. However, it is important to point out that my theory recognizes creativity as a fundamental and intrinsic human characteristic. In a very real sense, all human beings are creative and all are potentially members of the creative class. It is just that 38 million people – roughly 30 percent of the workforce – are fortunate enough to be paid to use their creativity in their work.

In my research I have discovered a number of trends that are indicative of the new geography of creativity. These are some of the patterns of the creative class:

- The creative class is moving away from traditional corporate communities, working class centers, and even many Sunbelt regions to a set of places I call "creative centers."
- The creative centers tend to be the economic winners of our age. Not only do they have high concentrations of creative-class people, they have high concentrations of creative economic outcomes, in the form of innovations and high-tech industry growth. They also show strong signs of overall regional vitality, such as increases in regional employment and population.
- The creative centers are not thriving due to traditional economic reasons such as access to natural resources or transportation routes. Nor are they thriving because their local governments have gone bankrupt in the process of giving tax breaks and other incentives to lure business. They are succeeding largely because creative people want to live there. The companies follow the people – or, in many cases, are started by them. Creative centers provide the integrated ecosystem or habitat where all forms of creativity – artistic and cultural, technological and economic – can take root and flourish.
- Creative people are not moving to these places for traditional reasons. The physical attractions that most cities focus on – sports stadiums, freeways, urban malls, and tourism-and-entertainment districts that resemble theme parks – are irrelevant, insufficient, or actually unattractive to many creative-class people. What they look for in communities are abundant high-quality experiences, an openness to diversity of all kinds, and, above all else, the opportunity to validate their identities as creative people.

THE NEW GEOGRAPHY OF CREATIVITY

These shifts are giving rise to powerful migratory trends and an emerging new economic geography. In the leading creative centers, the creative class makes up more than 35 percent of the workforce, regions such as the greater Washington, DC, region, the Raleigh-Durham area, Boston, and Austin. But despite their considerable advantages, large regions have not cornered the market as creative-class locations. In fact, a number of smaller regions have some of the highest creative-class concentrations in the nation – notably college towns such as East Lansing, Michigan, and Madison, Wisconsin.

At the other end of the spectrum are regions that are being bypassed by the creative class. Among large regions, Las Vegas, Grand Rapids, and Memphis harbor the smallest concentrations of the creative class. Members of the creative class have nearly abandoned a wide range of smaller regions in the outskirts of the South and Midwest. In small metropolitan areas such as Victoria, Texas, and Jackson, Tennessee, the creative class comprises less than 15 percent of the workforce. The leading centers for the working class among large regions are Greensboro, North Carolina, and Memphis, Tennessee, where the working class makes up more than 30 percent of the workforce. Several smaller regions in the South and Midwest are veritable working-class enclaves with 40 to 50 percent or more of their workforce in the traditional industrial occupations. These places have some of the most minuscule concentrations of the creative class in the nation. They are symptomatic of a general lack of overlap between the major creative-class centers and those of the working class. Of the 26 large cities where the working class comprises more than one-quarter of the population, only one, Houston, ranks among the top 10 destinations for the creative class.

Las Vegas has the highest concentration of the service class among large cities, 58 percent, while West Palm Beach, Orlando, and Miami also have around half of their total workforce in the service class. These regions rank near the bottom of the list for the creative class. The service class makes up more than half the workforce in nearly 50 small and medium-size regions across the country. Few of them boast any significant concentrations of the creative class, save as vacationers, and offer little prospect for upward mobility. They include resort towns such as Honolulu and Cape Cod. But they also include places like Shreveport, Louisiana, and Pittsfield, Massachusetts. For these places that are not tourist destinations, the economic and social future is troubling to contemplate.

Places that are home to large concentrations of the creative class tend to rank highly as centers of innovation and high-tech industry. Three of the top five large creative-class regions are among the top five high-tech regions. Three of the top five large creative class regions are also among the top five most innovative regions (measured as patents granted per capita). And, the *same five* large regions that top the list on the Talent Index (measured as the percentage of people with a bachelor's degree or above) also have the highest creative-class concentration: Washington, DC, Boston, Austin, the Research Triangle, and San Francisco. The statistical correlations comparing creative-class locations to rates of patenting and high-tech industry are uniformly positive and statistically significant.

TECHNOLOGY, TALENT, AND TOLERANCE

The key to understanding the new economic geography of creativity and its effects on economic outcomes lies in what I call the 3Ts of economic development: *technology, talent, and tolerance.* Creativity and members of the creative class take root in places that possess all three of these critical factors. Each is a necessary but by itself insufficient condition. To attract creative people, generate innovation, and stimulate economic development, a place must have all three. I define tolerance as openness, inclusiveness, and diversity to all ethnicities, races, and walks of life. Talent is defined as those with a bachelor's degree and above. And technology is a function of both innovation and high-technology concentrations in a region. My focus group and interview results indicate that talented individuals are drawn to places that offer tolerant work and social environments. The statistical analysis validates not only the focus group results, but also indicates strong relationships between technology, tolerance, and talent.

The 3Ts explain why cities such as Baltimore, St. Louis, and Pittsburgh fail to grow despite their deep reservoirs of technology and world-class universities: they are unwilling to be sufficiently tolerant and open to attract and retain top creative talent. The interdependence of the 3Ts also explains why cities such as Miami and New Orleans do not make the grade even though they are lifestyle meccas: they lack the required technology base. The most successful places – the San Francisco Bay area, Boston, Washington, DC, Austin, and Seattle – put all 3Ts together. They are truly creative places.

My colleagues and I have conducted a great deal of statistical research to test the creative capital theory by looking at the way these 3Ts work together to power economic growth. We found that talent or creative capital is attracted to places that score high on our basic indicators of diversity – the Gay, Bohemian, and other indexes. It is not because high-tech industries are populated by great numbers of bohemians and gay people; rather, artists, musicians, gay people, and members of the creative class in general prefer places that are open and diverse. Such low entry barriers are especially important because, today, places grow not just through higher birth rates (in fact virtually all U.S. cities are declining on this measure), but by their ability to attract people from the outside.

As we have already seen, human capital theorists have shown that economic growth is closely associated with concentrations of highly-educated people. But few studies have specifically looked at the relationship between talent and technology, between clusters of educated and creative people and concentrations of innovation and high-tech industry. Using our measure of the creative class and the basic Talent Index, we examined these relationships for the 49 regions with more than one million people and for all 206 regions for which data are available. As well as some well-known technology centers, smaller college and university towns rank high on the Talent Index – places such as Santa Fe, Madison, Champaign-Urbana, State College, and Bloomington, Indiana. When I look at the subregional level, Ann Arbor (part of the Detroit region) and Boulder (part of the Denver region) rank first and third, respectively.

These findings show that both innovation and high-tech industry are strongly associated with locations of the creative class and of talent in general. Consider that 13 of the top 20 high-tech regions also rank among the top 20 creative-class centers, as do 14 of the top 20 regions for high-tech industry. Furthermore, an astounding 17 of the top 20 Talent Index regions also rank in the top 20 of the creative class. The statistical correlations between Talent Index and the creative-class centers are understandably among the strongest of any variables in my analysis because creative-class people tend to have high levels of education. But the correlations between the Talent Index and working-class regions are just the opposite – negative and highly significant – suggesting that working-class regions possess among the lowest levels of human capital.

Thus, the creative capital theory says that regional growth comes from the 3Ts of economic development, and to spur innovation and economic growth a region must have all three of them.

THE ROLE OF DIVERSITY

Economists have long argued that diversity is important to economic performance, but they have usually meant the diversity of firms or industries. The economist John Quigley, for instance, argues that regional economies benefit from the location of a diverse set of firms and industries. Jacobs long ago highlighted the role of diversity of both firms and people in powering innovation and city growth. As Jacobs saw it, great cities are places where people from virtually any background are welcome to turn their energy and ideas into innovations and wealth.

This raises an interesting question. Does living in an open and diverse environment help to make talented and creative people even more productive; or do its members simply cluster around one another and thus drive up these places' creativity only as a byproduct? I believe both are going on, but the former is more important. Places that are open and possess *low entry barriers* for people gain creativity advantage from their ability to attract people from a wide range of backgrounds. All else being equal, more open and diverse places are likely to attract greater numbers of talented and creative people – the sort of people who power innovation and growth.

LOW BARRIERS TO ENTRY

A large number of studies point to the role of immigrants in economic development. In *The Global Me*, the *Wall Street Journal* reporter Pascal Zachary argues that openness to immigration is the cornerstone of innovation and economic growth. He contends that America's successful economic performance is directly linked to its openness to innovative and energetic people from around the world, and attributes the decline of once prospering countries, such as Japan and Germany, to the homogeneity of their populations.

My team and I examined the relationships between immigration or percent foreign born and high-tech industry. Inspired by the Milken Institute study, we dubbed this the Melting Pot Index. The effect of openness to immigration on regions is mixed. Four out of the top 10 regions on the Melting Pot Index are also among the nation's top

10 high-technology areas; and seven of the top 10 are in the top 25 high-tech regions. The Melting Pot Index is positively associated with the Tech-Pole Index statistically. Clearly as University of California at Berkeley researcher Annalee Saxenian argues, immigration is associated with high-tech industry. However, immigration is not strongly associated with innovation. The Melting Pot Index is not statistically correlated with the Innovation Index, measured as rates of patenting. Although it is positively associated with population growth, it is not correlated with job growth. Furthermore, places that are open to immigration do not necessarily number among the leading creative-class centers. Even though 12 of the top 20 Melting Pot regions number in the top 20 centers for the creative class, there is no significant statistical relationship between the Melting Pot Index and the creative class.

THE GAY INDEX

Immigrants may be important to regional growth, but there are other types of diversity that prove even more important statistically. In the late 1990s, the Urban Institute's Gary Gates, along with the economists Dan Black, Seth Sanders, and Lowell Taylor, used information from the U.S. Census of Population to figure out where gay couples located. He discovered that particular cities were favorites among the gay population.

The U.S. Census Bureau collects detailed information on the American population, but until the 2000 Census it did not ask people to identify their sexual orientation. The 1990 Census allowed couples that were not married to identify as "unmarried partners," different from "roommates" or "unrelated adults." By determining which unmarried partners were of the same sex, Gates identified gay and lesbian couples. The Gay Index divides the percentage of coupled gay men and women in a region by the percentage of the population that lives there and thus permits a ranking of regions by their gay populations. Gates later updated the index to include the year 2000.

The results of our statistical analysis on the gay population are squarely in line with the creative capital theory. The Gay Index is a very strong predictor of a region's high tech industry concentration. Six of the top 10 1990 and five of the top 10 2000 Gay Index regions also rank among the nation's top 10 high-tech regions. In virtually all of our statistical analyses, the Gay Index did better any than other individual measure of diversity as a predictor of high-tech industry. Gays not only predict the concentration of high-tech industry, they also predict its growth. Four of the regions that rank in the top 10 for high-technology growth from 1990–1998 also rank in the top 10 on the Gay Index in both 1990 and 2000. In addition, the correlation between the Gay Index (measured in 1990) and the Tech-Pole Index calculated for 1990–2000 increases over time. This suggests that the benefits of diversity may actually compound.

There are several reasons why the Gay Index is a good measure for diversity. As a group, gays have been subject to a particularly high level of discrimination. Attempts by gays to integrate into the mainstream of society have met substantial opposition. To some extent, homosexuality represents the last frontier of diversity in our society, and thus a place that welcomes the gay community welcomes all kinds of people.

THE BOHEMIAN INDEX

As early as the 1920 studies by Robert Park, sociologists have observed the link between successful cities and the prevalence of bohemian culture. Working with my Carnegie Mellon team, I developed a new measure called the Bohemian Index, which uses Census occupation data to measure the number of writers, designers, musicians, actors, directors, painters, sculptors, photographers, and dancers in a region. The Bohemian Index is an improvement over traditional measures of amenities because it directly counts the producers of the amenities using reliable Census data. In addition to large regions, such as San Francisco, Boston, Seattle, and Los Angeles, smaller communities such as Boulder and Fort Collins, Colorado; Sarasota, Florida; Santa Barbara, California; and Madison, Wisconsin, rank rather highly when all regions are taken into account.

The Bohemian Index turns out to be an amazingly strong predictor of everything from a region's high-technology base to its overall population and employment growth. Five of the top

10 and 12 of the top 20 Bohemian Index regions number among the nation's top 20 high-technology regions. Eleven of the top 20 Bohemian Index regions number among the top 20 most innovative regions. The Bohemian Index is also a strong predictor of both regional employment and population growth. A region's Bohemian presence in 1990 predicts both its high-tech industry concentration and its employment and population growth between 1990 and 2000. This provides strong support for the view that places that provide a broad creative environment are the ones that flourish in the Creative Age.

TESTING THE THEORIES

Robert Cushing of the University of Texas has undertaken to systematically test the three major theories of regional growth: social capital, human capital, and creative capital. His findings are startling. In a nutshell, Cushing finds that social capital theory provides little explanation for regional growth. Both the human capital and creative capital theories are much better at accounting for such growth. Furthermore, he finds that creative communities and social capital communities are moving in opposite directions. Creative communities are centers of diversity, innovation, and economic growth; social capital communities are not.

Cushing went to great pains to replicate Putnam's data sources. He looked at the surveys conducted by a team that, under Putnam's direction, did extensive telephone interviewing in 40 cities to gauge the depth and breadth of social capital. Based on the data, Putnam measured 13 different kinds of social capital and gave each region a score for attributes like "political involvement," "civic leadership," "faith-based institutions," "protest politics," and "giving and volunteering." Using Putnam's own data, Cushing found very little evidence of a decline in volunteering. Rather, he found that volunteering was up in recent years. People were more likely to engage in volunteer activity in the late 1990s than they were in the 1970s. Volunteering by men was 5.8 percent higher in the five-year period 1993–1998 than it had been in the period 1975–1980. Volunteering by women was up by 7.6 percent. A variety of statistical tests confirmed these results, but Cushing did not stop there. He then combined this information on social capital trends with independent data on high-tech industry, innovation, human capital, and diversity. He added the Milken Institute's Tech-Pole Index, the Innovation Index, and measures of talent, diversity, and creativity (the Talent Index, the Gay Index, and the Bohemian Index). He grouped the regions according to the Tech-Pole Index and the Innovation Index (their ability to produce patents).

Cushing found that regions ranked high on the Milken Tech-Pole Index and Innovation Index ranked low on 11 of Putnam's 13 measures of social capital. High-tech regions scored below average on almost every measure of social capital. High-tech regions had less trust, less reliance on faith-based institutions, fewer clubs, less volunteering, less interest in traditional politics, and less civic leadership. The two measures of social capital where these regions excelled were "protest politics" and "diversity of friendships." Regions low on the Tech-Pole Index and the Innovation Index were exactly the opposite. They scored high on 11 of the 13 Putnam measures but below average on protest politics and diversity. Cushing then threw into the mix individual wages, income distribution, population growth, numbers of college-educated residents, and scientists and engineers. He found that the high-tech regions had higher incomes, more growth, more income inequality, and more scientists, engineers, and professions than their low-tech, but higher social capital counterparts. When Cushing compared the Gay and Bohemian Indexes to Putnam's measures of social capital in the 40 regions surveyed in 2000, the same basic pattern emerged: Regions high on these two diversity indexes were low on 11 of 13 of Putnam's categories of social capital. In Cushing's words, "conventional political involvement and social capital seem to relate negatively to technological development and higher economic growth." Based on this analysis, Cushing identified four distinct types of communities. The analysis is Cushing's; the labels are my own.

- *Classic Social Capital Communities.* These are the places that best fit the Putnam theory – places such as Bismarck, North Dakota; rural South Dakota; Baton Rouge, Louisiana; Birmingham, Alabama; and Greensboro, Charlotte, and

Winston-Salem, North Carolina. They score high on social capital and political involvement but low on diversity, innovation, and high-tech industry.

- *Organizational Age Communities.* These are older, corporate-dominated communities such as Cleveland, Detroit, Grand Rapids, and Kalamazoo. They have average social capital, higher-than-average political involvement, low levels of diversity, and low levels of innovation and high-tech industry. They score high on my Working Class Index. In my view, they represent the classic corporate centers of the organizational age.
- *Nerdistans.* These are fast-growing regions such as Silicon Valley, San Diego, Phoenix, Atlanta, Los Angeles, and Houston, lauded by some as models of rapid economic growth but seen by others as plagued with sprawl, pollution, and congestion. These regions have lots of high-tech industry, above-average diversity, low social capital, and low political involvement.
- *Creative Centers.* These large urban centers, such as San Francisco, Seattle, Boston, Chicago, Minneapolis, Denver, and Boulder, have high levels of innovation and high-tech industry and very high levels of diversity but lower than average levels of social capital and moderate levels of political involvement. These cities score highly on my Creativity Index and are repeatedly identified in my focus groups and interviews as desirable places to live and work. That's why I see them as representing the new creative mainstream.

In the winter of 2001, Cushing extended his analysis to include more than three decades of data for 100 regions. Again he based his analysis on Putnam's own data sources: the 30-year time series collected by DDB Worldwide, the advertising firm, on activities such as churchgoing, participation in clubs and committees, volunteer activity, and entertaining people at home. He used these data to group the regions into high and low social capital communities and found that social capital had little to do with regional economic growth. The high social capital communities showed a strong preference for "social isolation" and "security and stability" and grew the least – their defining attribute being a "close the gates" mentality. The low social capital communities had the highest rates of diversity and population growth.

Finally, Cushing undertook an objective and systematic comparison of the effect of the three theories – social capital, human capital, and creative capital – on economic growth. He built statistical models to determine the effect of these factors on population growth (a well-accepted measure of regional growth) between 1990 and 2000. To do so, he included separate measures of education and human capital; occupation, wages, and hours worked; poverty and income inequality; innovation and high-tech industry; and creativity and diversity for the period 1970–1990.

Again his results were striking. He found no evidence that social capital leads to regional economic growth; in fact the effects were negative. Both the human capital and creative capital models performed much better, according to his analysis. Turning first to the human capital approach, he found that while it did a good job of accounting for regional growth, "the interpretation is not as straightforward as the human capital approach might presume." Using creative occupations, bohemians, the Tech-Pole Index, and innovations as indicators of creative capital, he found the creative capital theory produced formidable results, with the predictive power of the Bohemian and Innovation Indexes being particularly high. Cushing concluded that the "creative capital model generates equally impressive results as the human capital model and perhaps better."

DIRECTIONS FOR FUTURE RESEARCH

The nature and function of the city is changing in ways and dimensions we could scarcely have expected even a decade or two ago. For much of the past century the city has been viewed as a center for physical production and trade, industrialization, and the agglomeration of finance, service, and retail activities. Our theories of the city are all based on the basic notion of the city as an arena for production and largely based on activities that take place in the city during the daylight hours. Similarly, our theories of community are largely based on notions of the tightly knit community of the past, a community defined by strong ties – a conceptual theme that has been revived by the work of Robert Putnam and widespread interest in social capital both inside academe and in public policy circles. But, as this article has tried to show, the past decade has seen a sweeping transformation in the nature and functions of cities and communities. My own field research, as well as the research of others, has shown the preference for weak ties and quasi-anonymity. Social capital is at best a limited theory of community – one that fits uneasily with many present-day realities. Magically invoking it will not somehow recreate the stable communities characterized by strong ties and commitments of the past. On this score, the key is to understand the new kinds of communities – communities of interest – that are emerging in an era defined by weak ties and contingent commitments. Much more research is needed on these and related issues.

Our theories of cities, neighborhoods, and urban life are undergoing even more sweeping transformation. Sociologists such as Terry Clark, Richard Lloyd, and Leonard Nevarez have been dissecting the new reality of the city as a center for experience, lifestyle, amenities, and entertainment. This shift is not only noticeable in cites such as San Francisco, Seattle, and Boston, which have long been centers of culture and lifestyle, but in older industrial cities such as Chicago, which have been dramatically transformed into centers of entertainment, experience, and amenity. According to Clark, entertainment has replaced manufacturing and even services as Chicago's "number 1" industry. Understanding the city as an arena for consumption, for entertainment, and for amenities – a city that competes for people as well as for firms, a city of symbols and experiences, the city at night – is a huge research opportunity for sociology, geography, and related disciplines.

At the organizational level, there is a great need for research on the factors that motivate creative people and how organizations and workplaces can adapt. We are at the very infancy of organizational and workplace experiments on how to motivate creative people. Recent experiments with open office design, flexible schedules, and various accoutrements are only the very beginning. Research on the psychology of creativity by Teresa Amabile, Robert Sternberg, and others shows that creativity is an intrinsically motivated process and further suggests that the use of extrinsic rewards, such as financial incentives, may actually be counterproductive to motivating creative work. This suggests that both academic economists and professional managers have gone off in the wrong direction, particularly during the 1990s, with the use of stock options and other forms of equity compensation to motivate creative workers. A great deal more research is needed on the intrinsic factors that motivate creative workers and, even more importantly, on the characteristics and factors associated with organizations and workplaces that can best motivate and enhance creative work.

Turning now to larger macrosocietal questions. The past several decades have seen a dramatic shift in the underlying nature of advanced capitalist economies, from a traditional industrial-organizational system based on large factories and large corporate office towers, and premised on economies of scale and the extraction of physical labor, to newer, emergent systems based on knowledge, intellectual labor, and human creativity. Understanding the underlying dynamics of the system, the social structures on which it rests, the kinds of workplace transformations it is setting in motion, and its effects on community as well as city form, structure, and function is a tremendous opportunity for research. In *The Rise of the Creative Class* I try to identify some of these underlying changes and the structural transformations they have set in motion as workplaces, lifestyles, and communities all begin to adapt and evolve in light of these deep economic and social shifts. That work is my best first pass, but there is much, much more to do.

A final and critically important avenue for research is to begin to get a handle on the downsides, tensions, and contradictions of this new Creative Age – and there are many. One that I am exploring with my team is rising inequality. Our preliminary investigations suggest that inequality is increasing at both the inter-regional and intra-regional scales. At the inter-regional level, increased inequality appears to be a consequence of what we have come to call the "new great migration" as creative-class people relocate to roughly a dozen key creative regions nationwide. Other preliminary research, for example, by Robert Cushing at the University of Texas, suggests that Austin is importing high-skill creative-class people and exporting lower-skill individuals. The same pattern appears to hold for other creative centers. Inequality is also on the rise within regions. Preliminary research I have conducted with Kevin Stolarick indicates that inequality is highest in creative centers such as San Francisco and Austin. Then there is the question of the relationship between knowledge-based, creative capitalism and new types of workplace injury. At the turn of the century, during the explosion of industrial capitalism, there was great incidence and, later, great concern over physical injuries in the workplace. Eventually, after much examination and policy debate, there emerged mechanisms like OSHA to reduce physical injury in the workplace. In the Creative Age, when the mind itself becomes the mode of production, so to speak, the nature of workplace injury has changed to what I term "mental injury." Sociologists and social psychologists have much to offer in identifying the factors associated with the increasing incidence of anxiety disorder, depression, substance abuse, and other forms of mental injury, and their relationship to creative work. I often make an analogy to Charlie Chaplin's *Modern Times*: The creative-class worker racing frantically to keep up with e-mails, telephone calls, and other aspects of information overload resembles Chaplin's assembly-line worker frantically trying to keep up with the assembly line – but the creative-class worker has to do this on a 24/7, around-the-clock-basis. A great deal of research needs to be done on the incidence of mental injury and its relationship to new ways of working.

[. . .]

"Cultures of Circulation and the Urban Imaginary: Miami as Example and Exemplar"

from *Public Culture* (2005)

Edward LiPuma and Thomas Koelble

Editors' Introduction

With three million people per day entering and leaving its airport, Miami is truly a world city and a critical site for the production and circulation of information, services, and capital. The city holds the strongest connections with Latin American and Caribbean nations, as reflected in the diversity of its communities, culture, commerce, and politics. In this reading, LiPuma and Koelble consider Miami the exemplar global city for the circulation of goods, people, services, and capital. They ask how is it that, despite the transitory nature of flows and circulation, Miami has internalized its overlapping and sometimes conflicting ties with other cultures and maintains a recognized and even stable place identity. Their answer relies on their use of the concept of the urban imaginary. An urban imaginary is a narrative of a city's past, present, and future that neatly presents a coherent, stable, and idealized identity of place. In order to do so, the urban imaginary must artfully represent certain aspects of city life while concealing others. In the case of Miami, its urban imaginary makes intelligible the flows, migrations, and movements that have come to define its everyday reality. The imaginary Miami seeks to normalize the often chaotic nature of the city's vast diversity and represent a stable, ideal sense of identity that is, in fact, reflective of certain interests while excluding others.

Edward LiPuma is Professor of Anthropology at the University of Miami and his areas of research focus include labor migration into South Florida and cultural integration. Thomas Koelble is Academic Director and Professor of Business Administration at the University of Cape Town. His research has focused on the global economy and local democracy, economic decision making, social democracy, and identity politics and democratic theory.

ARRIVALS AND DEPARTURES

The scene is played out each and every day at the Miami airport. A porter loads a trolley with the suitcases of an inbound passenger from Bogotá, Colombia; Caracas, Venezuela; or other points south. The trolley – loaded six or seven hard-bound cases high – is jostled ever so slightly, but the tremor begins, and after seemingly hesitating for a second, the luggage crashes to the hard tile floor. As they hit, they make a deep, hollow, empty sound: the acoustics of empty Samsonite waiting to be filled. Their owner has come to Miami on a routine shopping trip, purchasing goods sometimes made in the Americas but most often in China, India, or Europe. Specialty malls (such as Sawgrass Mills or the appropriately named Mall of the Americas) cater to this transnational trade, their marketing campaigns expressly aimed at a cosmopolitan class. Other people arrive to undergo medical procedures, to stash monies earned elsewhere, to secure temporary dollar-paying employment, or simply to ferry products back and forth through the wired channels of an underground economy that is everywhere visible. Even on such an everyday level, Miami functions as the nucleus for global circulations of goods, people, services, and capital. Hyperaware of its position in the cosmology of hemispheric capitalism, the Miami-Dade Board of Tourism has marketed the city aggressively and internationally as the region's world-class shopping, banking, and entertainment experience. And though Miami has one of the highest crime rates in the Americas, its shopping malls are privatopias secured by hidden cameras, high-tech surveillance, and armed guards.

Within the context of the multiple and overlapping cultures of circulation that interconnect the Americas, this essay seeks to develop a notion of the urban imaginary, conceptualizing it as a culturally imaginary space that is created in and through the relationship between these forms of circulation and the practices of stabilization that seek to objectify the city as a totality. The issue is how the postmodern city can imagine and represent itself as a totality – as an enframed, territorialized space of events, ethnicities, landmarks, and representations – when it must internalize persons, histories, and economic realities originating elsewhere as a condition of its reproduction. How is the social imaginary of a city engendered when the circulations that define that city and give it a recognized identity depend on necessarily fluid and transversal spaces and a temporality that is intrinsically connected to temporalities elsewhere? In what way, for example, can we consider Miami a city in standard modernist terms when there is a constant circulation of large numbers of people between Haiti and Miami and between Cuba and Miami and when the histories of Haiti and Cuba are grasped as intrinsic to the city's ongoing history? And, it is not simply people from Cuba and Haiti, but also from Colombia, Venezuela, Nicaragua, Honduras, Jamaica, and elsewhere that are – each in his or her own way – internalized.

[...]

One way of conceptually locating a global city such as Miami is to exhume the raw numbers. To give some idea of the level of circulation, note that on any given day the metropolitan Miami airports offer approximately fifty nonstop commercial passenger flights to Colombia alone and an even larger number of flights by commercial cargo transporters and corporate aircraft ferrying goods and company personnel. An equally impressive number of flights are routed to Venezuela, Brazil, Argentina, Mexico, and other countries. The city of San Salvador has nine direct passenger flights daily, Port-au-Prince seven, and Managua a dozen. Some 3 million people enter and leave Miami every month, about half of whom the government classifies as international travelers. More than 8 million tons of declared goods pass through the port of Miami every year (U.S. Department of Commerce 2003) and perhaps an equal amount of undeclared and illegal commodities, including, of course, cocaine and other illicit drugs. While these official raw numbers ornament the story of the multiple flows of persons, practices, and goods and also underline the extent to which Miami is both a globally directed city and the site of multiethnically based encounters, they only hint at the centrality of circulation to its self-construction.

Another way of conceptually locating Miami as a global city is to see it as a self-established spatial node, a place where governance, policing, transport services (e.g., aviation and shipping), communication networks, power generation, and other forms of infrastructure create a totality that serves as an address or base for the circulation and accumulation of capital and as an operational hub for the exchange of commodities, services, and people across local, inter-American, and global markets. Appearing in the academic literature as world city theory, this approach encourages an analysis of the relationship among cities, interurban networks, and the globalizing economy. There has, accordingly, been a continuous stream of research on a dozen or so cities – such as New York, London, Hong Kong, Los Angeles, Amsterdam, Paris, Singapore, Mexico City, Frankfurt, São Paulo, Miami, and Toronto – that attempts to connect the trajectory of urban economies to the global forces that increasingly impinge upon them. In particular, analysts have focused on the effects of the decline of production-based capitalism on the social and political economy of the city (epitomized by deindustrialization), the fracturing of the domestic labor market, and the eruption of what analysts have called the "informal sector." In Miami, the term "informal sector" has a euphemistic tone because it constitutes a substantial, ubiquitous, and dynamic dimension of the economy. In consequence there are absolutely no reliable (or, perhaps, even possible) statistics that profile the actual, operating economy, most importantly because it is embedded in trans-American networks where the medium of exchange is cash (usually U.S. dollars in large denominations) that moves across private and public circulatory networks. Miami also combines an assortment of spatial and temporal expressions of modern capitalism, from that of primary extraction to technofinancial enterprise. As magnets for the absorption and exhalation of the rural political economies, they incarnate not only contemporary versions of previous stages of capitalism (especially the rural versions of racialized capitalism exemplified by inflows of black Haitian immigrants who harvest Florida's sugarcane and Mayan Indians from the Yucatán who harvest vegetables and fruit in Homestead) but also the epistemologies carried along by these circulations. So, for example, many of the adult Haitians speak only Creole, practice a mixture of vodun and Catholicism, and think of their life here as part of an extended stopover on the way to a better life back home.

Although world city theory locates cities such as Miami on the map of global processes, its tendency is to think of them as if they were stable enclosed spaces that organize interurban circulations and global flows. Its method is to focus on the emergent connectivities between established spaces (arguing, for example, that they are increasingly linked by technologically amplified flows of information and capital). If, however, we follow the lead of world cities theory and take Miami as an exemplar of the postmodern city, we have little choice but to reverse the ontological polarity of this mode of theorization and argue that cities like Miami are, in the most generative sense, pivotal sites for, and products of, circulations. For Miami, this encompasses the cultures of circulations of people via immigration, tourism, business travel, and numerous forms of temporary residence, from vacationing second home owners to Mayan migrant workers. It also includes the cultures of circulations of capital and financial instruments, goods, and services of every type imaginable plus ideologies, images, institutions, and, of course, multiple languages. If we reverse the theoretical polarity, Miami appears to function as a kind of circulatory hub, an infrastructure defined as much by the dispositions and habituses of its residents as by its institutions and physical infrastructure for the movement and translation of people and cultural forms. This intricate ensemble of flows appears less as a simple circulatory system than as an infinite set of overlapping loops, a multicentric series of processes that cross-cut different scales, some intensely local, others transversally American in the hemispheric sense of the term. Framed within its cultures of circulation, Miami seems to function as an infrastructural platform for the flow of cultural forms through superimposed spatial planes that have literally no beginning or end. It is a central part of a transurban archipelago of Latin and Caribbean cities that continually incorporate and consume one another, so that each appears as a kind of plurality as opposed to a totality. The city here is global precisely because it is not an external yet possessed presence, capitalizing on the concreteness of certain forms, such as the city skyline, to produce an easy and uncontested image of stability. Here, the attempt to theorize the city always runs up against the reality that it is built up more on disjunctures than conjunctures and comprised of a composite of shifting identities and mobile forms. The increasingly important point is that the circulatory fragmentation of Miami is overdetermined because its states and scales of fragmentation are the very condition of its global extensionality. So within the city there is no primary scale – that is, a city government that oversees a fixed space composed of stable residents who identify with the city – nor is the city a part of a primary scale for globalizing process but rather is presently defined – like a cross-currency floating rate note – by its shape-shifting relationships to global, hemispheric, national, regional, and municipal levels. There is here a hierarchy of scales, in which the scales are defined more by their interface and the forms of connectivity that bring them into relation than by legal and political systems that often seem to be one step behind in recognizing and regulating their realities. This is, of course, particularly true with respect to the illicit goods that flow daily in and out of the hub – a contraband that includes not only the infamous illegal drugs but an equally substantial flow of repackaged, gray-market, prescription drugs.

The other analytical approach used to understand cities such as Miami is to begin with the sociological notion that the city is the site of juxtaposed ethnicities. This appears in the notion/ideology that Miami is a bicultural Anglo-Cuban city composed of ethnic enclaves or that these ethnic groups resist assimilation into the mainstream culture of Anglo America. This account, which replicates one of the perspectival views given by Miamians, conceives ethnicity as a naturally pre-existing identity that the urban environment then modifies and channels. These theorizations seek to grasp the "inner" urban processes by recourse to the notion of an ethnically diverse city of competing ethnic groups. On this view, what makes a person a Miamian is distinct from what imbues that person with an ethnic identity. But the problem with the idea of a bicultural or even multicultural city is not only that there are many cultures shaded by shifting gradients of difference and distance or that cultures overlap and interpenetrate through the circulation of people ("intermarriages" are unexceptional), it is that the deeply implicit habitus of ethnicity understands multicultural urbanism as a set of radiating lines that link levels of organization and spaces in an uninterrupted and unmediated way. The interethnic relationships among Anglos, Cubans, Colombians, African Americans, Nicaraguans, Jamaicans, and Haitians (to name only a few) are constituted in and through the urban imaginary and enframing of metropolitan life, a pluralistic imaging that extends back into Cuba, Colombia, and Haiti plus parts of the United States (the North for Anglos and the South for African Americans). In local usage, which has its own notions of political correctness, ethnic identities are rarely hyphenated: people are Anglos, Cubans, Colombians, blacks, Haitians, and so on; terms like Cuban American or Haitian American are mostly reserved for official communications with the United States. When, for example, in 1990, a Colombian-born policeman, William Lorenzo, was indicted for two counts of manslaughter, he raised nearly a quarter of a million dollars for his defense by appealing to Colombian nationals living in not only Miami but Bogotá and Cali.

These two common approaches to grasping Miami, world cities theory, and the sociology of ethnicity converge in their treatment of Miami as a stable, bounded space – a move that raises to a theoretical presupposition the city's own ideology of itself. They provide essentially accurate reports about the ideology of stabilization, which, in adopting the categories proposed by the city (in its commission reports, promotional materials, and so on), replicate its forms of misrecognition. But from this theoretical standpoint, there is no way to determine the relationship between the cultures of circulation and the creation of a Miami urban imaginary that organizes the global flows of persons, ideas, and images and, more importantly, is instrumental in defining the meaning and socio-structural implications of these flows (e.g., the meaning of ethnic identity). So the most significant feature of this transformation is that urban regions – or, more precisely, regions of urbanity – are emerging as the elementary spatial structures for the cultures of the circulation that define the new millennium. Miami's integration and trajectory within the globalizing economy is as the primary intermediation between Latin America, the Caribbean, the United States, and the global economy.

LOCATING THE CITY OF MIAMI

Miami must be the archetype of the (post)modern city; for, in the sense in which it is globally and usually understood, there is no city of Miami. In the overall urban scheme, the actually incorporated city of Miami is a medium-sized municipality whose distinctive feature is a downtown business area of glassy skyscrapers and hotels surrounded by the urban litter of decaying buildings, ruined and abandoned factories, and tropical slums composed of tightly bunched, run-down, single-family houses. What the world visualizes as the city of Miami is the web of cities and unincorporated municipalities that constitute Miami-Dade County – the name recently changed from Dade to Miami-Dade in recognition of this reality. In local conversation, the entire urban corridor from Fort Lauderdale to Miami is often treated as the single site of multiple levels of cultural and economic circulation.

If Los Angeles and more recently Phoenix represent the last great modern attempts at city creation, defined by their relentless progress toward the incorporation of outlying communities into a bounded and hierarchical structure, Miami is truly postmodern (though perhaps never modern) in that it comprises a mosaic of "cities" and urbanized spaces in which neither the sum nor its parts is sovereign or hierarchically organized.

[. . .]

Miami is not based on a principle of assimilation. The transparent but unspoken concept is of permanent plurality and the internalization of others as moments or dimensions of itself. Rather than the image of immigrants arriving in a new world, then slowly, generationally "fitting in" by sloughing off those parts of their traditions that do not suit their new environment, an altogether different paradigm emerges. The imagined city is imagined ideologically as a fixed, stable, enclosed space with its own indelible character and identity; in the case of Miami, its multiple lines, temporalities, and magnitude of circulation breed an ideology of the retention of culture – as though the constant circulations, intermixings, and reter-ritorializations of Nicaraguans, Colombians, Venezuelans, El Salvadorians, Haitians, or Cubans, along with internally differentiated Anglo and African American communities, have not powerfully redetermined all of them. If the modernist ideology is that of a minority assimilation that, as studies have repeatedly demonstrated, is only ever partly realized, the ideology here is of the transportability and continuity of culture, not least because culture itself is seen as having a fundamental socioracial component. The result is the birth of a notion of the composite city, a permeated space made up of heterogeneous and incommensurable elements that seldom ascend to the level of parts (of some larger whole) other than as part of a discursive program that constitutes itself by ignoring this reality.

[. . .]

CONCEPTUALIZING MIAMI: ETHNICITY OR ETHNI-CITY?

On one of Miami's promotional brochures there is a jubilant multiethnic-looking face, the model a dark-skinned Latina with certain Anglo features, the physiognomistic image at once suggesting a hybridization of African, Latin, and Anglo cultures. Numerous advertisements in the yellow pages, wishing to suggest that their businesses cater to the full range of ethnic groups, also feature such hybrid images – often accomplished by blending digital photographs of a range of ethnic faces. Other similarly intended advertisements suggest the same message by juxtaposing images of what they imagine to be the baseline socioracial categories. Among other things, this shows the extent to which Miami's deepest realities emanate from the production of appearances, which is fitting for a city whose primary principle of difference and group identity – racialized ethnicity – appears only in a most unreflexive and deeply misrecognized form.

Miami is the southernmost city of the United States, but it is certainly not a southern city. Southern U.S. cities have always been defined by a certain insularity: a barely suppressed disdain and diffidence toward what was distant and different, coupled with a thoroughgoing segregation built up firstly along racial lines and secondarily along class lines and always situated within a history that is close at hand, a process of remembering and memorializing the past inscribed in an institutional history of battlefields, buildings, and pioneers; and an inculcated history of dispositions toward northerners, foreigners, and distant intimates, mainly African Americans and American Indians. It is the disfiguring configuration of these features that gives southern cities marked identities, a form of identity making that is very remote from that practiced in Miami. And so there is the joke that as one flies north and west from Miami to the truly deep southern cities of the Florida panhandle, the pilot reminds the passengers when landing to turn back their watches one century.

At least from a modern perspective, Miami seems to be spiritually unclaimed, to lack a substantive line of identity formation. It seems unclaimed spiritually in the sense that there appears to be no animating principle through which people identify with the city and with their coresidents. One response to this vacuum is that Miami will develop a spirituality as it matures, as it integrates the third great wave of immigration and other transformative processes into the urban fabric. Another response is that it is creatively inventing new forms of spirituality in which there is a continual infiltration and internalization of globally circulating images, identities, and ideologies, especially between the Americas. There are numerous ways in which those who govern Miami attempt to stabilize these circulations and, at least symbolically, consolidate a plethora of cities and diffuse unincorporated areas in order to engender an urban imaginary – from the way Miami represents itself in the commercial media and public spheres to the way it negotiates its relationships with state and federal governments. But the most critical counterweight to circulation and plurality is the notion of ethnicity, which serves as the modernist trope for the social organization of interurban group relationships. Thus the ideology of (racialized) ethnicity as enunciated by politicians, the media, and community leaders imagines an urban landscape of permanently identifiable and fixed groups, each arising from a set of primordial cultural attachments.

The referencing of ethnicity in the media and everyday speech is done with such matter-of-factness and is so much a part of the unquestioned object language of urban conversation that it appears as though speakers were simply reading the objects off of the urban landscape much as they identify the palms and saw grass that define Miami's tropical environs. The city of Miami attempts to continually enact this ideology, to imbue it with substance, less with the results of institution building than with an intricate symbolic staging designed to spotlight ethnicity. When Nelson Mandela visited Miami in 1990 to thank the supporters who had remained loyal to the struggle against apartheid, Miami politicians and activists snubbed him because, in their narrowed vision, he embodied black and African, as opposed to Cuban and American, political causes; in Carl Goldfarb's (1990: 2) words, "They replayed their parts from past ethnic controversies, like wooden horses on a merry-go-round, unable to escape their ideological harnesses." What could not be hidden was that Miami was home to a liberal expatriate South African community that expressly opposed the Afrikaner regime and its racial world-view, a community that occupied a position in socioglobal space that was as geographically distant from South Africa as it was politically near to its public sphere and the subversive discourse of democracy that was nurtured there.

The majority of Latin and Caribbean persons who inhabit Miami do not fit into the classifications that have evolved from the modern city. As we have tried to indicate, they are not in any simple way immigrants, exiles, tourists, business travelers, or migrant workers – all categories whose resonance derives from a prefigured relationship between societies of origin and those of destination. The circulation of people does not simply interlink South Florida with South America and the Caribbean, it engenders a connectivity in which their flows determine and define both the forms that circulate transversally and the terminals themselves, so that a new and heightened habitus of interdependence becomes the foreground of social action and position taking. At the beginning of the city's expansion through immigration, for example, just after the Cubans had arrived and the hope of an immediate reconquest still vibrated the air(waves), the horizons of homeland and exile as implacable oppositions fought to impose what seemed to be mutually exclusive visions of the collective future, only to lead to a horizon in which "here" and "there" could be deictically centered in both Cuba and America. Here's an example that only barely suggests rather than captures all the possibilities and improvisations. Carlos and a host of others like him, including a growing number of his relatives, oscillate between Miami and a small village on the outskirts of Santiago, Cuba, spending multimonth stretches in each. In both locations, Carlos makes his living in the home improvement business, though he also transports other products back and forth, maintaining a house and family in Cuba and another in Miami. What is different about his experience in Miami from his experience in Cuba is the contact he has with other people from Latin America, many of whom are also involved in lives that pluralize the places they inhabit. Or, consider Marisa, a converted evangelical Christian from a remote mountain area of Honduras who cleans house for a number of wealthy families, taking as part of her compensation their cast-off (but often hardly worn) clothes. She brings these clothes back to her village and distributes them among the villagers, thereby improving the social capital of her mother and other relatives and transforming dramatically the conditions of her periodic return.

The contradiction between the city as a site of circulations and the city as a fixed platform for the interaction of ethnic groups is itself an expression of the contradictory conceptions of ethnicity that are necessary to sustain the Cuban community's complex relationship to its once native landscape. On the one hand, the Cuban community must imagine that its Cuban identity is so pure and natural that it transcends space, existing as much in Miami as formerly in Havana; on the other hand, it must imagine this identity as so ephemeral and politically made that those who have never left, who still live in Havana, have ceased to be truly Cuban because they have embraced Castro – they have fallen from a state of ethnic grace by consorting with the devil. Accordingly, as traitors to their own identity, they are not worthy or fit to step on the Cuban soil of Miami. It takes so much political labor to foreground ethnicity for Miami Cubans; it uses up so much political capital, yet is so necessary precisely because the transnational and transurban flows of the circulatory system always threaten to dethrone the ideological image of ethnicity as an identity that derives from an earlier temporality and previous territoriality, such as the imagined space-time of prerevolutionary Cuba. Indeed, ethnicity is so significant a baseline of social identity and ground for the construction of Miami's urban space, and its importance as a generative scheme for placing and representing others is seemingly so much greater than its measurable material effects, precisely because ethnicity allows for the provisional enclosure and constitution of the parts, the ethnic communities and groups, that, in the modernist view, society assembles into the totality called the city. It is for this exact reason that Miami seems, if not to have invented, then to have elevated to unprecedented heights, the deeply contradictory form of what might be called the transnational ethnic secular ritual.

Every ethnic community in Miami has its own day, its own ritual festival, its own celebratory forms. At the surface level, these celebrations of identity in America appear to parallel a variety of ethnically inspired days, parades, and crusades. For those from northeastern cities of the United States, St. Patrick's Day and Columbus Day should come easily to mind. The purpose of the holiday is to present, by parade, for example, the contributions and accomplishments of that ethnic community to the overall city. In Miami, the production and centrality of difference motivate these rituals, and so they unfold primarily along another set of axes that appear skewed to those more accustomed to the ethnic rites of the older cities. Along one line, the festivals underline the distinctiveness and insularity of the ethnic group; those who attend are divided up into the congregation of ethnic souls and those who are simply onlookers. Such a ritual works by objectively presupposing the very state of affairs, the existence of an enclosed and cohesive ethnic group, that it creatively enacts. These rites of ethnic identity are central to the core experience of Miami because they provisionally stabilize and reaffirm the reality of ethnic communities in the face of highly deterritorializing and culturally dissembling forms of circulation. Indeed, on another axis, these celebrations and fiestas invariably have a transnational dimension to them. Each ethnic festival, for example, includes a contingent of musicians and political figures who have come from or (in the case of Cuba) are thought to embody the spirit of the homeland. All this serves to emphasize that the community's primary cultural political attachment is overseas, that its place in the world and location in urban space is defined in reference to its country of origin.

If what are called "white Cubans" constitute the model minority in Miami's self-imagination, then black Haitian immigrants are the refuse of a nation beset by grinding poverty and perpetual political turbulence. This opposition is part of a larger, ideologically infused attempt to stabilize the city by establishing a socioracial order. No matter their varying social histories or positions in economic space, the urban narrative assumes that all Haitians are fundamentally the same. This discourse serves to attach, through racialization, internally diverse communities to uniform experiences, social geographies, and cultural beliefs. The basic project has been to try to stabilize the city ethnically by establishing a continuum from white Anglos and Cubans to black African Americans and Haitians. The basic theme is that Miamians can identify each of these socioracial groups in terms of where they originated from, the space in the city that constitutes their ethnic spirit and where that spirit is reproduced in its purest form (e.g., Little Havana), and, of course, their color. The notion – which is frequently filtered through the positive image of ethnic harmony and the virtues of combining the cultural talents of so many races – is of the city as a composite of these groups in and for themselves and in relation to one another.

[...]

CONCLUSION

To put the case in the modern negative, which, from the standpoint of (post)modern circulation, is a positive attribute, Miami cannot be known by its internal coherence and thus its closure and territorialization. Its residents do not therefore become part of the city by becoming part of a community; and, to hear people talk, it is abundantly clear that living in Hialeah, Coral Gables, Kendall, or Miami Lakes does not engender an identity. In this sense, global cities, such as Miami, have relations with other cities, such as Havana, Mexico City, and Bogotá, that cannot be uncritically mapped as forms of external connectivity; for the city already contains these relations as webs of internal linkages and differences. In this sense, external relations produce similar effects to internal ones, meaning that there is no concerted effort to construct or demarcate the city with respect to specifications beyond it. At one level, the city continually mimics the constraints of urban lifeways by pushing them to excess, in this way exposing and deflating their capacity for constraint and limitations. As an exemplar of global cities, Miami appears at once as a continuous series of border-crossing circulations and as a very recognizable entity in the contemporary space-time.

That Miami is so recognizable, indeed, overexposed on the world stage, involves more than constant nominalization. It turns on the creation of images of totality on a number of symbolic registers. The quest is to create a complex spatial image, composed of other nested images, that temporarily arrests the numerous flows, extensions, and relations that extend outward even as they wind back, in order to provisionally stabilize a territorialized space of concretized forms. Certain cultural forms bring the imaginary of the city into existence by presupposing its reality as a condition of their own, though many of them, such as the skyline, tropical botanical gardens, and the long stretches of beach, have only the most symbolic connection to any territorialized state. The city's project of self-totalization centers on canonizing these images as self-referential emblems of Miami, so that seen from any distance, whether from abroad or from its own communities, to consume them is to presuppose the existence of the city. These enactments of the city turn on the interaction between self-reflexivity and circulation in that the objectification of the city creates the performative basis for the cultures of circulation that have come to define Miami's urbanity. Critical aspects of the construction of the city reside in the performative ideology of territorializing, reading, writing, vocalizing, and visualizing the city. To be attentive to this discourse, which is disseminated translocally through almost every medium (sometimes foregrounded, as in commercials lauding Miami, most often as background to other communications) is to identify the contemporary space-time of Miami and to identify with the audience addressed by this conversation; it is also to become aware that it is necessary to apprehend the city as a totality because others think of it that way. They act, desire, and conceptualize the future on that basis. In other words, the existence of Miami as a social imaginary depends on the ongoing and self-reflexive reincorporation of performativity – a process that provisionally stabilizes and reaffirms the reality of the city in the face of the transgressions, internalizations, and pluralities inherent in the multicentric, multilevel circulations that make Miami what it is. From this viewpoint, the fetish of the city is less an act of misrecognition than the sum of those acts of shared imagination through which the mutuality and performativity of people's actions constitutes the city as a quasi-collective agent. Nothing exemplifies this more than the words of those, such as locally elected politicians, who, in nominating themselves to speak on behalf of the city, perform true social magic in presupposing both the life of the bounded city as the authority for their position taking and the unbounded authority of their position to imagine the city. The city as a social totality thus appears as a homogeneous agency that subsumes individuals who are in principle unique; at the same time, transcendent homogeneous agencies are seen as mutually differentiating.

[...]

Like those once backwater locales that technologically jumped from no phones to wireless phones, skipping landlines altogether, Miami has so swiftly become postmodern because it never concretized the modern or evolved much of a vested institutional interest in its preservation, only truly developing from orange groves and plantations over the past forty years. Miami, because of its extraordinary mobility, never fully experienced the modernist city's dialectics of trespassing, then reestablishing fixed boundaries, the political aim to create the forms of separation and difference that help to ensure consent in the quest to reproduce an image of the status quo in a space defined by connectivity and transformation. One result of its uncommon past and present is that, even when viewed from the most modernist perspective possible, the panoptic view from nowhere, Miami appears to be composed of transversal, fluid, partially overlapping topographies, as though, like a set of overlaid transparencies, there are many superimposed cities occupying what appears to be the same nominal space.

"Staying Vietnamese: Community and Place in Orange County and Boston"

from *City and Community* (2005)

Karin Aguilar-San Juan

Editors' Introduction

Cities abound with diverse neighborhoods that offer up their own distinctive attractions – be they ethnic restaurants and shops or outdoor markets, street fairs, or music venues. Neighborhoods are places that reflect the cultural forms of the people who inhabit and visit them. But that statement is only partially correct, as Karin Aguilar-San Juan shows us in this reading on two Vietnamese neighborhoods in the United States. Space is not simply a "container" that houses a community but is an active element in the creation of community itself. Place, she argues, is a constitutive element of the community-building process. In order to demonstrate the relationship between place and community, Aguilar-San Juan focuses upon place-making activities that influence, and in turn are influenced by, community building in two very different Vietnamese enclaves: Little Saigon in southern California's Orange County and Fields Corner in Boston.

Orange County encompasses an enormous region stretching from Los Angeles to San Diego, comprised of a number of multinucleated cities and planned communities marked by suburban sprawl. The growth of Little Saigon reflects the sprawling, (auto)mobile quality of southern California. By the late 1970s the town of Westminster was described as Vietnamese, as evidenced by a sizable number of Vietnamese groceries, restaurants, and newspapers. Throughout the 1980s this community grew larger, leading to the development of an Asian-themed shopping mall. Little Saigon's location clearly influenced its path of development as an ethnic community. Plans are in the works to further develop Little Saigon as a business district primarily serving Vietnamese to a tourist district for southern Californians from a variety of ethnic backgrounds. So, too, does the development of the Fields Corner Vietnamese community reflect the sociospatial realities of dense, older neighborhoods in northeastern U.S. cities. Here, the Vietnamese community is enmeshed in (and even overlaps with) other enclave communities – a pattern reflective in many older immigrant districts. In Fields Corner, community development emerges in conjunction with other, adjacent "brother" communities to advance common causes, such as housing, livable wages, and street crime reduction. Aguilar-San Juan contrasts this neighborhood-oriented approach to community building from that of Little Saigon's, emphasizing the importance of place as an explanatory variable.

Karin Aguilar-San Juan teaches American Studies and Asian American Studies at Macalester College in St. Paul, Minnesota. She is the Editor of *The State of Asian America: Activism and Resistance in the 1990s* (South End Press, 1994). She has studied Vietnamese communities across the United States and teaches courses on community and identity among Asian Americans.

The arrival of Vietnamese refugees to U.S. shores in 1975, their forced dispersal, and their subsequent creation of distinct and recognizable Vietnamese American places in various metropolitan areas over the ensuing decades presents a number of puzzling theoretical issues for sociologists who study cities and immigration. Well into the second generation, Vietnamese and Vietnamese American places, communities, and identities persist despite expectations that their being scattered by federal resettlement policy would have lessened the possibility of hanging on to "Vietnamese-ness" (Vo, 2000). These places are often seen primarily as business districts or "ethnic enclaves," but they are also sites for the creation of social networks, aggregating devices, anchors for identity, and representations of culture. Importantly, these places make it possible for Vietnamese American communities to "stay Vietnamese" despite dispersal and suburbanization.

Spatial-assimilation theory continues to be a central component of the ethnicity paradigm, encapsulating the ecological tradition of the Chicago School by equating residential mobility – specifically suburbanization – to acculturation and to an advanced phase of "structural" assimilation. To the proponents of spatial-assimilation theory, the appearance of immigrant groups in suburbia not at the end of their trajectories of assimilation but at, or near, the beginning is seen as an enigmatic "hallmark" event. Recent scholarship on Vietnamese refugees and immigrants, though extremely valuable on its own terms, tends to accommodate and further spatial-assimilation theory by treating space as an empty vessel into which immigrants are poured rather than a medium through and against which communities are established.

This article illuminates further the relationship between place and community for Vietnamese Americans. I focus on space not only as a "container" for the Vietnamese American population but also as an active element in the creation of Vietnamese American community. A key tenet of the socio-spatial approach is that spatial formations are contested – rather than natural, inevitable, or "empty" of structural relations of power or inequality.

As a result of a collective history of war, displacement, and exile, Vietnamese American community building has become, in large part, a concerted effort at recreating a "homeland." In many important ways, Vietnamese American place making attempts to remedy displacement by reconstructing a Vietnamese "cosmos" within various metropolitan regions across the United States. Vietnamese American place making may be understood, at least in part, as a recuperative activity responding to traumatic global and historical events.

This paper focuses on place making as the meso-level transformation of existing places in order to construct and produce new places that strengthen and proliferate certain aspects of community. Place making in the context of the present study means assembling the features of place so that specific forms of community are bolstered or promoted, while others are diminished or extinguished. Place making requires taking apart and reformulating location, material form, or representation in order to modify, destroy, or rebuild place. When tied to the demands of community building, place making becomes ever more complex.

Three types of place-making activities influence, and in turn are influenced by, community building: *territorializing*, *regulating*, and *symbolizing*. Each activity puts a unique spin on place by fitting together its three dimensions in unique ways.

[. . .]

A TALE OF TWO COMMUNITIES

This article tells a tale of two Vietnamese American communities. To the untrained eye, they appear similar. They are comprised of Vietnamese refugees and immigrants, in addition to growing numbers of U.S.-born Vietnamese Americans. They contain gift shops, jewelry stores, groceries, bakeries, restaurants, nail salons, and doctors' offices that cater to a Vietnamese or Vietnamese American clientele. They are recognizable to Vietnamese and non-Vietnamese alike as culturally distinct places.

Upon closer examination, the two communities are much more different than they are similar, particularly when it comes to three commonly examined features of community: population size, geographic distribution, and institutional completeness. The Vietnamese population of the Los Angeles–Riverside–Orange County CMSA is more than seven times the Vietnamese population of the Boston–Worcester–Lawrence CMSA. In southern

California, nearly 75% of the Vietnamese population lives in five adjacent cities: Westminster (the site of Little Saigon), Santa Ana, Garden Grove, Fountain Valley, and Anaheim. In Boston, Vietnamese are scattered throughout the city, with four nonadjacent neighborhoods – South Dorchester (the site of Fields Corner), North Dorchester, Allston-Brighton, and East Boston – holding the bulk of the population. However, the residential core lies within neighborhood boundaries of South Dorchester.

LITTLE SAIGON: THE HEART OF VIETNAMESE AMERICA

Home to Mickey Mouse and a center of the John Birch Society, Orange County, California, has long represented for many scholars and journalists the pinnacle of suburbia: a bastion of conservativism, affluence, white privilege, and uncontrolled sprawl. A region encompassing nearly 800 square miles but only 3 million people, Orange County is a conglomeration of about 27 dispersed and relatively sparsely populated municipalities linked by a freeway system extending from Los Angeles (to the north) and San Diego (to the South).

In its spatial design, Orange County embodies the antithesis of the Chicago School's ecological version of community. Orange County owes its decentered, multinucleated, regional sprawl not to the random movement of untold numbers of individuals but to a *conscious and deliberate scheme by experts and elites* to construct a series of planned communities and specialty zones catering to for-profit enterprise. The Irvine Company has played a key role in shaping Orange County's land uses since the late 1880s, promoting a model of community that clearly and without apology prioritizes profit maximization over social or environmental concerns.

In 1975, the city of Westminster contained a small, mostly white working-class community living among used auto body shops and rambling farms of strawberries and chili peppers. The formation of Little Saigon began with the arrival of first-wave refugees directly from Camp Pendleton. The bulk of these refugees were channeled toward Orange County after resettlement agencies in San Francisco (to the north), Los Angeles, and San Diego (to the south) complained to federal officials that their cities were overburdened with other immigrant groups. By the late 1970s, enterprising families had marked Westminster as a Vietnamese place by establishing a handful of Vietnamese groceries, restaurants, and *Nguoi Viet* (now one of several Vietnamese-language newspapers published in the region). This early migration set the stage for later efforts to build community and place.

As a result of a second influx of mostly ethnic Chinese from Vietnam in the 1980s, an ethnic Chinese economy grew in Westminster, providing goods and services, as well as employment for residents of Little Saigon who were otherwise cut out of the primary economy (Gold, 1994). Through the 1980s, Westminster's real estate market attracted foreign investors from Hong Kong, Taiwan, and Singapore (Weiss, 1994). Chain migration and secondary migration brought new residents, workers, and entrepreneurs from Vietnam and from other regions of the United States to Orange County's Vietnamese district.

The Chinese Vietnamese ethnic economy fuels Little Saigon, which is recognizable to outsiders for its Vietnamese-language signage, distinctive architecture, Vietnamese restaurants, bakeries, and cafes. In 1987, Frank Jao – Little Saigon's "premier developer" known for his biography as a rags-to-riches entrepreneur – put up the Asian Garden Mall. The two-story, privately owned mall houses dozens of jewelers selling gold and jade, gift and clothing shops, and several beef noodle shops and food stalls. Flanked in the front by three larger-than-life statues representing Prosperity, Longevity, and Good Health, the edifice is widely acknowledged as the "center" or "heart" of Little Saigon.

However, shopping is not the only thing that happens in Little Saigon. On any given day, the Vietnamese Community of Orange County (VNCOC) is packed with dozens of youth and elders studying English, preparing for citizenship classes, or just playing mahjong with friends. In 1989, the refugee resettlement support agency moved into its own building in Santa Ana, about 5 miles from the Asian Garden Mall. Vietnamese gather in many other areas, including the many nearby strip malls offering Vietnamese coffee, food, and services; the multimillion dollar Vietnamese Catholic Center (where Vietnamese priests are not allowed to hold their own services); the Westminster Cemetery (which has a separate Asian section); and recently, the Westminster War Memorial.

[. . .]

Little Saigon's business leaders recognize that in order to survive and thrive, the business district needs to attract foreign investment and to cultivate a wider and more affluent clientele. In the late 1990s, Orange County's Vietnamese Chamber of Commerce worked to promote tourism in Little Saigon by offering guided tours to business people and by organizing trade missions to Southeast Asia. For the Chamber, the business district is the Vietnamese community. Nhat Le, then director of the Chamber, told me, "Whatever is good for this area is good for us." When I interviewed him, Mr. Nhat and his colleagues were actively pursuing discussions with non-Vietnamese Orange County business leaders to include Little Saigon as the exotic food source of an official tourist triangle, composed of Little Saigon, Disneyland, and Knott's Berry Farm. Le Pham, an editor and journalist, put it this way: "The most important business in Little Saigon in the future should be eating . . . The basic tourist concession is a triangle of Disneyland in Anaheim, Knott's Berry Farm, and Little Saigon . . . with the Queen Mary in Long Beach far in the distance." According to Mr. Le, "Along Bolsa will be a good place for a hundred of eating places. It's look like Waikiki." Along with restaurants for tourists, the plans would include a posh, five-star hotel.

Transforming Little Saigon into a tourist paradise might be the best way to spur economic development, but doing so also brings to a heavy boil a conflict that might otherwise simmer quietly beneath the surface. The conflict revolves around the role of ethnic Chinese in Little Saigon. As a business district, Little Saigon thrives because of its ethnic Chinese economy (Gold, 1994). Timothy Le, a Chinese-Vietnamese entrepreneur whose family owns several major groceries in Little Saigon, estimates that as much as 90% of the businesses in Little Saigon are run by Chinese. As he put it, "It's always the Chinese doing more business than the Vietnamese. Vietnamese, true Vietnamese, are better in politics." In 1997, Frank Jao tried to build a suspended mini-mall across Bolsa Avenue called "Harmony Bridge." His proposal turned out to be rancorous: Community leaders pointed to certain design elements of the bridge and to the ethnicity of the developer himself and charged that Jao and the bridge were "too Chinese." Mai Cong, an elder community leader, told the press, "It's the uniqueness [of Little Saigon] that attracts tourist to come, not diversity, [which] is boring." Meanwhile, the former mayor of Westminster, Charles Smith, extolled the bridge as a structure that "would have helped promote Little Saigon as a tourist destination and as the cultural and economic capital of the Vietnamese free world."

Taking their complaints all the way to City Hall, Jao's detractors argued that the bridge's architectural references to China and to Chinese culture would destroy their community because these references threaten the "ethnic integrity" of Little Saigon (Dizon, 1996; Fisher, 1994; Pope, 1996). One group of Vietnamese American students analyzed the conflict over Harmony Bridge as yet another disturbing example of Little Saigon's cultural inauthenticity (Nguyen et al., 1996). In community newsletters, much ink was spilled over the precise elements of architectural design – dragons, birds, flowers, color, and slope of the roof – that were described as "truly Vietnamese" as opposed to Chinese. Those Chinese elements were further described as offensive due to China's 1,000-year-long history of colonizing Vietnam. In the end, the protestors prevented the bridge from being built. Adding to the bridge's collapse may have been the lack of financial backing by the city of Westminster. Jao wanted to use municipal bonds to fund construction, but those monies are not supposed to fund private commercial developments.

[. . .]

SHAKING HANDS WITH GI JOE

While some leaders concern themselves with the economic vitality and cultural authenticity of Little Saigon, others make it their task to deal with public memories of the past, especially of the Vietnam War. Staging events and designing monuments that commemorate the War from a refugee/U.S. Ally perspective help to carve out places that challenge the image of South Vietnamese soldiers as "losers" and contribute to a positive rendition of Vietnamese "character and tradition" in Little Saigon. In 1996, Van T. Do, in his mid-20s, and Oai Ta, around 50, described to me the community's plans to raise hundreds of thousands of dollars for a Vietnam War memorial including a statue of GI Joe and a Vietnamese soldier shaking hands. Once an engineer but now the elected president of a Vietnamese community organization, Mr. Oai sees himself as having no other job than "leader of the community." In his opinion the "Vietnam syndrome" means Americans are obsessed about having lost the War. Making common ground with Americans, then, means telling them "we lose too." Plans became reality in September 2002, when a bronze sculpture about 10 feet high and weighing several tons was delivered to Westminster. The monument shows two soldiers standing side-by-side, according to the artist, "working together." The statue was unveiled on April 27, 2003, near the 28th anniversary of the fall of Saigon.

The Vietnam War Memorial in Westminster puts into concrete form one of the most pressing concerns of at least some segments of the Vietnamese American community: to be seen in friendly and equal terms as people who are deserving of moral, emotional, and financial support because of their position as U.S. Allies during the War. Some groups feel closer to the past than others. Older refugees, especially those who were detained in Vietnam for several years after the War, harbor particularly bitter feelings about the past. Ex-detainees see first-wave refugees who escaped immediately after the War as being too quick to move on toward trade relations with Vietnam, or to invite any kind of cultural exchange with Vietnam. Because the public memories of the War are so deeply entrenched into U.S. culture, the Vietnamese American community as a whole must contend with the impact of the War regardless of their many internal differences.

While most Vietnamese residential areas in Orange County are not architecturally distinct, many important commercial buildings including the Asian Garden Mall are distinctively Asian if not specifically Vietnamese. The yellow and red-striped flag of South Vietnam is a common sight throughout Little Saigon, displayed along major avenues and inside shops and stalls. Together with the flags, statues of dragons and other mythical animals, sloping tile roofs, commercial billboards in Vietnamese, and street signage, including the freeway signs announcing Little Saigon, help to give the place its cultural and political identity.

Orange County's cultural and political economy dictates many of the salient features of Vietnamese American place making, including efforts to make Little Saigon part of a tourist triangle and the design of the Westminster War Memorial. Knowing that the long-term growth of the business district means cultivating an affluent, suburban clientele, business leaders hope to turn Little Saigon into a place that welcomes and even caters to a larger, non-Vietnamese public. Portraying the War as an event where Vietnamese and U.S. soldiers stood at each other's sides as equals creates a version of history that corroborates the region's emphasis on capitalism and free trade.

FIELDS CORNER: BOSTON'S VIETNAMESE VILLAGE

In stark contrast to southern California – Orange County in particular – Boston is a tightly packed network of old villages replete with references to the past (Brown and Tager, 2000). An old city built on land recovered from the sea, Boston spans 47 square miles and is home to slightly more than half a million people . . . Often left out of Boston's historical narrative are Blacks and Chinese, whose long-standing neighborhoods also represent migrant communities.

The suburbs, rather than the city center, have led Boston's tangled web of growth. Tracing back to the end of the Civil War, this pattern took shape as the upper- and middle- classes moved to "streetcar suburbs," filling the spaces in between long-established centers of industry. The city's elaborate patchwork of 20 neighborhoods connected by a mass transit system and federally subsidized highways poses an unusual decentralized urban formation. The spatial divide between the haves and have-nots among Boston's social classes and racial groups also takes a scattered, labyrinthine shape.

[. . .]

In 1975, the first Vietnamese refugees used Boston's Chinatown as a port of entry. They were settled in Chinatown by federal and local agencies at a time when the city was still reeling from the turmoil of the War, busing, and school desegregation. These events, along with Boston's severe economic decline, exacerbated tensions throughout the city and often forced Chinese Americans (Boston's dominant Asian American group at the time) into shaky "pan-ethnic" coalitions with Southeast Asians, especially Vietnamese. According to one expert on Boston's Chinatown, refugee relocation agencies turned to Chinatown for help partly out of ignorance ("all Asians are alike") and partly because Chinatown's leaders saw it in their financial and political interests to help the new refugees. In 1980, 60% of Chinatown residents were Asian or Pacific Islander; of those, 23% were living below poverty. Pulling in new refugees meant receiving federal and state funds that would stimulate Chinatown's economy.

Later, as each Southeast Asian group voiced its own cultural and linguistic requirements, separate mutual assistance agencies were created in specific neighborhoods outside of Chinatown. By the mid-1980s, rent increases chased Vietnamese out of Brighton and into Fields Corner. Slowly, Vietnamese-owned businesses in Chinatown followed their clientele into Fields Corner, too. In the late 1980s, second-wave refugees (ethnic Chinese merchants, "boat people," and ex-political detainees) arrived, settling in both Chinatown and Fields Corner.

[. . .]

While Vietnamese are not the largest immigrant or minority group in Fields Corner, their place-making activities make them the most visible and most powerful group. Their move into Fields Corner has been dramatic.

Community leaders worked hard to develop a strategy for community development that would go beyond refugee resettlement support services to meet the community's long-term needs for social and political resources and recognition in the city. Younger (under 35 years old) leaders formed the Vietnamese American Initiative for Development (Viet-AID), a community development corporation based in Fields Corner. Viet-AID took as their mission to spur the growth of the Vietnamese American community by creating affordable housing, jobs, and senior services, and child care within the context of a multicultural, multiracial neighborhood.

Soon Viet-AID put forward the idea for a Vietnamese American community center – an actual building owned and operated by Vietnamese that would house events and social services. Some elders, key figures in the community, objected to the idea and refused to support it publicly, mostly out of a perceived threat to their seniority. They did not like the fact of younger people choosing a direction that they, as elders, might have to follow. Anh Thach, a lawyer in his early 30s and a key voice among Vietnamese Americans in Boston, told me that the older generation clings to an "exile concept of community." This concept reflects a simple desire among elders to "establish something for their own."

In part because many Vietnamese refugees first landed in Chinatown, and because Chinatown has been reenergized by Chinese-Vietnamese entrepreneurship, the distinction between Boston's two Asian neighborhoods – Fields Corner and Chinatown – requires constant attention from Vietnamese American community leaders.

In making their mark on Boston, Vietnamese American leaders have had to deal with a landscape replete with historical references to the Vietnam War. In 1981, Dorchester's Vietnam veterans put up a memorial to the sixty-two Dorchester residents who gave their lives in the Vietnam War. Two years later, 62 "hero squares" were dedicated on Veterans Day. Lining Dorchester Avenue, the hero squares stand juxtaposed with Vietnamese-owned storefronts; together they offer a powerful reminder of the impact of the War.

Vietnamese veterans would like to find common ground with U.S. Veterans. Referring to the hero squares, one Vietnamese veteran I interviewed told the press, "It is something that meshes together. When we came here, we tried to do something for our new country, and something to honor [those Americans who served in Vietnam]. We'd like to find what we can do to honor our own people." But common ground is hard to find.

[. . .]

Vietnamese American community building and place making in Boston occur on a multifaceted and rapidly changing backdrop. Since its incorporation, Boston seems to have laid out the welcome mat for immigrants, harboring them in neighborhoods known for their ethnic networks and institutions. The Vietnamese American community has traveled a circuitous and unexpected path from Chinatown to

Fields Corner. It may be too early to predict in which neighborhoods Boston's Vietnamese will finally land. However, in cementing their relationship as a group to Fields Corner by building an actual community center there in 2002, leaders have not followed the spatial-assimilation patterns of Boston's 19th-century European immigrants.

DISCUSSION: RECREATING A VIETNAMESE HOMELAND

Migration and settlement help to set the stage for Vietnamese American community formation, but without purposeful attempts at place making, Little Saigon and Fields Corner might not exist as they now do. Striving to reconstruct a Vietnamese "cosmos" in each region, Vietnamese American leaders territorialize, regulate, and symbolize place. Bit by bit, Vietnamese American leaders in both regions reconstruct mini-homelands that reflect and reinforce the cultural and political identities of their constituencies, thus building community.

Transferring Vietnamese American experiences onto the physical landscape saturates place with "Vietnamese-ness." However, the symbolization of Vietnamese American places increasingly reflects the hybridized realities of Vietnamese American community building. For example, a restaurant in Boston's Chinatown was once named "Pho Bolsa," making plain reference to Orange County's Vietnamese "Main Street." If the prototype for Vietnamese American place making becomes Orange County's Little Saigon but not the Saigon in Vietnam, what might that presage for a concept of authentic or original identity among Vietnamese Americans?

As they piece together the places that will focus and strengthen their respective communities, Vietnamese American leaders in Orange County and Boston must deal with the already existing territorialization, regulation, and symbolization of place in each region. As a result, the sites which emerge as centers of Vietnamese American social life in Orange County and Boston differ greatly in their scope and character. In Orange County, Little Saigon provides an intense and vigorous site for Vietnamese American community activity not only for local residents, but also for suburbanized Vietnamese who drive hundreds of miles just to shop or eat there. Territorialization and symbolization have moved along over the years, with the result that Little Saigon is widely referred to as "the capital of Vietnamese America." In Boston, Fields Corner provides only a small fraction of the shops, restaurants, and services that are found in Little Saigon. Territorialization and symbolization are nascent in Boston, with the result that Fields Corner is a much weaker instance of Vietnamese American place making than is Little Saigon.

To be sure, compositional factors such as population size also influence the noticeable variations in Vietnamese American community and place between the two regions. In Orange County, Vietnamese Americans dominate the Asian American landscape in sheer numbers. Their power derives partly from size and affluence; Vietnamese American leadership activities appear to be cast in wholly ethnic-specific terms and addressed exclusively toward the needs and interests of Vietnamese Americans. In Boston, a smaller and less affluent population means Vietnamese Americans must be team players in pan-ethnic Asian American networks and organizations if their own ethnic-specific needs and interests are to be met. Several of the leaders I interviewed are in fact key voices in those Asian American formations.

Specific differences in leaders' strategies for handling issues such as spatial assimilation, economic development, cultural authenticity, and public memories of the Vietnam War indicate the powerful sway of compositional and spatial factors on Vietnamese American community building and place making. These differences may be further explained in terms of territorialization, regulation, and symbolization. For example, a pre-existing Chinatown creates an obstacle for Vietnamese American community building in Boston. Since younger leaders succeeded in planning, designing, and constructing a new edifice in Fields Corner known simply as the "Vietnamese American community center," the symbolism of Fields Corner as a Vietnamese place has intensified. Yet Boston's leaders continue to struggle with the territorialization and regulation of place because they do not fully control the multicultural and multiracial composition of the community center, the neighborhood, or the city. They have also not built their own monument to the War, instead responding in a haphazard fashion to the city's dominant, though conflicted, War memories. While Vietnamese American place making in Boston is still in its infancy, demographic, industrial, or physical changes in Boston's landscape may propel Vietnamese American territorialization and regulation into other phases in the future.

Place making in Orange County's Little Saigon is in many ways exemplary for its size and scope, but place making there is not without its own peculiar dilemmas. For instance, Orange County's Vietnamese Americans reside far from Monterey Park, southern California's suburban Chinatown. While Vietnamese American leaders appear to have accomplished the territorialization and regulation of their own place, they struggle nevertheless with the symbolization of Little Saigon because Chinese-Vietnamese entrepreneurship constitutes a foundational element of Little Saigon's ethnic economy. As part of doing business with the Vietnamese American community, some of Little Saigon's busiest entrepreneurs regularly conduct business in Chinese with other Chinese in the southern California region. As business leaders proceed to design strategies for attracting tourism to the district, debates over Little Saigon's "authenticity" will inevitably proliferate. Worse, symbolization may take on ideological tones as community leaders who are driven by anticommunist sentiments view business trade with Vietnam as suspicious, and therefore "inauthentic," activity. While these complex dynamics are present also in Boston's Vietnamese American community, in Orange County the conflicts they engender are much larger and more destabilizing to leadership.

The mechanisms leading to successful Vietnamese American place making in Orange County and Boston may be summarized in general terms. Territorialization extracts a distinctive group identity out of a bland, homogeneous postsuburban landscape on one hand, and a tightly packed urban web on the other. Mainly through its intense and recognizable Vietnamese business district, Little Saigon gives physical structure to "Vietnamese-ness" in Orange County. In a much subtler fashion, Fields Corner announces the presence of Vietnamese in Boston by dotting the street with Vietnamese-owned shops and by constructing a new, two-story building known as the Vietnamese American community center in the middle of Fields Corner.

Regulation monitors the cultural and ideological boundaries of place. To my knowledge, Vietnamese American leaders' efforts to supervise and control what goes on in place do not currently receive the formal backing of the state. Protesting the "too Chinese" mall/bridge in Little Saigon indicates one group effort to police the cultural content of place. The bridge was never built, but Chinese-Vietnamese will continue to provide the backbone of Little Saigon's ethnic economy. Protesting the shadow of Ho Chi Minh in the Boston gas tank rainbow is an example of a group effort to police the ideological content of place. The rainbow was never removed, and most likely Vietnamese Americans in Boston will have to coexist with public memories of the War they view as hostile or disrespectful. To the extent that Vietnamese American place making is successful in both regions, these places mesh with the prevailing landscape of power. Neither group exhibits much capacity for independent regulation.

Symbolization hooks meaning to place. These symbols reflect and reinforce the overall agenda of place making and community building. Establishing places that anchor Vietnamese communities on U.S. soil requires complex narratives that portray Vietnamese as upstanding individuals who deserve full citizenship. Plans to extend the function of Little Saigon from business district for Vietnamese to tourist district for southern California indicate the eagerness of the Vietnamese Americans to partner with the region's larger business community. Whether Little Saigon is truly Vietnamese or actually somewhat Chinese is a symbolic battle that is subordinate to the plan to put Little Saigon on the map with Mickey Mouse. In Fields Corner, Vietnamese American leaders are taking on the issues and concerns of "brother" communities in the neighborhood: affordable housing, livable wages, clean and well-lit streets, safety, park benches. In so doing, they embrace a neighborhood-oriented approach to community building and put some of their own issues such as their feelings about Boston's public memories of the War on the back burner. Vietnamese American place making is slowed here in comparison to Orange County.

[. . .]

CONCLUSION

[...]

Places such as Little Saigon and Fields Corner territorialize, regulate, and symbolize the Vietnamese American community. In establishing places of significance and distinction, leaders are responding to the already existing landscape by finding ways to fit into, on top of, or in between the places that make up Orange County and Boston. In different ways, both places contradict the predictions of spatial-assimilation theory.

The creation of Vietnamese American places allow even the most assimilated and suburbanized Vietnamese Americans to "stay Vietnamese." In truth, staying Vietnamese is about generating new ways of being Vietnamese in America, not exporting old traditions and culture directly from Vietnam and transplanting them whole in America. These new Vietnamese ways often give rise to issues that are not present in Vietnam.

Federal resettlement policy dispersed refugees in order to hasten their incorporation into mainstream society, vastly underestimating their capacity to recongregate and rebuild communities. However, once many people arrive in one physical location, communities do not simply arise. This article shows that by assembling and organizing the various elements of place, leaders purposefully turn groups of people into communities with a "sense of solidarity and significance." In future research, place must be treated as a constitutive element of the community-building process.

PART SEVEN

Regulation and Rights in Urban Space

Plate 11 Urban protest in New York City, USA January 28–30, 2002 by Larry Towell. During protests at the World Economic Forum, a protestor holds a "Make the Global Economy Work for Working Families" sign. Reproduced by permission of Magnum Photos.

INTRODUCTION TO PART SEVEN

Cities are both sites of incredible ethnic, racial, and cultural differences and social and economic power – the latter, concerned with risks of social disorder, tends to control and shape the former. While certain forms of control and regulation, such as zoning laws, are obvious, others are not. In recent decades, municipal leaders, business owners, planners, and city elites have developed innovative physical and administrative ways to regulate people and places. Regulation, as the readings in the first half of this section attest, raises concerns among many urbanists, as certain populations are systematically denied access to areas within the city and, consequently, full participation in civil society.

In the first reading, Sally Engle Merry outlines how order is constructed through the regulation of spaces (as opposed to individuals, per se) in contemporary urban society. Merry claims cities have embraced a largely unnoticed tactic of spatial governmentality, in which particular urban areas are defined as "off-limits" to certain behaviors deemed undesirable and, consequently, to those social groups who practice them. She demonstrates how mechanisms of crime control are deployed as a means of spatial exclusion and to manage social difference in order to regulate the everyday lives of city dwellers, especially the urban poor. With both continued urban inequality and the need to protect investments in the upscale redevelopment of city centers, cities have embraced "zero-tolerance" policies to regulate their cores, effectively banishing unwanted populations from the streets.

The second reading, by Setha Low, picks up on this issue of regulation as it pertains to urban public spaces. Public spaces include the streets, sidewalks, parks, "open areas," and plazas that are accessible and open to everyone – they constitute the foundational locations of ideal democracy and civic virtue in the city. But in recent decades the meaning of "public" – in terms of both control of the space itself and who should be allowed to use it and how – has been contested. Beginning in the late 1980s and continuing since, urban public space has largely been redefined to reflect the interests of risk-averse homeowners, elites, store owners, corporations, local governments, and planners. Broadly speaking, their interests lie in the regulation of public space to promote predictability over chance, comfort over uneasiness (picture homeless people populating benches in a park), and sameness over social difference. As Low indicates in her study of Lower Manhattan, the management and oversight of public spaces are increasingly privatized, leading to the promotion of certain ideal uses and users (such as consumers) and the exclusion of certain groups, such as locals (acting as non-consumers), youth subcultures, and the urban poor. Hence, while the search for order, comfort, and security remains the palatable interest of most city dwellers, taken to its extreme (as privatization of public space may entail) it comes at the cost of excluding others, reducing their rights and opportunities to use public space, and further rendering them invisible.

The dichotomies of inclusion/exclusion and visibility/invisibility are fully explored in Teresa Caldeira's study of fortified residential enclaves or "gated communities." Caldeira focuses on São Paulo, Brazil, where walled communities are status symbols for the wealthy and acquisitive middle classes. Picture landscaped and sprawling estates or condominium complexes hidden behind tall stucco walls painted in pastel colors, covered with vines, nearly totally disconnected from the surrounding city. While its residents may enjoy its amenities and hypersecurity, the fortified enclave raises some disturbing questions about the public life of cities. When Caldeira turns her eye to gated communities in Los Angeles she sees similar issues of status enhancement and security consciousness, coupled with the demands of the wealthy and upper middle class to isolate themselves from the urban masses. Caldeira's study addresses the important fact that, for those with financial means, privatization and the free market allow for exclusion and invisibility to be readily reproduced.

The drive toward the privatization of not only public spaces and gated neighborhoods but also of hospitals, prisons, and even public schools, is a hallmark feature of contemporary governance for cash-strapped cities seeking ways to balance budgets, rebuild neighborhoods, and develop downtowns. In a global era of tough competition for investment capital, most cities have embraced a pro-growth, pro-market style of governance that cedes traditional municipal power to private interests and, in turn, further disenfranchises city dwellers with respect to control over the decisions that shape their city. Scholars refer to this form of governance philosophy as neoliberalism. In the fourth reading in this section, Christopher Mele traces the outcomes and impacts of neoliberal urban redevelopment for Chester, Pennsylvania, a small city south of Philadelphia caught in the seemingly inescapable trap of deindustrialization and suburban flight. Chester is somewhat unique in that the city redefined itself as both a "waste magnet" and a "destination" for gamblers and soccer fans. Mele sees these efforts as reflecting two phases of neoliberalism – an earlier period of "rolling back," when local governments eliminated regulations and oversight in often desperate efforts to attract development, and a later "rolling out," when quasi-government agencies and private actors court investment capital to develop large-scale projects such as casinos and sports stadiums. As a consequence, Chester's waterfront – although a quirky hodgepodge of mixed uses – depends on the exclusion of the majority of its own poor, predominantly black residents. This situation in which redevelopment hinges on social regulation and exclusion is an attribute of neoliberal urbanism.

The final two readings address the consequences of stepped-up efforts at regulation and social exclusion and ask who has the right to the city. The question of rights stems from the 1968 publication of *Le Droit à la ville*, by the renowned critical urban theorist Henri Lefebvre. Influenced by the era of social protest, Lefebvre advocated a form of urban politics from a position of the inherent rights of the inhabitants. This rights-based politics would both challenge the state and corporate control over the shape, design, and use of the city and, more importantly, allow the fulfillment of the city's promise of self-determination by its inhabitants. Free from the nearly exclusive economic dictates that dominate the purpose of the city, the right to the city unleashes the possibilities of spaces for encounters and exchanges by a host of different social groups, cultures, and classes. In the reading entitled "Spaces of Insurgent Citizenship," James Holston presents one version of this open-ended vision of a very different kind of city. Holston connects the dominating and restraining aspects of urban life to the politics and power of urban design, indicating how modernism has shaped both the city's built environment and the minds (and imaginations) of those who live there. But fissures and cracks abound as urban dwellers remake and reimagine their environment in ways meaningful to their experiences (and outside the intentions of planners and developers). In this, Holston finds hope, as these insurgent forms of citizenship offer promise roughly along the lines of Lefebvre's earlier call to action.

Finally, David Harvey begins his short essay by pointing out that Lefebvre's "right to the city" is "not merely a right to access to what already exists, but a right to change it after our heart's desire." Harvey first reminds us of the simple fact that most of us live in societies where the rights of private property and profit making prevail over any other notions of rights. Indeed, the ordinariness of that statement reveals how deeply the precepts of late capitalism have become "normalized." What we perceive as natural is actually socially constructed. In order, then, for any kind of emancipatory urban politics to take hold, Harvey contends we must first acknowledge that the kind of city we want cannot be separated from the kind of society we want to (re)invent. The "right to the city," therefore, is a fundamental human right: the right to remake the city in a different image. Reimagining – and, more importantly, remaking – the city offers the opportunity to change ourselves. What kind of cities do we wish to create? What kind of people do we want to be? By asserting our right to address the first question, we answer the second.

"Spatial Governmentality and the New Urban Social Order: Controlling Gender Violence through Law"

from *American Anthropologist* (2001)

S. E. Merry

Editors' Introduction

Both private and public spaces in contemporary cities are highly regulated. Mounted cameras, sensors, and other advanced surveillance technologies allow police and private security firms to monitor the use (and users) of streets, parks, sidewalks, and shopping malls. Certain forms of architecture and exterior designs discourage certain behaviors, such as the gathering and forming of groups, while encouraging others, such as shopping and other forms of consuming. The regulation of particular spaces – drug-free zones, panhandler-free zones, gun-free zones, and even teenager-free zones (shopping malls at certain hours) – adds another layer of social control to the urban landscape. These regulatory methods diminish the likelihood or risk of so-called undesirable behaviors by spatially excluding certain activities – and actors – in designated places. Cities have become especially adept at managing risk by anticipating problems and preventing them *spatially* – as reflected in the concept, spatial governmentalilty. In this reading, Sally Merry shows us how spatial governmentality differs from other forms of regulation that tend to focus on managing individuals or social groups, rather than the spaces they live in, work, or occupy. Spatial regulation includes curfews, allowing people into a space at specific times only, and prohibiting behavior, such as prostitution or alcohol consumption, in certain areas. As Merry demonstrates, spatial regulation is aimed at those deemed incapable of self-management. Hence, managing spaces and other traditional forms of regulation, such as imprisonment of offenders, work hand in hand as effective strategies of urban governance. As Merry notes, these strategies are the result of decades of economic inequalities that divide urban populations in the United States. In American cities, spatial strategies such as "gated communities" exclude the poor from access to the wealthy. In this reading, Merry examines spatial control and how it relates to gender violence in the town of Hilo, Hawai'i. The focus of security and spatial regulations enacted in Hilo is to protect potential victims and to reduce the risk of danger, not to reform those who commit crimes. In this case, self-management training programs for offenders and a spatially based system of deterrence increased the public safety of women in Hilo.

Sally Engle Merry is a Professor in the Department of Anthropology and the Institute for Law and Society at New York University. She is the author of *Human Rights and Gender Violence*, which examines how international human rights laws about gender violence apply to local problems. She has published articles on women's human rights, violence against women, and the process of localizing human rights. Her book *Colonizing Hawai'i: The Cultural Power of Law* (Princeton, 2000) received the 2001 J. Willard Hurst Prize from the Law and Society Association.

Although modern penality is largely structured around the process of retraining the soul rather than corporal punishment, as Foucault argued in his study of the emergence of the prison (1979), recent scholarship has highlighted another regime of governance: control through the management of space. New forms of spatially organized crime control characterize contemporary cities, from the explosion of gated communities to "prostitution free zones" as a regulatory strategy for the sex trade to "violence free zones" as a way of diminishing communal conflict in India. Spatialized strategies have been applied to the control of alcohol consumption and the regulation of smoking. In the 1970s, concerns about fear of crime in the United States expanded from a focus on catching offenders to removing "incivilities" in public spaces. This meant creating spaces that appeared safe to urbanites by removing people who looked dangerous or activities that seemed to reveal social disorder such as homeless people or abandoned trash. New community-policing strategies toward youths emphasize moving potentially criminal youths to other areas rather than prosecuting them.

These are all examples of new regulatory mechanisms that target spaces rather than persons. They exclude offensive behavior from specified places rather than attempting to correct or reform offenders. The regulation of space through architectural design and security devices is generally understood as a complement to disciplinary penality but fundamentally different in its logic and technologies. While disciplinary mechanisms endeavor to normalize the deviant behavior of individuals, these new mechanisms focus on governing populations as a whole. They manage risks by anticipating problems and preventing them rather than punishing offenders after the incident. Governance through risk management means mitigating harms rather than preventing transgressions. It is future oriented and focuses on prevention, risk minimization, and risk distribution.

A focus on managing risks rather than enforcing moral norms has transformed police practices in recent years (Ericson and Haggerty 1997, 1999; O'Malley 1999a: 138–139). This approach seeks to produce security rather than to prevent crime – to reduce the risk of crime rather than to eliminate it. Order is defined by actuarial calculations of tolerable risk rather than by consensus and conformity to norms (Simon 1988). New policing strategies seek to diminish risks through the production of knowledge about potential offenders (Ericson and Haggerty 1997). In general, modern democratic countries have experienced a pluralizing of policing, which joins private and community-based strategies that focus on protection of space with public strategies that detect and punish offenders (Bayley and Shearing 1996).

New mechanisms of social ordering based on spatial regulation have been labeled spatial governmentality (Perry 2000; Perry and Sanchez 1998). They differ substantially from disciplinary forms of regulation in logic and techniques of punishment. Disciplinary regulation focuses on the regulation of persons through incarceration or treatment, while spatial mechanisms concentrate on the regulation of space through excluding offensive behavior. Spatial forms of regulation focus on concealing or displacing offensive activities rather than eliminating them. Their target is a population rather than individuals. They produce social order by creating zones whose denizens are shielded from witnessing socially undesirable behavior such as smoking or selling sex. The individual offender is not treated or reformed, but a particular public is protected. The logic is that of zoning rather than correcting.

Spatialized regulation is always also temporal as well. Regulations excluding offensive behavior usually specify time as well as place. Systems such as curfews designate both where and when persons can appear. Spatial regulations may interdict particular kinds of persons from an area only during certain times, such as business hours, or prohibit behavior, such as drinking, only after a certain time at night. Spatial regulation may cover all periods of time, but it is typically targeted to some specified part of the day. It may also be imposed only for a limited duration, as in the case of the stay-away court orders discussed below.

Although spatial forms of governmentality are not exclusively urban, they have taken on particular importance in modern cities. In addition to features of size, scale, heterogeneity, and anonymity, many contemporary cities are characterized by sharp economic inequalities, major differences in levels of development, global labor and capital flows, and a shift to neoliberal forms of governance. As states endeavor to govern more while spending less, they have adopted mechanisms that build on individual self-governance and guarded spaces. They establish areas to which only people seen as capable of self-governance have access and incarcerate those who cannot be reformed. People are encouraged to participate in their own self-governance, whether by voluntarily passing through metal-detector machines in airports or organizing into community watch brigades.

[. . .]

Within the neoliberal regime of individual responsibility and accountability, populations are divided between those understood as capable of self-management and those not. Managing spaces and incarcerating offenders are therefore complementary rather than opposing strategies. These complementary strategies are the product of the vast economic inequalities dividing urban populations in the United States. In American cities, spatial strategies are typically used by the wealthy to exclude the poor, while those who fail to respect these islands of safety are incarcerated. Private organizations pursue similar strategies, developing systems of private policing and governance that parallel those of the state. This transformation seems to be characteristic of cities outside the United States as well. Indeed, contemporary urbanism is shaped not only by features of size, scale, and anonymity but also by globally produced inequalities and transnationally circulating notions of governance.

[. . .]

The expansion of spatial governmentality diminishes the scope of collective responsibility for producing social order characteristic of governance in the modern state. Some persons are defined as hopeless, deserving exclusion rather than correction and reintegration. The collectivity takes less responsibility for the excluded. Prisons are increasingly seen as holding pens rather than places of education, training, and reform.

Although spatial governmentality is generally described as a system that provides safety for those who can afford it while abandoning the poor to unregulated public spaces, in this article I describe a different use of spatial governance: the spatial exclusion of batterers from the life space of their victims. This is not an instance of creating a collective safe space but, instead, of protecting a person by prohibiting access to her home or workplace. This approach emphasizes the safety of the victim rather than the punishment or reform of the offender. Unlike the more recognized uses of spatial governance, this initiative endeavors to protect poor women as well as rich women. It represents the use of spatial systems of governance that benefit more than the wealthy and privileged. Like other forms of spatial governmentality, however, this regime typically controls the disadvantaged rather than the privileged. People subject to restraining orders for gender violence are typically poor men very similar to, and often identical with, those generally controlled by the forms of spatial governmentality developed by the wealthy. It is not that these are the only men who batter, but these are usually the only ones who end up in the restraining order process.

The use of spatial control in gender violence situations is relatively new. It took a powerful social movement many years to develop this legal protection for battered women. Punishing batterers for the crime of assault is an old practice; providing legal restrictions on their movements to create a safe space for victims is much newer. A concerted social movement of feminist activists beginning in the late 1960s argued for the applicability of protective orders for such situations. Commonly referred to as temporary restraining orders (TROs), these orders supplement more conventional strategies for punishing batterers. TROs are court orders that require the person who batters (usually but not always male) to stay away from his victim (usually but not always female) under penalty of criminal prosecution. In the United States, protective orders were used for domestic abuse situations beginning in the 1970s, about the same time as refuges and shelters were being promoted by the battered-women's movement.

[. . .]

The article is based on a decade of ethnographic research in a town in Hawai'i, a place with a distinctive colonial past and plantation legacy but a thoroughly American legal system and feminist movement against battering. Its courts follow mainland U.S. patterns in their reliance on spatial processes for protecting battered women as well as in their approaches to punishing and reforming batterers. The town, Hilo, has 45,000 inhabitants and serves as the hub of a large agricultural region dotted with vast sugar cane plantations, the recent collapse of which has exacerbated problems of unemployment and poverty. Although it lacks the anonymity of larger cities, Hilo shares the wide economic disparities of contemporary American cities. Farmers, plantation workers, part-time construction workers, homeless people living on the beach, welfare families, professors, judges, and county officials jostle one another in the streets, stores, and offices of Hilo and its environs. Although dispersed rural communities often do not use shelters or courts to handle gender violence, Hilo is sufficiently urban to rely heavily on the law and formal organizations such as shelters. Moreover, Hilo is influenced by changing conceptions of governance from the U.S. mainland. This analysis of spatial approaches to wife battering shows the importance of spatial modes of governance in contemporary urban life and reveals the extent to which these new forms of governance are circulating from one local place to another.

THEORIZING SPATIAL GOVERNMENTALITY

The concept of spatial governmentality derives from Foucault's elaboration of the notion of governmentality, a neologism that incorporates both government and rationality (1991). Governmentality refers to the rationalities and mentalities of governance and the range of tactics and strategies that produce social order. It focuses on the "how" of governance (its arts and techniques) rather than the "why" (its goals and values). Techniques of governmentality are applied to the art of governing the self as well as that of governing society.

Considerable research on governmentality has delineated a rough historical sequence from eighteenth-century mechanisms that act primarily on the body, such as exile or dramatic physical punishment, to a modern, nineteenth-century system of social control that relies on reforming the soul of the individual and normalizing rule breakers, to a late-twentieth-century postmodern form of social control that targets categories of people using actuarial techniques to assess the characteristics of populations and develops specific locales designed for prevention rather than the normalization of offenders . . . The modern system relies on disciplinary technologies to forge the modern subject at work as well as in the family. The postmodern system is premised on a postindustrial subjectivity of consumption, choice, introspection about feelings, and flexibility.

At the same time, there has been an elaboration of mechanisms that promote security by diminishing risks. Risk-based techniques such as social insurance, workers' compensation, and income tax are examples of security-focused technologies of governance. They offer more efficient ways of exercising power since they tolerate individual deviance but produce order by dividing the population into categories organized around differential degrees of risk. Risk-based approaches fall within the sphere of neoliberal techniques of governance, which Valverde et al. describe as the downloading of risk management to individuals and families, "responsibilization," empowerment, and consumer choice (1999: 19). Responsibilization involves the inculcation and shaping of responsibility for good health and good order within the home, the family, and the individual by means of expert knowledges (Rose 1999: 74).

Many forms of governmentality have a spatial component. Foucault recognized a critical role for spatial ordering in his analysis of systems of discipline in the nineteenth century, but he saw its role largely as a frame for ordering and confining bodies and as a structure of surveillance (1979). In contemporary society, spatialized forms of ordering are connected to the recent intensification of consumption as a mode of identity formation along with neoliberal approaches to government. In contemporary cities, there is increasing focus on managing the spaces people occupy rather than managing the people themselves. Systems of providing security through the private regulation of spaces reveal the emergence of postcarceral forms of discipline that do not focus on individualized soul training. Instead, these new forms of regulation depend on creating spaces characterized by the consensual, participatory governance of selves. These systems rely on selves who see themselves as choice-making consumers, defining themselves through the way they acquire commodities and choose spouses, children, and work. In liberal democracies of the postwar period, citizens are to regulate themselves, to become active participants in the process rather than objects of domination. Citizen subjects are educated and solicited into an alliance between personal objectives and institutional goals, creating "government at a distance."

Disney World and the shopping mall represent locales for such participatory regulation in which the self is made and makes itself in ways structured by the private regulation of the space. These forms of regulation rely on the state only minimally and are largely maintained through private security forces. The space itself creates expectations of behavior and consumption. These systems are not targeted at reforming the individual or transforming his or her soul; instead they operate on populations, inducing cooperation without individualizing the object of regulation. Private control lacks a moral conception of order and is concerned only with what works; it is preventative rather than punishing. This shift to an instrumental focus means a move away from concern with individual reform to control over opportunities for breaches of order. Spatial governmentality works not by containing disruptive populations but by excluding them from particular places. The shopping mall, the prototype of spatial governmentality, is also the product of market-based technologies for shaping and controlling identity and behavior. As subjects become consumers, the autonomous citizens regulate themselves through organizing their lives around the market. The individual invested with rights is replaced by the individual who defines himself or herself by consumption. This control is promotive rather than reactive, voluntary rather than coercive, based more on choice than constraint. Power appears to disappear behind individual choice. Systems of private regulation are backed by formal legal processes, which will remove those who cannot govern themselves.

Thus, the newer systems coexist with morally reformist carceral systems, each defined by whom and what it excludes. The prison system survives and expands along with nonpunitive systems, which manage the opportunities for behaviors rather than the behaviors themselves. Outside the space marked by the absence of penal power, there is a world of unemployed, insane, socially marginal people subjected to the penal power of police and prisons. As in the control of gender-based violence, spatial forms of ordering operate against a backdrop of punishment.

PUNISHMENT/THERAPY/SAFETY: APPROACHES TO GENDER VIOLENCE

From a governmentality perspective, there are three distinct forms of governance: punishment, discipline, and security. One is based on punishing offenders, one on reforming offenders through therapy and training, and one on keeping offenders away from victims through spatial separation. All three are used in dealing with gender-based violence in cities in North America. In this section, I describe each form of governmentality as it has been developed to control wife abuse in the United States and show how it works in practice in the particular context of Hilo, Hawai'i. In wife-battering cases, the dominant mode of punishment is incarceration. Reform depends on a range of services such as batterer intervention programs, control of alcohol and drug use, parenting classes, counseling, and the ongoing supervision of a probation officer. Security is produced by spatial systems such as civil protective orders, which require batterers to stay away from their victims. A detailed analysis of the operation of each of these mechanisms indicates fundamental differences in the logic of each as well as intersections in practice. Spatial forms of governance require punishment as a last resort, while they also are connected to efforts to reform batterers.

Punishment/prison

Punishment is targeted to a particular act rather than the character of the offender or the plight of the victim. It seeks to deter future offenses with the fear of punishment. In the past, forms of public punishment were designed not to reform the offender but, as Foucault argues, to express the will of the sovereign (1979). These punishments were tailored to the offense itself and included flogging and public torture. In the modern period, punishment is largely deprivation of property (fines) and deprivation of liberty (prison). Incarceration is justified by the possibility of reform even though it is generally acknowledged that prisons fail to reform.

Beginning in the 1970s, feminist activists pressed for a greater use of punishment in gender violence cases, advocating mandatory police arrests, no-drop prosecution, and mandatory incarceration.

Historically, under the legal doctrine of coverture, the family had been defined as a private sphere under the authority of the husband rather than the state. Although coverture was generally eliminated by the late nineteenth century in the United States, its legacy is a reluctance to intervene legally in the family in ways that challenge male authority. The law intervenes in gender violence incidents less readily than in other cases of assault. Until recently, violence within families was treated as a social problem reflective of poverty rather than as a criminal offense. As late as 1973, a prosecutor working in the District of Columbia bemoaned the lack of punishment batterers received through the law and the total absence of services to which batterers or their victims could be referred.

Gender violence cases did appear in court in the past in Hilo, and offenders were generally found guilty and fined, but the numbers were small. Between 1971 and 1976, there were between one and nine cases of gender violence in Hilo courts every year and between 1980 and 1986, fewer than twenty a year. By 1998 the number had increased 25 times to 538 cases a year. Calls to the police for help increased fivefold from 500 a year in 1974 to 2,500 in 1994 while the population doubled. Fragmentary data from other parts of the United States reveal a similar staggering growth in the number of criminal cases of domestic violence in the courts since the mid-1980s. These changes are a result of demands for a more activist police force and mandatory arrest policies along with no-drop prosecution.

This increase in cases has not translated into a significant increase in punishment. Instead, it has served to funnel offenders into an array of services and subject them to ongoing supervision by the courts. A 1995 study of 140 domestic violence arrests in 11 jurisdictions found that only 44 made it to conviction, plea, or acquittal, and of these, only 16 served any time. In Hawai'i, a new spouse abuse statute passed in 1986 mandated 48 hours of incarceration for a person convicted of battering. Judges and prosecutors in Hilo say that it is very common for men to escape jail time by pleading to a lesser charge, such as third degree assault, with the stipulation that the offender receive probation and attend a batterer intervention program – but not do jail time. It is also common for cases to be dismissed altogether because the victim refuses to testify or the defendant leaves the island. In Hilo, many cases are not prosecuted, but convictions usually lead to punishment . . . Thus, the major intervention of the court is reform through social services backed by the threat of punishment. The same people are often involved in both criminal and civil proceedings.

Discipline/reform

Disciplinary techniques work on persons rather than actions, seeking to reform them through rehabilitation and repentance. Disciplinary systems incorporate a broad range of therapeutic and group discussion techniques ranging from batterer's intervention programs to alcoholics-anonymous-style self-help meetings. Some are designed to reform by forcing the body to follow an orderly sequence of activities in work and everyday life, while others reform through introspection and insight, requiring consent from the subject of transformation. Prison reform models from the early nineteenth century already incorporated these two approaches to discipline: one was based on habituation of the body and coordination with the machinery of production while the other developed skills of self-management and self-control and promoted autonomy and integrity. These two forms continued to provide alternative models of discipline throughout the nineteenth and early twentieth centuries, but the latter came to predominate. In the late twentieth century, the criminal justice system in the United States has increasingly turned to introspective forms of discipline and self-management.

In the 1990s, this model dominated batterer reform efforts in Hilo as well as in the rest of the United States. This model focused on undermining the cultural support for male privilege and violence against women by exploring men's feelings and beliefs and encouraging men to analyze their own behavior during battering events. Violence against women was understood as an aspect of patriarchy. A dominant feature of group discussions was changing beliefs about men's entitlement to make authoritative decisions and back them up with violence.

Men convicted of spouse abuse or under a TRO were required to attend the Alternatives to Violence (ATV) program started in Hilo in 1986. ATV offers violence control training for men and a support group for women. Men are required to attend weekly two-hour group discussions for six months. In groups of 10 to 15 men and 2 facilitators, participants talk about their use of violence to control their partners. Discussions stress the importance of egalitarian relations between men and women and the value of settling differences by negotiation rather than by force. The men are taught that treating their partners with respect rather than violence will win them a more loving, trusting, and sexually fulfilling relationship and forge warmer relations with their children.

If men fail to attend the program, the staff informs their probation officers. Those whose attendance is a stipulation of a criminal spouse abuse conviction face revocation of their probation. Those required to attend as a condition of a TRO are guilty of contempt of court – a criminal offense – and their case is sent to the prosecutor. In practice, these men are typically sent back to ATV rather than receiving a jail sentence or other criminal penalty, but the threat of jail time is frequently articulated by judges during court hearings. Thus, attendance at this psychoeducational program is enforced by the threat of prison. The program emphasizes training in self-management of violence, but failure to accomplish this task results in the return to a regime of punishment, at least in theory. In practice, nonparticipants are typically sent back to the program. Only after new violations are they sent to jail.

[...]

The courts occasionally referred batterers to one of several alternative approaches to gender violence in Hilo. The most common were family therapy, Christian pastoral counseling, and an indigenous Native Hawaiian model of healing and conflict resolution. These alternatives incorporate quite different ideologies of gender and marriage than feminist programs. Yet all use techniques of self-management and self-reflection similar to those used at ATV, techniques that are characteristic of the technology of governance in the present period. Batterers, too, are to be reformed through these technologies of the self.

Security/spatial mechanisms

Security techniques are those that seek to minimize the harm wreaked by offenders by containing or diminishing the risks they pose to others. They focus on protecting victims or potential victims and spreading the cost of harms to a larger group through insurance systems. Security technologies assess risks, anticipate and prevent risks, and analyze factors that produce risk. Their target is an entire population rather than particular individuals, and the goal is not reform but security for the population as a whole. Foucault sees security as a specific principle of political method and practice capable of being combined with sovereignty and discipline. The method of security deals in a series of possible and probable events, calculates comparative costs, and, instead of demarcating the permissible and forbidden, specifies a mean and possible range of variation. Sovereignty works on a territory, discipline focuses on the individual, and security addresses itself to a population. From the eighteenth century on, security is increasingly the dominant component of modern governmental rationality.

[...]

Security systems are engaged in reducing danger not by reforming individuals who are threatening but by predicting who might be dangerous and either preventing and neutralizing that danger or spreading it evenly among the population. Some forms of criminal behavior, such as drug use, are currently being subjected to harm-minimization strategies designed to diminish the harm that this behavior imposes on individuals and the wider population instead of using disciplinary strategies.

In the domain of gender violence, security techniques are designed to protect victims instead of seeking to reform offenders. They did not emerge in the field of gender violence until the battered-women's movement of the 1970s.

[...]

In Hawai'i, a law providing for Ex-Parte Temporary Restraining Orders for victims of domestic violence was passed in 1979. Thus, the use of protective orders for domestic violence represented a new legal mechanism developed in the 1970s, which disseminated rapidly across the U.S.

This is the most innovative feature of contemporary American efforts to diminish wife battering. It is fundamentally a spatial mechanism since it simply separates the man and the woman. Shelters, which provide places of refuge for battered women, are similarly novel inventions of the battered-women's movement of the 1970s, although they build on older patterns of safe houses and helpful neighbors and relatives. Neither of these interventions makes an effort to reform the batterer, but seek only to keep him away from the victim.

CONCLUSIONS

Although Hilo is a small town, its changing practices of managing wife battering parallel those of big cities and exemplify shifting forms of governance in contemporary industrialized cities. In Hilo, as in many larger cities, responsibility for control of violence against women has shifted from kin and neighbors to the state. It is the law, rather than the family, to which these battered women turn. Such a decision is not easy and is often discouraged by kin and friends. Yet, the skyrocketing number of complaints shows that the turn to the law is happening in Hilo as well as elsewhere in the United States.

In Hilo as well as in large industrial cities, processes of spatial governmentality are shaped by inequalities linked to class and ethnicity. Yet, spatial governmentality does not simply increase the control over the poor, but also can increase the safety of all women. Many who write about risk society fear that it is a slide into a big brother state, but there may be possibilities for these new mechanisms when they are democratically distributed. It took a protracted struggle led by a powerful social movement to develop and implement a legal innovation that benefits poor women. Many judges still question its validity as a legal procedure, and police are often lax in enforcing it. Overworked prosecutors ignore TRO violations. Yet, the creation and implementation of this spatial mechanism of governmentality at least reveals the possibility of more democratic forms of spatial governance for the protection of vulnerable populations.

[. . .]

The adoption of new forms of spatial governmentality is part of a complex reconfiguration of governance in the postmodern world. These changes are fostered by globalization. Globalization distributes not only commodities and images, but also modes of governance. The invention of the TRO for gender violence was quickly followed by its rapid spread through the United States. There is now a global diffusion of batterer intervention programs, no-drop policies, and restraining orders. Practices in Hilo are brought by activists from other parts of the country, while judges and officials concerned about controlling gender violence face budget pressures found elsewhere in North America to do less and accomplish more. The Hilo judiciary has, like many other U.S. jurisdictions, focused on developing self-management training for offenders in conjunction with spatially based systems of deterrence in place of more costly systems of punishment. Along with neoliberal approaches to governance, these new technologies of spatial governmentality now circulate globally within cities large and small.

REFERENCES

Foucault, Michel (1979) *Discipline and Punish: The Birth of the Prison*. New York: Vintage.

Foucault, Michel (1991) "Governmentality," in Graham Burchell, Colin Gordon, and Peter Miller (eds.), *The Foucault Effect: Studies in Governmentality*, pp. 87–105. Chicago: University of Chicago Press.

"The Erosion of Public Space and the Public Realm"

from *City and Society* (2006)

Setha Low

Editors' Introduction

Public space has always been a crucial feature of the urban landscape. It is the necessary venue in which people from various class, ethnic, racial, and cultural backgrounds "rub shoulders," learn from each other, and socially interact. The public realm is meant to be unbounded and an expansive space for social interaction, political action, and the free exchange of ideas. The prolonged erosion of public space – through government regulations or increased private sector control – is of utmost concern to urbanists. In this reading, Setha Low first summarizes the condition of public space in New York City in the post-9/11 years. Previously open spaces closed after 9/11. Many New Yorkers feel comfortable with the number of surveillance cameras within public spaces as the price paid for so-called enhanced safety. As Low explains, privatization is the driving force behind the shrinkage of public spaces. Public/private partnerships have become more prevalent and decisions over how the space is to be used and by whom is increasingly predetermined by an elite set of actors. The World Trade Center, and in particular the use of the space at Ground Zero, is an example of this. The redevelopment of the site could have provided an opportunity for a public space that could respond to citizens' feelings and concerns, but instead private sector involvement prevailed. Low interviewed residents of nearby Battery Park City and found their ideas contrasted sharply from media and governmental representations of the reconstructed site. For example, residents felt too many memorials in their community spaces challenged the neighborhood's vibrancy. Low's point is that public space should remain part of the public sector, especially given the many ethnic and cultural groups who live in and visit New York. Public space, Low feels, could be used to promote diversity in gender, class, culture, nationality, and ethnicity. Low ends the discussion by looking at how this diversity can be promoted.

Setha Low is Professor of Environmental Psychology, Geography, Anthropology, and Women's Studies, and Director of the Public Space Research Group at The Graduate Center, City University of New York, where she teaches courses and trains Ph.D. students in the anthropology of space and place, urban anthropology, culture and environment, and cultural values in historic preservation.

INTRODUCTION

In New York City, we are losing public space and the democratic values it represents when we need it most. People went to Washington Square Park and Union Square after 9/11, and later to protest the Iraq war and mourn the dead soldiers. But during the Republican Convention, Central Park was closed to protesters because of the cost of re-seeding the lawn. What does this closure of the most symbolic of public spaces portend?

Nancy Fraser defines the public realm as an unbounded, expansive space of social interaction, free exchange of ideas, and political action that influences governmental practice (Kohn 2004). Without the encounters that occur in public space, the public realm contracts.

PUBLIC SPACE IN NEW YORK CITY

In the 1960s, William H. Whyte set out to find out why some New York City public spaces were successes, filled with people and activities, while others were empty, cold and unused. He found that only a few places were attracting daily users and saw this decline as a threat to urban civility. He advocated for viable places where people could meet and relax and his recommendations were implemented by the New York City Planning department to transform the city (Whyte 1980).

In this century, we are facing a different kind of threat to public space – not one of disuse, but of patterns of design, management, and systems of ownership that reduce diversity. In some cases these designs are a deliberate program to reduce the number of undesirables, and in others, a by-product of privatization, commercialization, historic preservation and poor planning and design. Both sets of practices reduce the vitality and vibrancy of the spaces and reorganize it to welcome only tourists and middle-class people.

Further, the obsession with security since September 11th has closed previously open spaces and buildings. Long before the World Trade Center bombings, insecurity and fear of others had been a centerpiece of the post-industrial American city. But New Yorkers are now overreacting by barricading themselves, reducing their sense of community, openness, and optimism.

Before 9/11, when designers talked about security issues they meant reducing vandalism, creating defensible spaces, and moving homeless people and vagrants to other locations (Sipes 2002). With the enhanced fear of terrorism, though, familiar physical barriers such as bollards, planters, security gates, turnstiles, and equipment for controlling parking and traffic are now reinforced by electronic monitoring tactics – such as metal detectors, surveillance cameras and continuous video recording (Speckhardt and Dowdell 2002). Before September 11th, the idea that New Yorkers would agree to live their lives under the gaze of surveillance cameras or real time police monitoring seemed unlikely. Yet the New York Civil Liberties Union has found more than 2,397 cameras trained on public spaces (Tavernise 2004). What was once considered "Big Brother" technology and an infringement of civil rights is now a necessary safety tool with little, if any, examination of the consequences.

PRIVATIZATION OF PUBLIC SPACE

Private interests take over public space in countless ways. Neil Smith, Don Mitchell and I have documented how sealing off a public space by force, redesigning it, and then opening it with intensive surveillance and policing is a precursor to private managementt (Low 2006). Restricting access and posting extensive restrictions further privatizes its use. For example, the interior public space of the Sony Atrium does not allow people in with excessive amounts of shopping bags or shopping carts. Napping is forbidden. At Herald Square in front of Macy's, the 34th Street Partnership has put up a list of rules prohibiting almost everything including sitting on the seat-height, planting walls. Gated communities exclude the public with fencing and guards, especially when there is a public amenity – such as a lake or walking trail – inside (Low 2003). Policing and other forms of surveillance insure that street vendors are strictly confined or banished to marginal areas, while malls and shopping centers have guards and 24 hour video surveillance to protect their facilities.

These physical tactics are bolstered by economic strategies in which public goods are controlled by a private corporation or agency. For example,

Business Improvement Districts can tax local businesses and retail establishments to provide policing, trash removal, and street renovation accompanied by imposed restrictions on the use of public sidewalks, pocket parks and plazas. Conservancies and public/private partnerships also blur public/private distinctions when the city grants decision-making powers to private citizens who then raise money to run what was formerly a publicly-funded park.

Gated communities employ a different set of regulatory practices connected with regional and municipal planning. Incorporation, incentive zoning, and succession and annexation recapture public goods and services including taxpayers' money and utilize these funds to benefit private housing developments. These strategies mislead taxpayers and channel money into amenities the public cannot use and contribute only to the maintenance of private communities. This shift toward privatization of land use controls is an impoverishment of the public realm as well as access to public resources.

THE WORD TRADE CENTER AS PUBLIC SPACE

There is an inherent tension between the meanings of the World Trade Center site created by dominant political and economic players, and the significance of the area for those who live near it. Most of the media reporting has been on the construction of a memorial space for an imagined national and global community of visitors who identify with its broader, state-produced meanings. But New Yorkers' meanings are as much a part of memorialization as the political machinations and economic competition for rental space and architectural status. In response, I have been studying what local Battery Park City residents say about the aftermath of 9/11 and recording their feelings about what they would like to see built at Ground Zero to expand and contest media and governmental representations of the design.

For New York Governor Pataki, New York City Mayor Bloomberg, and the architectural critic Paul Goldberger, the site plan and memorial space design is emotionally evocative. But for local residents, children, and the overall fabric of New York public spaces, it offers little to solve the problems – much less the feelings of fear and insecurity – of those who live and work downtown. For example, residents of Battery Park City say that they would not like to live in a cemetery, and feel that there are already too many memorials in their community spaces. They would like greater economic vitality, more people and businesses to enliven their neighborhood. Almost half of the pre-9/11 residents left shortly after the tragedy, and those who stayed still feel afraid and vulnerable. The current architectural and memorial designs do not take into consideration any of the residents' concerns elicited through interviewing. Sadly, the memorial space dominates the Battery Park City side of the site, while the retail and commercial space that the neighborhood needs is included within the outer ring of tall office buildings. And the sunken expanse of memorial space is not perceived by residents or children as a "safe" or "secure" space, even though it is defended by walls and a sunken, inaccessible site. So even at Ground Zero, we are losing the opportunity for a public space that could respond to citizens' feelings and concerns.

With globalization this trend of increased barricading and surveillance accompanied by privatization is intensifying. Immigrants, the mainstay of the U.S. economy, have again become the feared "other." Privatization, surveillance, and restrictive management have created an increasingly inhospitable environment for immigrants, local ethnic groups, and culturally diverse behaviors. If this trend continues, it will eradicate the last remaining spaces for democratic practices, places where a wide variety of people of different gender, class, culture, nationality, and ethnicity intermingle peacefully.

How can we integrate our diverse communities and promote social tolerance in this new political climate? One way is to make sure that our urban public spaces, where we all come together, remain public in the sense of providing a place for everyone to relax, learn, and recreate, and open so that we have places where interpersonal and intergroup cooperation and conflict can be worked out in a safe and public forum.

Based on twenty years of ethnographic research on parks, historic sites, and beaches, the Public Space Research Group has developed a series of principles that encourage, support, and maintain cultural diversity in public space that are presented in *Rethinking Urban Parks: Public Space and Cultural Diversity* (Low, Taplin and Scheld 2005). They include principles similar to William H. Whyte's rules for small urban spaces that promote their social viability, but in this case, these rules promote and/or maintain the "public" in urban open spaces. The principles are not applicable in all situations, but are meant as guidelines for empowered citizen decision-making in park planning, management, and design for the future.

1. If people are not represented in urban parks, historic national sites and monuments, and more importantly if their histories are erased, they will not use the park.
2. Access is as much about economics and cultural patterns of park use as circulation and transportation, thus income and visitation patterns must be taken into consideration when providing access for all social groups.
3. The social interaction of diverse groups can be maintained and enhanced by providing safe, spatially adequate "territories" for everyone within the larger space of the overall site.
4. Accommodating the differences in the ways social class and ethnic groups use and value public sites is essential to making decisions that sustain cultural and social diversity.
5. Contemporary historic preservation should not concentrate on restoring the scenic features without also restoring the facilities and diversions that attract people to the park.
6. Symbolic ways of communicating cultural meaning are an important dimension of place attachment that can be fostered to promote cultural diversity.

These principles for promoting and sustaining cultural diversity in urban parks and heritage sites are just a beginning, but they are a way for us to start to address the landscape of fear. The important point to be made, however, is that it is not just the landscape that we should be looking at, but the regulations, laws, and policies; restricted uses; paranoia; and citizen compliance.

"Fortified Enclaves: The New Urban Segregation"

from *Public Culture* (1996)

Teresa P. R. Caldeira

Editors' Introduction

In this selection, Teresa Caldeira examines fortified enclaves – what are sometimes referred to as "gated communities" – in São Paulo, Brazil and Los Angeles. Walled communities are an increasing presence in cities and suburbs around the world and Caldeira is interested in what these communities mean to the idea of the city as a space of opportunity, chance, flow and movement, and, of course, social diversity.

In São Paulo, fortified enclaves are popular among the middle and upper classes for two different reasons. Walled or gated communities are popular among those who can afford them because they provide "total security" from the real and imagined dangers of the outside world. In addition to risk-conscious businessmen and their families and corporate elites, enclaves appeal to middle-classes fearful of street crime and the chaotic character of city streets. Enclaves provide a sense (perhaps illusion) of control and protection over one's surroundings. A second reason behind the proliferation of walled communities is status. Enclaves are physically separate and socially exclusive hence they confer a higher status among those who live within them (there is a prestige ranging among enclaves, as well). Given enclaves offer numerous amenities in private facilities (sport, leisure, etc.) and household and personal services, they are seen as spaces of privilege. As Caldeira also points out, class separation is a form of social distinction.

What effects do fortified enclaves have on the character of city life? As part of the built environment, enclaves contribute neither to public space nor freedom of mobility within the city. Walls, fences, and gates impede the flow of traffic, pedestrian and otherwise; enclaves serve no greater purpose for nonresidents who are forced to move around them. Second, the social, cultural, and economic life of enclaves is self-contained and inward-looking. Enclaves do not gesture – either architecturally or sociospatially – to the streets and neighborhoods around them. Caldeira writes: "In other words, the relationship they establish with the rest of the city and its public life is one of avoidance; they turn their backs on them."

Los Angeles shares with São Paulo a fascination and preoccupation with walled communities. Two memorable television images of middle-class homeowners during the 1992 riots were the walled communities with its residents safely intact and the residents of neighborhoods with street access erecting makeshift barricades and gates and manning the entrances with shotguns. Caldeira draws on Mike Davis's extensive work on the architecture of fear found in his book, *City of Quartz* (London: Verso, 1990). For Davis, walled communities are "class warfare at the level of the street." They tend to reinforce social inequality and spatial segregation and contribute nothing positive to the public life of the greater city. In fact, with their friendly, inviting names and pastel-colored walls, gated communities "normalize" inequality and segregation.

In the end, Caldeira finds very little to celebrate with walled communities. As anti-public spaces, they minimize the contact among persons of different class, ethnic, national, and racial backgrounds, and contribute to a greater fear of the "Other." Urban fear and its underlying causes are not addressed – they are concretized and memorialized. Whatever contact middle classes have with the poor and ethnic and racial minorities is limited to consumption, she argues. Difference is encountered in brief exchanges at ethnic markets or restaurants or in contact with mostly minority maids and groundskeepers. Either way, contact is minimal, controlled, and on the terms of wealthier, more powerful whites.

Teresa Caldeira is Associate Professor of Anthropology at the University of California, Irvine. Her book, *City of Walls: Crime, Segregation, and Citizenship in São Paulo* (Berkeley, CA: University of California Press, 2000) won the American Ethnological Society Senior Book Prize in 2001. She has been a Visiting Scholar at the University of São Paulo and the Center for the Study of Violence (NEV). In addition to her research on urban segregation, she studies the impact of technological change on domestic spaces and generational relations in Brazil.

In the last few decades, the proliferation of fortified enclaves has created a new model of spatial segregation and transformed the quality of public life in many cities around the world. Fortified enclaves are privatized, enclosed, and monitored spaces for residence, consumption, leisure, and work. The fear of violence is one of their main justifications. They appeal to those who are abandoning the traditional public sphere of the streets to the poor, the "marginal," and the homeless. In cities fragmented by fortified enclaves, it is difficult to maintain the principles of openness and free circulation that have been among the most significant organizing values of modem cities. As a consequence, the character of public space and of citizens' participation in public life changes.

In order to sustain these arguments, this chapter analyzes the case of São Paulo, Brazil, and uses Los Angeles as a comparison. São Paulo is the largest metropolitan region (it has more than sixteen million inhabitants) of a society with one of the most inequitable distributions of wealth in the world. In São Paulo, social inequality is obvious. As a consequence, processes of spatial segregation are also particularly visible, expressed without disguise or subtlety. Sometimes, to look at an exaggerated form of a process is a way of throwing light onto some of its characteristics that might otherwise go unnoticed. It is like looking at a caricature. In fact, with its high walls and fences, armed guards, technologies of surveillance, and contrasts of ostentatious wealth and extreme poverty, contemporary São Paulo reveals with clarity a new pattern of segregation that is widespread in cities throughout the world, although generally in less severe and explicit forms.

[. . .]

BUILDING UP WALLS: SÃO PAULO'S RECENT TRANSFORMATIONS

The forms producing segregation in city space are historically variable. From the 1940s to the 1980s, a division between center and periphery organized the space of São Paulo, where great distances separated different social groups: the middle and upper classes lived in central and well-equipped neighborhoods and the poor lived in the precarious hinterland. In the last fifteen years, however, a combination of processes, some of them similar to those affecting other cities, deeply transformed the pattern of distribution of social groups and activities throughout the city. São Paulo continues to be a highly segregated city, but the way in which inequalities are inscribed into urban space has changed considerably.

[. . .]

São Paulo is today a city of walls. Physical barriers have been constructed everywhere – around houses, apartment buildings, parks, squares, office complexes, and schools. Apartment buildings and houses that used to be connected to the street by gardens are now everywhere separated by high fences and walls and guarded by electronic devices and armed security men. The new additions frequently look odd because they were improvised in spaces conceived without them, spaces designed to be open. However, these barriers are now fully integrated into new projects for individual houses, apartment buildings, shopping areas, and work spaces. A new aesthetic of security shapes all types of constructions and imposes its new logic of surveillance and distance as a means for displaying status, and it is changing the character of public life and public interactions.

[. . .]

Fortified enclaves represent a new alternative for the urban life of these middle and upper classes. As such, they are codified as something conferring high status. The construction of status symbols is a process that elaborates social distance and creates means for the assertion of social difference and inequality.

[. . .]

Closed condominiums are supposed to be separate worlds. Their advertisements propose a "total way of life" that would represent an alternative to the quality of life offered by the city and its deteriorated public space. Condominiums are distant, but they are supposed to be as independent and complete as possible to compensate for it; thus the emphasis on the common facilities they are supposed to have that transform them into sophisticated clubs. The facilities promised inside closed condominiums seem to be unlimited – from drugstores to tanning rooms, from bars and saunas to ballet rooms, from swimming pools to libraries.

In addition to common facilities, São Paulo's closed condominiums offer a wide range of services. Some of the services (excluding security) are psychologists and gymnastic teachers to manage children's recreation, classes of all sorts for all ages, organized sports, libraries, gardening, pet care, physicians, message centers, frozen food preparation, housekeeping administration, cooks, cleaning personnel, drivers, car washing, transportation, and servants to do the grocery shopping. If the list does not meet your dreams, do not worry, for "everything you might demand" can be made available. The expansion of domestic service is not a feature of Brazil alone. As Saskia Sassen has shown in the case of global cities, high-income gentrification requires an increase in low-wage jobs; yuppies and poor migrant workers depend on each other. In São Paulo, however, the intensive use of domestic labor is a continuation of an old pattern, although in recent years some relationships of labor have been altered, and this work has become more professional.

The multiplication of new services creates problems, including the spatial allocation of service areas. The solutions for this problem vary, but one of the most emblematic concerns the circulation areas. Despite many recent changes, the separation between two entrances – in buildings and in each individual apartment – and two elevators, one labeled "social" and the other "service," seems to be untouchable; different classes are not supposed to mix or interact in the public areas of the buildings. Sometimes the insistence on this distinction seems ridiculous, because the two elevators or doors are often placed side by side instead of being in separate areas. As space shrinks and the side-by-side solution spreads, the apartments that have totally separate areas of circulation advertise this fact with the phrase, "social hall independent from service hall." The idea is old: class separation as a form of distinction.

Another problem faced by the new developments is the control of a large number of servants. As the number of workers for each condominium increases, as many domestic jobs change their character, and as "creative services" proliferate for members of middle and upper classes who cannot do without them, so also do the mechanisms of control diversify. The "creative administrations" of the new enclaves in many cases take care of labor management and are in a position to impose strict forms of control that would create impossible daily relationships if adopted in the more personal interaction between domestic servants and the families who employ them. This more "professional" control is, therefore, a new service and is advertised as such. The basic method of control is direct and involves empowering some workers to control others. In various condominiums, both employees of the condominium and maids and cleaning workers of individual apartments (even those who live there) are required to show their identification tags to go in and out of the condominium. Often they and their personal belongings are searched when they leave work. Moreover, this control usually involves men exercising power over women.

The middle and upper classes are creating their dream of independence and freedom – both from the city and its mixture of classes and from everyday domestic tasks – on the basis of services from working-class people. They give guns to badly paid working-class guards to control their own movement in and out of their condominiums. They ask their badly paid "office boys" to solve all their bureaucratic problems, from paying their bills and standing in all types of lines to transporting incredible sums of money. They also ask their badly paid maids – who often live in the *favelas* on the other side of the condominium's wall – to wash and iron their clothes, make their beds, buy and prepare their food, and frequently care for their children all day long. In a context of increased fear of crime in which the poor are often associated with criminality, the upper classes fear contact and contamination, but they continue to depend on their servants. They can only be anxious about creating the most effective way of controlling these servants, with whom they have such ambiguous relationships of dependency and avoidance, intimacy and distrust.

Another feature of closed condominiums is isolation and distance from the city, a fact that is presented as offering the possibility of a better lifestyle. The latter is expressed, for example, by the location of the development in "nature" (green areas, parks, lakes), and in the use of phrases inspired by ecological discourses. However, it is clear in the advertisements that isolation means separation from those considered to be socially inferior, and that the key factor to assure this is security. This means fences and walls surrounding the condominium, guards on duty twenty-four hours a day controlling the entrances, and an array of facilities and services to ensure security – guardhouses with bathrooms and telephones, double doors in the garage, and armed guards patrolling the internal streets. "Total security" is crucial to "the new concept of residence." In sum, to relate security exclusively to crime is to fail to recognize all the meanings it is acquiring in various types of environments. The new systems of security not only provide protection from crime but also create segregated spaces in which the practice of exclusion is carefully and rigorously exercised.

[...]

ATTACKING MODERN PUBLIC SPACE

The new residential enclaves of the upper classes, associated with shopping malls, isolated office complexes, and other privately controlled environments represent a new form of organizing social differences and creating segregation in São Paulo and many other cities around the world. The characteristics of the Paulista enclaves that make their segregationist intentions viable may be summarized in four points. First, they use two instruments to create explicit separation: on the one hand, physical dividers such as fences and walls; on the other, large empty spaces creating distance and discouraging pedestrian circulation. Second, as if walls and distances were not enough, separation is guaranteed by private security systems: control and surveillance are conditions for internal social homogeneity and isolation. Third, the enclaves are private universes turned inward with designs and organization making no gestures toward the street. Fourth, the enclaves aim at being independent worlds that proscribe an exterior life, which is evaluated in negative terms. The enclaves are not subordinate either to public streets or to surrounding buildings and institutions. In other words, the relationship they establish with the rest of the city and its public life is one of avoidance: they turn their backs on them. Therefore, public streets become spaces in which the elite circulate by car and poor people circulate on foot or by public transportation. To walk on the public street is becoming a sign of class in many cities, an activity that the elite is abandoning. No longer using streets as spaces of sociability, the elite now want to prevent street life from entering their enclaves.

Private enclaves and the segregation they generate deny many of the basic elements that previously constituted the modern experience of public life: primacy of streets and their openness; free circulation of crowds and vehicles; impersonal and anonymous encounters of the pedestrian; unprogrammed public enjoyment and congregation in streets and squares; and the presence of people from different social backgrounds strolling and gazing at those passing by, looking at store windows, shopping, and sitting in cafes, joining political demonstrations, or using spaces especially designed for the entertainment of the masses (promenades, parks, stadiums, exhibitions).

[...]

Contemporary fortified enclaves use basically modernist instruments of planning, with some notable adaptations. First, the surrounding walls: unlike examples of modernist planning, such as Brasilia, where the residential areas were to have no fences or walls but to be delimited only by expressways, in São Paulo the walls are necessary to demarcate the private universes. However, this demarcation of private property is not supposed to create the same type of (nonmodernist) public space that characterizes the industrial city. Because the private universes are kept apart by voids (as in modernist design), they no longer generate street corridors. Moreover, pedestrian circulation is discouraged and shopping areas are kept away from the streets, again as in modernist design. The second adaptation occurs in the materials and forms of individual buildings. Here there are two possibilities. On the one hand, buildings may completely ignore the exterior walls, treating facades as their backs. On the other, plain modernist facades may be eliminated in favor of ornament, irregularity, and ostentatious materials that display the individuality and status of their owners. These buildings reject the glass and transparency of modernism and their disclosure of private life. In other words, internalization, privacy, and individuality are enhanced. Finally, sophisticated technologies of security assure the exclusivity of the already isolated buildings.

[. . .]

Instead of creating a space in which the distinctions between public and private disappear – making all space public as the modernists intended – the enclaves use modernist conventions to create spaces in which the private quality is enhanced beyond any doubt and in which the public, a shapeless void treated as residual, is deemed irrelevant. This was exactly the fate of modernist architecture and its "all public space" in Brasilia, a perversion of initial premises and intentions. The situation is just the opposite with the closed condominiums and other fortified enclaves of the 1980s and 1990s. Their aim is to segregate and to change the character of public life by bringing to private spaces constructed as socially homogeneous environments those activities that had been previously enacted in public spaces.

Today, in cities such as São Paulo, we find neither gestures toward openness and freedom of circulation regardless of differences nor a technocratic universalism aiming at erasing differences. Rather, we find a city space whose old modern urban design has been fragmented by the insertion of independent and well-delineated private enclaves (of modernist design) that pay no attention to an external overall ordination and are totally focused on their own internal organization. The fortified fragments are no longer meant to be subordinated to a total order kept together by ideologies of openness, commonality, or promises of incorporation. Heterogeneity is to be taken more seriously: fragments express radical inequalities, not simple differences. Stripped of the elements which in fact erased differences such as uniform and transparent facades, modernist architectural conventions used by the enclaves are helping to insure that different social worlds meet as infrequently as possible in city space – that is, that they belong to different spaces.

In sum, in a city of walls and enclaves such as São Paulo, public space undergoes a deep transformation. Felt as more dangerous, fractured by the new voids and enclaves, broken in its old alignments, privatized with chains closing streets, armed guards, guard dogs, guardhouses, and walled parks, public space in São Paulo is increasingly abandoned to those who do not have a chance to live, work, and shop in the new private, internalized, and fortified enclaves. As the spaces for the rich are enclosed and turned inside, the outside space is left for those who cannot afford to go in. A comparison with Los Angeles shows that this new type of segregation is not São Paulo's exclusive creation and suggests some of its consequences for the transformation of the public sphere.

SÃO PAULO, LOS ANGELES

Compared to São Paulo, Los Angeles has a more fragmented and dispersed urban structure. São Paulo still has a vivid downtown area and some central neighborhoods, concentrating commerce and office activities, which are shaped on the model of the corridor street and, in spite of all transformations, are still crowded during the day. Contemporary Los Angeles is "polynucleated and decentralized." And its renovated downtown, one of the city's economic and financial centers, does not have much street life: people's activities are contained in the corporate buildings and their under- and overpass connections to shopping, restaurants, and hotels. São Paulo's process of urban fragmentation by the construction of enclaves is more recent than Los Angeles's, but it has already changed the peripheral zones and the distribution of wealth and economic functions in ways similar to that of the metropolitan region of Los Angeles. According to Ed Soja, the latter is a multicentered region marked by a "peripheral urbanization," which is created by the expansion of high-technological, post-Fordist industrialization and marked by the presence of high-income residential developments, huge regional shopping centers, programmed environments for leisure (theme parks, Disneyland), links to major universities and the Department of Defense, and various enclaves of cheap labor, mostly immigrants. Although São Paulo lacks the high-technology industries found in Los Angeles, its tertiarization and distribution of services and commerce are starting to be organized according to the Los Angeles pattern.

Although we may say that São Paulo expresses Los Angeles's process of economic transformation and urban dispersion in a less explicit form, it is more explicit and exaggerated in the creation of separation and in the use of security procedures. Where rich neighborhoods such as Morumbi use high walls, iron fences, and armed guards, the West Side of Los Angeles uses mostly electronic alarms and small signs announcing "Armed Response." While São Paulo's elites clearly appropriate public spaces – closing public streets with chains and all sorts of physical obstacles, installing private armed guards to control circulation – Los Angeles elites still show more respect for public streets. However, walled communities appropriating public streets are already appearing in Los Angeles, and one can wonder if its more discreet pattern of separation and of surveillance is not in part associated with the fact that the poor are far from the West Side, while in Morumbi they live beside the enclaves. Another reason must surely be the fact that the Los Angeles Police Department – although considered one of the most biased and violent of the United States – still appears very effective and nonviolent if compared to São Paulo's police. São Paulo's upper classes explicitly rely on the services of an army of domestic servants and do not feel ashamed to transform the utilization of these services into status symbols, which in turn are incorporated in newspaper advertisements for enclaves. In West Los Angeles, although the domestic dependence on the services of immigrant maids, nannies, and gardeners seems to be increasing, the status associated with employing them has not yet become a matter for advertisement. In São Paulo, where the local government has been efficient in approving policies to help segregation, upper-class residents have not yet started any important social movement for this purpose. But in Los Angeles residents of expensive neighborhoods have been organizing powerful homeowner associations to lobby for zoning regulations that would maintain the isolation their neighborhoods now enjoy.

Despite the many differences between the two cases, it is also clear that in both Los Angeles and São Paulo conventions of modernist city planning and technologies of security are being used to create new forms of urban space and social segregation. In both cities, the elites are retreating to privatized environments, which they increasingly control, and are abandoning earlier types of urban space to the poor and their internal antagonisms. As might be expected given these common characteristics, in both cities we find debates involving planners and architects in which the new enclaves are frequently criticized but are also defended and theorized. In São Paulo, where modernism has been the dominant dogma in schools of architecture up to the present, the defense of walled constructions is recent and timid, using as arguments only practical reasons such as increasing rates of crime and of homelessness. Architects tend to talk about walls and security devices as an unavoidable evil. They talk to the press, but I could not find academic articles or books on the subject. In Los Angeles, however, the debate has already generated an important literature, and both criticism and praise of "defensible architecture" are already quite elaborated.

One person voicing the defense of the architectural style found in the new enclaves is Charles Jencks. He analyzes recent trends in Los Angeles architecture in relation to a diagnosis of the city's social configuration. In his view, Los Angeles's main problem is its heterogeneity, which inevitably generates chronic ethnic strife and explains episodes such as the 1992 uprising. Since he considers this heterogeneity as constitutive of Los Angeles's reality, and since his diagnosis of the economic situation is pessimistic, his expectation is that ethnic tension will increase, that the environment will become more defensive and that people will resort to nastier and more diverse measures of protection. Jencks sees the adoption of security devices as inevitable and as a matter of realism. Moreover, he discusses how this necessity is being transformed into art by styles that metamorphose hard-edged materials needed for security into "ambiguous signs of inventive beauty" and "keep out" and design facades with their backs to the street, camouflaging the contents of the houses. For him, the response to ethnic strife is "defensible architecture and riot realism." The "realism" lies in architects looking at "the dark side of division, conflict, and decay, and represent[ing] some unwelcome truths." Among the latter is the fact that heterogeneity and strife are here to stay, that the promises of the melting pot can no longer be fulfilled.

Jencks targets ethnic heterogeneity as the reason for Los Angeles's social conflicts and sees separation as a solution. He is not bothered by the fact that the intervention of architects and planners in the urban environment of Los Angeles reinforces social inequality and spatial segregation. He also does not interrogate the consequences of these creations for public space and political relationships. In fact, his admiration of the backside-to-the-street solution indicates a lack of concern with the maintenance of public streets as spaces that embed the values of openness and conviviality of the heterogeneous masses.

But Los Angeles's defensible architecture also has its critics. The most famous of them is Mike Davis, whose analysis I find illuminating, especially for its thinking about the transformations in the public sphere. For Davis, social inequality and spatial segregation are central characteristics of Los Angeles, and his expression "Fortress L.A." refers to the type of space being presently created in the city.

[. . .]

For Davis, the increasingly segregated and privatized Los Angeles is the result of a clear master plan of postliberal (i.e., Reagan–Bush Republican) elites, a theme he reiterates in his analysis of the 1992 riots. To talk of contemporary Los Angeles is, for Davis, to talk of a new "class war at the level of the built environment" and to demonstrate that "urban form is indeed following a repressive function in the political furrows of the Reagan–Bush era. Los Angeles, in its prefigurative mode, offers an especially disquieting catalogue of the emergent liaisons between architecture and the American police state."

Davis's writing is marked by an indignation fully supported by his wealth of evidence concerning Los Angeles. Nevertheless, sometimes he collapses complex social processes into a simplified scenario of warfare that his own rich description defies. Despite this tendency to look at social reality as the direct product of elite intentions, Davis elaborates a remarkable critique of social and spatial segregation and associates the emerging urban configuration with the crucial themes of social inequality and political options. For him, not only is there nothing inevitable about "fortress architecture," but it has, in fact, deep consequences for the way in which public space and public interactions are shaped.

My analysis of São Paulo's enclaves coincides with Davis's analysis of Los Angeles as far as the issue of public space is concerned. It is clear in both cases that the public order created by private enclaves of the "defensible" style has inequalities, isolation, and fragmentation as starting points. In this context, the fiction of the overall social contract and the ideals of universal rights and equality that legitimated the modern conception of public space vanish. We should ask, then, if there is already another political fiction organizing inequalities and differences at the societal level, and how best to conceive this new configuration as the old modern model loses its explanatory value. If social differences are brought to the center of the scene instead of being put aside by universalistic claims, then what kind of model for the public realm can we maintain? What kind of polity will correspond to the new fragmented public sphere? Is democracy still possible in this new public sphere?

PUBLIC SPHERE: INEQUALITIES AND BOUNDARIES

People attach meanings to the spaces where they live in flexible and varying ways, and the factors influencing these readings and uses are endless. However, cities are also material spaces with relative stability and rigidity that shape and bound people's lives and determine the types of encounters possible in public space. When walls are built up, they form the stage for public life regardless of the meanings people attach to them and regardless of the multiple "tactics" of resistance people use to appropriate urban space.

In this essay, I have been arguing that in cities where fortified enclaves produce spatial segregation, social inequalities become quite explicit. I have also been arguing that in these cities, residents' everyday interactions with people from other social groups diminish substantially, and public encounters primarily occur inside protected and relatively homogeneous groups. In the materiality of segregated spaces, in people's everyday trajectories, in their uses of public transportation, in their appropriations of streets and parks, and in their constructions of walls and defensive facades, social boundaries are rigidly constructed. Their crossing is under surveillance. When boundaries are crossed in this type of city, there is aggression, fear, and a feeling of unprotectedness; in a word, there is suspicion and danger. Residents from all social groups have a sense of exclusion and restriction. For some, the feeling of exclusion is obvious as they are denied access to various areas and are restricted to others. Affluent people who inhabit exclusive enclaves also feel restricted; their feelings of fear keep them away from regions and people identified in their mental maps of the city as dangerous.

Contemporary urban segregation is complementary to the issue of urban violence. On the one hand, the fear of crime is used to legitimate increasing measures of security and surveillance. On the other, the proliferation of everyday talk about crime becomes the context in which residents generate stereotypes as they label different social groups as dangerous and therefore as people to be feared and avoided. Everyday discussions about crime create rigid symbolic differences between social groups as they become aligned with either good or evil. In this sense, they contribute to a construction of inflexible separations as city walls do. Both enforce ungiving boundaries. In sum, one of the consequences of living in cities segregated by enclaves is that while heterogeneous contacts diminish, social differences are more rigidly perceived, and proximity to people from different groups is considered dangerous, thus emphasizing inequality and distance.

Nevertheless, the urban environment is not the only basis of people's experiences of social differences. In fact, there are other arenas in which differences tend to be experienced in almost opposite ways, offering an important counterpoint to the experience of the urban environment. This is the case of the perceptions of social difference forged through the intensification of communication networks and mass media (international news, documentaries about all types of lives and experiences), through mass movements of populations, through tourism, or through the consumption of ethnic products (food, clothes, films, music). In these contexts, boundaries between different social universes become more permeable and are constantly crossed as people have access to worlds not originally their own.

Thus, the perception and experience of social differences in contemporary cities may occur in quite distinct ways. Some tame social differences, allowing their appropriation by various types of consumers. Other experiences, such as those of emerging urban environments, characterized by fear and violence, magnify social differences and maintain distance and separateness. If the first type of experience may blur boundaries, the second type explicitly elaborates them. Both types of experience constitute the contemporary public sphere, but their consequences for public and political life are radically different. On the one hand, the softening of boundaries may still be related to the ideals of equality of the liberal-democratic polity and may serve as the basis of claims of incorporation. The tamed differences produced to be consumed do not threaten universalist ideals, and in their peculiar way put people into contact. On the other hand, the new urban morphologies of fear give new forms to inequality, keep groups apart, and inscribe a new sociability that runs against the ideals of the modern public and its democratic freedoms. When some people are denied access to certain areas and when different groups are not supposed to interact in public space, references to a universal principle of equality and freedom for social life are no longer possible, even as fiction. The consequences of the new separateness and restriction for public life are serious: contrary to what Jencks thinks, by making clear the extension of social inequalities and the lack of commonalities, defensible architecture and planning may only promote conflict instead of preventing it.

Among the conditions necessary for democracy is that people acknowledge those from different social groups as co-citizens, that is, as people having similar rights. If this is true, it is clear that contemporary cities that are segregated by fortified enclaves are not environments that generate conditions conducive to democracy. Rather, they foster inequality and the sense that different groups belong to separate universes and have irreconcilable claims. Cities of walls do not strengthen citizenship; rather, they contribute to its corrosion. Moreover, this effect does not depend either on the type of political regime or on the intentions of those in power, since the architecture of the enclaves by itself entails a certain social logic.

Discussions about cities such as Los Angeles, London, or Paris, that is, cities populated by people from the most diverse cultural origins, commonly invoke the theme of the limits of modern citizenship based on affiliation to a nation-state. One might rethink the parameters of citizenship in those cities and suggest that the criterion for participation in political life could be local residence rather than national citizenship. Moreover, it would be possible to argue that this local participation is increasingly necessary to make those cities livable and to improve the quality of life of the impoverished population, increasingly consisting of immigrants. The contrast between this alternative political vision and the reality of fortified cities allows for at least two conclusions, one pessimistic and one more optimistic.

The pessimist would say that the direction of new segregation and the extension of social separation already achieved would make impossible the engagement of a variety of social groups in a political life in which common goals and solutions would have to be negotiated. In this view, citizenship in cities of walls is meaningless. The optimistic interpretation, however, would consider that the change in the criteria for admission to political life, and the consequent change in status of a considerable part of the population, would generate a wider engagement in the search for solutions to common problems and would potentially bridge some distances. There are many reasons to be suspicious of such optimism: studies of homeowner associations in Los Angeles remind us how local democracy may be used as an instrument of segregation. However, the boom of social movements in São Paulo after the mid-1970s suggests a cautious optimism. Where excluded residents discover that they have rights to the city, they manage to transform their neighborhoods and to improve the quality of their lives. That fortified enclaves in part counteracted this process should not make us abandon this qualified optimism. The walls were not able to totally obstruct the exercise of citizenship, and poor residents continue to expand their rights.

NOTE

In Brazil in 1989, the proportion of income in the hands of the poorest 50 percent of the population was only 10.4 percent. At the same time, the richest 1 percent had 17.3 percent of the income. Data are from the National Research by Domicile Sample (PNAD) undertaken by the Brazilian Census Bureau. The distribution of wealth has become more inequitable since the early 1980s.

"Casinos, Prisons, Incinerators, and Other Fragments of Neoliberal Urban Development"

from *Social Science History* (2011)

Christopher Mele

Editors' Introduction

Neoliberalism is a sociopolitical ideology that advocates strict adherence to principles of private, "free" markets and entrepreneurialism to guide social and economic solutions to longstanding social problems. In cities, neoliberal social policy has led to either the elimination or dramatic transformation of urban governance over public housing, urban renewal, and federal assistance to cities and the introduction of market-based solutions, such as public–private partnerships in urban development. In this selection, Christopher Mele examines the emergence of neoliberal governance strategies around urban fragmentation within the city of Chester, an older, former industrial city in southeastern Pennsylvania. Chester is a poor city by any measure. Its current population is half of what it was in 1950, it ranks high in state levels of street gang activity, assaults, and homicides, bad schools, and poor community health. Yet Chester's recent past reveals some rather peculiar efforts to "turn the city around." Chester's waterfront landscape now includes empty, overgrown lots, a few factories, a state prison, office space in a renovated power plant, a municipal trash incinerator, a gambling casino, and a major league soccer stadium. Mele explains how this odd assortment of enclaves reflects two waves of neoliberal urban development: "roll-back neoliberalism," defined as a period of "chasing smokestacks," and "roll-out neoliberalism," in which governments use public resources and governance powers to promote private-sector-initiated development, such as stadiums and casinos. Both waves have had little or no positive impact upon the lives of the city's marginalized residents. Based on the case of Chester, Mele suggests that neoliberal urban development is not only incapable of resolving the extreme racial and class disparities within older industrial cities, but it furthers the development of exclusionary spaces that, in turn, enhance urban inequalities.

Neoliberal urban development, as a set of governance practices and regulations intended to valorize cities as sites for capital accumulation, brings about increasing social polarization and uneven spatial development at a number of scales within metropolitan areas, often producing "a patchwork quilt of islands of relative affluence struggling to secure themselves in a sea of spreading decay." The appearance of these self-contained, insular enclaves is born of a neoliberal ideology that vilifies comprehensive planning efforts to overcome segregation and celebrates market-based solutions to urban economic development.

Los Angeles is often cited as the exemplar of "fragmented urbanism" but older, former industrial cities have similarly participated in enclave-oriented redevelopment. State power, in particular, has become instrumental to the private market creation of isolated spaces of luxury shopping, entertainment, and housing vigorously disconnected from poor, mostly minority neighborhoods in cities like Baltimore or Camden, New Jersey. The neoliberal push toward state-facilitated market solutions, inter-city competition for capital, and public–private development partnerships has compelled municipalities to institute new spatial strategies that make room for development through enhanced racial and class segregation and exclusion. Within the past few decades, state and municipal authorities have deployed new mechanisms of urban governance that relinquish to private corporations administrative and legal control over how certain areas in the city are developed and used (or "consumed") and by whom – effectively rendering such spaces "cities within cities."

Although much of the literature on neoliberalism is theoretical in scope, recent work has addressed the sociohistorical dimensions of how modes of neoliberal regulation and governance practices take shape. The recent emergence of governance strategies around urban fragmentation within older, former industrial cities serves as a window through which to view the actual and often contradictory workings of neoliberalism, along with the social costs of their implementation.

DISJOINTED DEVELOPMENT IN CHESTER, PENNSYLVANIA

Chester, Pennsylvania is a small city (4.8 square miles) of 36,000 persons located on the Delaware River, just 15 miles south of Philadelphia. Chester is the state's oldest city; it was settled by Swedes and Finns in the 1640s and was briefly home to William Penn, who spent a year there (in part to settle a land dispute) before relocating upriver to present-day Philadelphia. It remained a minor port town until the end of the nineteenth century, when shipbuilding and industrial manufacturing fueled a robust population growth of immigrant workers and a vibrant commercial downtown. Its twentieth century history of industrial decline, suburban outmigration, and racial unrest rings similar to any number of mid-sized cities in the Northeast and the Midwest. Today, Chester is bleak and brutally poor. Its population is half its peak of the 1950s and the majority of its present-day residents are black families (75.7% of total population) whose annual incomes are far below the state's median level. The city's landscape remains, for the most part, desolate – comprised of dilapidated rowhouses, empty storefronts, and abandoned lots. There are a handful of corner grocers (but no supermarket), an auto shop, a pharmacy, and a number of bars, taverns, and liquor stores. It has long ranked top in the state for street gang activity, assaults and homicides, poorly-performing schools, and sagging community health levels – most of which are clustered in the confined space of Chester's poorest neighborhoods. Reflective of a downward spiral since the early 1970s, most Delaware Valley residents have low regard for Chester: they seldom visit it, take pains to avoid driving through it, and actively reproduce its negative reputation (helped along by memories of racial unrest and routine news reports of homicides and unemployment rates). From a casual visitor's perspective, Chester is (tragically) unremarkable – it could be any deindustrialized U.S. city.

Yet Chester's story offers up something peculiar. The eight-lane wide Interstate 95 cuts through the city, marking (if not making) the northeastern boundary of Chester's ghetto. The boundary to the southwest is the city's waterfront and location of its once bustling shipbuilding and factory district, marked by Route 291. Chester's short stretch of waterfront boasts an odd and striking assortment of non-descript warehouses and small factories, garbage-strewn and overgrown lots, an international paper products plant, a state prison, a handful of Colonial-era buildings, Class-A office space in a renovated power plant, a municipal trash incinerator, Harrah's casino and race track, a professional soccer stadium, and plans for an upscale housing, retail and office complex. To the north of Chester lie more plants and factories and an international airport, to the south, one of the largest oil refineries on the east coast. Together with its ghetto, Chester is a city of fragments, separated by widened roads and empty lots, with each of the different enclaves sustaining its own function, economy and demographic, ranging from incarceration, bare subsistence and legalized gambling to noxious industry and insurance underwriting and, if built, high-end consumption.

There is little interaction among the users of these discrete micro-worlds. Other than the ghetto, with its predominantly poor and black residents, the enclaves are the intended domains of visitors-of-sorts: gamblers, prisoners, white-collar workers, truck drivers, laborers, and soccer fans. Gamblers drive along the gloomy Industrial Highway to get to Harrah's – a fortress-like and brightly-lit complex of restaurants, shops, a race track, a parking garage and a 2,700-slot machine gaming floor. Harrah's stands in stark contrast to the surrounding smokestacks, empty lots, and, curiously, a state prison (juxtaposing the jail cell and the slot machine floor and forced temperance and indulgent excess). The nearby waste incinerator is testament to the rampant fiscal desperation of the 1980s and 1990s; the noise of garbage trucks and plumes of smoke remind even the most ardent city boosters this is still Chester. A very short distance south along the waterfront, the cluster of renovated office space, a major league soccer stadium, and ambitious plans for upscale shopping, condominiums, and apartments is named "Rivertown" – a place name which has no historical connection to Chester whatsoever, as its developers gladly intend.

On the surface, the city's turn to the promise of gambling, prison construction, waste management, and a professional sports arena reflect separate and discrete development strategies, each linked to national and international trends in urban economic development (and reflected in their own scholarly literatures on gaming, mass incarceration, and environmental racism). But there is something more here, made obvious by the fact that these different fragments of urban development coexist in a confined space within a small city. And while Chester's curious effort at revitalization certainly reflects its unique and specific history, the underlying forces behind these developments are best understood as localized manifestations of neoliberal urbanism.

The hodgepodge, seemingly disjointed urban landscape of today's Chester is the result of two chronological waves of neoliberal governance strategies. In the first and earlier wave, local government sought to make the most of the city's deindustrialized expanse, an increasingly obsolete labor force, and lack of economic development options by allowing, if not welcoming, private companies to locate noxious industries and unwanted institutions – incinerators and a prison – in former industrial parcels adjacent to poor, black neighborhoods. This first wave of "chasing smokestacks" may be conceived as a localized version of "roll-back neoliberalism" and is largely defined by (1) the active dismantling of urban land use policies rooted in an earlier period of comprehensive physical and social planning and, concurrently, (2) a readiness among governments to use public resources and governance powers to promote private-sector initiated development. "Roll-back neoliberalism" took root in the late 1970s and early 1980s, when states and municipal governments faced federal cutbacks and were compelled to turn to market-based solutions to address long-standing urban problems. An ensuing period of "roll-out neoliberalism" began in the 1990s, when states and local governments developed policies to aggressively attract private capital. Faced with the limitations of chasing smokestacks in an increasingly postindustrial political economy, Chester officials actively courted amusement/consumption-oriented development – casinos and stadiums – in former industrial spaces bordering poor, black communities. State and local governments have "rolled-out" a variety of site-specific development incentive programs, including property tax relief, favorable land use policies, and direct capital assistance, to aggressively assist in the private urban redevelopment of certain targeted areas within the city and not others.

Today, the result of both waves is a landscape of disconnected fragments of redevelopment that coexist awkwardly. Chester's recent past, I contend, provides a supportive case for conceiving neoliberalism as temporally dynamic and continually adaptive – an illustration of both "roll-back" and "roll-out" neoliberal urban development.

CHESTER'S BRIEF RISE AND LONG FALL

Like many older industrial cities located adjacent to water, Chester's industrial rise was tied to shipping; raw materials were delivered to manufacturing plants along the Delaware River and finished goods were loaded on to ships for export. In the late nineteenth and early twentieth century, the riverfront housed numerous light manufacturing industries, producing dyes, artificial silk, men's and women's worsteds and wools, and metal syringes. By the 1920s, Chester's waterfront emerged as a premier location for heavy industry along the eastern United States. Dozens of firms manufactured brick, chemicals, and steel castings and iron tubing, employing tens of thousands of immigrant laborers from Europe. The Sun Shipbuilding and Dry Dock Company (where both Harrah's casino and the state prison now sit) was the city's largest industrial concern at the time. Like other industrial-based cities, the reliance on manufacturing output left Chester susceptible to vagaries in the national economy, including the Great Depression. As in the case of many industrial cities, the twin postwar processes of deindustrialization and suburbanization produced long-term sociospatial effects for Chester, many of which remain visible throughout the city today. Chester's manufacturing economy disappeared incrementally in the decades after World War II. Larger firms relocated to cheaper and bigger parcels in the immediate suburbs, in the South, or, eventually, overseas; no new firms replaced them. As industries along the river shut down, waterfront parcels lay abandoned, polluted, and vacant. Department stores, restaurants, and small commercial businesses vanished from the adjacent downtown business district. Between 1967 and 1972, for example, Chester lost 199 non-industrial establishments (a loss of 1,487 jobs) and 32 industrial facilities (a loss of 3,000 jobs).

As in many industrial cities, factory closings and the accompanying decline in industrial employment occurred roughly in tandem with major demographic changes: mainly, the arrival of a black workforce and the exodus of white residents. Blacks comprised only about 12% of Chester's total population between 1900 and 1920; by 1940 it was 17%. By 1970, blacks comprised 45% of the total population but the city's increasingly large proportion of blacks was primarily an effect of staggering overall population losses, due largely to the departure of whites. The city's population was estimated at more than 80,000 in 1945. However, that number quickly fell after the War, leveling off at nearly 66,000 in 1950, 63,000 in 1960, and falling to 56,000 by 1970.

[...]

Conditions within Chester worsened in the 1970s and 1980s. The city's manufacturing base collapsed, employment levels plummeted and the downtown core and adjacent run-down working-class neighborhoods witnessed rising crime levels, worsening educational opportunities, and increased poverty and destitution. Chester today is a much smaller and poorer city than it was fifty years ago. Located in one of the most affluent counties in Pennsylvania, the city is a pocket of severe deprivation. The number of people currently living in Chester (36,000) is nearly half that of 1950 (66,000). Chester's population is 75.7% black, 5% Latino, and the remaining 18% is white. Twenty-eight percent of the city's residents live below the official poverty level (three times the state average) and the majority of poor residents are clustered in the ghetto adjacent to the former manufacturing district along the waterfront. Over half of the city's residents are renters. Chester's unemployment rate runs double the state average. Thirty-two percent of its residents over the age 25 have not earned a high school diploma.

ROLL-BACK NEOLIBERALISM: CHASING SMOKESTACKS AND PRISONS

Federal subsidies and leadership in urban policy receded by the first half of the 1980s, compelling state and municipal governments to devise different governance arrangements specifically suited to growing local political economic needs and circumstances. Despite differences, emerging forms of urban governance shared a common dimension: municipal policy priorities moved away from redistributive concerns based on expanding social welfare rights towards creating an amenable pro-business government culture to attract increasingly footloose private capital. This ideological-practical shift in governance priorities initially took a form of "rolling-back" – a gradual unraveling of urban spatial regulations and a defunding and dismantling of social welfare protections.

Historically, Chester has long seen its share of environmental problems from industrial production and shipbuilding bleeding into adjacent neighborhoods of working-class row homes. Environmental conditions in Chester's poor and minority neighborhoods worsened in the late 1980s and early 1990s as limited efforts at local urban redevelopment took shape in toxic form. Legislators and government officials partnered with investment firms and private corporations to transform a large portion of Chester's waterfront into a "waste magnet," allowing former industrial land to be developed as a specialized zone for various trash treatment and waste processing plants. By the early 1990s, Chester's waste zone would include a household and light industrial waste incinerator, a wastewater and sewage treatment facility, a medical waste treatment facility, and three waste processing plants.

[...]

Chester's waste zone is an enclave carved out for a specific use through a public–private initiative that generates private profit and some public revenues (through municipal and county operating fees) at substantial community costs. The story of how the waste zone came to be exemplifies "roll-back" neoliberal urban governance. Local, county and state governments fast-tracked land use variances and created a business-friendly environment with little environmental oversight for an industry not typically associated with sustainable urban economic development. But there is money to be made in waste management and the emergence of Chester's waste zone exemplifies a dependence upon public–private partnerships characteristic of neoliberal governance.

The plan for a concentrated zone of waste industries appealed to city and county officials: it held out the ultimately elusive promise of local job creation and offered reuse of mostly contaminated former industrial space. With few revenue-generating development alternatives on hand or looming, officials began to envision Chester's rundown waterfront as an obvious magnet for the waste treatment industry. Local government enthusiasm handed the private waste business consortium a profitable return: where other communities took legislative steps to force out waste treatment plants, Chester welcomed them.

[...]

Shortly after the approval of the incinerator, other waste management corporations approached the city and county with plans to build additional facilities. In a similar fees-based revenue sharing plan with the city, Delaware County officials approved construction of an infectious and chemotherapeutic medical waste autoclave adjacent to the trash-to-steam incinerator. Once built, the plant was the largest medical waste autoclave in the US (it closed in 1996). At the peak of operation in 1993, its owner, Thermal Pure Systems, processed three times the amount of medical waste produced in the entire state of Pennsylvania. Chester and Delaware County officials gave similar approval to plans and permits to construct two contaminated soil remediation sites (incinerators) but these facilities never opened.

[...]

Cooperation between government and the private sector excluded any mechanisms for meaningful citizen participation and dissent. As the number of existing and planned facilities in the waste zone grew, however, residents became increasingly organized. Fed up with noise and odors and increasing concerns over health consequences, residents living adjacent to the waste zone formed the community-based organization, Chester Residents Concerned for Quality Living (CRCQL), in 1992. CRCQL sued Pennsylvania's Department of Environmental Protection (PADEP) in May 1996, mounting a novel legal claim of discrimination under the Civil Rights Act of 1964. The lawsuit held that the state discriminated against the black community when PADEP did not consider the racial makeup of Chester or the number of existing noxious facilities before approving permits for a fifth waste treatment plant. The district court ruled in favor of PADEP and the company that had applied for a permit to open a contaminated soil incinerator, Soil Remediation Services. In 1997, The 3rd Circuit Court of Appeals overturned that ruling in favor of the residents. The U.S. Supreme Court was to hear the case but dismissed it after Soil Remediation Services withdrew its application for a permit to operate.

CRCQL disbanded shortly after the 1998 ruling but the effects of its protest and legal actions were lasting: The movement succeeded in drawing media and legal attention to Chester's toxic cluster of waste treatment facilities. In 1994, CRCQL successfully pressured Chester's city council to pass an ordinance that required new companies wishing to locate in Chester to certify that they would not contribute to a net increase in pollution levels in the area. The ordinance proved instrumental to a shift away from toxic development in the late 1990s, as described in the later section on "roll-out" neoliberalism.

In addition to chasing smokestacks, Chester officials turned to another unconventional form of urban development that promised to deliver "jobs and community development": housing prisoners. In the past two decades, poor, mostly rural communities across the US have often turned to prison construction and operation as a subsidized solution to local economic development problems and unemployment. There is little evidence that prisons stimulate direct growth or generate positive spillover effects within communities. To the contrary, the construction and operation of a new prison seems to hinder other development possibilities and tends to reinforce local preexisting patterns of spatial class and racial inequalities.

ROLL-OUT NEOLIBERALISM: THE PROMISE OF CASINOS, SPORTS STADIUMS AND UPSCALE DEVELOPMENT

In their periodization of neoliberalism, Peck and Tickell (2002) conceive governance strategies as continually adapting to circumstances on the ground and responding to both opportunities and challenges as cities and nation-states compete for private capital. In their conceptual ideal, "roll-out" neoliberalism – characterized by proactive and novel forms of governance – succeeded "roll-back" neoliberalism in the mid 1990s. Governments increase alliances and partnerships with the private sector and, in effect, institutionalize governance through and by the market. Cities no longer simply accommodate private development requests; they often act as the driving force of private redevelopment.

Although created by local governments, with few exceptions local development authorities operate outside the framework of direct local government control; they are accountable less to constituents and their representatives than to the successful achievement of a generalized directive to increase local economic development. Development authorities routinely leverage private investment capital around particular development projects but they do much more: They initiate development schemes with private developers, coordinate city, county, state, and federal government assistance, assemble real estate parcels and promote a city's use of eminent domain and similar takeover powers when necessary, co-author project designs, assure fast-tracking of local permits and approvals, secure capital by offering government coffers as collateral (bond sales), obtain public subsidies through tax abatements, and deliver infrastructure improvements (sewers, roads, sidewalks). Local investment and business communities strongly supported the administrative consolidation of development planning and assistance outside the rules-driven policy oversight of municipal government. Development authorities are not charged as impartial intermediaries between citizens and corporations but as representatives of business interests that, in turn, become city interests.

Chester followed the lead of larger cities and created the Chester Economic Development Authority (CEDA) in 1994 to serve as a quasi-non-governmental administrative agent for economic and community development programs and activities. The authority works directly with city agencies in seeing a redevelopment project through from inception to completion. CEDA benefits from its political legitimacy as a municipal creation and has been granted a broad mandate to define and execute development project choices. The authority also oversees state economic and community development funds and two key development incentive programs generated by the state and municipal governments: the Keystone Opportunity Zone (KOZ) program, which resembles enterprise zones in place in most U.S. states, and the Local Economic Revitalization Tax Assistance Act (LERTA), a state tax incentive program.

The Keystone Opportunity Zone (KOZ) Program is a state-initiated tax abatement and incentive program for private redevelopment of specific zones of abandoned, unused, or underutilized land and buildings throughout Pennsylvania. KOZs are managed by local development authorities, like CEDA, which follow state guidelines to locate eligible districts within a city, seek private developers, and assist in the design of projects. The KOZ program provides full tax forgiveness until 2013 for municipal, school district, and county real estate taxes. Corporations are also exempt from taxes on earned income/net profits, business gross receipts, sales and use, wage and net profits, realty use and occupancy. KOZs appeal to large private development corporations because they offer large tracts of "shovel-ready" land for development, free of any existing cultural, social, legal, or political privileges, claims or entitlements, along with considerable tax savings. Delaware County and the City of Chester have offered additional tax incentives to corporate relocation, development planning and construction in KOZs. The Chester municipal government opted to implement the state's Local Economic Revitalization Tax Assistance Act (LERTA) for companies involved in the revitalization of deteriorated properties in specific areas in the city. LERTA allows private companies to defer increases in real estate taxes on the value of the improvements over ten years.

Chester's KOZs favor a donut-shaped pattern of development, in which newer, upscale sites ring the poorest residential neighborhoods and the defunct central business district. The zones are also not contiguous, leading toward fragmented redevelopment. Newer developments, then, are disconnected from the urban core, allowing – if not encouraging – developers to define their projects as distinct from the city proper. The location of Chester's KOZs facilitates what many developers consider as significant as the subsidies and tax-breaks: a project's physical separation from the ghetto. A project's success hinges upon the degree to which it is isolated and protected from the ghetto and its reputation as a dangerous space. Two recent KOZ developments along the Chester waterfront – Harrah's casino and racetrack and the Rivertown soccer stadium and mixed-use complex – best illustrate how physical separation from the ghetto makes social and cultural boundaries feasible as well.

[. . .]

The $400 million Harrah's Chester casino and racetrack opened in 2007. The casino features a 100,000 square foot gaming floor with 2,700-slot and video poker machines, 5/8-mile harness racetrack with 1,500-seat grandstand, a buffet, 24-hour restaurant and a 300-seat clubhouse dining area. The racetrack-casino operates 24 hours a day and brings in about 8,500 visitors to Chester every day – the majority of whom live outside the city. The location of the casino development qualified for KOZ designation. As a result, Harrah's benefited from a range of state and local tax breaks through 2013, including local property tax relief. Harrah's saved over $8 million in taxes during the construction phase of the $429 million complex alone. If the casino's owners should decide to sell it before 2013, they pay no tax on capital gains.

When Harrah's announced its intention to build and operate the casino in a Chester KOZ, many county and state legislators were troubled by the idea of gambling companies profiting from the tax-free zones. Yet the state law that produced KOZ stipulates the tax incentives be attached to the land itself, rather than specific kinds of taxpaying corporations or particular types of industries. CEDA was instrumental in bringing Harrah's to Chester. The agency worked with the casino industry and state officials to overcome legal restrictions and helped wage a public relations campaign to assuage concern about gambling and the negative effects of a casino's presence to the community. CEDA, Chester's mayor, and other city leaders portrayed Harrah's as Chester's best and last hope for breaking with its recent past and downbeat reputation as a poor, crime-ridden city. In response to the casino's critics, CEDA predicted that Harrah's Chester would create a multiplier effect in the adjacent deprived neighborhoods, attracting new restaurants, gas stations and numerous other tourist-driven businesses. While no new businesses have appeared nearby, Harrah's reportedly uses local vendors for some of its restaurant supplies. The company also endowed a $200,000 college scholarship for Chester students in 2008. CEDA correctly reported that the casino would generate close to $10 million in annual revenue for the city (roughly 20% of Chester's annual budget). Prior to the casino's opening, Harrah's sponsored job fairs within Chester to recruit for over 750 positions.

While city leaders unabashedly predicted community-wide benefits from the casino, others questioned the anticipated spill-over effects based on the experience of gambling and urban development in Atlantic City, where the impact has been overwhelmingly negative.

[. . .]

Just south of Harrah's is "Rivertown," a multi-year, multi-million-dollar private development of a large office complex, a public access pedestrian walkway along the Delaware River, and a soccer stadium for the Philadelphia major league soccer franchise. Developers first took advantage of the 90-acre KOZ designation in the purchase and complete renovation of the former Chester Station Power Plant into an office and recreation complex called the Wharf at Rivertown. The power plant stood empty for decades and its owner, Philadelphia Electric Company (PECO) had planned to demolish it until the site earned KOZ status in 1999. In October 2000, Preferred Real Estate Investments Inc. purchased the plant from PECO for $1 with the provision that the new owner tackle interior environmental cleanup. Chester's mayor – seeing the Wharf as a showcase of the city's potential – helped Preferred secure nearly $1.1 million in public subsidies. The state provided $2.6 million in grants and loans to Preferred for infrastructure development, land reclamation, and installation of fiber-optic cable. Then-Senator Rick Santorum sought and won $15 million in federal assistance, mainly for a direct highway link between Rivertown and the Commodore Barry Bridge that allows workers (as well as casino and stadium visitors) to bypass downtown Chester and avoid seeing or interacting with its residents.

[. . .]

In March 2008, Philadelphia was awarded the league's sixteenth franchise with its home stadium in Chester's Rivertown. The development of the Rivertown soccer stadium and planned residential and commercial district is a public–private endeavor. Keystone Sports and Entertainment LLC, a conglomerate enterprise of operating partners and the soccer franchise among others, constructed the 18,634-seat stadium. Private investor funding for the stadium hinged on a $30 million bond from Delaware County, $10 million from the Delaware River Port Authority for waterfront improvements, and an anticipated $25 million in state aid (and an additional $400 million slated for future on-site mixed-use development). Delaware County led negotiations over lease options and acquired the land for the stadium. The second phase of the development calls for 180 town homes, 225 apartment units, 42,000 square feet of retail space, 435,000 square feet of new office space and a 200,000 square feet convention center. A community boat marina and spacious greenways complete the developer's plans for the 70 acre site.

[. . .]

CONCLUSION

Drawing on the work of Peck and Tickell (2002), Chester's disjointed redevelopment landscape is best explained not as a result of haphazard or bad city planning but as of the result of two successive waves of neoliberal urbanism: "roll-back" and "roll-out" governance arrangements that favor exclusionary redevelopment. In chronological terms, municipal governments increased alliances and partnerships with the private sector to redefine and institutionalize redevelopment as market-driven and territorial- and project-based. Following, cities no longer simply accommodated private development needs and requests; they developed new policies that initiate and drive private redevelopment using an arsenal of tax cuts and direct subsidies. The programs and policies identified as "roll-back" and "roll-out" together mark a neoliberal mode of governance: administrative and fiscal control is largely ceded to private developers acting in conjunction with local development agencies, public accountability and participation is limited, and development is targeted to sublocal enclaves rather than the entire city.

The consequences of neoliberal urban development for the majority of poor, disenfranchised residents are mostly negative. Discursive efforts to include poor minority residents in Chester's "renaissance" appear routinely on websites, such as "Chester Yes" and "Kickstart Chester," and promotional brochures that accompany announcements of new development projects. Substantial revenue flows from Harrah's Casino into city coffers. Yet the casino's presence has generated little tangible effect on either the surrounding streets of dilapidated homes and empty lots or the job opportunities of Chester's residents. City and county officials and Rivertown's developers have touted the future benefits of the soccer stadium and the future residential–office–shopping complex. After years of waiting for the casino's spill-over benefits to materialize, many locals are skeptical. While some residents are hopeful the economic investment in the waterfront will pay, others question whether the benefits of redevelopment will include them.

Neoliberal urban governance suggests two related meanings of the "dividing line" between the fragmented developments along the waterfront and Chester's adjacent ghetto. First, the developments are such that most of Chester's residents are excluded. The normative ideal of a city as a space that accommodates social, class, racial, sexual, and gender differences is rendered untenable when urban landscapes are carved up into self-contained fragments that are designed for specific social groups and types of behaviors and not (many) others. The fragmented city, therefore, allows for a delimited set of specialized statuses and subjectivities for users of each of its different enclaves or zones at particular times: the gambler in the casino, the sports fan at the soccer stadium, the white-collar worker in the office complex, and the marginalized prisoner and the ghetto residents in their respective carceral and carceral-like spaces. State support for enclave-oriented private redevelopment (and subsequent disregard for more holistic, community-wide development) reinforces and normalizes longstanding patterns of class exclusion and racial segregation. Consequently, deep-rooted urban social and economic problems that pertain to large segments of the city are no longer addressed, nor are there public efforts to attempt to solve them. Instead, disagreeable conduct and unwelcome activities are zoned out of exclusionary spaces where regulation of social behavior (and, in turn, subjects) is privatized. Access to and from specialized enclaves, such as the casino or the stadium, free movement within their adjacent "public" spaces, and the range of social activities prescribed to occur within such spaces are monitored and controlled; private security, surveillance cameras, and other regulatory mechanisms effectively weed out not only generally offensive behavior but many other activities (loitering, organizing, non-consuming) that fall outside a zone's stated purpose.

Rhetorically, city leaders and developers have stressed that the casino, the soccer stadium and the proposed residential and entertainment complex will spur development in the surrounding city (across the dividing line of Route 291). But the mechanisms that now govern redevelopment – project-specific public–private ventures with highly specific funding sources – makes a spillover effect improbable. KOZs, for example, focus on increasing economic activity within their boundaries, not the entire city. There are no development incentives for small shop keepers or owners of small parcels surrounding Harrah's or Rivertown. Indeed, neoliberal urban development is structurally incapable of stitching together the disparate enclaves it produces. Rather, the opposite is in effect. Extreme disparities in wealth and income are not bridged by exclusionary enclave development; they are further legitimized, naturalized, and made to appear inevitable.

REFERENCE

J. Peck and A. Tickell (2002) "Neoliberalising Space," *Antipode* 34(3), 380–404.

"Spaces of Insurgent Citizenship"

from Leonie Sandercock (ed.) *Making the Invisible Visible* (1998)

James Holston

Editors' Introduction

The shape and form of many cities today are largely the products of a modernist era in which technology, rational planning, and the state's careful governance of urban space would eventually rid society of many social ills, including overcrowding, density, and the unregulated development of the built environment. Explicit in this modernist notion of city development is the core belief that planning and governance can change society for the better by envisioning an alternative better future. As James Holston tells us, the idea of planning is "central to the identity of the modern state: it motivates political authorities to attempt to create and legitimate new kinds of public spheres, with new subjects and subjectivities for them." So, as Holston informs us, rational planning not only shapes the built environment of cities but attempts to influence and form the identities or subjectivities of those who live there. But this effort to define citizenship is not complete as opposition to modernist urbanism produces opportunities for an insurgent citizenship. By insurgent, Holston is referring to the opposition to the modernism that, in effect, "absorbs citizenship into a plan of state building." Insurgent citizenship embodies alternative futures to the modernist notion of normative order. This is because they are new metropolitan forms of social interaction not yet absorbed by the modernist doctrine. Holston recognizes the importance of including the ethnographic present into city planning. By this he means "the possibilities for chance encountered in existing social conditions." Holston's audience includes planners, whom he advises to engage with the new, or insurgent, ideas that emerge from grassroots mobilization and everyday practices that empower, parody, and even derail state agendas. Examples cited include transnational networks of immigrant urban dwellers, Queer Nation, ganglands, employee-owned factories, and zones of a "new racism." These examples are labeled insurgent because they introduce into the cities new identities that may not coincide with existing histories and planning agendas.

James Holston is a Professor and Socio-Cultural House Head in Sociocultural Anthropology at University of California, Berkeley. His areas of interest include cities and citizenship, planning and architecture, and urban ethnography. He is author of *Insurgent Citizenship: Disjunctions of Democracy and Modernity in Brazil* (Princeton, NJ: Princeton University Press, 2008) and *Cities and Citizenship* (Durham, NC: Duke University Press, 1999).

Cities are plugged into the globe of history like capacitors: they condense and conduct the currents of social time. Their layered surfaces, their coats of painted stucco, their wraps of concrete register the force of these currents both as wear and as narrative. That is, city surfaces tell time and stories. Cities are full of stories in time, some sedimented and catalogued; others spoorlike, vestigial, and dispersed. Their narratives are epic and everyday; they tell of migration and production, law and laughter, revolution and art. Yet, although obvious, their registry is never wholly legible because each foray into the palimpsest of city surfaces reveals only traces of these relations. Once lived as irreducible to one another, they are registered as part of the multiplicity and simultaneity of processes that turn the city into an infinite geometry of superimpositions. Their identities, modes, forms, categories, and types recombine in the gray matter of streets. City narratives are, as a result, both evident and enigmatic. Knowing them is always experimental.

[. . .]

I want to argue that one of the most urgent problems in planning and architectural theory today is the need to develop a different social imagination – one that is not modernist but that nevertheless reinvents modernism's activist commitments to the invention of society and to the construction of state. I suggest that the sources of this new imaginary lie not in any specifically architectural or planning production of the city but rather in the development of theory in both fields as an investigation into what I call the spaces of insurgent citizenship – or insurgent spaces of citizenship, which amounts to the same thing. By insurgent, I mean to emphasize the opposition of these spaces of citizenship to the modernist spaces that physically dominate so many cities today. I also use it to emphasize the opposition of these spaces of citizenship to the modernist political project that absorbs citizenship into a plan of state building and that, in the process, generates a certain concept and practice of planning itself. At the heart of this modernist political project is the doctrine – also clearly expressed in the tradition of civil or positivist law – that the state is the only legitimate source of citizenship rights, meanings, and practices. I use the notion of insurgent to refer to new and other sources and to their assertion of legitimacy.

THE ALTERNATIVE FUTURES OF MODERNISM

The spaces of an insurgent citizenship constitute new metropolitan forms of the social not yet liquidated by or absorbed into the old. As such, they embody possible alternative futures. It is important to distinguish this concept of the possible from the fundamentally different idea of alternative futures inherent in modernist planning and architectural doctrine. Both express the basic paradigm of modernity that emphasizes that alternative futures are indeed possible. But the insurgent and the modernist are competing expressions, which I will distinguish as ethnographic and utopian, respectively. In modern architecture and urban design, manifestos have called for the state to assert the priority of collective interests over private interests by imposing on the chaos of existing cities the construction of a new type of city based on its master plans. But that model derives in turn from the pervasive ideal of modernity that the state, usually in the form of a national government, can change society and manage the social by imposing an alternative future embodied in the plans. In this sense, the project of modernist planning is to transform an unwanted present by means of an imagined future. Whether in the form of urban design or applied social science, this idea of planning is central to the identity of the modern state: it motivates political authorities to attempt to create and legitimate new kinds of public spheres, with new subjects and subjectivities for them. The instruments of these initiatives define not only the development agenda of the state but also its accredited liberal profession and social sciences – architecture, urban design, demography, bureaucratic administration, sociology, criminology, epidemiology, and so forth – by means of which governments try to forge new forms of collective association and personal habit as the basis of propelling their societies into a proclaimed future.

This ideology of planning is utopian not because it is critical of the present or because it has as its objective the disruption of taken-for-granted norms. It shares these characteristics with the ethnographic mode I propose. Rather, it is utopian because its notion of alternative futures is based on absent causes and its methods on a theory of total decontextualization.

[. . .]

[It] also relies on a subjective transformation of existing conditions. In this case, borrowing from other avant-garde movements of the early twentieth century, it uses techniques of shock to force a subjective appropriation of the new social order inherent in its plans. These techniques emphasize decontextualization, defamiliarization, and dehistoricization. Their central premise of transformation is that the new architecture/urban design would create set pieces within existing cities that would subvert and then regenerate the surrounding fabric of denatured social life. El Lissitzky explained this premise concisely in 1929: "The introduction of new building types into the old fabric of the city affects the whole by transforming it" (1970: 52). It is a viral notion of revolution, a theory of decontextualization in which the radical qualities of something totally out of context infect and colonize that which surrounds it. This something may be a single building conceived as an instance of the total plan, that is, as a fragment of its radical aesthetics and social practices. Or it may be an entire city designed as an exemplar, as in the case of Brasilia. Either way, the radical fragment is supposed to create new forms of social experience, collective association, perception, and personal habit. At the same time, it is supposed to preclude those forms deemed undesirable by negating previous social and architectural expectations about urban life.

This use of decontextualization ultimately springs from the conviction that it is possible to extract antithetically from existing conditions an absent ideal as a new positive entity. This extraction is achieved, in other words, through subjective synthesis. Such synthesis is reached through the shock of defamiliarization during which the subject identifies with the ideal in the dialectic as the means necessary to bridge the now evident gap between his or her local and splintered situation and the proposed future plenitude.

[. . .]

However, if few promises for change have captured the world's imagination to a greater degree than this idealist project of alternative futures, few have yielded greater perversity. A fundamental dilemma inevitably dominates this project if it is to have any substance beyond the imaginary world of plans. It is one inherent in all forms of planning – both as urban design and as applied social science – that propose an alternative future based on absent totalities: the necessity of having to use what exists to achieve what is imagined destroys the utopian difference between the two that is the project's premise. Worse, examples such as Brasilia show that attempts to maintain the plan in spite of the corrosive effects of this utopian paradox exacerbate the very conditions that generate the desire for change. Perversely, they tend to turn the project into an exaggerated version of what its planners wanted to preclude in the first place.

[. . .]

The imagined future raises a further dilemma. On the one hand, it always runs the risk of the utopian paradox I just described: either it remains without substance and thus disconnected from the conditions that generate a desire for it; or, in gaining history, it exacerbates the very issues it intends to negate. On the other hand, a second conclusion is also apparent: without a utopian factor, plans remain locked in the prison house of unacceptable existing conditions. Is not the elimination of the desire for a different future as oppressive as the modernist perversion of it? To exclude the imaginary and its inherently critical perspective in that way is to condemn planning to accommodations of the status quo, and I reject such paralysis. Hence, a difficult question remains: if the notion of alternative futures is both indispensable and yet, in its utopian form, perverse, what kind of intervention in the city could construct a sense of emergence without imposing a teleology that disembodies the present in favor of a utopian difference?

INSURGENT CITIZENSHIP

My criticism of modernist planning is not that it presupposes a nonexistent egalitarian society or that it dreams of one. To deny that dream is also to conceal or encourage a more totalitarian control of the present. It is rather that modernist planning does not admit or develop productively the paradoxes of its imagined future. Instead, it attempts to be a plan without contradiction, without conflict. It assumes a rational domination of the future in which its total and totalizing plan dissolves any conflict between the imagined and the existing society in the imposed coherence of its order. This assumption is both arrogant and false. It fails to include as *constituent* elements of planning the conflict, ambiguity, and interdeterminacy characteristic of actual social life. Moreover, it fails to consider the unintended and the unexpected as part of the model. Such assumptions are common to master plan solutions generally and not only to those in urban planning. Their basic feature is that they attempt to fix the future – or the past, as in historical preservation – by appealing to the precedents that negate the value of present circumstance. The crucial question for us to consider, therefore, is how to include the ethnographic present in planning, that is, the possibilities for change encountered in existing social conditions.

[...]

This frame of complementary perspectives offers an important suggestion for thinking about a new production of the city. If modernist planning relies on and builds up the state, then its necessary counteragent is a mode of planning that addresses the formations of insurgent citizenship. Planning theory needs to be grounded in these antagonistic complements, both based on ethnographic and not utopian possibility: on one side, the project of state-directed futures which can be transformative but which is always a product of specific politics; and, on the other, the project of engaging planners with the insurgent forms of the social which often derive from and transform the first project but which are in important ways heterogeneous and outside the state. These insurgent forms are found both in organized grassroots mobilization and in everyday practice that, in different ways, empower, parody, derail, or subvert state agendas. They are found, in other words, in struggles over what it means to be a member of the modern state – which is why I refer to them with the term citizenship. Membership in the state has never been a static identity, given the dynamics of global migrations and national ambitions. Citizenship changes as new members emerge to advance their claims, expanding its realm, and as new forms of segregation and violence counter these advances, eroding it. The sites of insurgent citizenship are found at the intersection of these processes of expansion and erosion.

These sites vary with time and place. Today, in many cities, they include the realm of the homeless, networks of migration, neighborhoods, or Queer Nation, constructed peripheries in which the poor build their own homes in precarious material and legal conditions, ganglands, fortified condominiums, employee-owned factories, squatter settlements, suburban migrant labor work camps, sweatshops, and the zones of so-called new racism. They are sites of insurgence because they introduce into the city new identities and practices that disturb established histories. These new identities and the disturbances they provoke may be of any social group, elite or subaltern. Their study views the city as not merely the container of this process but as its subject as well – a space of emergent identities and their social organization. It concentrates on practices that engage the problematic nature of belonging to society. It privileges such disturbances, emergences, and engagements because it is at the fault lines of these processes that we perceive the dynamism of society. This perception is quite different, however, from a sociological accretion of data, and its register includes the litter and not only the monuments of urban experience.

The dynamism and its perception are the theoretical objectives of a planning linked to insurgent forms of the social. It differs from the modernist objectives of planning because it aims to understand society as a continual reinvention of the social, the present, and the modern and their modes of narrative and communication. What planners need to look for are the emergent sources of citizenship – and their repression – that indicate this invention. They are not hard to find in the wake of this century's important processes of change: massive migration to the world's major cities, industrialization and deindustrialization, the sexual revolution, democratization, and so forth. The new spaces of citizenship that result are especially the product of the compaction and reterritorialization in cities of so many new residents with histories, cultures and demands that disrupt the normative and assumed categories of social life. This disruption is the source of insurgent citizenship and the object of a planning theory that includes the ethnographic present in its constitution.

The distinction between formal and substantive citizenship is useful in identifying this object because it suggests how the forms of insurgent citizenship appear as social practice and therefore how they may be studied. Formal citizenship refers to membership in a political community – in modern history, preeminently, the nation-state. Substantive citizenship concerns the array of civil, political, and social rights available to people.

[...]

In many cases formal citizenship is neither a necessary nor a sufficient condition for substantive citizenship. In other words, although in theory full access to rights depends on membership, in practice that which constitutes citizenship substantively (rights and duties) is often independent of its formal status. Indeed, it is often inaccessible to those who are formal citizens (e.g., the native poor), yet available to those who are not (e.g., legally resident "aliens"). These kinds of problems challenge the dominant notion of citizenship as national identity and the historic role of the nation-state as the preeminent form of modern political community.

But in so doing, they indicate a new possibility that could become an important focus for urban planning: they suggest the possibility of multiple citizenships based on the local, regional, and transnational affiliations that aggregate in contemporary urban experience. Although this possibility represents a significant change in the recent history of citizenship, it is not a new arrangement. Multiple and overlapping jurisdictions predominated in Europe until the triumph of national citizenship obliterated other forms, among them the urban citizenships that organized so many regions of the ancient and the premodern world. The modern state explicitly competed with the city for the primary affiliation of its citizens. Moreover, it usurped their differences, replacing the local management of history with the national. That is, the state recognized local diversity under the banner of national heritage. One of the most widely shared projects of modern states, this nationalization of diversity legitimates a singular state citizenship as the best condition for securing a society of plural cultural identities. But the recent worldwide manipulation of "rights to difference" movement profoundly challenges this claim. Their new ethnocultural politics and violence are in large part a response to the perceived failures of a singular national citizenship. In this reevaluation, the local and the urban reappear as the crucial sites for articulating not only new fanaticisms and hooliganisms but also the new transnational and diasporic identities. If planning theory, as I suggest, can conceptualize this collision between state citizenship and these insurgent alternatives, planning practice can respond to this articulation first by expressing its heterogeneity – the social condition we actually live in – and then by developing some of the ethnographic possibilities that are, by definition, embedded in heterogeneous conditions.

[...]

PLANNING THE ETHNOGRAPHICALLY POSSIBLE

In this essay, I have raised the problem of developing a new social imagination in planning and architecture. I have suggested that when citizenship expansions and erosions focus on urban experience, they constitute an insurgent urbanism that informs this development in several ways. First, they present the city as both the text and the context of new debates about fundamental social relations. In their localism and strategic particularism, these debates valorize the constitutive role of conflict and ambiguity in shaping the multiplicity of contemporary urban life. In a second sense, this heterogeneity works against the modernist absorption of citizenship into a project of state: building, providing alternative, possible sources for the development of new kinds of practices and narratives about belonging to and participating in society. This "working against" defines what I called an insurgent citizenship; and its spatial mode, an insurgent urbanism. This insurgence is important to the project of rethinking the social in planning because it reveals a realm of the possible that is rooted in the heterogeneity of lived experience, which is to say, in the ethnographic present and not in the utopian futures.

But in advocating a move to the ethnography of the present, I do not suggest that planning abandon the project of state building that modernist doctrine defines and that is basic to the notion of modernity itself. Excessive attention to the local has its own dangers. Although I argue, for example, that ethnographic investigation is the best way to establish the terms by which residents participate in the planning of their communities, such participation can be paradoxical: residents across the economic spectrum will often decide, by the most democratic of processes, to segregate their communities "from the evil outside," closing, fortifying, and privatizing their spaces in relation to those deemed outsiders. Hence, throughout the United States, it is common to find home-owner associations trying to use the powers and privileges of democratic organization to exclude and discriminate. Local enactments of democracy may thereby produce antidemocratic results.

The lesson of this paradox is that planning needs to engage not only the development of insurgent forms of the social, but also the resources of the state to define, and occasionally impose, a more encompassing conception of right than is sometimes possible to find at the local level. An example of this transformative power of the state comes from the conflict over legal segregation in the southern United States during the 1960s when the federal government eventually intervened in local affairs and acted against local authorities. Above all, planning needs to encourage a complementary antagonism between these two engagements. It needs to operate simultaneously in two theaters, so to speak, maintaining a productive tension between the apparatus of state-directed futures and the investigation of insurgent forms of the social embedded in the present.

In developing the latter as the counter of the former, planners and architects engage a new realm of the possible with their professional practice. But this realm requires a different kind of practice, different in both objective and method, and this difference amounts to reconceptualization of the fields. In terms of methods, I mean to emphasize those of an urban ethnographer – or of a detective, which are similar: methods of tracing, observing, decoding, and tagging, at one moment of the investigation, and those of reconstructing, identifying, presenting, and rearticulating, at another. Both the trace and the reconstruction compose this engagement with the ethnographic present.

[. . .]

"The Right to the City"

from *International Journal of Urban and Regional Research* (2003)

David Harvey

Editors' Introduction

Cities are not simply the products of human decisions and actions; cities shape most aspects of our daily lives. But to what degree does the average city dweller have any say over how the city provides for her needs? Who has the right to the city? By the "right to the city," Harvey is referring to claims made by social groups possessing unequal power to shape and have say over the process of urbanization (as discussed more fully in the earlier reading by Harvey). From the onset of the earliest cities, urbanization has been a class phenomenon, with the wealthier classes controlling the shape and form of the urban landscape. Throughout modern history, the right to the city has been entwined with property rights and related economic interests tied to the accumulation of capital. But the free market has (re)produced only greater disparities between socioeconomic classes, as well as producing monopoly interests with greater levels of power. Today, the right to the city is held largely in the hands of political and economic elites who are in the position to shape the city according to their own interests. Harvey rejects the inevitability of this condition and calls for a different right to the city to be asserted. His prescriptions are briefly outlined in the reading. Changes can be brought to the city, but they require political organization and demonstration (sometimes brought to the streets of the city itself). With social action, city dwellers come to realize the significance of the right to the city trumps the right to accumulate capital (and reproduce inequality) held among elites. To create or imagine this new city Harvey implores that we must not just reimagine the city itself but the politics and economics within the city.

The city, the noted urban sociologist Robert Park once wrote, is:

> man's most consistent and on the whole, his most successful attempt to remake the world he lives in more after his heart's desire. But, if the city is the world which man created, it is the world in which he is henceforth condemned to live. Thus, indirectly, and without any clear sense of the nature of his task, in making the city man has remade himself.

The right to the city is not merely a right of access to what already exists, but a right to change it after our heart's desire. We need to be sure we can live with our own creations (a problem for every planner, architect and utopian thinker). But the right to remake ourselves by creating a qualitatively different kind of urban sociality is one of the most precious of all human rights. The sheer pace and chaotic forms of urbanization throughout the world have made it hard to reflect on the nature of this task. We have been made and remade without knowing exactly why, how, wherefore and to what end. How then can we better exercise this right to the city?

The city has never been a harmonious place, free of confusions, conflicts, violence. Only read the history of the Paris Commune of 1871, see Scorsese's fictional depiction of gangs of New York in the 1850s, and think how far we have come. But then think of the violence that has divided Belfast, destroyed Beirut and Sarajevo, rocked Bombay, even touched the "city of angels". Calmness and civility in urban history are the exception not the rule. The only interesting question is whether outcomes are creative or destructive. Usually they are both: the city is the historical site of creative destruction. Yet the city has also proven a remarkably resilient, enduring and innovative social form.

But whose rights and whose city? The communards of 1871 thought they were right to take back "their" Paris from the bourgeoisie and imperial lackeys. The monarchists who killed them thought they were right to take back the city in the name of God and private property. Both Catholics and the Protestants thought they were right in Belfast as did Shiv Sena in Bombay when it violently attacked Muslims. Were they not all equally exercising their right to the city? "Between equal rights", Marx once famously wrote, "force decides". So is this what the right to the city is all about? The right to fight for one's heart's desire and liquidate anyone who gets in the way? It seems a far cry from the universality of the UN Declaration on Human Rights. Or is it?

Marx, like Park, held that we change ourselves by changing our world and vice versa. This dialectical relation lies at the root of all human labor. Imagination and desire play their part. What separates the worst of architects from the best of bees, he argued, is that the architect erects a structure in the imagination before materializing it upon the ground. We are, all of us, architects, of a sort. We individually and collectively make the city through our daily actions and our political, intellectual and economic engagements. But, in return, the city makes us. Can I live in Los Angeles without becoming a frustrated motorist?

We can dream and wonder about alternative urban worlds. With enough perseverance and power we can even hope to build them. But utopias these days get a bad rap because when realized they are often hard to live with. What goes wrong? Do we lack the correct moral and ethical compass to guide our thinking? Could we not construct a socially just city?

But what is social justice? Thrasymachus in Plato's *Republic* argues that each form of government enacts the laws with a view to its own advantage so that the just is the same everywhere, the advantage of the stronger. Plato rejected this in favor of justice as an ideal. A plethora of ideal formulations now exist. We could be egalitarian, utilitarian in the manner of Bentham (the greatest good of the greatest number), contractual in the manner of Rousseau (with his ideals of inalienable rights) or John Rawls, cosmopolitan in the manner of Kant (a wrong to one is a wrong to all), or just plain Hobbesian, insisting that the state (Leviathan) impose justice upon reckless private interests to prevent social life being violent, brutal and short. Some even argue for local ideals of justice, sensitive to cultural differences. We stare frustratedly in the mirror asking: "which is the most just theory of justice of all?" In practice, we suspect Thrasymachus was right: justice is simply whatever the ruling class wants it to be.

Yet we cannot do without utopian plans and ideals of justice. They are indispensable for motivation and for action. Outrage at injustice and alternative ideas have long animated the quest for social change. We cannot cynically dismiss either. But we can and must contextualize them. All ideals about rights hide suppositions about social processes. Conversely, social processes incorporate certain conceptions of rights. To challenge those rights is to challenge the social process and vice versa. Let me illustrate.

We live in a society in which the inalienable rights to private property and the profit rate trump any other conception of inalienable rights you can think of. This is so because our society is dominated by the accumulation of capital through market exchange. That social process depends upon a juridical construction of individual rights. Defenders argue that this encourages "bourgeois virtues" of individual responsibility, independence from state interference, equality of opportunity in the market and before the law, rewards for initiative, and an open market place that allows for freedoms of choice. These rights encompass private property in one's own body (to freely sell labor power, to be treated with dignity and respect and to be free from bodily coercions), coupled with freedoms of thought, of expression and of speech. Let us admit it: these derivative rights are appealing. Many of us rely heavily upon them. But we do so much as beggars live off the crumbs from the rich man's table. Let me explain.

To live under capitalism is to accept or submit to that bundle of rights necessary for endless capital accumulation. "We seek", says President Bush as he goes to war, "a just peace where repression, resentment and poverty are replaced with the hope of democracy, development, free markets and free trade". These last two have, he asserts, "proved their ability to lift whole societies out of poverty". The United States will deliver this gift of freedom (of the market) to the world whether it likes it or not. But the inalienable rights of private property and the profit rate (earlier also embedded, at US insistence, in the UN declaration) can have negative, even deadly, consequences.

Free markets are not necessarily fair. "There is", the old saying goes, "nothing more unequal than the equal treatment of unequals". This is what the market does. The rich grow richer and the poor get poorer through the egalitarianism of exchange. No wonder those of wealth and power support such rights. Class divisions widen. Cities become more ghettoized as the rich seal themselves off for protection while the poor become ghettoized by default. And if racial, religious and ethnic divisions cross-cut, as they so often do, with struggles to acquire class and income position, then we quickly find cities divided in the bitter ways we know only too well. Market freedoms inevitably produce monopoly power (as in the media or among developers). Thirty years of neoliberalism teaches us that the freer the market, the greater the inequalities and the greater the monopoly power.

Worse still, markets require scarcity to function. If scarcity does not exist, then it must be socially created. This is what private property and the profit rate do. The result is much unnecessary deprivation (unemployment, housing shortages, etc.) in the midst of plenty. Hence, the homeless on our streets and the beggars in the subways. Famines can even occur in the midst of food surpluses.

The liberalization of financial markets has unleashed a storm of speculative powers. A few hedge funds, exercising their inalienable right to make a profit by whatever means, rage around the world, speculatively destroying whole economies (such as that of Indonesia and Malaysia). They destroy our cities with their speculations, reanimate them with their donations to the opera and the ballet while, like Kenneth Lay of Enron fame, their CEOs strut the global stage and accumulate massive wealth at the expense of millions. Is it worth the crumbs of derivative rights to live with the likes of Kenneth Lay?

If this is where the inalienable rights of private property and the profit rate lead, then I want none of it. This does not produce cities that match my heart's desire, but worlds of inequality, alienation and injustice. I oppose the endless accumulation of capital and the conception of rights embedded therein. A different right to the city must be asserted.

Those that now have the rights will not surrender them willingly: "Between equal rights, force decides". This does not necessarily mean violence (though, sadly, it often comes down to that). But it does mean the mobilization of sufficient power through political organization or in the streets if necessary to change things. But by what strategies do we proceed?

No social order, said Saint-Simon, can change without the lineaments of the new already being latently present within the existing state of things. Revolutions are not total breaks but they do turn things upside down. Derivative rights (like the right to be treated with dignity) should become fundamental and fundamental rights (of private property and the profit rate) should become derivative. Was this not the traditional aim of democratic socialism?

There are, it turns out, contradictions within the capitalist package of rights. These can be exploited. What would have happened to global capitalism and urban life had the UN declaration's clauses on the derivative rights of labor (to a secure job, reasonable living standards and the right to organize) been rigorously enforced?

But new rights can also be defined: like the right to the city which, as I began by saying, is not merely a right of access to what the property speculators and state planners define, but an active right to make the city different, to shape it more in accord with our heart's desire, and to remake ourselves thereby in a different image.

The creation of a new urban commons, a public sphere of active democratic participation, requires that we roll back that huge wave of privatization that has been the mantra of a destructive neoliberalism. We must imagine a more inclusive, even if continuously fractious, city based not only upon a different ordering of rights but upon different political-economic practices. If our urban world has been imagined and made, then it can be reimagined and remade. The inalienable right to the city is worth fighting for. "City air makes one free", it used to be said. The air is a bit polluted now. But it can always be cleaned up.

ILLUSTRATION CREDITS

PART ONE

1 A transatlantic arriving in the harbor, New York City, 1959 by Henri Cartier-Bresson. Reproduced by permission of Magnum Photos.
2 Skateboarder and biker. Reproduced by permission of Corbis.

PART TWO

3 Manhattan, 1959 by Henri Cartier-Bresson. In the upper part of the city, around 103rd Street, some 61 blocks from Times Square, slums are being torn down with ruthless speed to make way for low cost housing projects such as these seen against the skyline. Reproduced by permission of Magnum Photos.

PART THREE

4 East 100th Street, Spanish Harlem, New York City, USA, 1966, by Bruce Davidson. Reproduced by permission of Magnum Photos.
5 Homeless in São Paulo, Brazil, 2003 by Ferdinando Scianna. Reproduced by permission of Magnum Photos.

PART FOUR

6 Commuters in Hong Kong MTR Train, by Minjoo Oh.
7 Gay Pride Parade, New York City, USA, 1998 by Nikos Economopoulos. Reproduced by permission of Magnum Photos.

PART FIVE

8 Chinatown, New York City, USA, 1998 by Chien-Chi Chang. A newly arrived immigrant eats noodles on a fire escape. Reproduced by permission of Magnum Photos.

PART SIX

9 Jazz Musician, New York City, 1958 by Dennis Stock. Reproduced by permission of Magnum Photos.

10 Times Square, Manhattan, New York City, 1999 by Raymond Depardon. Reproduced by permission of Magnum Photos.

PART SEVEN

11 Urban protest in New York City, January 28–30, 2002 by Larry Towell. During protests at the World Economic Forum, a protestor holds a "Make the Global Economy Work for Working Families" sign. Reproduced by permission of Magnum Photos.

COPYRIGHT INFORMATION

PART ONE

PART TWO

LOGAN and MOLOTCH John Logan and Harvey Molotch, "The City as a Growth Machine," excerpts from *Urban Fortunes: The Political Economy of Place* (1987). Berkeley, CA: University of California Press: pp. 1, 17–20, 23–25, 32, 52–62. © 1987 The Regents of the University of California.

SQUIRES Gregory Squires, "Partnership and the Pursuit of the Private City," in *Urban Life in Transition*, Mark Gottdiener and Chris Pickvance (eds.) (1991). Newbury Park; London: Sage: 1–12. Reprinted by permission of Sage Publications, Inc. Books.

DEAR Michael Dear, "Los Angeles and the Chicago School: Invitation to a Debate," *City and Community* 1, 1 (March): 5–28. © 2002 American Sociological Association. Reprinted by permission of John Wiley and Sons.

BRENNER and THEODORE Neil Brenner and Nik Theodore, "Cities and the Geographies of 'Actually Existing Neoliberalism'," *Antipode* 34, 3: 349–379 (2002). © 2002, Editorial Board of *Antipode*. Reprinted by permission of John Wiley and Sons.

DREIER, MOLLENKOPF, and SWANSTROM Peter Dreier, John Mollenkopf, and Todd Swanstrom, "Metropolitics for the Twenty-First Century," in *Place Matters*. University Press of Kansas (2001).

REES and WACKERNAGEL William Rees and Mathis Wackernagel, "Urban Ecological Footprints: Why Cities Cannot Be Sustainable – And Why They Are a Key to Sustainability," *Urban Ecology* (2008), originally published in *Environmental Impact Assessment Review* 16: 223–248. © 1996. Reprinted by permission of Elsevier.

PART THREE

DU BOIS W. E. B. Du Bois, "The Environment of the Negro" (1900), *The Philadelphia Negro.* Public Domain.

WACQUANT and WILSON Loïc J. D. Wacquant and William Julius Wilson, "The Cost of Racial and Class Exclusion in the Inner City," *Annals of the American Academy of Political and Social Science* 501: 8–26. © 1989, American Academy of Political and Social Science. Reprinted by permission of Sage Publications, Inc.

MASSEY and DENTON Douglas S. Massey and Nancy A. Denton, excerpts from *American Apartheid: Segregation and the Making of the Underclass* (1993). Cambridge, MA: Harvard University Press. © 1993 by the President and Fellows of Harvard College.

PORTES and MANNING Alejandro Portes and Robert D. Manning, "The Immigrant Enclave: Theory and Empirical Examples," in *Competitive Ethnic Relations*, Susan Olzak and Joane Nagel (eds) (1986). Orlando, FL: Academic Press. Reprinted by permission of Elsevier Online.

LIN and ROBINSON Jan Lin and Paul Robinson, "Spatial Disparities in the Expansion of the Chinese Ethnoburb of Los Angeles," *Geojournal* 64: 51–61 (2005).

DUNCAN James S. Duncan, "Men Without Property: The Tramp's Classification and Use of Urban Space," *Antipode* 10: 24–34 (1983). © 1978. Reprinted by permission of John Wiley and Sons.

ELLIOTT and PAIS James R. Elliott and Jeremy Pais, "Race, Class, and Hurricane Katrina: Social Differences in Human Responses to Disaster," *Social Science Research* 35: 295–321 (2006). © 2006. Reprinted by permission of Elsevier.

PART FOUR

MARKUSEN Ann R. Markusen, "City Spatial Structure, Women's Household Work, and National Urban Policy," *Signs* (Spring 1981): S23–S43. Reprinted by permission of University of Chicago Press.

GILBERT Melissa R. Gilbert, "'Race,' Space, and Power: The Survival Strategies of Working Poor Women," *Annals of the Association of American Geographers* 88, 4: 595–621. © 1998, Association of American Geographers. Reprinted by permission of Taylor & Francis Ltd., online.

ADLER and BRENNER Sy Adler and Johanna Brenner, "Gender and Space: Lesbians and Gay Men in the City," *International Journal of Urban and Regional Research* 16, 1: 24–34. © 1992, Joint Editors and Blackwell Publishing. Reprinted by permission of Blackwell Publishing.

DONHAM Donald L. Donham, "Freeing South Africa: The 'Modernization' of Male–Male Sexuality in Soweto," *Cultural Anthropology* 13, 1: 3–21. © 1998 American Anthropological Association. Reprinted by permission of University of California Press.

TANI Sirpa Tani. "Whose Place is this Space? Life in the Street Prostitution Area of Helsinki, Finland," *International Journal of Urban and Regional Research*, 26, 2 (June 2002): 343–359. © 2002. Reprinted by permission of John Wiley and Sons.

PART FIVE

FRIEDMANN John Friedmann, "The World City Hypothesis," *Development and Change* 17: 69–83 (1986). © 1982. Reprinted by permission of John Wiley and Sons.

SASSEN Saskia Sassen, "Whose City Is It? Globalization and the Formation of New Claims," *Public Culture* 8, 2: 205–223. © 1996. All rights reserved. Reprinted by permission of Duke University Press.

YEOH and CHANG Brenda S. A. Yeoh and T. C. Chang, "Globalising Singapore: Debating Transnational Flows in the City," *Urban Studies* 38, 7: 1025–1044 (2001). © 2001 Urban Studies Journal Ltd. Reprinted by permission of Sage Publications.

STOLLER and McCONATHA Paul Stoller and Jasmin Tahmaseb McConatha, "City Life: West African Communities in New York," *Journal of Contemporary Ethnography* 30, 6: 651–677. © 2001. Reprinted by permission of Sage Publications.

LEVITT Peggy Levitt, "Social Remittances: Migration Driven Local-level Forms of Cultural Diffusion," *International Migration Review* 32, 4 (Winter): 926–948. © 1998. Reprinted by permission of Blackwell Publishing Ltd.

PART SIX

ZUKIN Sharon Zukin, "Whose Culture? Whose City?" in *The Cultures of Cities.* Cambridge, MA: Blackwell Publishing: 1–48. © 1995. Reprinted by permission of John Wiley and Sons.

FLORIDA Richard Florida, "Cities and the Creative Class," *City and Community* 2, 1 (March 2003): 3–19. © 2003 American Sociological Association. Reprinted by permission of John Wiley and Sons.

LIPUMA and KOELBLE Edward LiPuma and Thomas Koelble, "Cultures of Circulation and the Urban Imaginary: Miami as Example and Exemplar," *Public Culture* 17, 1: 153–179. © 2005. All rights reserved. Reprinted by permission of Duke University Press.

AGUILAR-SAN JUAN Karin Aguilar-San Juan, "Staying Vietnamese: Community and Place in Orange County and Boston," *City and Community* 4, 1 (March): 37–65. © 2005. Reprinted by permission of John Wiley and Sons.

PART SEVEN

MERRY S. E. Merry, "Spatial Governmentality and the New Urban Social Order: Controlling Gender Violence through Law," *American Anthropologist* 103, 1: 16–30 (2001). Reprinted by permission of American Anthropological Association.

LOW Setha Low, "The Erosion of Public Space and the Public Realm," *City and Society* 18, 1: 43–49 (2006). Reprinted by permission of American Anthropological Association.

Index

Page numbers in italics indicate figures
Page numbers in bold indicate tables